# THE ALKALOIDS

## Chemistry and Biology

## VOLUME 63

# THE ALKALOIDS

## Chemistry and Biology

*Edited by*

Geoffrey A. Cordell

*College of Pharmacy*
*University of Illinois at Chicago*
*Chicago, Illinois*

**VOLUME 63**

ELSEVIER

Amsterdam • Boston • Heidelberg • London• Oxford •
Paris • San Diego • San Francisco • Singapore • Sydney • Tokyo
Academic Press is an imprint of Elsevier

ACADEMIC PRESS

Academic Press is an imprint of Elsevier
84 Theobald's Road, London WC1X 8RR, UK
Radarweg 29, PO Box 211, 1000 AE Amsterdam, The Netherlands
The Boulevard, Langford Lane, Kidlington, Oxford OX5 1GB, UK
30 Corporate Drive, Suite 400, Burlington, MA 01803, USA
525 B Street, Suite 1900, San Diego, CA 92101-4495, USA

First edition 2006



ISBN-13: 978-0-12-469563-4
ISBN-10: 0-12-469563-9
ISSN: 1099-4831

For information on all Academic Press publications
visit our website at books.elsevier.com

Printed and bound in USA

06 07 08 09 10 10 9 8 7 6 5 4 3 2 1

# CONTENTS

## Bisindole Alkaloids

TOH-SEOK KAM AND YEUN-MUN CHOO

# CONTRIBUTORS

*Numbers in parentheses indicate the pages on which the authors' contributions begin.*

JAUME BASTIDA (87), Departament de Productes Naturals, Facultat de Farmàcia, Universitat de Barcelona, 08028 Barcelona, Spain

YEUN-MUN CHOO (181), Department of Chemistry, University of Malaya, 50603 Kuala Lumpur, Malaysia

PETER J. FACCHINI (1), Department of Biological Sciences, University of Calgary, Calgary, AB, Canada

TOH-SEOK KAM (181), Department of Chemistry, University of Malaya, 50603 Kuala Lumpur, Malaysia

RODOLFO LAVILLA (87), Parc Científic de Barcelona, Universitat de Barcelona, 08028 Barcelona, Spain

DANIEL G. PANACCIONE (45), Division of Plant and Soil Sciences, West Virginia University, Morgantown, WV 26506-6108, USA

CHRISTOPHER L. SCHARDL (45), Department of Plant Pathology, University of Kentucky, Lexington, KY 40546-0312, USA

PAUL TUDZYNSKI (45), Institut für Botanik, Westfälische Wilhelms Universität Münster, Münster D-48149, Germany

FRANCESC VILADOMAT (87), Departament de Productes Naturals, Facultat de Farmàcia, Universitat de Barcelona, 08028 Barcelona, Spain

# PREFACE

This volume of *The Alkaloids: Chemistry and Biology* is comprised of four very different chapters; a reflection of the diverse facets that comprise the study of alkaloids today. As awareness of the global need for natural products which can be made available as drugs on a sustainable basis increases, so it has become increasingly important that there is a full understanding of how key metabolic pathways can be optimized. At the same time, it remains important to find new biologically active alkaloids and to elucidate the mechanisms of action of those that do show potentially useful or novel biological effects.

Facchini, in Chapter 1, reviews the significant studies that have been conducted with respect to how the formation of alkaloids in their various diverse sources are regulated at the molecular level.

The history of the ergot alkaloids and their biological effects is a very rich and fascinating one. In Chapter 2, advances in the biosynthetic pathway of these alkaloids are discussed, the genes involved are reviewed, and their biological and clinical significance are presented by Schardl, Panaccione, and Tudzynski.

In Chapter 3, Bastida, Lavilla, and Viladomat provide a classical review of the isolation, structure elucidation, biosynthesis, synthesis, and biology of the *Narcissus* alkaloids. This group of alkaloids is receiving a great deal of attention at the present time because of the significant biological activities observed for some of these metabolites.

It has been over 25 years since the broad area of the bisindole alkaloids was reviewed in this series. In the concluding chapter, Kam and Choo have taken on this formidable task and provide a very comprehensive summation of tremendous advances in the structural, synthetic, and biological aspects of more than 30 basic alkaloid types represented.

Geoffrey A. Cordell
*University of Illinois at Chicago*

—CHAPTER 1—

# REGULATION OF ALKALOID BIOSYNTHESIS IN PLANTS

PETER J. FACCHINI

*Department of Biological Sciences, University of Calgary, Calgary, Alberta, Canada*



## I. Introduction

Several key technical breakthroughs over the last half-century have contributed to an impressive advancement in our understanding of alkaloid biosynthesis in plants. The use of radiolabeled precursors in the 1950s allowed the elucidation of several biogenic pathways. The widespread application of plant cell cultures during the 1970s provided a rich source of biosynthetic enzymes and encouraged work on the elucidation of signal transduction mechanisms that activate alkaloid pathways. The introduction of molecular techniques in the 1990s prompted the isolation of numerous molecular clones involved in alkaloid biosynthesis (*1*), which have been used to determine the tissue-specific localization of alkaloid biosynthetic enzymes and gene transcripts, and functionally analyze the corresponding promoters. Recent applications of genomics-based technologies, such as expressed sequence tag (EST) databases, DNA microarrays, and proteome analysis, have shown the potential to accelerate the discovery of new components and mechanisms involved in the assembly and function of plant alkaloids. Our emerging ability to investigate alkaloid metabolism from a combined biochemical, molecular, cellular, and physiological perspective has greatly improved our appreciation for the complex regulation of diverse biosynthetic pathways. Unlike other types of secondary metabolites, the different structural categories of alkaloids have unique biosynthetic origins. This review will focus on recent advances in our understanding of the metabolic regulation involved in the biosynthesis of six groups of alkaloids – benzylisoquinoline, monoterpenoid indole, tropane, purine, pyrrolizidine, and quinolizidine alkaloids – for which biosynthetic and regulatory genes have been reported and characterized.

THE ALKALOIDS, Vol. 63
ISSN: 1099-4831
DOI: 10.1016/S1099-4831(06)63001-0

## II. Alkaloid Biosynthetic Pathways

### A. BENZYLISOQUINOLINE ALKALOIDS

Many benzylisoquinoline alkaloids (BAs) are used as pharmaceuticals due to their potent pharmacological activity, which is often an indication of their biological function. For example, the effectiveness of morphine as an analgesic, colchicine as a microtubule disrupter, and (+)-tubocurarine as a neuromuscular blocker suggests that these alkaloids function as herbivore deterrents. The antimicrobial properties of sanguinarine and berberine suggest that they confer protection against pathogens (*2*). BAs occur mainly in basal angiosperms, including the members of Ranunculaceae, Papaveraceae, Berberidaceae, Menispermaceae, and Magnoliaceae.

BA biosynthesis begins with the conversion of tyrosine into both dopamine and 4–hydroxyphenylacetaldehyde by a lattice of decarboxylations, *ortho*-hydroxylations, and deaminations (*3*). The aromatic amino acid decarboxylase (TYDC) that converts tyrosine and dopa into their corresponding amines has been purified, and several cDNAs have been cloned (Scheme 1) (*4–6*).

A family of ~15 genes, which can be divided into two subgroups based on sequence identity, encodes TYDC in opium poppy (*Papaver somniferum*) (*5*). The catalytic properties of the various protein isoforms are similar despite the differential developmental and inducible expression of the *TYDC* gene family (*5,7–9*). Dopamine and 4–hydroxyphenylacetaldehyde are condensed by norcoclaurine synthase (NCS) to yield (*S*)-norcoclaurine (Scheme 1), the central precursor to all BAs in plants (*10,11*). Owing to the inability of NCS to discriminate between 4–hydroxyphenylacetaldehyde and 3,4–dihydroxyphenylacetaldehyde, and the nonspecificity of the subsequent methyltransferase reactions, it was originally thought that the tetrahydroxyBA (*S*)-norlaudanosoline was the precursor (*12*). However, only norcoclaurine has been found to occur in plants. NCS has been purified (*13,14*) and corresponding molecular clones have been isolated from *Thalictrum flavum* (*15*) and *P. somniferum* (*16*). NCS is ancestrally related to the pathogenesis-related (PR) 10 and Bet v 1 protein families, which include the enzyme (HYP1) from St. John's wort (*Hypericum perforatum*) responsible for the biosynthesis of the bioactive naphthodianthrone hypericin (*17*). The novel catalytic functions of NCS and HYP1 define a new class of plant secondary metabolic enzymes.

(*S*)-Norcoclaurine is converted into (*S*)-reticuline by a 6–*O*-methyltransferase (6OMT), an *N*-methyltransferase (CNMT), a P450 hydroxylase (CYP80B1), and a 4′-*O*-methyltransferase (4′OMT) (Scheme 1) (*18–21*). 6OMT and 4'OMT were purified (*18*), and corresponding cDNAs isolated and characterized from *Coptis japonica* (*22*), *T. flavum* (*23*), and *P. somniferum* (*24–26*). Each enzyme exhibits unique substrate specificity and a different reaction mechanism despite extensive homology. A previously reported family of catechol *O*-methyltransferases that accepts (*S*)-norcoclaurine as a substrate is probably not involved in alkaloid biosynthesis (*27*). CNMT has also been purified (*28*), and the corresponding cDNA isolated from *C. japonica* (*29*), *T. flavum* (*23*), and *P. somniferum* (*24*). The aromatic ring hydroxylation involved in the conversion of (*S*)-norcoclaurine into (*S*)-reticuline was once thought to proceed *via* a nonspecific phenol oxidase (*30*). However, a P450-dependent monooxygenase (CYP80B1) was shown to have a lower $K_m$ for (*S*)-*N*-methylcoclaurine than the

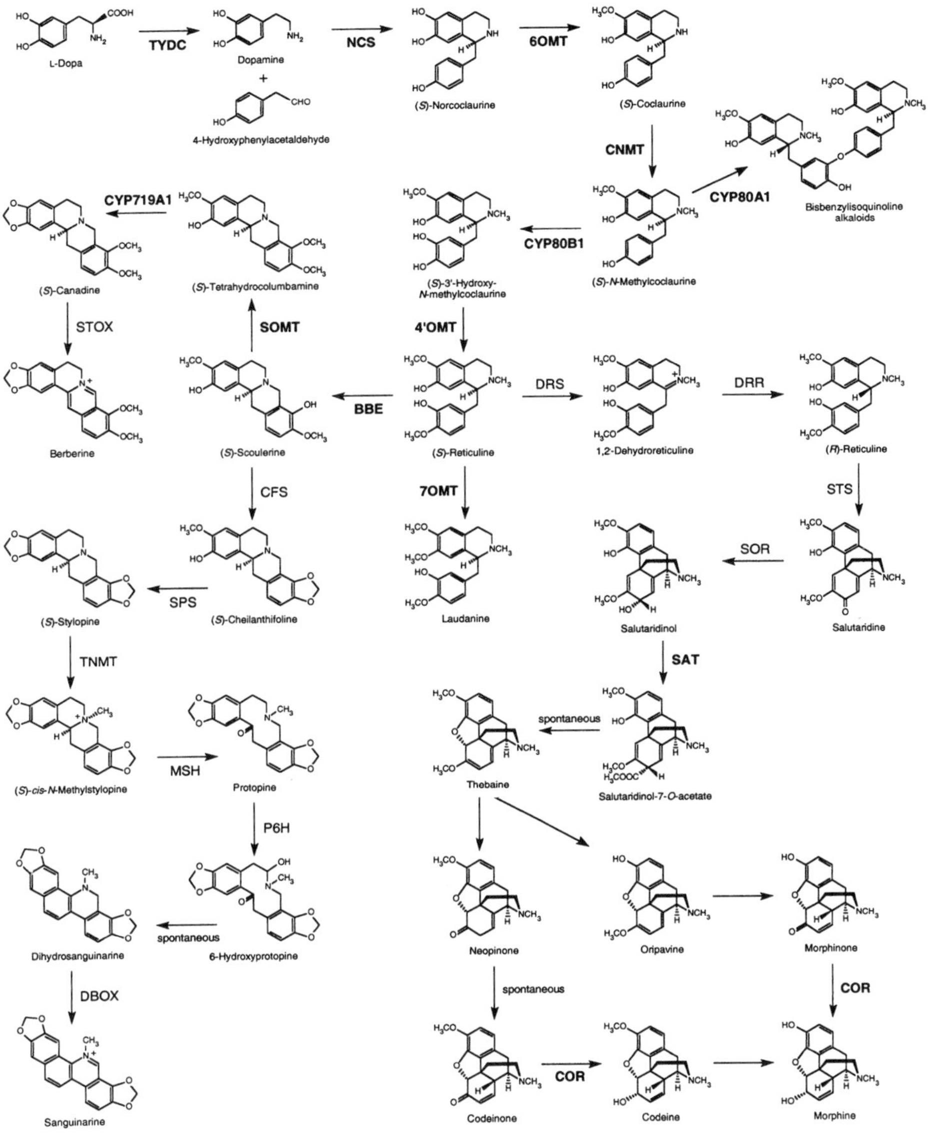


Scheme 1. Biosynthesis of the benzylisoquinoline alkaloids berberine, morphine, and sanguinarine. Enzymes for which corresponding molecular clones have been isolated are shown in bold. Abbreviations: 4′OMT, 3′-hydroxy-*N*-methylcoclaurine 4′-*O*-methyltransferase; 6OMT, norcoclaurine 6-*O*-methyltransferase; 7OMT, reticuline 7-*O*-methyltransferase; BBE, berberine bridge enzyme; CFS, cheilanthifoline synthase; CNMT, coclaurine *N*-methyltransferase; COR, codeinone reductase; CYP719A1, canadine synthase; CYP80A1, berbamunine synthase; CYP80B1, *N*-methylcoclaurine 3′-hydroxylase; DBOX, dihydrobenzophenanthridine oxidase; DRR, 1,2-dehydroreticuline reductase; DRS, 1,2-dehydroreticuline synthase; MSH, *N*-methylstylopine 14-hydroxylase; NCS, norcoclaurine synthase; P6H,

*(contd)*

phenolase and is now accepted as the enzyme catalyzing the conversion of (*S*)-*N*-methylcoclaurine into (*S*)-3'-hydroxy-*N*-methylcoclaurine (Scheme 1) (*21*). Molecular clones encoding CYP80B1 have been isolated from California poppy (*Eschscholzia californica*) (*21*), *P. somniferum* (*31,32*), and *T. flavum* (*23*).

(*S*)-Reticuline pathway intermediates also serve as precursors to the bis-benzylisoquinoline alkaloids, such as (+)-tubocurarine (Scheme 1). A phenol-coupling P450-dependent oxidase (CYP80A1) was purified, and the corresponding cDNA was isolated from *Berberis stolonifera* (*33,34*). CYP80A1 couples two molecules of (*R*)-*N*-methylcoclaurine or one each of (*R*)- and (*S*)-*N*-methylcoclaurine to form (*R,R*)-guattegaumerine or (*R,S*)-berbamunine, respectively (Scheme 1). Phenyl ring substitutions, regiospecificity, number of ether linkages, and monomer stereospecificity add additional diversity to the bis-benzylisoquinoline alkaloids. A cytochrome P450 reductase (CPR) was purified from opium poppy, and corresponding cDNAs were isolated from opium poppy and *E. californica* (*35*).

(*S*)-Reticuline is a branch-point intermediate in the biosynthesis of most BAs and many substituted derivatives are produced. For example, a molecular clone encoding (*R,S*)-reticuline 7–*O*-methyltransferase (7OMT), which catalyzes the conversion of (*R,S*)-reticuline into laudanine, has been identified (*25*). Much of the work has focused on branch pathways leading to the benzophenanthridine (e.g., sanguinarine), protoberberine (e.g., berberine), and morphinan (e.g., morphine and codeine) alkaloids (*36*). A multitude of relevant enzymes have been isolated, many have been purified, and an impressive number of corresponding cDNAs have been cloned (*36*). The first committed step in benzophenanthridine and protoberberine alkaloid biosynthesis involves the conversion of (*S*)-reticuline into (*S*)-scoulerine by the berberine bridge enzyme (BBE) (Scheme 1). BBE was purified from *Berberis beaniana* (*37*), corresponding cDNAs were cloned from *B. stolonifera*, *E. californica*, and *T. flavum* (*23,38,39*), the recombinant enzyme was characterized (*40,41*), and *BBE* genes were isolated from *P. somniferum* and *E. californica* (*42,43*). Interestingly, BBE belongs to the same FAD-dependent oxidoreductase family as $\Delta^1$-tetrahydrocannabinolic acid synthase, which is involved in the biosynthesis of the psychoactive compound $\Delta^1$-tetrahydrocannabinol in *Cannabis sativa* (*44*).

Benzophenanthridine alkaloid biosynthesis requires the conversion of (*S*)-scoulerine into (*S*)-stylopine by two P450-dependent oxidases, (*S*)-chelanthifoline synthase (CFS) and (*S*)-stylopine synthase (SPS), resulting in the formation of two methylenedioxy bridges (*45,46*), (*S*)-Stylopine is *N*-methylated by tetrahydroprotoberberine-*cis-N*-methyltransferase, which has been isolated from *E. californica* and *Corydalis vaginans* cells (*47*), and purified from *Sanguinaria canadensis* cultures (*48*). A P450-dependent monooxygenase, (*S*)-*cis-N*-methylstylopine 14–hydroxylase (MSH), then catalyzes the formation of protopine (*49*). Another P450-dependent enzyme, protopine-6–hydroxylase (PPH) followed by a spontaneous intramolecular rearrangement converts protopine into dihydrosanguinarine (*50*). The oxidation of dihydrosanguinarine to sanguinarine occurs *via* dihydrobenzophenanthridine

protopine 6-hydroxylase; SAT, salutaridinol 7-*O*-acetyltransferase; SOMT, scoulerine 9-*O*-methyltransferase; SOR, salutaridine:NADPH 7-oxidoreductase; SPS, stylopine synthase; STOX, (*S*)-tetrahydroxyprotoberberine oxidase; STS, salutaridine synthase; TNMT, tetrahydroprotoberberine *cis-N*-methyltransferase; TYDC, tyrosine decarboxylase.

oxidase (DBOX) (*51*), a cytosolic enzyme purified from *S. canadensis* cultures (*52*). Two other species-specific enzymes, dihydrochelirubine-12–hydroxylase and 12–hydroxydihydrochelirubine-12–*O*-methyltransferase, catalyze the final two steps in the biosynthesis of macarpine, the most highly oxidized BA found in nature (*53*).

In some plants, (*S*)-scoulerine is methylated, rather than oxidized, to yield (*S*)-tetrahydrocolumbamine (Scheme 1). The reaction is catalyzed by scoulerine-9–*O*-methyltransferase (SOMT) (*54*), which was purified, and the corresponding cDNA isolated, from *C. japonica* (*55,56*) and *T. flavum* (*23*). The P450-dependent enzyme canadine synthase (CYP719A) was detected in members of the genera *Coptis* and *Thalictrum* and shown to catalyze methylenedioxy bridge formation in (*S*)-tetrahydrocolumbamine (*57*), but not in the quaternary alkaloid columbamine (*58*), showing that berberine biosynthesis cannot proceed *via* columbamine as once proposed. Molecular clones for CYP719A1 have been isolated from *C. japonica* (*59*) and *T. flavum* (*23*). (*S*)-Canadine, also known as (*S*)-tetrahydroberberine, is oxidized to berberine by either (*S*)-canadine oxidase (CDO) or (*S*)-tetrahydroprotoberberine oxidase (STOX) (*60*). These enzymes catalyze the same reaction, but their biochemical properties are distinct. STOX from the genus *Berberis* is a flavinylated protein with a broad substrate range, whereas CDO from the genera *Coptis* and *Thalictrum* contains iron, proceeds *via* a different mechanism, and prefers (*S*)-canadine as a substrate.

Conversion of (*S*)-reticuline into its (*R*)-epimer is the first committed step in morphinan alkaloid biosynthesis in certain species. The still poorly characterized enzymes 1,2–dehydroreticuline synthase and 1,2–dehydroreticuline reductase catalyze the stereospecific reduction of 1,2–dehydroreticuline to (*R*)-reticuline (*61,62*). Intramolecular carbon–carbon phenol coupling of (*R*)-reticuline by the P450-dependent enzyme salutaridine synthase (STS) results in the formation of salutaridine (*63*). The cytosolic enzyme, salutaridine: NADPH 7–oxidoreductase (SOR), found in *Papaver bracteatum* and *P. somniferum* reduces salutaridine to (7*S*)-salutaridinol (*64*). Conversion of (7*S*)-salutaridinol into thebaine requires closure of an oxide bridge between C-4 and C-5 by acetyl coenzyme A: salutaridinol-7–*O*-acetyltransferase (SAT). The enzyme was purified from opium poppy cell cultures and the corresponding cDNA isolated (Scheme 1) (*65,66*). In the last steps of morphine biosynthesis, codeinone is produced from thebaine and then reduced to codeine, which is finally demethylated to yield morphine. Codeinone reductase (COR), which reduces (–)-codeinone to (–)-codeine, has been purified and the corresponding cDNA isolated from *P. somniferum* (Scheme 1) (*67,68*).

## B. MONOTERPENOID INDOLE ALKALOIDS

Monoterpenoid indole alkaloids are found mainly in the Apocynaceae, Loganiaceae, and Rubiaceae. Many have potent biological activities (*69*), including the cerebral metabolism enhancing alkaloids of periwinkle (*Vinca minor*), the antimalarial and cardiotonic alkaloids of cinchona (*Cinchona ledgeriana*), and the anticancer agents from Chinese "happy" tree (*Camptotheca acuminata*). The well-known central nervous stimulants strychnine and yohimbine are also monoterpenoid indole alkaloids. Perhaps the most important from a health perspective are the anticancer agents vincristine and vinblastine from the Madagascar periwinkle (*Catharanthus roseus*). The importance of *C. roseus* as a source of anticancer medicines has prompted intensive research on alkaloid biosynthesis in this plant.

Monoterpenoid indole alkaloids consist of an indole moiety provided by tryptamine and a terpenoid component derived from the iridoid glucoside secologanin (Scheme 2). Molecular clones for both the α-and β-subunits of anthranilate synthase (AS), which catalyze the first committed reaction of the indole pathway, have been isolated from *C. acuminata* (*70,71*). Comparison of the two differentially regulated genes encoding the AS α-subunit showed that the spatial and developmental expression of only one paralleled that of the β-subunit gene, and the pattern of camptothecin accumulation. Thus, the indole and monoterpenoid indole alkaloid pathways appear coordinately regulated through the duplication of specific genes, such as that encoding the AS α-subunit. Tryptophan is converted into tryptamine by tryptophan decarboxylase (TDC), which is encoded by a single gene in *C. roseus* (*72–74*), and by two autonomously regulated genes in *C. acuminata* (*75*). A molecular clone for TDC was also reported from *Ophiorrhiza pumila* (*76*).

Secologanin is formed from precursors derived from the triose phosphate/pyruvate pathway (*77*). Two cDNAs encoding the enzymes 1–deoxy-D-xylulose 5–phosphate reductoisomerase (DXR) and 2C-methyl-D-erythritol 2,4–cyclodiphosphate synthase (MECS) of the 2C-methyl-D-erythritol 4–phosphate (MEP) pathway were isolated from *C. roseus* (*78*). The corresponding gene transcripts were induced in *C. roseus* cell cultures producing monoterpenoid indole alkaloids. The first committed step in secologanin biosynthesis is the hydroxylation of geraniol to 10–hydroxygeraniol (*79,80*). A novel P450-dependent monooxygenase (CYP76B6) is specific for the C-10 position of geraniol and exhibits similar affinity for nerol, the *cis*-isomer of geraniol. The enzyme was purified and shown to contain FMN and FAD as cofactors (*81*), and the corresponding cDNA was isolated (Scheme 2) (*82*). Conversion of loganin into secologanin represents the last step in the pathway and is catalyzed by another P450-dependent enzyme (CYP72A1) for which the corresponding cDNA has also been reported (Scheme 2) (*83,84*). The enzyme 3–hydroxy-3–methylglutaryl coenzyme A reductase (HMGR), which is involved in the biosynthesis of mevalonate, was cloned and characterized from *C. roseus* (*85*) and *C. acuminata* (*86,87*). The differential expression of *HMGR* genes in response to wounding and methyl jasmonate (MeJA) was suggested to contribute to the regulation of terpenoid indole alkaloid (TIA) biosynthesis. However, the formation of secologanin *via* the non-mevalonate terpenoid pathway (*77*) indicates that the correlation between *HMGR* expression and TIA accumulation is coincidental.

Tryptamine and secologanin condense to form strictosidine, the common precursor to all monoterpenoid indole alkaloids, by strictosidine synthase (STR) (Scheme 2). STR cDNAs have been isolated from *Rauwolfia serpentina*, *C. roseus,* and *O. pumila* (*88–91*). Crystallization and preliminary X-ray analysis of STR from *R. serpentina* has also been reported (*92*). Strictosidine is deglucosylated by strictosidine β-D-glucosidase (SGD), which has been purified (*93*), and the corresponding cDNA isolated from *C. roseus* (*94*) and *R. serpentina* (*95*) cell cultures (Scheme 2). STR and SGD have also been crystallized and preliminary X-ray analyses have been performed (*96,97*). Deglucosylated strictosidine is converted *via* several unstable intermediates into 4,21–dehydrogeissoschizine. Although many monoterpenoid indole alkaloids are produced from 4,21–dehydrogeissoschizine, the enzymology of the branch pathways leading to catharanthine and most other products is poorly understood. However, the final steps of vindoline biosynthesis from tabersonine have been characterized in considerable detail (Scheme 2).

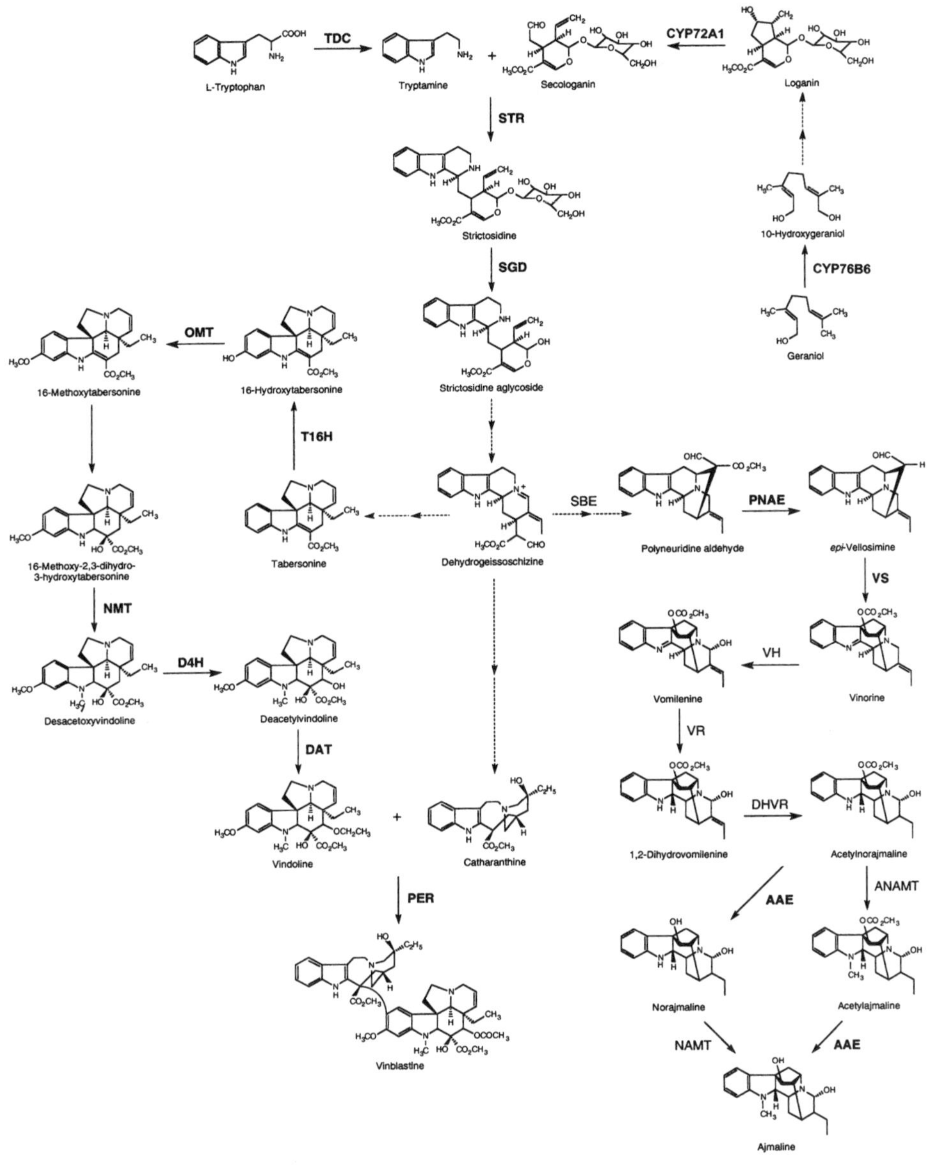


Scheme 2. Biosynthesis of the monoterpenoid indole alkaloids vinblastine and ajmaline. Enzymes for which corresponding molecular clones have been isolated are shown in bold. Abbreviations: AAE, acetylajmalan esterase; ANAMT, acetylnorajmalan methyltransferase; D4H, desacetoxyvindoline 4–hydroxylase; DAT, deacetylvindoline 4–*O*-acetyltransferase; DHVR, dihydrovomilenine reductase; PER, peroxidase; CYP76B6, geraniol 10–hydroxylase; NAMT, norajmalan methyltransferase; NMT, *N*-methyltransferase; OMT, *O*-methyltransferase; PNAE, polyneuridine aldehyde esterase; SBE, sarpagan bridge enzyme; SGD, strictosidine $\beta$-D-glucosidase; CYP72A1, secologanin synthase; STR, strictosidine synthase; T16H, tabersonine 16–hydroxylase; TDC, tryptophan decarboxylase; VH, vinorine hydroxylase; VR, vomilenine reductase; VS, vinorine synthase.

The first of the six steps in the conversion of tabersonine into vindoline is catalyzed by tabersonine 16–hydroxylase (T16H) (Scheme 2), which was characterized as a P450-dependent monooxygenase (*98*). A cDNA able to hydroxylate tabersonine at the C-16 position was cloned using a strategy based on the light activation of the enzyme in *C. roseus* cultures (*99*). P450 sequences amplified with heme-binding domain-specific primers were hybridized to RNA from induced and noninduced cells. One cDNA encoding a P450-dependent enzyme (CYP71D12), showed induction kinetics consistent with T16H activity and was capable of converting tabersonine into 16–methoxytabersonine when expressed as a translational fusion with CPR (*99*). The next three steps in vindoline biosynthesis consist of 16–*O*-methylation, hydration of the 2,3–double bond, and *N*-methylation of the indole-ring nitrogen (*100–102*). An *O*-methyltransferase that methylates 16–*O*-demethyl-4–*O*-deacetylvindoline has been reported (*103*). Initially, two consecutive hydroxylations at the C-3 and C-4 positions were proposed to follow the 16–hydroxylation of tabersonine (*103*). However, the isolation of an *N*-methyltransferase (NMT) specific for the indole-ring nitrogen of 16–methoxy-2,3–dihydro-3–hydroxytabersonine showed that the *O*-methylation step precedes *N*-methylation, and that 16–hydroxytabersonine is the natural substrate of the *O*-methyltransferase (*102,104*). The enzyme involved in hydrating the 2,3–double bond has not been reported. The final two steps in vindoline biosynthesis are catalyzed by a 2–oxoglutarate-dependent dioxygenase (D4H), which hydroxylates the C-4 position of desacetoxyvindoline (*105,106*) and an acetylcoenzyme A-dependent *O*-acetyltransferase (DAT) specific for the 4–OH of deacetylvindoline (Scheme 2) (*107–109*). Molecular clones encoding D4H and DAT have been isolated (*110,111*). Vindoline is ultimately coupled to catharanthine by a non-specific peroxidase (PER) to yield vinblastine (*112,113*). A cDNA encoding a protein homologous to DAT was also isolated from *C. roseus* roots and the enzyme shown to function as minovincinine-19–hydroxy-*O*-acetyltransferase (*114*).

Progress has also been made in the isolation and characterization of enzymes involved in ajmaline biosynthesis in *R. serpentina* (Scheme 2). Conversion of strictosidine-derived intermediates under the control of the P450-dependent sarpagan bridge enzyme opens the pathway to the sarpagan-type alkaloids. A specific methylesterase of the $\alpha/\beta$-fold superfamily of hydrolases, polyneuridine aldehyde esterase (PNAE), converts the intermediate polyneuridine aldehyde into 16–*epi*-vellosimine (*115,116*). A member of the BAHD superfamily of acyl-transfer enzymes, vinorine synthase (VS) catalyzes the acetylation of 16–*epi*-vellosimine (*117*). Molecular clones for PNAE (*115*) and VS (*118*) have been isolated and the crystal structure of VS has been solved (*119,120*). Vinorine is then hydroxylated by the P450-dependent enzyme vinorine hydroxylase (VH) to form vomilenine (*121*), which is subsequently reduced in two steps by the NADPH-dependent enzymes vomilenine reductase (VR) and 1,2–dihydrovomilenine reductase (DHVR) (*122,123*). VR might play an important regulatory role in the ajmaline pathway as the acetyl group introduced by VS can be removed after reduction of the indolenine double bond. In contrast, deacetylation at the indolenine stage (i.e. vinorine or vomilenine) would lead to spontaneous ring opening and formation of sarpagan-type alkaloids, such as 16–*epi*-vellosimine and vellosimine. A molecular clone has been reported for acetylajmalan esterase (AAE), a highly specific enzyme that hydrolyzes the 17–*O*-acetyl group of ajmalan-type alkaloids ultimately leading to ajmaline (*124,125*). *R. serpentina* cultures normally accumulate the side-product raucaffricine rather than ajmaline. The reutilization of raucaffricine for ajmaline production involves raucaffricine-*O*-$\beta$-glucosidase (RG).

The cDNA for RG was isolated and shown to encode a protein with homology to SGD (*126*).

## C. TROPANE ALKALOIDS AND NICOTINE

Tropane alkaloids possess an 8–azabicyclo[3.2.1]octane nucleus and are found in plants of three families, Solanaceae, Erythroxylaceae, and Convolvulaceae. Although several tropane alkaloids are used as legitimate pharmaceuticals most are known for their toxicity. Atropine, the racemic form of (+)- and (–)-hyoscyamine, is used to dilate the pupil during opthalmological examinations and to treat poisoning by organophosphorous insecticides, nerve gas, and the toxic principles of the mushroom *Amanita muscaria*. Scopolamine continues to be used to prevent motion sickness and, in combination with morphine, as a sedative during labor. Peruvian coca (*Erythroxylum coca*) is cultivated for the production of cocaine, which is used licitly and illicitly as a local anesthetic in ophthamology, as a central nervous system stimulant, and as drug to improve physical endurance and reduce appetite. Calystegines possess strong therapeutic potential as potent inhibitors of $\beta$-glucosidase and are found primarily in the Convolvulaceae, which includes sweet potato (*Ipomoea batatas*). They also occur in potato (*Solanum tuberosum*) and other members of the Solanaceae family.

The *N*-methyl-$\Delta^1$-pyrrolinium cation is the last common intermediate in both tropane alkaloid and nicotine biosynthesis (Scheme 3). *N*-Methyl-$\Delta^1$-pyrrolinium cation formation begins with the decarboxylation of ornithine and arginine by ornithine decarboxylase (ODC) and arginine decarboxylase (ADC), respectively. Putrescine is formed directly from ornithine by ODC, whereas agmatine and *N*-carbamoylputrescine serve as intermediates when putrescine is formed *via* ADC. An ODC cDNA isolated from *Datura stramonium* showed similarity to ADCs and other ODCs (*127*). ADC cDNAs have been isolated from oat and tomato (*128,129*). Arginine appears to supply most of the putrescine for alkaloid biosynthesis (*130*).

Tropane alkaloid biosynthesis begins with the *S*-adenosylmethionine (SAM) -dependent *N*-methylation of putrescine by the enzyme putrescine *N*-methyltransferase (PMT) (Scheme 3) (*131*), cDNAs of which have been isolated from *Atropa belladonna* and *Hyoscyamus niger* (*132*), and the calystegine-producing plant *S. tuberosum* (*133*). *N*-Methylputrescine is oxidatively deaminated by a diamine oxidase to 4–aminobutanal, which undergoes spontaneous cyclization to form the reactive *N*-methyl-$\Delta^1$-pyrrolinium cation. Although the enzymes involved are unknown, the *N*-methyl-$\Delta^1$-pyrrolinium cation is thought to condense either with acetoacetic acid to yield hygrine as a precursor to the tropane ring, or with nicotinic acid to form nicotine.

A key branch point in tropane alkaloid biosynthesis is the conversion of tropinone into either tropine or pseudotropine (or $\psi$-tropine), which possess opposite stereochemistry at the 3–hydroxyl position. Two different NADPH-dependent enzymes catalyze the stereospecific reduction of tropinone: the 3–carbonyl of tropinone is reduced to the 3$\alpha$-hydroxy group of tropine by tropinone reductase I (TR-I) and to the 3$\beta$-hydroxy group of pseudotropine by tropinone reductase II (TR-II). Genes encoding both TR-I and TR-II have been identified in the tropinone

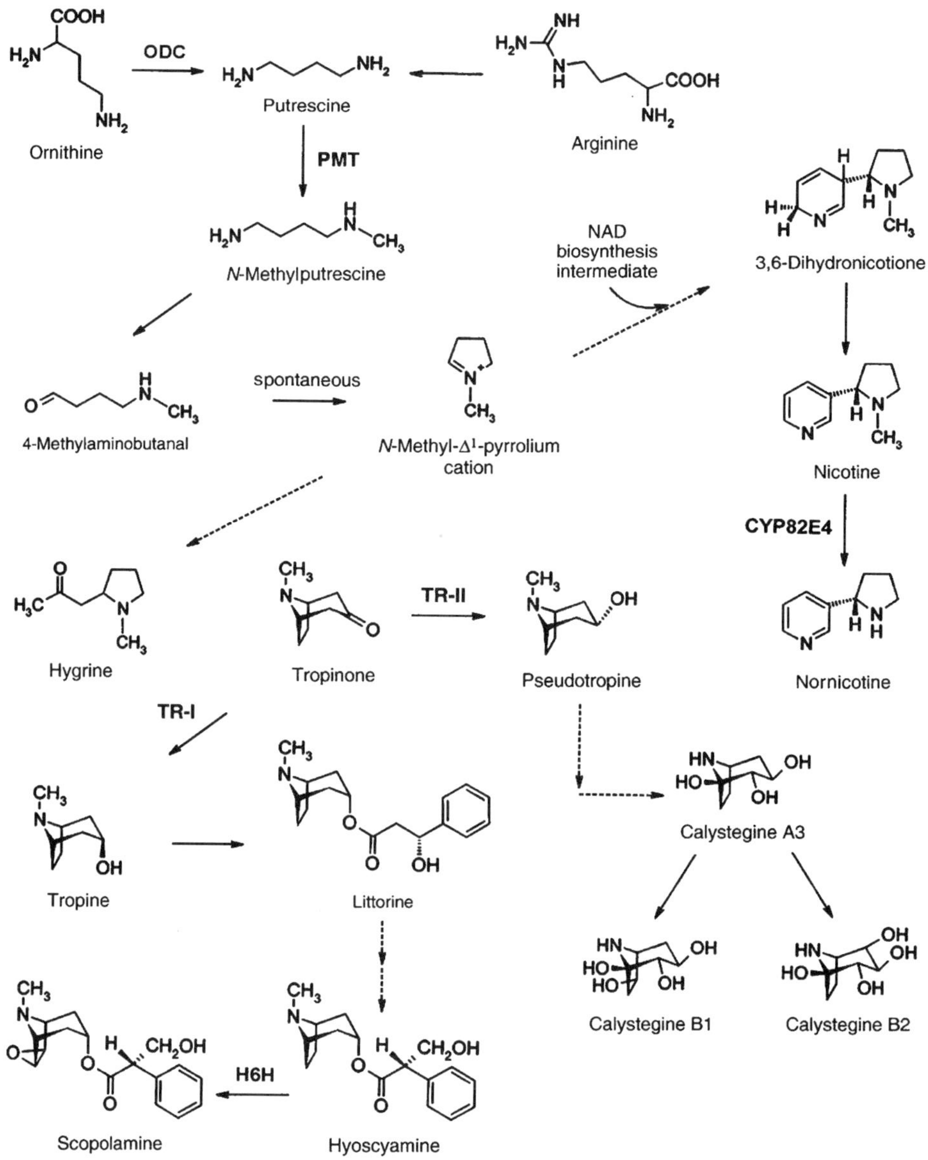


Scheme 3. Biosynthesis of the tropane alkaloids hyoscyamine and scopolamine, the calystegines, and nicotine. Molecular clones have been isolated for the enzymes shown. Abbreviations: CYP82E4, nicotine *N*-demethylase; H6H, hyoscyamine 6$\beta$-hydroxylase; ODC, ornithine decarboxylase; PMT, putrescine *N*-methyltransferase; TR-I, tropinone reductase-I; TR-II, tropinone reductase-II.

alkaloid-producing species *A. belladonna* and *H. niger*, but not in tobacco, which accumulates nicotine rather than tropane alkaloids (*134–136*). Hyoscyamine and scopolamine are derived from tropine *via* TR-I, whereas TR-II yields pseudotropine, which is converted into calystegines and other nortropane alkaloids (*137,138*). Crystallization of TR-I and TR-II revealed the structural basis for the different reaction stereospecificities in the same protein fold of these highly related enzymes (*139*).

Hyoscyamine is produced by the condensation of tropine and the phenylalanine-derived intermediate tropic acid, although the pathway is not fully understood (*140,141*). Hyoscyamine can be further converted into its epoxide scopolamine by 6$\beta$-hydroxylation of the tropane ring followed by intramolecular epoxide formation *via* removal of the 7$\beta$-hydrogen (Scheme 3). Both reactions are catalyzed by a 2–oxoglutarate-dependent dioxygenase, hyoscyamine 6$\beta$-hydroxylase (H6H), for which cDNAs have been isolated from *H. niger* (*142*), *A. belladonna* (*143*), and *Anisodus tanguticus* (*144*). Only scopolamine-producing plants exhibit H6H activity (*145*). The distribution of hyoscyamine in the Solanaceae is wider than that of scopolamine suggesting that only certain phylogenetic lineages acquired the gene encoding H6H (*146*).

The active principle in most members of the genus *Nicotiana* is nicotine, which exerts physiological effects, including addiction, by acting on nicotinic acetylcholine receptors. Nicotine is composed of a pyridine ring joined to an *N*-methylpyrrolidine ring. The pyridine ring is derived from quinolinic acid and the pyrrolidine ring, as with the tropane alkaloids, comes from putrescine. The nicotine biosynthetic pathway consists of at least eight enzymatic steps *via* the *N*-methyl-$\Delta^1$-pyrrolinium cation (Scheme 3), which is thought to condense with an intermediate from the nicotinamide adenine dinucleotide (NAD) pathway (either 1,2–dihydropyridine or nicotinic acid) to form 3,6–dihydronicotine. This reaction is apparently catalyzed by the still poorly defined enzyme nicotine synthase (*147*). Genes encoding PMT, the first committed step in the biosynthesis of the *N*-methyl-$\Delta^1$-pyrrolinium cation, and quinolinate phosphoribosyltransferase (QPRtase), the enzyme responsible for the conversion of quinolinic acid into nicotinic acid, have been cloned and characterized from tobacco (*148,149*). The final step in nicotine biosynthesis is the removal of a proton from the pyridine ring by a predicted 1,2–dihydronicotine dehydrogenase enzyme. Nornicotine, produced by the *N*-demethylation of nicotine, is an undesirable alkaloid because it serves as the precursor of the carcinogen *N′*-nitrosonornicotine during the curing and processing of tobacco. A molecular clone was recently isolated that encodes nicotine *N*-demethylase (CYP82E4), a P450-dependent monooxygenase that catalyzes the conversion of nicotine into nornicotine (*150*).

## D. PURINE ALKALOIDS

Three groups of plants that produce purine alkaloids can be identified on the basis of the types of alkaloids that accumulate: caffeine-producing plants such as coffee, tea, and mate (*Ilex paraguariensis*) (*151*); theobromine-producing plants represented by cacao, cocoa tea (*Camellia ptilophylla*), and *C. irrawadiensis* (*152*); and methyluric acid-producing plants including *Coffea dewevrei*, *C. liberica*, *C. excelsa* and kucha (*Camellia assamica*). Caffeine, the world's most widely used pharmacologically active substance, is primarily ingested in coffee, tea, and cola soft drinks (*153*). As an antagonist of endogenous adenosine receptors, caffeine causes vasoconstriction and increases blood pressure. However, the acute and chronic risks and benefits of caffeine use are not fully understood.

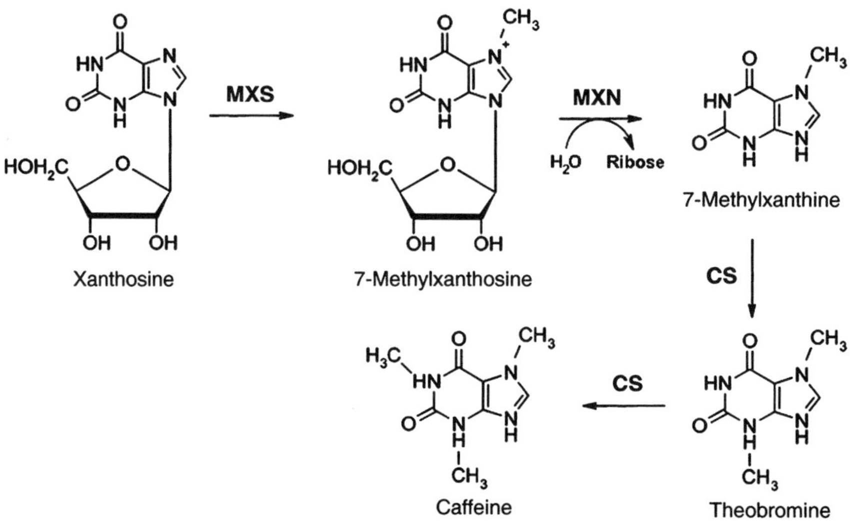


Scheme 4. Biosynthesis of the purine alkaloids caffeine and theobromine. Molecular clones have been isolated for all enzymes shown. Abbreviations: CS, caffeine synthase; MXN, 7–methylxanthosine nucleosidase; MXS, 7–methylxanthosine synthase.

Caffeine is produced from xanthosine *via* three distinct *N*-methylations (Scheme 4) (*154–158*). Partially purified enzyme extracts from tea (*Camellia sinensis*) and coffee (*Coffea arabica*) were shown to exhibit all three activities, suggesting either that the NMT steps in caffeine biosynthesis are catalyzed by a single enzyme, or by multiple enzymes with similar properties (*159*). Three SAM-dependent NMT activities are found in young tea leaves, but are absent in fully developed leaves (*160*). However, a specific NMT purified from coffee was active only toward 7–methylxanthine and theobromine (*161*). An NMT catalyzing the methylation of methylxanthines and designated as caffeine synthase (CS) was purified from tea (*162*). CS catalyzes two consecutive methylations involved in the conversion of 7–methylxanthine into caffeine, but is inactive toward xanthosine indicating that the first methylation proceeds *via* a different enzyme. Heterologous expression of the CS cDNA showed that the enzyme was active toward 7–methylxanthine, paraxanthine, and theobromine (Scheme 4) (*162*). However, a cDNA was recently isolated from coffee encoding a methyltransferase (XMT) that catalyzes the conversion of 7–methylxanthine into theobromine (Scheme 4) (*163*). XMT did not accept other biosynthetic intermediates, suggesting that caffeine is synthesized *via* several independent methyltransferases in coffee.

## E. PYRROLIZIDINE ALKALOIDS

Pyrrolizidine alkaloids are the leading plant toxins that have deleterious effects on the health of humans and animals. Over 360 different pyrrolizidine alkaloids are found in approximately 3% of the world's flowering plants. These noxious natural products are primarily restricted to the Boraginaceae (many genera), Asteraceae (tribes Senecionae and Eupatoriae), Fabaceae (mainly the genus *Crotalaria*), and

Orchidaceae (nine genera). Additional pyrrolizidine alkaloid-producing plants are scattered throughout six other unrelated families (*164*).

Most pyrrolizidine alkaloids are esters of basic alcohols known as necine bases. The most frequently studied pyrrolizidine alkaloids are formed from the polyamines, putrescine and spermidine, and possess one of three common necine bases: retronecine, heliotridine, and otonecine. Putrescine is utilized exclusively as a substrate in secondary metabolism, whereas spermidine is a universal cell-growth factor involved in many physiological processes in eukaryotes. Spermidine biosynthesis begins with the decarboxylation of SAM by SAM decarboxylase (*165*). The aminopropyl group is then transferred from decarboxylated SAM to putrescine by spermidine synthase to form spermidine (Scheme 5). Putrescine can be produced from ornithine by ODC. However, putrescine is derived from the arginine–agmatine pathway in pyrrolizidine alkaloid-producing plants due to the absence of ODC activity (*166*).

Homospermidine, the first pathway-specific intermediate in pyrrolizidine alkaloid biosynthesis, is formed from putrescine and spermidine by homospermidine

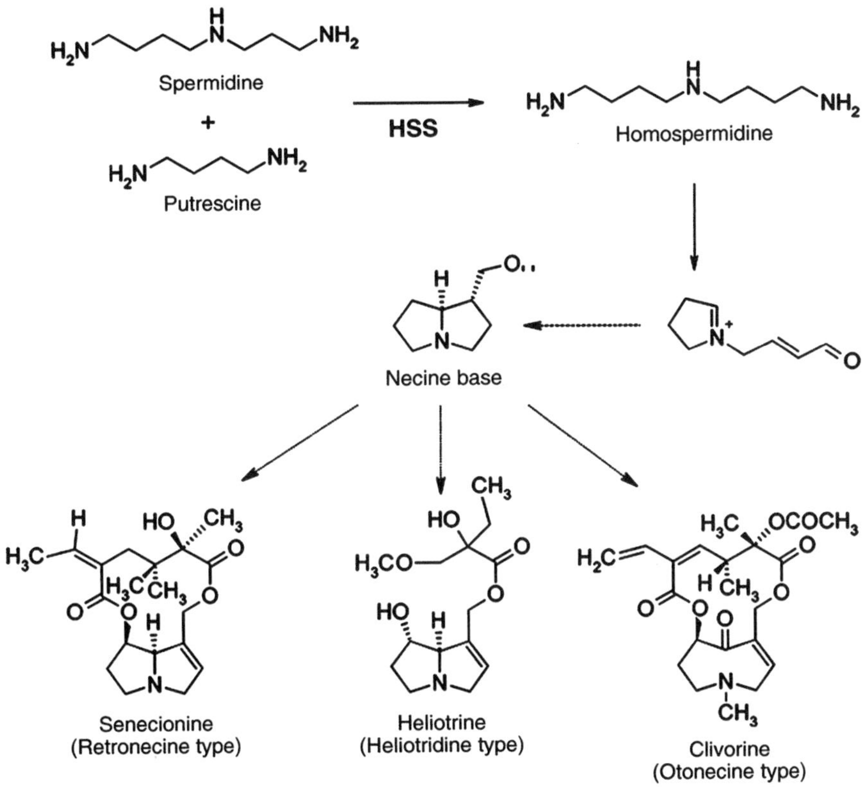


Scheme 5. Biosynthesis of pyrrolizidine alkaloids. A molecular clone has been isolated for only one biosynthetic enzyme. Abbreviation: HSS, homospermidine synthase.

synthase (HSS) (*167*). Homospermidine formation is the only known example of a functional moiety of spermidine used in a secondary metabolic pathway. The necine base moiety is formed from homospermidine *via* consecutive oxidative deaminations (*167*) and subsequently converted into senecionine (Scheme 5). Molecular clones encoding HSS have been cloned from several pyrrolizidine alkaloid-producing plants (*168*) and best characterized from *Senecio vernalis* (*167,169*) and *Eupatorium cannabinum* (*170*).

HSS exhibits strong sequence similarity to deoxyhypusine synthase (DHS), a ubiquitous enzyme responsible for the activation of eukaryotic initiation factor 5A (eIF5A) (*171*). DHS catalyzes the transfer of an aminobutyl moiety from spermidine to a lysine residue of eIF5A. In contrast, HSS does not accept eIF5A as a substrate (*172*) except, possibly, in rare cases (*173*). However, HSS and DHS both catalyze the formation of homospermidine by the aminobutylation of putrescine, although this reaction is generally not catalyzed by DHS *in vivo* (*167,171*). HSS is thought to have evolved from DHS after duplication of the single-copy *DHS* gene in at least four independent phylogenetic lineages, including the genera *Senecio* and *Eupatorium*, both of which are part of the Asteraceae family (*168*). The product of these duplications appears to have lost the ability to interact with eIF5A, but has retained HSS activity. HSS is a well-documented example of the evolutionary recruitment of a primary metabolic enzyme into a secondary metabolic pathway (*174*).

## F. QUINOLIZIDINE ALKALOIDS

Quinolizidine alkaloids are mainly distributed in the family Fabaceae and are, in particular, produced by members of the genus *Lupinus*. The seeds of *Lupinus* spp. are rich in protein and have long been used by humans as food and livestock feed. In developing countries, cultivated or wild lupins are often the major source of protein in the human diet, and are also used to feed domestic animals because the plants can be grown in poor soil not suitable for the cultivation of traditional crops. Quinolizidine alkaloids are generally toxic compounds that must be removed by soaking seeds in water prior to consumption. The development of commercial cultivars of "sweet" varieties of *L. angustifolius* with low quinolizidine alkaloid content has expanded the potential use of lupin as a food source. However, quinolizidine alkaloids are involved in the defense response in plants and consequently, sweet lupin varieties are less resistant to pathogens and herbivores (*175*).

The biochemistry and molecular biology of quinolizidine alkaloid biosynthesis have not been fully characterized. Quinolizidine alkaloids are formed from lysine *via* lysine decarboxylase (LDC), where cadaverine is the first detectable intermediate (Scheme 6). Biosynthesis of the quinolizidine ring is thought to arise from the cyclization of cadaverine units *via* an enzyme-bound intermediate (*176*). LDC and the quinolizidine skeleton-forming enzyme have been detected in chloroplasts of *L. polyphyllus* (*177*). Once the quinolizidine skeleton has been formed, it is modified by dehydrogenation, hydroxylation, or esterification to generate the diverse array of alkaloid products.

Lupins produce both tetracyclic (e.g., lupanine) and bicyclic quinolizidine alkaloids (i.e., lupinine), which typically accumulate as esters of tiglic acid, *p*-coumaric

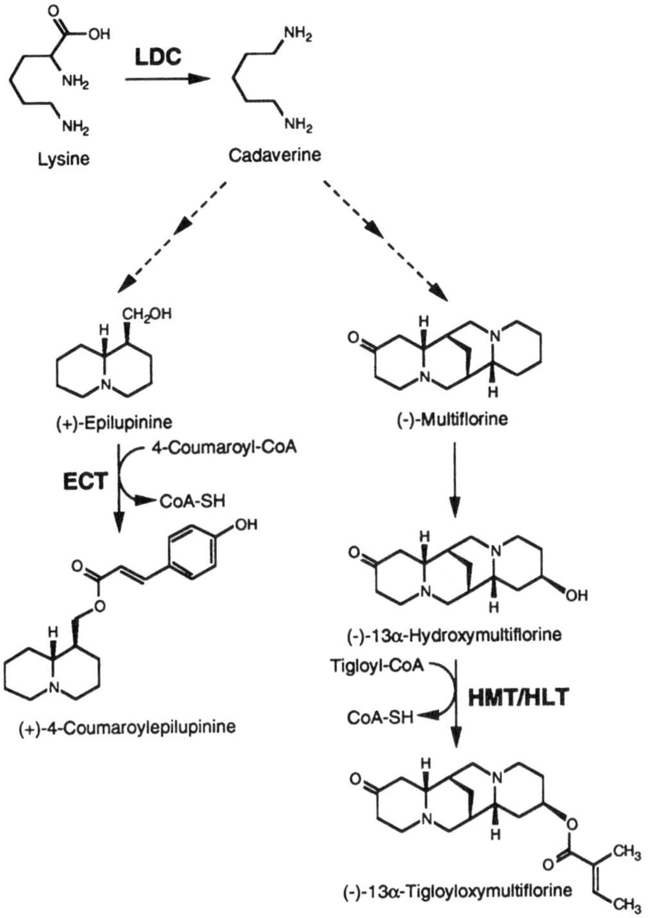


Scheme 6. Biosynthesis of the quinolizidine alkaloids (+)-4-coumaroylepilupinine and (–)-13α-tigloyloxymultiflorine. Molecular clones have been isolated for the enzymes shown. Abbreviations: ECT, *p*-coumaroyl-coenzymeA:(+)-epilupinine *O*-*p*-coumaroyltransferase; HMT/HLT, tigloyl-coenzymeA:(–)-13α-hydroxymultiflorine/(+)-13α-hydroxylupanine *O*-tigloyltransferase; LDC, lysine decarboxylase.

acid, acetic acid, and ferulic acid (*176*). The biological significance of these esters is not clear. Some species lack the acyltransferase required for ester formation and accumulate quinolizidine alkaloid aglycones. The substrate specificity of these acyltransferases largely defines the alkaloid profile of different lupin species (Scheme 6). Two acyltransferases involved in quinolizidine alkaloid ester biosynthesis have been characterized. The transfer of a tigloyl group from tigloyl-CoA to the 13α-hydroxyl of 13α-hydroxymultiflorine or 13α-hydroxylupanine is catalyzed by tigloyl-CoA:(–)-13α-hydroxymultiflorine/(+)-13α-hydroxylupanine *O*-tigloyltransferase (HMT/HLT) (*178*). Molecular cloning of HMT/HLT showed that the enzyme belongs to a unique subfamily of the BAHD superfamily of plant acyltransferases (*179*). A second acyltransferase, *p*-coumaroyl-CoA: (+)-epilupinine *O*-*p*-coumaroyltransferase (ECT) transfers *p*-coumaroyl to the hydroxyl moiety of (+)-epilupinine (*178*). Although bitter and sweet lupins exhibit distinct alkaloid accumulation profiles, there are no discernable differences in acyltransferase activity between such varieties.

## III. Regulation of Alkaloid Biosynthesis

### A. DEVELOPMENTAL AND ENVIRONMENTAL CONTROLS

#### *1. Light-Induced Regulation*

Several enzymes – TDC, STR, T16H, NMT, D4H, and DAT – involved in vindoline biosynthesis are developmentally controlled in *C. roseus* seedlings, and some are further regulated by light (*98,101,180*). However, the light activation of vindoline biosynthesis does not require cytomorphogenesis (*181*). Dark-grown *C. roseus* seedlings accumulate tabersonine and smaller amounts of other vindoline pathway intermediates (*100*). The transfer of etiolated seedlings to light caused the turnover of tabersonine, and other intermediates, to vindoline, suggesting that some enzymes in the pathway are regulated by light. The light-activation of T16H probably occurs at the transcriptional level (*98,99*). NMT and D4H activities also increased when etiolated seedlings were exposed to light (*105,180*). The *D4H* gene appears to be under multi-level developmental and light-mediated regulation in *C. roseus* because D4H activity was low in etiolated seedlings despite an abundance of transcripts (*110,182*). Exposure of etiolated seedlings to light caused a rapid increase in D4H activity without an increase in transcript levels (*110*). Despite the presence of only one *D4H* gene in *C. roseus*, several protein isoforms were detected in etiolated and light-grown seedlings, suggesting that D4H is affected by light at the post-transcriptional level (*182*). DAT activity was also found to increase in response to light, but the induction appears to occur only at the transcriptional level (*111,180*). Phytochrome was implicated in the activation of vindoline biosynthesis by light (*183*), and was shown to control *D4H* gene expression (*182*).

#### *2. Inducible Regulation*

Several alkaloid pathways are induced by treatment with fungal elicitors, heavy metal ions, UV radiation, and osmotic shock. Typically, work on the inducible regulation of alkaloid biosynthesis has relied on the use of cell cultures treated with fungal elicitors. For example, treatment of *C. roseus* cultures with a fungal elicitor typically increases the accumulation of tryptamine and monoterpenoid indole alkaloids, such as catharanthine (*184,185*). In contrast, the biosynthesis of vindoline and, thus, vinblastine is blocked in *C. roseus* cell cultures due, at least in part, to the transcriptional inactivation of the *DAT* gene (*186*). Fungal elicitors were shown to induce *TDC*, *STR*, and *SGD* gene expression, suggesting that elicitor-mediated signal transduction pathways are relatively short and activate pre-existing transcription factors (*90,94*). Fungal elicitors also induce jasmonic acid (JA) biosynthesis in *C. roseus* cell cultures (Fig. 1A) (*187,188*). Treatment of *C. roseus* seedlings with MeJA caused an increase in TDC, STR, D4H, and DAT activities and enhanced vindoline accumulation (*189*). TDC and STR transcript levels were also induced in *C. roseus* cultures treated with MeJA or the JA precursor $\alpha$-linolenic acid (*187*). MeJA also promoted the biosynthesis of tabersonine 6,7–epoxidase, which catalyzes the NADPH- and $O_2$-dependent selective epoxidation of tabersonine to

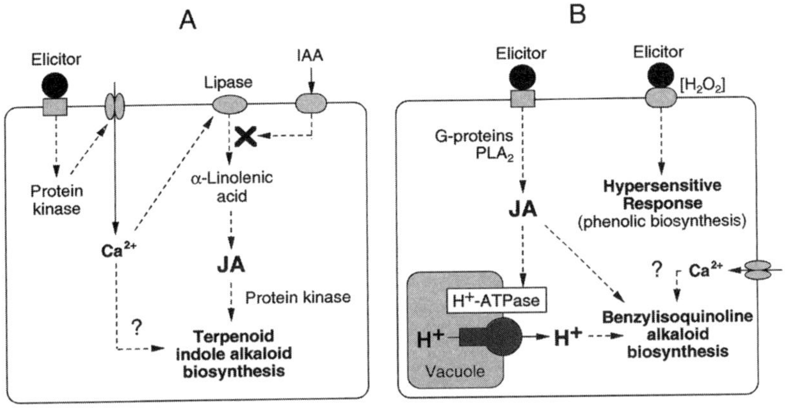


Fig. 1. Stimulus-response coupling models for the transcriptional activation of alkaloid biosynthesis in plant cell cultures. (A) Monoterpenoid indole alkaloids; (B) Benzylisoquinoline alkaloids. Abbreviation: JA, jasmonic acid.

lochnericine, in *C. roseus* root cultures (*190*). Inhibition by CO, clotrimazole, miconazole, and cytochrome c suggested that the enzyme was a P450-dependent monooxygenase. A JA biosynthetic inhibitor, diethyldithiocarbamic acid, blocked elicitor-induced JA formation, and TDC and STR activation. MeJA – and salicylic acid – also induced a fourfold increase in the accumulation of the alkaloid pilocarpine in *Pilocarpus jaborandi* seedlings (*191*).

## B. SIGNAL TRANSDUCTION

The elicitor-induced biosynthesis of JA and the MeJA-induced expression of *TDC* and *STR* were blocked by the protein kinase inhibitor K-252a, suggesting the involvement of protein phosphorylation in the signal transduction pathway (Fig. 1A). The JA-mediated induction of TDC and D4H in *C. roseus* seedlings involves transcriptional and post-translational controls (*192*). For example, MeJA treatment activated *TDC* expression and increased protein stability, but did not result in higher enzyme activity in light- or dark-grown seedlings. Similarly, exogenous MeJA induced D4H activity and protein levels, but only in light-grown seedlings.

In *C. roseus* cell cultures, the coordinate expression of several monoterpenoid indole alkaloid biosynthetic genes is induced by treatment with fungal elicitors. A yeast-derived elicitor was shown to activate the protein phosphorylation- and calcium flux-dependent production of reactive oxygen species (ROS) as a second messenger (*193*). However, ROS generation was not required for the elicitor-mediated induction of alkaloid biosynthetic genes, suggesting that the oxidative burst is coupled to the activation of other defense genes. The necessity of a functional octadecanoid pathway was shown in *C. roseus* cells cultured in an auxin-starved medium (*194*). Auxin inhibited monoterpenoid indole alkaloid accumulation and reduced *TDC* transcription (*73*), whereas alkaloid biosynthesis could be induced in an auxin-free medium (*195*). Exogenous MeJA restores the ability of cells grown in the presence of auxin to produce alkaloids (*194*). In cells cultured in auxin-free media, MeJA or JA treatment causes a further increase in alkaloid accumulation, whereas alkaloid production is reduced in auxin-starved cells treated with octadecanoid pathway inhibitors. These results suggest that JAs are produced in response to auxin depletion and function in coordinating biochemical events that lead to alkaloid biosynthesis (Fig. 1A). However, despite the role of JAs in linking physiological and environmental signals to alkaloid biosynthesis, JA treatment of etiolated seedlings did not enhance TDC activity, nor did it replace the light requirement for *D4H* expression (*192*). In plants, JAs might function only to modulate events already controlled by other mechanisms.

A role for the CaaX-prenyltransferases, protein farnesyltransferase (PFT) and type-I protein geranylgeranyltransferase (PGGT-I), in regulating early monoterpenoid indole alkaloid biosynthetic genes, including those encoding enzymes involved of the MEP pathway, was demonstrated by the RNA interference (RNAi)-mediated silencing of the corresponding genes in *C. roseus* cell cultures (*196*). The induced expression of a subset of genes by auxin depletion in wild-type cells was suppressed in a cell line with post-transcriptionally silenced *PFT* and *PGGT*-I genes. Protein targets for both PFT and PGGT-I appear necessary for the expression of several early alkaloid biosynthetic genes in *C. roseus*.

Leaf damage caused by herbivores also increased JA and nicotine levels in *Nicotiana sylvestris* roots (*197*). Exogenous MeJA induced nicotine production in roots, suggesting that JAs function in transferring the damage signal from the shoot to the root. MeJA also caused an increase in ODC, PMT, and *S*-adenosylmethionine synthetase (SAMS) transcript levels, and the accumulation of putrescine, *N*-methylputrescine, and nicotine in tobacco cell cultures (*198*). However, the induction of *SAMS* and *PMT* gene expression by MeJA was blocked by cycloheximide, whereas *ODC* expression was not, suggesting that multiple control mechanisms are involved in the MeJA-mediated regulation of nicotine biosynthesis. Recently, it was shown that insects could reduce the JA-induced accumulation of nicotine by regulating *PMT* gene expression through the activation of ethylene production (*199*). In contrast, a fusion of the tobacco *PMT3* gene promoter to the *β-glucuronidase* (*GUS*) reporter gene showed constitutive expression only in roots, but wound-inducible expression is localized around wound sites in leaves of transgenic tobacco (*200*). Jasmonate signaling did not appear to contribute to the transient, wound-induced expression of the *PMT:GUS* transgene.

Cell cultures of several Papaveraceae species accumulate macarpine, sanguinarine, and other benzophenanthridine alkaloids in response to treatment with fungal elicitors *via* the transcriptional activation of biosynthetic genes (*24,201–204*). In general, membrane-associated enzymes are induced by treatment with elicitor or MeJA (*205*). However, some cytosolic enzymes, such as TYDC (*206*), NCS (*13*), and DBOX were also induced in response to specific treatments (*207*). The induction of benzophenanthridine alkaloid biosynthesis in *E. californica* has been shown to occur at elicitor concentrations below the threshold required to stimulate events associated with the hypersensitive response, such as the production of phenolic compounds (Fig. 1B) (*208*). Phenolic production could be selectively blocked by catalase at higher elicitor concentrations, suggesting that alkaloid biosynthesis is coupled to a signal transduction pathway which is not mediated by ROS involved in activating the classic hypersensitive response. Uncoupled induction mechanisms for phenylalanine ammonia lyase (PAL), TYDC, and downstream enzymes of sanguinarine biosynthesis have also been demonstrated in opium poppy cell cultures (*206*). The elicitor-mediated activation of BA biosynthesis appears to require a transient decrease in cytosolic pH caused by an efflux of protons from the vacuole (Fig. 1(B)) (*208*). Artificial acidification of the cytosol induced alkaloid biosynthesis, but not the hypersensitive response, whereas a depletion of vacuolar protons blocked the increase in alkaloid accumulation. The induction of BA accumulation in *E. californica* cell cultures by yeast glycoproteins, MeJA, artificial acidification, or mild osmotic stress was shown to attenuate a subsequent response to the same stimulus – a phenomenon known as homologous desensitization (*209*). Yeast elicitor and artificial acidification resulted in mutual desensitization, whereas elicitor-treated cells remained responsive to MeJA and osmotic stress. Moreover, the transient acidification of the cytoplasm required for elicitor-mediated signal–response coupling was not detected after jasmonate treatment (*209*). Both jasmonate-dependent and jasmonate-independent, ΔpH-controlled signaling pathways, appear to be involved in the induction of alkaloid pathways in *E. californica* (Fig. 1B).

Phospholipase $A_2$ ($PLA_2$), a G-protein-controlled redox-dependent plasma membrane protein, might trigger the signal transduction pathway leading to the efflux of vacuolar protons (Fig. 1B) (*210*). The role of G-proteins in the induction of BA biosynthesis was also detected in *S. canadensis* cells treated with modulators of GTP-binding proteins and G-protein activators (*211*). The induction of BA biosynthesis also appears to depend on an external source of $Ca^{2+}$, suggesting that $Ca^{2+}$ and perhaps calmodulin participate in the signal transduction pathway (Fig. 1B) (*203*).

## C. PROMOTER ELEMENTS AND TRANSCRIPTIONAL REGULATORS

Recent studies have identified several promoter elements and transcription factors involved in the regulation of *TDC*, *STR1*, and *CPR*. Three regions involved in basal or elicitor-induced expression were detected from 160 to −37 in the *TDC* promoter (*212*). The region from −160 to −99 acted as the main transcriptional enhancer for basal expression, and two separate elicitor-responsive elements were found from −99 to −87, and from −87 to −37. *In vitro* binding of factors in tobacco and *C. roseus* nuclear protein extracts identified two binding activities that interact with multiple *TDC* promoter regions as GT-1 and 3AF1 (*213*). Mutagenesis of GT-1 binding sites did not affect basal or elicitor-induced expression, but reduced *TDC* promoter activation by UV light (*213*). Only UV-B wavelengths were shown to induce TIA accumulation and TDC expression in *C. roseus* suggesting the involvement of a specific receptor (*213,214*). Loss-of-function analysis revealed redundant UV-responsive elements in the *TDC* promoter between −99 and +198 (*214*).

Coordinate transcript accumulation suggests that *TDC*, *STR*, and *CPR* genes are regulated by common nuclear factors in response to elicitor treatment and UV light (*90,215*). Deletion of the *CPR* promoter to −366 eliminated the elicitor-inducible expression observed with a longer promoter (*216*). The −632 to −366 region of the *CPR* promoter was also shown to contain GT-1 binding sites. The main elicitor-responsive elements of the *STR1* promoter were identified from −339 to −145 (*217*). Again, GT-1 was shown to bind to this and other regions of the *STR1* promoter. Despite the interaction of GT-1 with *TDC, STR,* and *CPR* promoters, the *in vivo* role of this transcription factor in the regulation of gene expression is unknown.

Additional transcriptional regulators must also participate in the expression of *TDC, STR1, CPR,* and other genes involved in monoterpenoid indole alkaloid biosynthesis. A G-box motif at −105 binds G-box binding factors (GBFs) *in vitro*, but is not essential for the elicitor-induced expression of *STR1 in vivo* (*217*). This G-box element also interacts with tobacco nuclear factors and the GBF 3AF1 (*218*). A G-box tetramer, fused only to a TATA-box, confers seed-specific expression in transgenic tobacco, but requires the enhancer region from the CaMV promoter for expression in leaves. Thus, sequences flanking the G-box motif might be necessary for *STR1* promoter activity in different tissues (*218*).

Other binding factors (CrGBF-1 and CrGBF-2) that bind to the G-box motif in the *STR* promoter were isolated using a yeast one-hybrid screen (*219*). A GCC-box element from −100 to −58 of the *STR1* promoter was necessary and sufficient for JA- and elicitor-responsive expression (*220*). Two cDNAs encoding octadecanoid-derivative-responsive *C. roseus* APETALA2-domain (ORCA) proteins that bind the JA- and elicitor-responsive elements of *STR1* were isolated by yeast one-hybrid screening. The ORCA2 protein *trans*-activates the *STR1* promoter, and expression of the *ORCA2* gene is induced by elicitor and JA treatment of *C. roseus* cell cultures. In contrast, the *ORCA1* gene is constitutively expressed. The closely related *ORCA3* gene was cloned using T-DNA activation tagging and shown to enhance the expression of several monoterpenoid indole alkaloid biosynthetic genes – *TDC, STR, SGD, CPR,* and *D4H* – when overexpressed in *C. roseus* cultures (*221,222*). ORCA3 was shown to activate gene expression *via* an interaction with a JA-responsive promoter element (*223*). Yeast one-hybrid screening using the STR promoter as bait identified a MYB-like protein in *C. roseus* (CrBPF-1) with homology to a box P binding factor (BPF-1) (*224*). Unlike the ORCA transcription factors, CrBPF-1 was rapidly activated by elicitor, but not by JA. The induction of CrBPF-1 appears to depend on protein kinase activity and cytosolic calcium concentration (Fig. 1A). These data illustrate the existence of two distinct transcriptional regulators capable of directing *STR* expression. Yeast one-hybrid screening also identified three members (ZCT1, ZCT2, and ZCT3) of the Cys(2)/His(2)-type (transcription factor IIIA-type) zinc-finger protein family that bind to elicitor-responsive promoter elements of the *TDC* and *STR* genes of *C. roseus* (*225*). The ZCT proteins, which were transcriptionally induced by yeast elicitor and MeJA, repressed the activation of the *TDC* and *STR* genes and, in particular, the activating process of ORCA transcription factors on the *STR* promoter.

Deletion analysis has only begun to reveal the location of regulatory domains in BA biosynthetic gene promoters (*226*). The –393 to –287 regions of the *TYDC7*, and the −355 to −200 regions of the *BBE1* promoters were functionally required in a transient GUS assay system based on the microprojectile bombardment of opium poppy cell cultures. Time courses for the induction of TYDC7 and BBE1 mRNAs in wounded opium poppy cells were nearly identical to those for GUS activity in cells bombarded with promoter-*GUS* constructs when the regions from –393 to –287 in *TYDC7* or –355 to –200 in *BBE1* were present. These results suggest that the wound signal caused by the DNA-coated microcarriers induced wound-responsive regulatory elements located from –393 to –287 in *TYDC7*, and from –355 to –200 in *BBE1*. Functional analysis of the *BBE1* promoter from *E. californica* identified the region from –496 to –455 as necessary for basal activity (*43*). Comparison of this region to the –355 to –200 domain in opium poppy *BBE1* revealed 55% nucleotide identity within a 40–base pair domain (*226*).

## D. METABOLIC CHANNELS AND ENZYME COMPLEXES

It is generally assumed that many metabolic enzymes do not function in isolation, but interact physically, or are in close proximity with, other enzymes that participate in common pathways (*227,228*). Structures composed of interacting proteins involved in specific metabolic processes are known as metabolic complexes, metabolic channels, metabolons, or multi-enzyme complexes. A metabolic channel is a complex of two or more enzymes that facilitates the direct transfer of biosynthetic intermediates between the catalytic sites of sequential enzymes in a biosynthetic pathway. Theoretically, metabolic channels provide several advantages that promote efficient cellular metabolism. The direct transfer of a pathway-intermediate from one enzyme to another maintains a high local substrate concentration, which avoids the dilution of intermediates released into the cytoplasm. Enzyme complexes also eliminate competition from other enzymes for the same substrate, increase the stability of intermediates, and minimize the deleterious effects of cytotoxic intermediates. Several investigations of the interactions between pathway enzymes have hinted at the importance of metabolic channels in primary and secondary metabolism (*228,229*).

Recent studies have implicated a role for enzyme complexes in alkaloid biosynthetic pathways. Diamine oxidase, which participates in the deamination of methylated putrescine as an early step in nicotine biosynthesis, was purported to associate with *S*-adenosylhomocysteine hydroxylase, an enzyme involved in the formation of SAM (*230*). A metabolic channel in BA metabolism was suggested by unexpected results obtained from the RNAi-mediated silencing of *COR* genes in opium poppy (*231*). Although seven enzymatic steps occur between (*S*)-reticuline and codeinone (i.e., the substrate for COR), only (*S*)-reticuline accumulates at elevated levels. No intermediates between (*S*)-reticuline and morphine were detected. Removal of COR was suggested to disrupt a metabolic channel composed of morphinan branch pathway enzymes resulting in the accumulation of an alkaloid intermediate produced by enzymes that are not a part of the same complex. Interestingly, thebaine and oripavine, intermediates upstream of COR substrates, accumulate to high levels in some opium poppy cultivars; thus, an accumulation of morphinan branch pathway intermediates was expected.

Despite the detection of COR and other known morphinan branch pathway enzymes in opium poppy cell cultures (*63–65,67*), morphine and codeine are not produced. The inability of cultured cells to accumulate morphine might reflect the absence of other relevant enzymes or structural proteins that preclude the formation of a requisite metabolic channel. Nevertheless, the efficient operation of the sanguinarine pathway indicates that cultured opium poppy cells possess the essential regulatory, cellular, and metabolic components to support the biosynthesis and accumulation of highly substituted and cytotoxic BAs. The assembly of metabolic channels involved in the biosynthesis of (*S*)-reticuline and the benzophenanthridine branch pathway were suggested by the antisense RNA-mediated suppression of CYP80B1 and BBE in transgenic California poppy cell and root cultures (*232,233*). Substantially decreased CYP80B1 and BBE levels result in the overall suppression of BA biosynthesis. As reported for the RNAi-mediated silencing of COR (*231*), accumulation of the expected pathway intermediates, (*S*)-*N*-methylcoclaurine and (*S*)-reticuline, does not occur.

## E. CELLULAR COMPARTMENTALIZATION AND TRAFFICKING

Alkaloids generally accumulate in specific cell types due to their cytotoxicity and probable role in plant defense responses. For example, alkaloids are sequestered to isolated idioblasts and laticifers in *C. roseus* (*234,235*), root endodermis and stem cortex or pith in *T. flavum* (*236*), and laticifers in opium poppy (*237*). Within cells, alkaloids are often stored in discrete vesicles or the central vacuole (*238*). The cell type- and organelle-specific accumulation of alkaloids has prompted studies on the cellular and subcellular localization of the relevant biosynthetic pathways. The cell biology of alkaloid biosynthesis is remarkably diverse and complex (Fig. 2).

### 1. *Intercellular Trafficking*

Enzymes involved in the biosynthesis of monoterpenoid indole alkaloids have been localized to several different cell types in *C. roseus*. The compartmentalization of metabolic reactions to specialized cell types might, in part, explain the inability of de-differentiated *C. roseus* cell cultures to produce alkaloids. TDC and STR are most abundant in *C. roseus* roots, but also occur in photosynthetic organs (*90*). In contrast, T16H (*98*), D4H (*110*), and DAT (*111*) are restricted to young leaves and other shoot organs where vindoline biosynthesis occurs. *In situ* hybridization and immunocyto-chemical localization studies have shown that CYP72A1, TDC, and STR are localized to the epidermis of stems, leaves, and flower buds (Fig. 2B) (*84,235,239*). In roots, these enzymes occur in cells near the apical meristem. In contrast, D4H and DAT are associated with laticifers and idioblasts of shoot organs, but are absent from roots (Fig. 2B). Laticifers and idioblasts are distributed throughout the mesophyll in *C. roseus* leaves, and are often several cell layers away from the epidermis; thus, vindoline biosynthesis involves at least two distinct cell types and requires the inter-cellular translocation of a pathway intermediate. Moreover, gene transcripts encoding enzymes involved in the MEP pathway, along with CYP76B6, were co-localized to the internal phloem parenchyma of young *C. roseus* aerial organs (*240*). These results suggest the translocation of vindoline biosynthetic intermediates from the internal phloem to the epidermis and from the epidermis to laticifers and idioblasts. Interestingly, transcripts encoding protein S (CrPS) were also localized to the internal phloem parenchyma of *C. roseus* developing leaves (*241*). CrPS is a member of the $\alpha/\beta$-hydrolase superfamily and was suggested to play a role in a signal transduction pathway triggering monoterpenoid indole alkaloid biosynthesis. Another cDNA with homology to an E2 ubiquitin-conjugating enzyme was also proposed to participate in alkaloid biosynthesis in *C. roseus* solely on the basis of transcript accumulation characteristics (*242*). However, the identification of cell types involved in vindoline biosynthesis by laser-capture microdissection and reverse transcription-polymerase

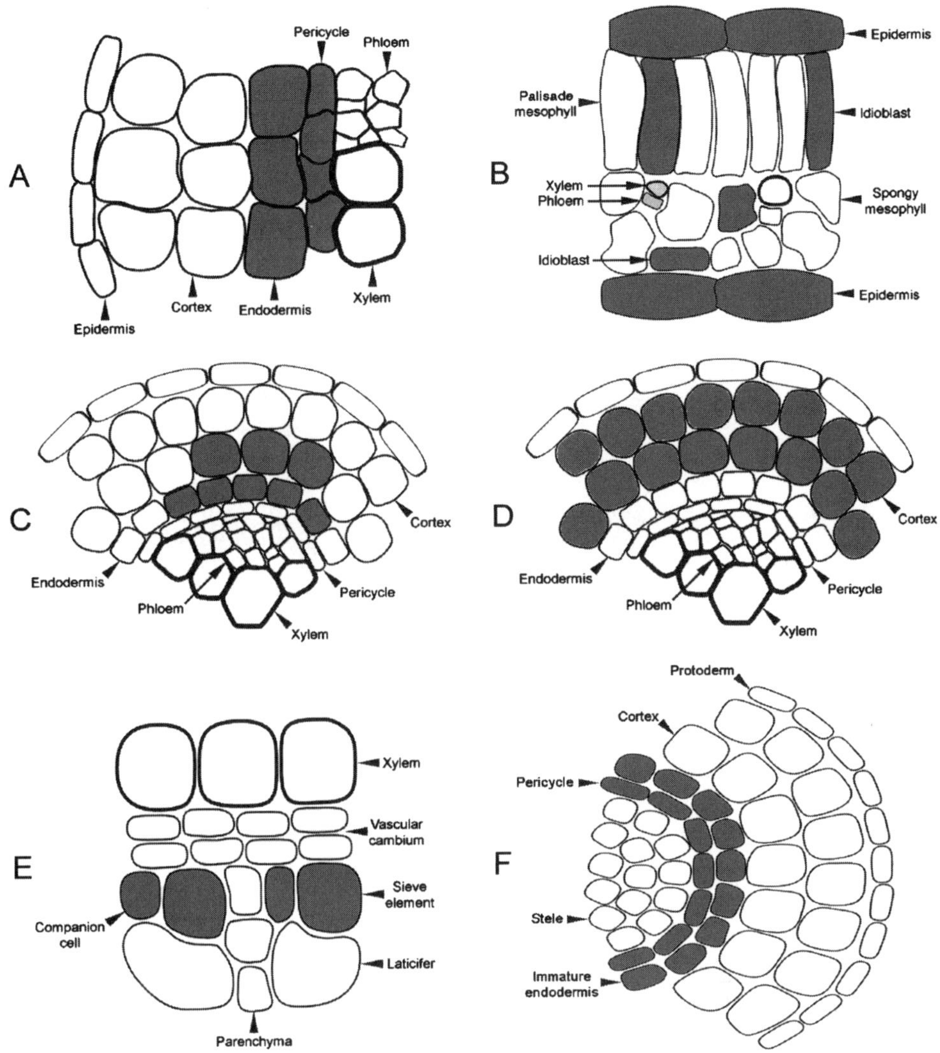


Fig. 2. Alkaloid biosynthetic pathways are associated with a diverse variety of cell types. The tissue-specific localization (shaded) of enzymes and/or gene transcripts are depicted for the biosynthesis of tropane alkaloids in *Atropa belladonna* and *Hyoscyamus niger* roots (A), monoterpenoid indole alkaloids in *Catharanthus roseus* leaves (B), pyrrolizidine alkaloids in *Senecio vernalis* roots (C), pyrrolizidine alkaloids in *Eupatorium cannabinum* roots (D), benzylisoquinoline alkaloids in *Papaver somniferum* vascular bundles (E), and protoberberine alkaloids in *Thalictrum flavum* roots (F).

chain reaction, and also by epidermal enrichment *via* carborundum abrasion, suggested that *C. roseus* leaf epidermis is fully and independently capable of producing 16-methoxytabersonine (Scheme 2) (*243*). The role of internal phloem in the supply of monoterpenoid precursors to alkaloid biosynthesis requires further evaluation.

PMT and H6H, which catalyze the first and last steps, respectively, in the biosynthesis of the tropane alkaloid scopolamine (Scheme 3), were localized to the pericycle in the roots of *A. belladonna* and *Hyoscyamus muticus* (Fig. 2A) (*132,145*). PMT also catalyzes the first step in nicotine biosynthesis (Scheme 3) and has been localized to the endodermis, outer cortex, and xylem in *N. sylvestris* (*244,245*). In contrast, TR-I, an intermediate enzyme in the tropane alkaloid pathway, resides in the endodermis and nearby cortical cells (Fig. 2A) (*135*); thus, intermediates of tropane alkaloid metabolism must also be transported between cell types. The biosynthesis and storage of acridone alkaloids were also associated with endodermis in *Ruta graveolens* (*246*).

In opium poppy, BA accumulation is restricted to the articulated laticifers found adjacent or proximal to sieve elements of the phloem (*204,237*). The cytoplasm of laticifers, known as latex, contains a full complement of cellular organelles in addition to many large vesicles to which alkaloids are sequestered (*237,238*). These vesicles were suggested to originate from dilations of the endoplasmic reticulum (ER) (*237,247*). Laticifers were long considered the site of BA biosynthesis and accumulation (*248–250*). However, the cellular localization of BA biosynthetic enzymes and gene transcripts showed that alkaloid synthesis does not involve laticifers (*7,251–253*). Initial *in situ* hybridization experiments demonstrated the localization of *TYDC* gene transcripts to the phloem, but not to laticifers (*7*). The morphinan pathway enzymes STS and SOR (Scheme 1) were also not detected in isolated latex (*63,64*). Seven other biosynthetic enzymes (i.e., 6OMT, CNMT, CYP80B1, 4′OMT, BBE, SAT, and COR) have been localized to sieve elements in opium poppy and their corresponding gene transcripts to associated companion cells (Fig. 2E) (*251*) (P. Facchini, unpublished results). BA biosynthetic enzymes synthesized in companion cells would be transported to sieve elements through pore-plasmodesmata units (Fig. 3). The localization of BA metabolism to phloem cells distinct from latificers predicts the intercellular transport of pathway intermediates or products, and breaks a long-standing paradigm in plant biology. Previously, sieve elements were not known to support complex metabolism, and were assumed to possess only a limited number of proteins required for cell maintenance and solute transport (*254*). However, the expanding physiological roles for sieve elements include the transport of macromolecules involved in systemic communication and the biosynthesis of JA, ascorbic acid, and defense-related compounds (*255*). Sieve elements clearly possess a previously unrealized biochemical potential.

*T. flavum* accumulates protoberberine alkaloids, such as the antimicrobial agent berberine. *In situ* RNA hybridization analysis revealed the cell type-specific expression of protoberberine alkaloid biosynthetic genes in roots and rhizomes of *T. flavum* (*23*). In roots, gene transcripts for all nine enzymes were localized to the pericycle – the innermost layer of the cortex – and adjacent cortical cells (Fig. 2F). In rhizomes, all biosynthetic gene transcripts were restricted to the protoderm of leaf primordia. The localization of biosynthetic gene transcripts was in contrast to the tissue-specific accumulation of protoberberine alkaloids. In roots, protoberberine alkaloids were restricted to endodermal cells upon the initiation of secondary growth, and distributed throughout the pith and cortex in rhizomes. Thus, the cell type-specific localization of protoberberine alkaloid biosynthesis and accumulation are temporally and spatially separated in *T. flavum* roots and rhizomes, respectively. Similarly, although berberine accumulates in the main roots of *C. japonica*, the low levels of SOMT activity and protein suggest that the main root is not the main site of

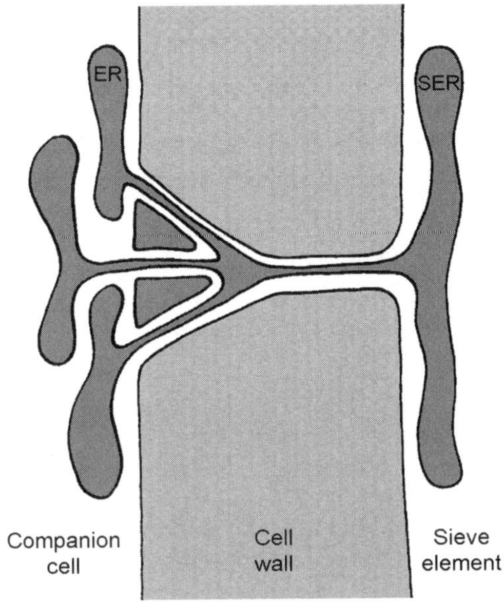


Fig. 3. Putative protein trafficking mechanism between companion cells and sieve elements. In opium poppy, BA biosynthetic enzymes synthesized in companion cells would be transported to sieve elements through pore-plasmodesma units. Abbreviations: ER, endoplasmic reticulum; SER, sieve element reticulum.

berberine biosynthesis (*256*). Indeed, the highest levels of SOMT are found in lateral roots and stems. Despite the close phylogenetic relationships between corresponding biosynthetic enzymes (*16,257*), different cell types are involved in the biosynthesis and accumulation of BAs in opium poppy, *T. flavum*, and perhaps *C. japonica*.

Pyrrolizidine alkaloids also share a common metabolic pathway in restricted, yet diverse taxa (*258*). In *Senecio* species, pyrrolizidine alkaloids are produced in actively growing roots as senecionine *N*-oxides, which are transported *via* the phloem to above-ground organs (*258*). Senecionine *N*-oxides are subsequently modified by one or two species-specific reactions (i.e., hydroxylation, dehydrogenation, epoxidation, or *O*-acetylation) that result in the unique pyrrolizidine alkaloid profile of different plants. Specific carriers are thought to participate in the phloem loading and unloading of senecionine *N*-oxides because plants that do not produce pyrrolizidine alkaloids are unable to translocate these polar intermediates (*259*). Pyrrolizidine alkaloids are spatially mobile, but do not show any turnover or degradation (*258*). Inflorescences appear to be the major sites of pyrrolizidine alkaloid accumulation in *S. jacobaea* and *S. vernalis*, with jacobine occurring in flowers.

HSS, the first committed pathway enzyme, was localized to the root endodermis and cortex adjacent to the phloem in *S. vernalis* (Fig. 2C), which might reflect a functional accommodation for systemic transport of pyrrolizidine alkaloids to the stem (*259*). However, HSS was found throughout the root cortex in *E. cannabinum* (Fig. 2D), which is a member of the Asteraceae family along with *S. vernalis* (*170*). In contrast to the general monophyletic origin of BA biosynthesis (*16*), pyrrolizidine alkaloid pathways appear to have evolved in at least four different angiosperm lineages (*168*). Two scenarios concerning the relationship between the cell type-specific localization and evolutionary origin of plant alkaloids have emerged. In the case of pyrrolizidine alkaloids, the differential localization of a key pathway enzyme was interpreted to support the polyphyletic origin of the biosynthetic pathway (*170*). In contrast, the monophyletic origin of BA biosynthesis (*16*) and differential cell type-specific localization of gene transcripts in *T. flavum* and *P. somniferum* implicates the migration of established pathways between cell types as a key feature of phytochemical evolution. Indeed, benzylisoquinoline, pyrrolizidine, and other alkaloid biosynthetic pathways might operate in cell types other than those already identified.

### 2. *Subcellular Localization of Biosynthetic Enzymes*

Many alkaloid biosynthetic enzymes have been localized to subcellular compartments other than the cytosol (*260*). Enzyme compartmentalization sequesters toxic alkaloids and pathway intermediates away from sensitive areas of the cell. The subcellular trafficking of biosynthetic intermediates might also create an important level of metabolic regulation. Understanding the subcellular compartmentalization of alkaloid pathways will show whether enzyme characteristics observed *in vitro* represent *bona fide* regulatory mechanisms *in vivo*.

Enzymes involved in vindoline biosynthesis have been localized to no fewer than five subcellular compartments in at least two distinct cell types. The conversion of tryptophan into tryptamine by TDC occurs in the cytosol (*261,262*). Since STR is localized in the vacuole, tryptamine must be transported across the tonoplast before being coupled to secologanin (*263*). CYP76B6 and CYP72A1, which catalyze the first and last steps in the conversion of geraniol into secologanin, are endomembrane-bound P450-dependent enzymes (*82,84*). CYP76B6 has also been putatively associated with provacuolar membranes (*264*). SGD, which catalyzes the deglucosylation of strictosidine, is potentially bound to the external side of the tonoplast (*262*), but *in vivo* localization studies showed that SGD is associated with the ER (*94*). The P450-dependent enzyme T16H, which catalyzes the 16–hydroxylation of tabersonine, is also associated with the ER (*98*). NMT is responsible for the third to last step in vindoline biosynthesis and was found to be associated with thylakoid membranes (*104,261*). The last two steps in vindoline biosynthesis, catalyzed by D4H and DAT, occur in the cytosol (*105,261*). Vindoline must then be channeled to the vacuole where the non-specific peroxidases that produce the bis-monoterpenoid indole alkaloids, such as vinblastine, are localized (*112*). Although T16H was localized to epidermis (*243*), it remains unclear whether NMT occurs in the epidermis, or in laticifers and idioblasts (Fig. 2B). The complex compartmentalization of the monoterpenoid indole alkaloid pathway suggests extensive subcellular trafficking of pathway intermediates. Although the vacuolar uptake of some TIAs, such as ajmalicine and serpentine, has been studied (*265*), the mechanisms involved in channeling pathway intermediates to specific subcellular compartments are poorly understood.

Several enzymes involved in BA biosynthesis were also associated with a subcellular compartment other than the cytosol. In the sanguinarine branch pathway, density gradient fractionation suggested the localization of BBE, CFS, SPS,

MSH, and P6H to microsomes with a density of 1.11–1.14 g $mL^{-1}$ (*46,49,60*). Except for BBE, these non-cytosolic enzymes are P450-dependent and are thus, integral endomembrane proteins. BBE is not an integral membrane protein, but is initially targeted to the ER and, subsequently, sorted to a vacuolar compartment (*266*).

Other soluble enzymes involved in berberine biosynthesis were also localized to endomembranes, including STOX (*60*), CDO (*57*), and columbamine *O*-methyltransferase (*267*). The association of these enzymes with a unique membrane fraction with a density of 1.14 g $mL^{-1}$ led to the speculation that distinct *alkaloid synthesizing vesicles* are found in certain cell types (*60*). Vesicles with the same density and containing various alkaloids and biosynthetic enzymes have been visualized within vacuole-like compartments (*60*). However, CYP80B1, BBE, and sanguinarine were co-localized to the ER in opium poppy cell cultures (*268*). Moreover, the induction of sanguinarine biosynthesis was shown to correlate with extensive dilations of the ER and the accumulation of an electron-dense flocculent material within both the ER and the vacuole (Fig. 4). Electron-dense, ER-derived vesicles appear to fuse with the central vacuole (*268,269*). As the optimum pH of BBE is ~8.8 (*37*), sanguinarine biosynthesis is unlikely to involve the vacuole; thus, it is possible that alkaloid metabolism occurs entirely in the ER. The proposed existence of a unique organelle, distinct from the ER, involved in BA biosynthesis (*60*) must now be reconsidered.

The apparent biosynthesis and accumulation of sanguinarine within a single cultured opium poppy cell is in contrast to the requirement for different cell types to facilitate biosynthetic gene expression, alkaloid biosynthesis, and product accumulation in the plant (*251*). The movement of proteins from companion cells to sieve

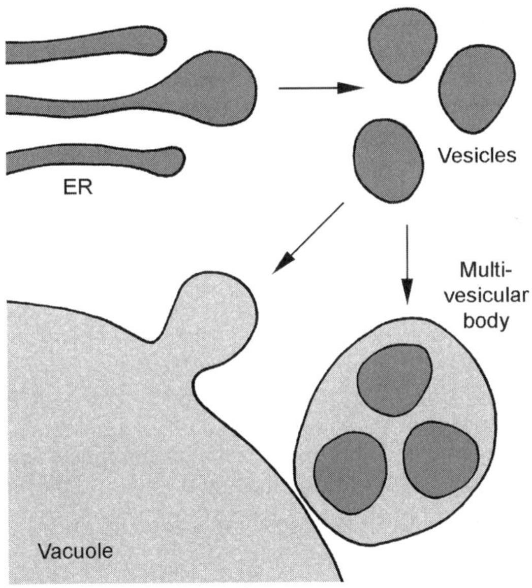


Fig. 4. Model showing the proposed changes in the morphology of the endoplasmic reticulum (ER), leading to the accumulation of sanguinarine in the central vacuole of cultured opium poppy cells treated with a fungal elicitor.

elements is well established (*254*). However, the mechanism involved in the translocation of alkaloids from sieve elements to laticifers is not known, but must occur either by apoplastic transport across plasma membranes and cell walls, or through symplastic connections between sieve elements and laticifers. An ATP-binding-cassette (ABC) multidrug-resistant transporter associated with xylem parenchyma and capable of transporting berberine and other low molecular weight molecules was isolated and characterized from *C. japonica* and *Thalictrum minus* (*270–274*). ABC transporters from cultured tobacco cells have also been associated with the translocation of tropane alkaloids and nicotine (*275*). The symplastic movement of alkaloids would require the presence of functional plasmodesmata between sieve elements and laticifers. The sequestration of alkaloids to vacuolar compartments within laticifers, and the localization of sanguinarine biosynthesis to the ER (*268*), suggest symplastic transport from the sieve element reticulum into the dilated ER of the laticifer through plasmodesmata desmotubules. The ER is contiguous through plasmodesmata and permits the movement of small molecules, such as dextrans. The endomembrane-mediated transport of alkaloids into laticifers is an intriguing option because alkaloid-containing vacuoles are derived from dilating ER cisternae in laticifers (*237,247,276*). The presence of functional plasmodesmata between sieve elements and laticifers would also eliminate the requirement for multiple membrane transporters. However, a vacuolar $H^+$/berberine antiporter stimulated by $Mg^{2+}$/ATP and $Mg^{2+}$/pyrophosphate was detected in cultured *C. japonica* cells (*277*). The transporter preferred berberine as a substrate, but also accepted other alkaloids. The role of such transporters in alkaloid metabolism remains to be determined.

Unique subcellular compartmentation is also present in quinolizidine alkaloid biosynthesis, which occurs in the mesophyll chloroplasts of some legumes (*177*). HMT/HLT activity is localized to the mitochondrial matrix, but not to chloroplasts where *de novo* quinolizidine alkaloid biosynthesis is thought to occur (*176*). ECT is present in an organelle distinct from the mitochondria and chloroplasts, but has not been unambiguously localized. Although the quinolizidine nucleus appears to be synthesized in the chloroplast, subsequent modifications can occur only after alkaloid intermediates are transported to the cytosol and mitochondria. Quinolizidine alkaloids appear to accumulate in vacuoles of epidermal cells where their defensive properties would be most effective.

## IV. Applications of Genomics to the Study of Alkaloid Biosynthesis

Genomic approaches, such as genome sequencing, EST and DNA microarray analysis, proteomics, and metabolic profiling, have accelerated new discoveries in the field of secondary metabolism. The availability of the complete Arabidopsis (*Arabidopsis thaliana*) genome sequence rapidly revolutionized plant biology research. Other notable genome sequencing projects have targeted rice, soybean, corn, and the model legume *Medicago truncatula*. However, sequence databases are not available for the genomes of species that produce complex alkaloids. Recently, however, ESTs have allowed the rapid identification of several genes involved in the secondary metabolism of non-model plants, avoiding the labor associated with traditional molecular biochemical methods. A new dimension was added to transcriptome analysis with the advent of DNA microarrays to monitor broad changes in gene expression patterns. Such high-throughput analysis offers the unprecedented ability to rapidly identify novel candidate genes on the basis of gene expression profile, to monitor changes in global gene expression patterns in species with complex secondary metabolic pathways, and to rapidly genotype mutagenized or transgenic plants.

The application of genomics to the study of alkaloid biosynthesis has been focused on BA metabolism in opium poppy (*25,26,278,279*) and *C. japonica* (*280*), monoterpenoid indole alkaloid metabolism in *C. roseus* (*91,281*), and nicotine metabolism in tobacco (*150,282*). An EST database containing transcripts present in cultured *C. japonica* cells was used to identify columbamine *O*-methyltransferase (*280*), and a DNA microarray-based approach was used to isolate nicotine *N*-demethylase from tobacco (*150*). A collection of 849 non-redundant opium poppy ESTs were generated using an amplified fragment-length polymorphism strategy (*26*). DNA microarray analysis of the differential expression among morphine-containing *P. somniferum* plants and eight other *Papaver* species that do not accumulate morphine revealed three cDNAs with higher expression in opium poppy, one of which encoded 4′OMT (*26*). DNA microarray analysis of the *top1* mutant variety of opium poppy, which accumulates thebaine and oripavine rather than codeine and morphine, revealed 10 down-regulated transcripts, although none encoded proteins likely to be responsible for the high-thebaine, low-morphine phenotype (*279*). In contrast, a DNA microarray was an effective tool to isolate a molecular clone encoding nicotine *N*-demethylase by comparing high-nicotine to high-nornicotine tobacco varieties (*150*).

Proteomics was used to identify proteins found in the exuded latex of opium poppy (*278*) and, subsequently, characterize 6OMT and 7OMT within the protein population (*25*). A proteomic approach was also undertaken to identify novel proteins involved in alkaloid biosynthesis during the growth cycle of *C. roseus* cell cultures (*281*). Eighty-eight protein spots were selected for identification by matrix-assisted laser desorption inoization tandem mass spectrometry (MALDI-MS/MS) of which 58 were identified, including two STR isoforms and tryptophan synthase. Several pioneering metabolomic analyses have recently been performed on various medicinal plant species, including *C. roseus* (*283*) and *Strychnos* spp. (*284*). Comprehensive metabolic profiling of various phytoplasma-treated *C. roseus* leaves by nuclear magnetic resonance (NMR) spectroscopy, followed by principal component analysis (PCA), identified compounds present at different levels in infected versus healthy tissues (*283*). Phytoplasma infection caused an increase in phenylpropanoid and monoterpenoid indole alkaloid pathway intermediates and products, including chlorogenic acid, loganic acid, secologanin, and vindoline. Another metabolomic study was conducted using liquid chromatography coupled with photodiode array detection and electrospray-ionization ion-trap mass spectrometry on methanol extracts from *O. pumila*, *Nothapodytes foetida*, and *C. acuminata* to compare the profile of camptothecin-related alkaloids (*76*). Considerable differences were detected in the constituents of various plants, hairy roots, and cell cultures. *In vivo* monitoring of alkaloid metabolism by two-dimensional cryo-NMR also has potential for metabolomic applications (*285*).

## V. Metabolic Engineering of Alkaloid Pathways

Metabolic engineering is broadly defined as the improvement of cellular activities by the manipulation of enzymatic, transport, or regulatory functions using recombinant DNA technology. Several alkaloid biosynthetic genes have been used to genetically alter the production of various secondary metabolites in plants. Transgenic plants with altered or novel enzyme activities have also become a powerful tool to study the metabolic control architecture of alkaloid pathways (*286*).

## A. MONOTERPENOID INDOLE ALKALOIDS

A *TDC* transgene was first introduced into *C. roseus* cells by infecting seedlings with an oncogenic strain of *Agrobacterium tumefaciens* (*287*). Tumorigenic calli expressing the *TDC* transgene showed increased TDC activity and tryptamine content, but alkaloid levels were not affected compared with wild-type controls. These results contradicted previous conclusions that tryptamine is a limiting substrate for monoterpenoid indole alkaloid biosynthesis in some tissues (*288*).

Nontumorigenic *C. roseus* cell cultures transformed with a *STR1* transgene showed 10–fold higher STR activity and accumulated higher levels of strictosidine and other monoterpenoid indole alkaloids, compared with wild-type cultures, but TDC activity was not affected (*289*). In contrast, high-TDC activity conferred by a *TDC* transgene introduced alone, or in combination with the *STR1* transgene, did not affect alkaloid accumulation. These results further suggest that STR catalyzes a rate-limiting step of alkaloid biosynthesis in *C. roseus* cell cultures (*289*). The influence of precursor availability on monoterpenoid indole alkaloid accumulation was investigated by feeding various concentrations and combinations of tryptamine and loganin to a transgenic *C. roseus* cell line overexpressing *STR1* (*290*). High rates of tryptamine synthesis were found to occur even when TDC activity was low. Moreover, efficient STR activity was possible even when the tryptamine pool was small. However, the overall formation of strictosidine was shown to require a sufficient supply of both secologanin and tryptamine, such that the efficient utilization of one depended on the availability of the other. Since precursor availability is a tissue-specific trait, different rate-limiting factors must exist in various cell lines, organs, and plants.

Expression of *C. roseus TDC* and *STR* transgenes resulting in high-constitutive TDC and STR activities was reported in transgenic *Cinchona officinalis* root cultures (*291*) and tobacco cell cultures (*292*). The transgenic tobacco cell line accumulated tryptamine and produced strictosidine when secologanin was added to the culture medium. These results demonstrate that two consecutive steps in the monoterpenoid indole alkaloid pathway can be cooperatively expressed in a foreign plant species that does not normally produce these metabolites. Examination of the integration frequencies and expression levels of *TDC* and *STR1* in tobacco showed that both transgenes were expressed in only 33% of the plants (*293*). Thus, the extensive phenotypic variation in alkaloid production in various transgenic tissues (*289*) is at least partly caused by gene silencing events affecting *TDC* and *STR1*.

Hairy roots of *C. roseus* transformed with a glucocorticoid-inducible, feedback-resistant chimeric *AS α-subunit* and *TDC* genes showed a sixfold increase in tryptamine levels that were not observed in roots transformed with the *TDC* gene alone (*294*). Transformed roots expressing only the chimeric *AS β-subunit* and/or the *AS α-subunit* genes showed substantial increase in the accumulation of tyrptophan and tyyptamine (*295–297*). However, increased monoterpenoid indole alkaloid accumulation was not observed, supporting the conclusion that although the feedback inhibition of AS by tryptophan regulates the production of tryptamine, this precursor normally does not limit alkaloid biosynthesis. *C. roseus* cell cultures transformed with a constitutive *TDC* gene also displayed high levels of tryptamine without increased alkaloid accumulation (*298*). However, feeding of loganin or secologanin resulted in the production of large quantities of specific monoterpenoid indole alkaloids.

## B. TROPANE ALKALOIDS, NICOTINE, AND OTHER POLYAMINE DERIVATIVES

Some of the earliest attempts to metabolically engineer plant secondary products involved polyamine-derived alkaloids. Transgenic *Nicotiana rustica* root cultures expressing an *ODC* gene from yeast produced higher levels of putrescine and nicotine (*299*). However, despite strong heterologous *ODC* expression, nicotine, putrescine, and *N*-methylputrescine levels increased only twofold, suggesting that ODC is not a rate-limiting step in nicotine biosynthesis. Overexpression of oat *ADC* in tobacco increased the accumulation of agmatine, the ADC reaction product (*300*). However, increased nicotine production was not detected, despite suggestions that the putrescine required for nicotine biosynthesis is generated *via* ADC rather than ODC (*301*). It is possible that the additional agmatine was not accessible to the nicotine pathway.

Hairy root cultures of *Datura metel* and *H. muticus* transformed with both the Ri plasmid and a constitutively regulated, chimeric tobacco *PMT* gene showed higher tropane alkaloid accumulation than wild-type controls (*302*). However, the scopolamine and hyoscyamine content of *D. metel* roots increased, but only the hyoscyamine level of *H. muticus* roots was elevated showing that differential results can be obtained with a common modification of the same alkaloid biosynthetic pathway in different systems. Similarly, hairy root cultures of a scopolamine-rich *Duboisia* hybrid expressing a constitutively regulated, chimeric tobacco *PMT* gene showed no change in tropane alkaloid levels compared with wild-type controls (*303*). The antisense-mediated down-regulation of PMT in transformed tobacco roots resulted in a corresponding reduction in nicotine accumulation (*304*). Several transformed root lines and leaves of some transgenic plants with reduced PMT levels showed higher anatabine accumulation. The increase in anatabine levels at the expense of nicotine was suggested to result from a relative excess of nicotinic acid, which is used to synthesize anatabine in the absence of a sufficient supply of 1–methyl-$\Delta^1$-pyrrolinium cation, the ultimate product of PMT activity. Over-expression of TR-I or TR-II in *A. belladonna* root cultures resulted in increased accumulation of the corresponding reaction products, tropine and pseudotropine, respectively (*305*). Higher pseudotropine availability led to an increased accumulation of calystegines. In contrast, elevated tropine production resulted in three- to fivefold higher tropane alkaloid levels, and a 30–90% decrease in calystegine accumulation compared with controls, demonstrating the role of TRS in regulating the ratio of tropine versus pseudotropine alkaloids.

The *H6H* gene from *H. niger* was constitutively expressed in the hyoscyamine-rich plant *A. belladonna* (*306*). Transgenic plants exhibiting high-H6H activity contained scopolamine almost exclusively. Similar results were obtained in transgenic root cultures of *A. belladonna* (*307*) and *H. muticus* (*308*). A large variation in TPA content was observed in root cultures of *H. muticus*, a species with a high capacity for tropane alkaloid accumulation, expressing *H6H*. One line produced over 100–fold more scopolamine than controls, but hyoscyamine conversion was still incomplete (*308*). Transgenic tobacco plants expressing a *H6H* were also capable of converting exogenous hyoscyamine and 6$\beta$-hydroxyhyoscyamine into scopolamine, which accumulated in leaves (*309*). Wild-type tobacco plants are unable to convert these intermediates into scopolamine. Tobacco and *H. muticus* root cultures transformed with a constitutively expressed, chimeric *H6H* gene were also capable of converting exogenous hyoscyamine into scopolamine (*310*). The addition of exogenous hyoscyamine caused changes in nicotine accumulation in the transgenic tobacco roots and up to 85% of scopolamine was secreted into the culture medium. The simultaneous over-expression of *PMT* and *H6H* in *H. niger* hairy root cultures resulted in the accumulation of up to ninefold more scopolamine than wild type and transgenic lines harboring individual transgenes (*311*). These results underscore the importance of targeting multiple rate-limiting enzymes in alkaloid pathways to achieve substantial increase in product yield.

Attempts to modulate tropane alkaloid or nicotine biosynthesis using other transgenes have achieved different degress of success. *A. belladonna* roots were transformed with an armed strain of *Agrobacterium rhizogenes* and a disarmed strain of *A. tumefaciens* harboring a construct containing the *rol A*, *B*, and *C* genes (*312*) or only the *rol C* gene (*313*). The accumulation of tropane alkaloids increased up to 27–fold in transgenic roots compared with wild-type controls. Tobacco root cultures transformed with a bacterial lysine decarboxylase *LDC* gene produced higher levels of cadaverine, the product of the LDC reaction, and the alkaloid anabasine, produced by the coupling of cadaverine and *N*-methylputrescine (*314*). The activity of LDC and the accumulation of cadaverine and anabasine were enhanced by fusing LDC to the rubisco small subunit transit peptide (*315*), which demonstrates the importance of directing foreign enzymes to appropriate subcellular compartments. Individual callus cultures transformed with a constitutively expressed, chimeric gene encoding SAMSynthetase displayed a range of phenotypes from reduced to increased enzyme activity and corresponding qualitative changes in alkaloid accumulation (*316*).

## C. BENZYLISOQUINOLINE ALKALOIDS

Procedures have been developed for the transformation of opium poppy plants (*317,318*), root cultures (*319,320*), and cell cultures (*321*), *E. californica* plants (*322*), root cultures (*233,319*), and cell cultures (*232*), and *T. flavum* plants (*323*). These transformation systems have provided the opportunity to alter the activity of individual enzymes of BA biosynthesis, and to examine the consequences of such modifications on the accumulation of pathway products and intermediates.

Antisense RNA- or RNAi-based post-transcriptional gene silencing (PTGS) has been used to modify alkaloid biosynthesis in both California poppy and opium poppy, especially when single genes or a family of closely related genes were involved (*231–233,324*). PTGS frequently results in plants or cell cultures with essentially knockout phenotypes when reliable transformation protocols are available. The antisense RNA-mediated suppression of BBE and CYP80B1 dramatically reduced the accumulation of BAs in transgenic California poppy cell and root cultures (*232,233*). Aromatic amino acid levels were also altered, suggesting that flux changes in BA metabolism resulted in the modulation of relevant primary metabolic pathways. Transgenic California poppy root cultures with increased levels of BBE displayed higher levels of sanguinarine and other benzophenanthridine alkaloids compared with controls (*233*). Transgenic opium poppy plants expressing an antisense-*BBE* gene also exhibited an altered ratio of latex alkaloids compared with wild-type plants (*324*). Interestingly, root alkaloid ratios remained unchanged, suggesting the role of enzymes other than BBE in the control of sanguinarine biosynthesis. The RNAi-mediated replacement of morphine with non-narcotic (*S*)-reticuline highlights the potential commercial value of metabolic engineering in opium poppy (*231*). A chimeric hairpin RNA construct designed to silence all members of the *COR* gene family resulted in the accumulation of (*S*)-reticuline, which has pharmacological value as a potential hair-growth stimulant, antibacterial, antimalarial, and anticancer agent.

Biosynthetic enzymes present at pathway junctions have also been targets for the manipulation of BA metabolism. SOMT is thought to control the ratio of coptisine to berberine and columbamine in *C. japonica* cells (*325*). Over-expression of this gene caused a 20% increase in enzyme activity, and elevated levels of berberine and columbamine. Expression of *C. japonica SOMT* in cultured California poppy cells, which lacks this enzyme, resulted in the production of columbamine, which is normally not produced. The introduction of SOMT appeared to redirect flux toward columbamine at the expense of sanguinarine. The transient RNAi-mediated silencing of the *SOMT* gene in *C. japonica* protoplasts was shown to substantially reduce SOMT activity as a potentially useful system to systematically evaluate the function of alkaloid biosynthetic genes (*326*).

## D. OTHER USES OF ALKALOID BIOSYNTHETIC GENES IN GENETIC ENGINEERING

Enzymes that operate at the interface between amino acid and alkaloid metabolism have been used to alter substrate and product pools in a variety of plants. These studies provide insight into the control architecture of amino acid biosynthesis and relevant physiological processes. Tryptamine accumulation in tobacco plants expressing heterologous *TDC* was proportional to the level of TDC activity (*287, 327,328*). Remarkably, plant growth was not affected despite the creation of a large sink for tryptophan, and a large pool of soluble tryptamine. Moreover, no significant difference was detected in the activity of key aromatic amino acid biosynthetic enzymes (*328*), and IAA levels were identical in high-tryptamine and control plants (*327*). Similarly, expression of a *TDC* transgene in cell and root cultures of *Peganum harmala* resulted in increased serotonin accumulation (*329*). Transformed tobacco expressing *TDC* was also reported to accumulate more tyramine compared with controls (*330*). The most plausible explanation for this result is that the increased demand for tryptophan up-regulates the shikimate pathway resulting in increased tyrosine biosynthesis and, consequently, higher tyramine production *via* an endogenous TYDC activity. Similarly, expression of a *TDC* transgene in cell and root cultures of *P. harmala* resulted in increased serotonin accumulation (*329*).

Introduction of *TDC* into *Brassica napus* (canola) resulted in the redirection of tryptophan into tryptamine, rather than indole glucosinolates (*331*). The indole glucosinolate content of seeds from transgenic plants was only 3% of that in wild-type seeds. In oilseeds, such as canola, the presence of indole glucosinolates decreases the value of the seed meal as an animal feed. This study is an elegant example of how a gene normally operating at an entry point in alkaloid biosynthesis can be used to divert metabolic flow and reduce undesirable product levels in crop species. In contrast, the introduction of *TDC* into potato altered the balance of substrate and product pools involved in phenylpropanoid metabolism (*332*). The redirection of tryptophan to tryptamine caused a decrease in wild-type levels of tryptophan, phenylalanine, and phenylalanine-derived phenolic compounds such as chlorogenic acid and lignin in transgenic potato tubers; thus, artificial metabolic sinks can also alter substrate availability even if the foreign gene operates outside the pathway involved in substrate supply.

Canola was also transformed with *TYDC* transgenes encoding TYDC isoforms from opium poppy (*333*). Plants expressing high levels of *TYDC1* showed fourfold higher TYDC activity, a 30% decrease in cellular tyrosine pools, a twofold increase in cell wall-bound tyramine, and a reduced cell wall digestibility compared with wild-type plants. This study supports the involvement of *TYDC* and tyramine in cell wall formation *via* the synthesis of hydroxycinnamic acid amides. The engineering of amide metabolism could also provide an effective strategy to reduce crop susceptibility to a broad spectrum of pathogens by decreasing cell wall digestibility. An increase in the wound-induced accumulation of hydroxycinnamic acid amides was also reported in tobacco expressing *TYDC* transgenes (*334–336*).

## VI. Future Prospects

Alkaloid biosynthetic pathways are under strict regulation in plants. So far, our restricted knowledge of the fundamental mechanisms involved in the control of alkaloid metabolism has limited our ability to harness the vast biotechnological potential of these important secondary pathways. For example, the use of plant cell cultures for the commercial production of pharmaceutical alkaloids has not become a reality despite decades of empirical research. The application of traditional and modern biochemical, molecular, and cellular techniques has revealed important clues about the reasons why *C. roseus* cultures accumulate tabersonine and catharanthine, but not vindoline or vinblastine, and why opium poppy cultures produce sanguinarine, but not codeine or morphine. The inability of dedifferentiated cells to accumulate certain metabolites was interpreted as evidence that the regulation of many alkaloid pathways is tightly coupled to the development of specific tissues. Recent studies have shown that alkaloid pathways are controlled at multiple levels, including cell type-specific gene expression, induction by light and elicitors, enzymatic controls, and other poorly understood metabolic mechanisms. Our ability to exploit the biosynthetic capacity of plants through, for example, metabolic engineering will require a thorough understanding of the mechanisms that allow a cell to produce specific alkaloids. Metabolic engineering strategies targeting transcriptional regulators involved in the control of multiple biosynthetic genes offer new opportunities to coordinately modulate the flux through specific alkaloid pathways (*337*). Advances in genomics will provide a more rapid and efficient means to identify new biosynthetic and regulatory genes involved in alkaloid pathways. The apparently unique features of alkaloid biosynthesis also provide intriguing targets for plant cell biology research, in general. Novel insights obtained using a combination of traditional and modern technologies, including biochemistry, molecular biology, cell biology, and genetic engineering, highlight the importance of a multifaceted approach in studying the regulation of alkaloid biosynthesis in plants. A thorough understanding of alkaloid metabolism in plants will also facilitate pathway reconstitution in microorganisms (*338*) and other biotechnological applications.

## Acknowledgments

I thank members of my research group, past and present, whose hard work and dedication have been instrumental in the success of our research endeavors. Work in my laboratory is supported by grants from the Natural Sciences and Engineering Research Council of Canada and the Canada Foundation for Innovation. I am also privileged to hold the Canada Research Chair in Plant Biotechnology.

## References

1. T. Hashimoto and Y. Yamada, *Curr. Opin. Biotechnol.* **14**, 163 (2003).
2. T. Schmeller, B. Latz-Brüning, and M. Wink, *Phytochemistry* **44**, 257 (1997).
3. M. Rueffer and M. H. Zenk, *Z. Naturforsch.* **42c**, 319 (1987).
4. I. A. Marques and P. E. Brodelius, *Plant Physiol* **88**, 47 (1988).
5. P. J. Facchini and V. De Luca, *J. Biol. Chem.* **269**, 26684 (1994).
6. I. E. Maldonado-Mendoza, M. López-Meyer, J. R. Galef, R. J. Burnett, and C. L. Nessler, *Plant Physiol* **110**, 43 (1996).
7. P. J. Facchini and V. De Luca, *Plant Cell* **7**, 1811 (1995).
8. P. J. Facchini, C. Penzes-Yost, N. Samanani, and B. Kowalchuk, *Plant Physiol* **118**, 69 (1998).
9. P. J. Facchini, K. L. Huber-Allanach, and L. W. Tari, *Phytochemistry* **54**, 121 (2000).
10. R. Stadler, T. M. Kutchan, S. Loeffler, N. Nagakura, B. Cassels, and M. H. Zenk, *Tetrahedron Lett.* **28**, 1251 (1987).
11. R. Stadler, T. M. Kutchan, and M. H. Zenk, *Phytochemistry* **28**, 1083 (1989).
12. M. Rueffer, H. El-Shagi, N. Nagahura, and M. H. Zenk, *FEBS Lett.* **129**, 5 (1981).
13. N. Samanani and P. J. Facchini, *Planta* **213**, 898 (2001).
14. N. Samanani and P. J. Facchini, *J. Biol. Chem.* **277**, 33878 (2002).
15. N. Samanani, D. K. Liscombe, and P. J. Facchini, *Plant J.* **40**, 302 (2004).
16. D. K. Liscombe, B. P. Macleod, N. Loukanina, O. I. Nandi, and P. J. Facchini, *Phytochemistry* **66**, 1374 (2005).
17. H. P. Bais, R. Vepachedu, C. B. Lawrence, F. R. Stermitz, and J. M. Vivanco, *J. Biol. Chem.* **278**, 32413 (2003).
18. F. Sato, N. Takeshita, J. H. Fitchen, H. Fujiwara, and Y. Yamada, *Phytochemistry* **32**, 659 (1993).
19. T. Frenzel and M. H. Zenk, *Phytochemistry* **29**, 3491 (1990).
20. T. Frenzel and M. H. Zenk, *Phytochemistry* **29**, 3505 (1990).
21. H. H. Pauli and T. M. Kutchan, *Plant J.* **13**, 793 (1998).
22. T. Morishige, T. Tsujita, Y. Yamada, and F. Sato, *J. Biol. Chem.* **275**, 23398 (2000).
23. N. Samanani, S. U. Park, and P. J. Facchini, *Plant Cell* **17**, 915 (2005).
24. P. J. Facchini and S. U. Park, *Phytochemistry* **64**, 177 (2003).
25. A. Ounaroon, G. Decker, J. Schmidt, F. Lottspeich, and T. M. Kutchan, *Plant J.* **36**, 808 (2003).
26. J. Ziegler, M. L. Diaz-Chavez, R. Kramell, C. Ammer, and T. M. Kutchan, *Planta* **222**, 458 (2005).

27. S. Frick and T. M. Kutchan, *Plant J.* **17**, 329 (1999).
28. K. B. Choi, T. Morishige, and F. Sato, *Phytochemistry* **56**, 649 (2001).
29. K. B. Choi, T. Morishige, N. Shitan, K. Yazaki, and F. Sato, *J. Biol. Chem.* **277**, 830 (2002).
30. S. Loeffler and M. H. Zenk, *Phytochemistry* **29**, 3499 (1990).
31. F. C. Huang and T. M. Kutchan, *Phytochemistry* **53**, 555 (2000).
32. M. Yu and P. J. Facchini, *Plant Physiol.* **122**, 1457 (2000).
33. R. Stadler and M. H. Zenk, *J. Biol. Chem.* **268**, 823 (1993).
34. P. F. X. Kraus and T. M. Kutchan, *Proc. Natl. Acad. Sci. USA* **92**, 2071 (1995).
35. A. Rosco, H. H. Pauli, W. Priesner, and T. M. Kutchan, *Arch. Biochem. Biophys.* **348**, 369 (1997).
36. P. J. Facchini, *Annu. Rev. Plant Physiol. Plant Mol. Biol.* **52**, 29 (2001).
37. P. Steffens, N. Nagakura, and M. H. Zenk, *Phytochemistry* **24**, 2577 (1985).
38. H. Dittrich and T. M. Kutchan, *Proc. Natl. Acad. Sci. USA* **88**, 9969 (1991).
39. W. M. Chou and T. M. Kutchan, *Plant J.* **15**, 289 (1998).
40. T. M. Kutchan, A. Bock, and H. Dittrich, *Phytochemistry* **35**, 353 (1994).
41. T. M. Kutchan and H. Dittrich, *J. Biol. Chem.* **270**, 24475 (1995).
42. P. J. Facchini, C. Penzes, A. G. Johnson, and D. Bull, *Plant Physiol.* **112**, 1669 (1996).
43. K. Hauschild, H. H. Pauli, and T. M. Kutchan, *Plant Mol. Biol.* **36**, 473 (1998).
44. S. Sirikantaramas, S. Morimoto, Y. Shoyama, Y. Ishikawa, Y. Wada, Y. Shoyama, and F. Taura, *J. Biol. Chem.* **279**, 39767 (2004).
45. W. Bauer and M. H. Zenk, *Tetrahedron Lett.* **30**, 5257 (1989).
46. W. Bauer and M. H. Zenk, *Phytochemistry* **30**, 2953 (1991).
47. M. Rueffer, G. Zumstein, and M. H. Zenk, *Phytochemistry* **29**, 3727 (1990).
48. B. R. O'Keefe and C. W. W. Beecher, *Plant Physiol.* **105**, 395 (1994).
49. M. Rueffer and M. H. Zenk, *Tetrahedron Lett.* **28**, 5307 (1987).
50. T. Tanahashi and M. H. Zenk, *Phytochemistry* **29**, 1113 (1990).
51. H. M. Schumacher and M. H. Zenk, *Plant Cell Rep.* **7**, 43 (1988).
52. H. Arakawa, W. G. Clark, M. Psenak, and C. J. Coscia, *Arch. Biochem. Biophys.* **299**, 1 (1992).
53. L. Kammerer, W. De-Eknamkul, and M. H. Zenk, *Phytochemistry* **36**, 1409 (1994).
54. S. Muemmler, M. Rueffer, N. Nagakura, and M. H. Zenk, *Plant Cell Rep.* **4**, 36 (1985).
55. F. Sato, T. Tsujita, Y. Katagiri, S. Yoshida, and Y. Yamada, *Eur. J. Biochem.* **225**, 125 (1994).
56. N. Takeshita, H. Fujiwara, H. Mimura, J. H. Fitchen, Y. Yamada, and F. Sato, *Plant Cell Physiol.* **36**, 29 (1995).
57. E. Galneder, M. Rueffer, G. Wanner, M. Tabata, and M. H. Zenk, *Plant Cell Rep.* **7**, 1 (1988).
58. M. Rueffer and M. H. Zenk, *Phytochemistry* **36**, 1219 (1994).
59. N. Ikezawa, M. Tanaka, M. Nagayoshi, R. Shinkyo, T. Sakaki, K. Inouye, and F. Sato, *J. Biol. Chem.* **278**, 38557 (2003).
60. M. Amman, G. Wanner, and M. H. Zenk, *Planta* **167**, 310 (1986).
61. W. De-Eknamkul and M. H. Zenk, *Phytochemistry* **31**, 813 (1992).
62. K. Hirata, C. Poeaknapo, and M. H. Zenk, *Phytochemistry* **65**, 1039 (2004).
63. R. Gerardy and M. H. Zenk, *Phytochemistry* **32**, 79 (1993).
64. R. Gerardy and M. H. Zenk, *Phytochemistry* **34**, 125 (1993).
65. R. Lenz and M. H. Zenk, *J. Biol. Chem.* **270**, 31091 (1995).
66. T. Grothe, R. Lenz, and T. M. Kutchan, *J. Biol. Chem.* **276**, 30717 (2001).
67. R. Lenz and M. H. Zenk, *Eur. J. Biochem.* **233**, 132 (1995).
68. B. Unterlinner, R. Lenz, and T. M. Kutchan, *Plant J.* **18**, 465 (1999).
69. T. J. C. Luijendijk, E. Vandermeijden, and R. Verpoorte, *J. Chem. Ecol.* **22**, 1355 (1996).

70. H. Lu and T. D. McKnight, *Plant Physiol.* **120**, 43 (1999).
71. H. Lu, E. Gorman, and T. D. McKnight, *Planta* **221**, 352 (2005).
72. V. De Luca, C. Marineau, and N. Brisson, *Proc. Natl. Acad. Sci. USA* **86**, 2582 (1989).
73. O. J. M. Goddijn, R. J. de Kam, A. Zanetti, R. A. Schilperoort, and J. H. C. Hoge, *Plant Mol. Biol.* **18**, 1113 (1992).
74. O. J. M. Goddijn, F. P. Lohman, R. J. de Kam, R. A. Schilperoort, and J. H. C. Hoge, *Mol. Gen. Genet.* **242**, 217 (1994).
75. M. López-Meyer and C. L. Nessler, *Plant J.* **11**, 1667 (1997).
76. Y. Yamazaki, H. Sudo, M. Yamazaki, N. Aimi, and K. Saito, *Plant Cell Physiol.* **44**, 395 (2003).
77. A. Contin, R. van der Heijden, A. W. Lefeber, and R. Verpoorte, *FEBS Lett.* **434**, 413 (1998).
78. B. Veau, M. Courtois, A. Oudin, J. C. Chenieux, M. Rideau, and M. Clastre, *Biochim. Biophys. Acta* **1517**, 159 (2000).
79. T. D. Meehan and C. J. Coscia, *Biochem. Biophys. Res. Commun.* **53**, 1043 (1973).
80. O. Schiel, I. Witte, and J. Berlin, *Z. Naturforsch.* **42c**, 1075 (1987).
81. K. M. Madyastha and C. J. Coscia, *J. Biol. Chem.* **254**, 2419 (1979).
82. G. Collu, N. Unver, A. M. Peltenburg-Looman, R. van der Heijden, R. Verpoorte, and J. Memelink, *FEBS Lett.* **508**, 215 (2001).
83. H. Yamamoto, N. Katano, A. Ooi, and K. Inoue, *Phytochemistry* **53**, 7 (2000).
84. S. Irmler, G. Schröder, B. St-Pierre, N. P. Crouch, M. Hotze, J. Schmidt, D. Strack, U. Matern, and J. Schröder, *Plant J.* **24**, 797 (2000).
85. I. E. Maldonado-Mendoza, R. J. Burnett, M. López-Meyer, and C. L. Nessler, *Plant Cell Tiss. Organ Cult.* **38**, 351 (1994).
86. R. J. Burnett, I. E. Maldonado-Mendoza, T. D. McKnight, and C. L. Nessler, *Plant Physiol.* **103**, 41 (1993).
87. I. E. Maldonado-Mendoza, R. M. Vincent, and C. L. Nessler, *Plant Mol. Biol.* **34**, 781 (1997).
88. T. M. Kutchan, N. Hampp, F. Lottspeich, K. Beyreuther, and M. H. Zenk, *FEBS Lett.* **237**, 40 (1988).
89. T. D. McKnight, C. A. Roessner, R. Devagupta, A. I. Scott, and C. L. Nessler, *Nucl. Acids Res.* **18**, 4939 (1990).
90. G. Pasquali, O. J. M. Goddijn, A. de Waal, R. Verpoorte, R. A. Schilperoort, J. H. C. Hoge, and J. Memelink, *Plant Mol. Biol.* **18**, 1121 (1992).
91. Y. Yamazaki, A. Urano, H. Sudo, M. Kitajima, H. Takayama, M. Yamazaki, N. Aimi, and K. Saito, *Phytochemistry* **62**, 461 (2003).
92. J. Koepke, X. Ma, G. Fritzsch, H. Michel, and J. Stöckigt, *Acta Crystallogr. Sect. D Biol. Crystallogr.* **61**, 690 (2005).
93. T. J. C. Luijendijk, L. H. Stevens, and R. Verpoorte, *Plant Physiol. Biochem.* **36**, 419 (1998).
94. A. Geerlings, M. Martinez-Lozano Ibañez, J. Memelink, R. van der Heijden, and R. Verpoorte, *J. Biol. Chem.* **275**, 3051 (2000).
95. I. Gerasimenko, Y. Sheludko, X. Ma, and J. Stöckigt, *Eur. J. Biochem.* **269**, 2204 (2002).
96. X. Ma, J. Koepke, G. Fritzsch, R. Diem, T. M. Kutchan, H. Michel, and J. Stöckigt, *Biochem. Biophys. Acta* **1702**, 121 (2004).
97. L. Barleben, X. Ma, J. Koepke, G. Peng, H. Michel, and J. Stöckigt, *Biochim. Biophys. Acta* **1747**, 89 (2005).
98. B. St-Pierre and V. De Luca, *Plant Physiol.* **109**, 131 (1995).
99. G. Schröder, E. Unterbusch, M. Kaltenbach, J. Schmidt, D. Strack, V. De Luca, and J. Schröder, *FEBS Lett.* **458**, 97 (1999).
100. J. Balsevich, V. De Luca, and W. G. W. Kurz, *Heterocycles* **24**, 2415 (1986).

101. V. De Luca, J. Balsevich, R. T. Tyler, U. Eilert, B. D. Panchuk, and W. G. W. Kurz, *J. Plant Physiol.* **125**, 147 (1986).
102. V. De Luca, J. Balsevich, R. T. Tyler, and W. G. W. Kurz, *Plant Cell Rep.* **6**, 458 (1987).
103. W. Fahn, E. Laussermair, B. Deus-Neumann, and J. Stöckigt, *Plant Cell Rep.* **4**, 337 (1985).
104. M. Dethier and V. De Luca, *Phytochemistry* **32**, 673 (1993).
105. E. De Carolis, F. Chan, J. Balsevich, and V. De Luca, *Plant Physiol.* **94**, 1323 (1990).
106. E. De Carolis and V. De Luca, *J. Biol. Chem.* **268**, 5504 (1993).
107. V. De Luca, J. Balsevich, and W. G. W. Kurz, *J. Plant Physiol.* **121**, 417 (1985).
108. W. Fahn, H. Gundlach, B. Deus-Neumann, and J. Stöckigt, *Plant Cell Rep.* **4**, 333 (1985).
109. R. Power, W. G. W. Kurz, and V. De Luca, *Arch. Biochem. Biophys.* **279**, 370 (1990).
110. F. A. Vázquez-Flota, E. De Carolis, A. M. Alarco, and V. De Luca, *Plant Mol. Biol.* **34**, 935 (1997).
111. B. St-Pierre, P. Laflamme, A. M. Alarco, and V. De Luca, *Plant J.* **14**, 703 (1998).
112. M. Sottomayor, M. Lopez-Serrano, F. DiCosmo, and A. Ros-Barcelo, *FEBS Lett.* **428**, 299 (1998).
113. M. Sottomayor and A. Ros Barcelo, *Protoplasma* **222**, 97 (2003).
114. P. Laflamme, B. St-Pierre, and V. De Luca, *Plant Physiol.* **125**, 189 (2001).
115. E. Dogru, H. Warzecha, F. Seibel, S. Haebel, F. Lottspeich, and J. Stöckigt, *Eur. J. Biochem.* **267**, 1397 (2000).
116. E. Mattern-Dogru, X. Ma, J. Hartmann, H. Decker, and J. Stöckigt, *Eur. J. Biochem.* **269**, 2889 (2002).
117. I. Gerasimenko, X. Ma, Y. Sheludko, R. Mentele, F. Lottspeich, and J. Stöckigt, *Bioorg. Med. Chem.* **12**, 2781 (2004).
118. A. Bayer, X. Ma, and J. Stöckigt, *Bioorg. Med. Chem.* **12**, 2787 (2004).
119. X. Ma, J. Koepke, A. Bayer, V. Linhard, G. Fritzsch, B. Zhang, H. Michel, and J. Stöckigt, *Biochim. Biophys. Acta* **1701**, 129 (2004).
120. X. Ma, J. Koepke, S. Panjikar, G. Fritzsch, and J. Stöckigt, *J. Biol. Chem.* **280**, 13576 (2005).
121. H. Falkenhagen and J. Stöckigt, *Z. Naturforsch.* **50c**, 45 (1995).
122. G. von Schumann, S. Gao, and J. Stöckigt, *Bioorg. Med. Chem.* **10**, 1913 (2002).
123. S. Gao, G. von Schumann, and J. Stockigt, *Planta Med.* **68**, 906 (2002).
124. L. Polz, H. Schübel, and J. Stöckigt, *Z. Naturforsch* **42c**, 333 (1987).
125. M. Ruppert, J. Woll, A. Giritch, E. Genady, X. Ma, and J. Stöckigt, *Planta* **222**, 888 (2005).
126. H. Warzecha, I. Gerasimenko, T. M. Kutchan, and J. Stöckigt, *Phytochemistry* **54**, 657 (2000).
127. A. J. Michael, J. M. Furze, M. J. C. Rhodes, and D. Burtin, *Biochem. J.* **314**, 241 (1996).
128. E. Bell and R. L. Malmberg, *Mol. Gen. Genet.* **224**, 431 (1990).
129. R. Rastogi, J. Dulson, and S. J. Rothstein, *Plant Physiol.* **103**, 829 (1994).
130. T. Hashimoto and Y. Yamada, *in* "Proceedings of the Seventh Annual Penn State Symposium in Plant Physiology" (B. K. Singh, H. E. Flores, and J. C. Shannon, eds.), p. 122. American Society of Plant Physiologists Press, Rockville, 1992.
131. N. Hibi, T. Fujita, M. Hatano, T. Hashimoto, and Y. Yamada, *Plant Physiol.* **100**, 826 (1992).
132. K. Suzuki, Y. Yamada, and T. Hashimoto, *Plant Cell Physiol.* **40**, 289 (1999).
133. O. Stenzel, M. Teuber, and B. Drager, *Planta* **223**, 200 (2006).
134. K. Nakajima, T. Hashimoto, and Y. Yamada, *Proc. Natl. Acad. Sci. USA* **90**, 9591 (1993).
135. K. Nakajima and T. Hashimoto, *Plant Cell Physiol.* **40**, 1099 (1999).

136. K. Nakajima, Y. Oshita, Y. Yamada, and T. Hashimoto, *Biosci. Biotechnol. Biochem.* **63**, 1756 (1999).
137. Y. Scholl, D. Hoke, and B. Drager, *Phytochemistry* **58**, 883 (2001).
138. Y. Scholl, B. Schneider, and B. Drager, *Phytochemistry* **62**, 325 (2003).
139. K. Nakajima, A. Yamashita, H. Akama, T. Nakatsu, H. Kato, T. Hashimoto, J. Oda, and Y. Yamada, *Proc. Natl. Acad. Sci. USA* **95**, 4876 (1998).
140. R. Duran-Patron, D. O'Hagan, J. T. Hamilton, and C. W. Wong, *Phytochemistry* **53**, 777 (2000).
141. A. Lanoue, M. Boitel-Conti, J. C. Portais, J. C. Laberche, J. N. Barbotin, P. Christen, and B. Sangwan-Norreel, *J. Nat. Prod.* **65**, 1131 (2002).
142. J. Matsuda, S. Okabe, T. Hashimoto, and Y. Yamada, *J. Biol. Chem.* **266**, 9460 (1991).
143. K. Suzuki, D. J. Yun, X. Y. Chen, Y. Yamada, and T. Hashimoto, *Plant Mol. Biol.* **40**, 141 (1999).
144. T. Liu, P. Zhu, K. D. Cheng, C. Meng, and H. X. He, *Planta Med.* **71**, 249 (2005).
145. T. Hashimoto, A. Hayashi, Y. Amano, J. Kohno, H. Iwanari, S. Usuda, and Y. Yamada, *J. Biol. Chem.* **266**, 4648 (1991).
146. T. Kanegae, H. Kajiya, Y. Amano, T. Hashimoto, and Y. Yamada, *Plant Physiol.* **105**, 483 (1994).
147. A. Katoh and T. Hashimoto, *Front. Biosci.* **9**, 1577 (2004).
148. N. Hibi, S. Higashiguchi, T. Hashimoto, and Y. Yamada, *Plant Cell* **6**, 723 (1994).
149. S. J. Sinclair, K. J. Murphy, C. D. Birch, and J. D. Hamill, *Plant Mol. Biol.* **44**, 603 (2000).
150. B. Siminszky, L. Gavilano, S. W. Bowen, and R. Dewey, *Proc. Natl. Acad. Sci. USA* **102**, 14919 (2005).
151. H. Ashihara and T. Suzuki, *Front Biosci.* **9**, 1864 (2004).
152. X. Q. Zheng, Y. Koyama, C. Nagai, and H. Ashihara, *J. Plant Physiol.* **161**, 363 (2004).
153. H. Ashihara and A. Crozier, *Trends Plant Sci.* **6**, 407 (2001).
154. B. H. Schulthess, P. Morath, and T. W. Baumann, *Phytochemistry* **41**, 169 (1996).
155. T. Suzuki, H. Ashihara, and G. R. Waller, *Phytochemistry* **31**, 2575 (1992).
156. H. Ashihara, F. M. Gillies, and A. Crozier, *Plant Cell Physiol.* **38**, 413 (1997).
157. H. Ashihara, A. M. Monteiro, F. M. Gillies, and A. Crozier, *Plant Physiol.* **111**, 747 (1996).
158. K. Misako and M. Kouichi, *Front. Biosci.* **9**, 1833 (2004).
159. M. Kato, T. Kanehara, H. Shimizu, T. Suzuki, F. M. Gillies, A. Crozier, and H. Ashihara, *Physiol. Plant.* **98**, 629 (1996).
160. N. Fujimori, T. Suzuki, and H. Ashihara, *Phytochemistry* **30**, 2245 (1991).
161. P. Mazzafera, G. Wingsle, O. Olsson, and G. Sandberg, *Phytochemistry* **37**, 1577 (1994).
162. M. Kato, K. Mizuno, T. Fujimura, M. Iwama, M. Irie, A. Crozier, and H. Ashihara, *Plant Physiol.* **120**, 579 (1999).
163. M. Ogawa, Y. Herai, N. Koizumi, T. Kusano, and H. Sano, *J. Biol. Chem.* **276**, 8213 (2001).
164. D. Ober, L. Gibas, L. Witte, and T. Hartmann, *Phytochemistry* **62**, 339 (2003).
165. G. Graser and T. Hartmann, *Planta* **211**, 239 (2000).
166. T. Hartmann, H. Sander, and R. Adolf, *Planta* **175**, 82 (1988).
167. D. Ober and T. Hartmann, *Proc. Natl. Acad. Sci. USA* **96**, 14777 (1999).
168. A. Reimann, N. Nurhayati, A. Backenkohler, and D. Ober, *Plant Cell* **16**, 2772 (2004).
169. D. Ober, R. Harms, and T. Hartmann, *Phytochemistry* **55**, 305 (2000).
170. S. Anke, D. Niemuller, S. Moll, R. Hansch, and D. Ober, *Plant Physiol.* **136**, 4037 (2004).
171. D. Ober and T. Hartmann, *J. Biol. Chem.* **274**, 32040 (1999).

172. D. Ober, R. Harms, L. Witte, and T. Hartmann, *J. Biol. Chem.* **278**, 12805 (2003).
173. N. Nurhayati and D. Ober, *Phytochemistry* **66**, 1346 (2005).
174. D. Ober and T. Hartmann, *Plant Mol. Biol.* **44**, 445 (2000).
175. M. A. Ruiz and A. Sotelo, *J. Agric. Food Chem.* **49**, 5336 (2001).
176. H. Suzuki, Y. Koike, I. Murakoshi, and K. Saito, *Phytochemistry* **42**, 1557 (1996).
177. M. Wink and T. Hartmann, *Plant Physiol.* **70**, 74 (1982).
178. K. Saito, H. Suzuki, S. Takamatsu, and I. Murakoshi, *Phytochemistry* **32**, 87 (1992).
179. T. Okada, M. Y. Hirai, H. Suzuki, M. Yamazaki, and K. Saito, *Plant Cell Physiol.* **46**, 233 (2005).
180. V. De Luca, J. A. Fernandez, D. Campbell, and W. G. W. Kurz, *Plant Physiol.* **86**, 447 (1988).
181. F. A. Vazquez-Flota, B. St-Pierre, and V. De Luca, *Phytochemistry* **55**, 531 (2000).
182. F. A. Vázquez-Flota and V. De Luca, *Plant Physiol.* **117**, 1351 (1998).
183. R. Aerts and V. De Luca, *Plant Physiol.* **100**, 1029 (1992).
184. U. Eilert, V. De Luca, F. Constabel, and W. G. W. Kurz, *Arch. Biochem. Biophys.* **254**, 491 (1987).
185. P. R. H. Moreno, R. van der Heijden, and R. Verpoorte, *Plant Cell Rep.* **12**, 702 (1993).
186. F. A. Vázquez-Flota, V. De Luca, M. Carrillo-Pech, A. Canto-Flick, and M. de Lourdes Miranda-Ham, *Mol. Biotechnol.* **22**, 1 (2002).
187. F. L. Menke, S. Parchmann, M. J. Mueller, J. W. Kijne, and J. Memelink, *Plant Physiol.* **119**, 1289 (1999).
188. M. J. Mueller, W. Brodschelm, E. Spannagl, and M. H. Zenk, *Proc. Natl. Acad. Sci. USA* **90**, 7490 (1993).
189. R. Aerts, D. Gisi, E. De Carolis, V. De Luca, and T. W. Baumann, *Plant J.* **5**, 635 (1994).
190. S. Rodriguez, V. Compagnon, N. P. Crouch, B. St-Pierre, and V. De Luca, *Phytochemistry* **64**, 401 (2003).
191. G. Avancini, I. N. Abreu, M. D. Saldana, R. S. Mohamed, and P. Mazzafera, *Phytochemistry* **63**, 171 (2003).
192. F. A. Vázquez-Flota and V. De Luca, *Phytochemistry* **49**, 395 (1998).
193. B. Pauw, B. van Duijn, J. W. Kijne, and J. Memelink, *Plant Mol. Biol.* **55**, 797 (2004).
194. P. Gantet, N. Imbault, M. Thiersault, and P. Doireau, *Plant Cell Physiol.* **39**, 220 (1998).
195. J. M. Mérillon, L. Ouelhazi, P. Doireau, J. C. Chénieux, and M. Rideau, *J. Plant Physiol.* **134**, 54 (1989).
196. V. Courdavault, M. Thiersault, M. Courtois, P. Gantet, A. Oudin, P. Doireau, B. St-Pierre, and N. Giglioli-Guivarc'h, *Plant Mol. Biol.* **57**, 855 (2005).
197. I. T. Baldwin, E. A. Schmelz, and T. E. Ohnmeiss, *J. Chem. Ecol.* **20**, 2139 (1994).
198. S. Imanishi, K. Hashizume, M. Nakakita, H. Kojima, Y. Matsubayashi, T. Hashimoto, Y. Sakagami, Y. Yamada, and K. Nakamura, *Plant Mol. Biol.* **38**, 1101 (1998).
199. R. A. Winz and I. T. Baldwin, *Plant Physiol.* **125**, 2189 (2001).
200. N. Sachan and D. L. Falcone, *Phytochemistry* **61**, 797 (2002).
201. U. Eilert, W. G. W. Kurz, and F. Constabel, *J. Plant Physiol.* **119**, 65 (1985).
202. H. M. Schumacher, H. Gundlach, F. Fiedler, and M. H. Zenk, *Plant Cell Rep.* **6**, 410 (1987).
203. G. B. Mahady and C. W. Beecher, *Phytochemistry* **37**, 415 (1994).
204. P. J. Facchini and D. A. Bird, *In vitro Cell. Dev. Biol. Plant* **34**, 69 (1998).
205. S. Blechert, W. Brodschelm, S. Hölder, L. Kammerer, T. M. Kutchan, M. J. Mueller, Z. Q. Xia, and M. H. Zenk, *Proc. Natl. Acad. Sci. USA* **92**, 4099 (1995).
206. P. J. Facchini, A. G. Johnson, J. Poupart, and V. De Luca, *Plant Physiol.* **111**, 687 (1996).
207. A. Ignatov, W. G. Clark, S. D. Cline, M. Psenak, R. J. Krueger, and C. J. Coscia, *Phytochemistry* **43**, 1141 (1996).

208. W. Roos, S. Evers, M. Hieke, M. Tschoepe, and B. Schumann, *Plant Physiol.* **118**, 349 (1998).
209. K. Farber, B. Schumann, O. Miersch, and W. Roos, *Phytochemistry* **62**, 491 (2003).
210. W. Roos, B. Dordschbal, J. Steighardt, M. Hieke, D. Weiss, and G. Saalbach, *Biochim. Biophys. Acta* **1448**, 390 (1999).
211. G. B. Mahady, C. Liu, and C. W. Beecher, *Phytochemistry* **48**, 93 (1998).
212. P. B. F. Ouwerkerk and J. Memelink, *Plant Mol. Biol.* **39**, 129 (1999).
213. P. B. F. Ouwerkerk, T. O. Trimborn, F. Hilliou, and J. Memelink, *Mol. Gen. Genet.* **261**, 610 (1999).
214. P. B. F. Ouwerkerk, D. Hallard, and J. Memelink, *Plant Mol. Biol.* **41**, 491 (1999).
215. A. H. Meijer, M. I. L. Cardoso, J. T. Voskuilen, A. de Waal, R. Verpoorte, and J. H. C. Hoge, *Plant J.* **4**, 47 (1993).
216. M. I. Lopes Cardoso, A. H. Meijer, S. Rueb, J. A. Machado, J. Memelink, and J. H. C. Hoge, *Mol. Gen. Genet.* **256**, 674 (1997).
217. G. Pasquali, A. S. Erven, P. B. Ouwerkerk, F. L. Menke, and J. Memelink, *Plant Mol. Biol.* **39**, 1299 (1999).
218. P. B. F. Ouwerkerk and J. Memelink, *Mol. Gen. Genet.* **261**, 635 (1999).
219. Y. Siberil, S. Benhamron, J. Memelink, N. Giglioli-Guivarc'h, M. Thiersault, B. Boisson, P. Doireau, and P. Gantet, *Plant Mol. Biol.* **45**, 477 (2001).
220. F. L. Menke, A. Champion, J. W. Kijne, and J. Memelink, *EMBO J.* **18**, 4455 (1999).
221. L. van der Fits and J. Memelink, *Science* **289**, 295 (2000).
222. L. van der Fits, F. Hilliou, and J. Memelink, *Transgenic Res.* **10**, 513 (2001).
223. L. van der Fits and J. Memelink, *Plant J.* **25**, 43 (2001).
224. L. van der Fits, H. Zhang, F. L. Menke, M. Deneka, and J. Memelink, *Plant Mol. Biol.* **44**, 675 (2000).
225. B. Pauw, F. A. Hilliou, V. S. Martin, G. Chatel, C. J. de Wolf, A. Champion, M. Pre, B. van Duijn, J. W. Kijne, L. van der Fits, and J. Memelink, *J. Biol. Chem.* **279**, 52940 (2004).
226. S. U. Park, A. G. Johnson, C. Penzes-Yost, and P. J. Facchini, *Plant Mol. Biol.* **40**, 121 (1999).
227. P. A. Srere, *Trends Biochem. Sci.* **25**, 150 (2000).
228. B. S. J. Winkel, *Annu. Rev. Plant Biol.* **55**, 85 (2004).
229. K. Jorgensen, A. V. Rasmussen, M. Morant, A. H. Nielsen, N. Bjarnholt, M. Zagrobelny, S. Bak, and B. L. Moller, *Curr. Opin. Plant Biol.* **8**, 280 (2005).
230. W. G. Heim and J. G. Jelesko, *Plant Mol. Biol.* **56**, 299 (2004).
231. R. S. Allen, A. G. Millgate, J. A. Chitty, J. Thisleton, J. A. Miller, A. J. Fist, W. L. Gerlach, and P. J. Larkin, *Nat Biotechnol.* **22**, 1559 (2004).
232. S. U. Park, M. Yu, and P. J. Facchini, *Plant Physiol.* **128**, 696 (2002).
233. S. U. Park, M. Yu, and P. J. Facchini, *Plant Mol. Biol.* **51**, 153 (2003).
234. U. Eilert, B. Wolters, and F. Constabel, *Can. J. Bot.* **64**, 1089 (1986).
235. B. St-Pierre, F. A. Vázquez-Flota, and V. De Luca, *Plant Cell* **11**, 887 (1999).
236. N. Samanani, E. C. Yeung, and P. J. Facchini, *J. Plant. Physiol.* **139**, 1189 (2002).
237. Å. Thureson-Klein, *Ann. Bot.* **34**, 751 (1970).
238. T. M. Kutchan, M. Rush, and C. J. Coscia, *Plant Physiol.* **81**, 161 (1986).
239. V. De Luca and B. St-Pierre, *Trends Plant Sci.* **5**, 168 (2000).
240. V. Burlat, A. Oudin, M. Courtois, M. Rideau, and B. St-Pierre, *Plant J.* **38**, 131 (2004).
241. D. Lemenager, L. Ouelhazi, S. Mahroug, B. Veau, B. St-Pierre, M. Rideau, J. Aguirreolea, V. Burlat, and M. Clastre, *J. Exp. Bot.* **56**, 1221 (2005).
242. Y. Siberil, M. Thiersault, G. Nepumoceno, P. Doireau, and P. Gantet, *J. Exp. Bot.* **53**, 149 (2002).
243. J. Murata and V. De Luca, *Plant J.* **44**, 581 (2005).
244. T. Shoji, K. Nakajima, and T. Hashimoto, *Plant Cell Physiol.* **41**, 1072 (2000).

245. T. Shoji, R. Winz, T. Iwase, K. Nakajima, Y. Yamada, and T. Hashimoto, *Plant Mol. Biol.* **50**, 427 (2002).
246. K. T. Junghanns, R. E. Kneusel, D. Gröger, and U. Matern, *Phytochemistry* **49**, 403 (1998).
247. C. L. Nessler and P. G. Mahlberg, *Am. J. Bot.* **64**, 541 (1977).
248. J. W. Fairbairn and G. Wassel, *Phytochemistry* **3**, 583 (1964).
249. J. W. Fairbairn, J. Palmer, and A. Paterson, *Phytochemistry* **7**, 2117 (1968).
250. J. W. Fairbairn and M. Djote, *Phytochemistry* **9**, 739 (1970).
251. D. A. Bird, V. R. Franceschi, and P. J. Facchini, *Plant Cell* **15**, 2626 (2003).
252. M. Weid, J. Ziegler, and T. M. Kutchan, *Proc. Natl. Acad. Sci. USA* **101**, 13957 (2004).
253. T. M. Kutchan, *Curr. Opin. Plant Biol.* **8**, 292 (2005).
254. K. J. Oparka and R. Turgeon, *Plant Cell* **11**, 739 (1999).
255. P. J. Facchini, D. A. Bird, R. Bourgault, J. M. Hagel, D. K. Liscombe, B. P. MacLeod, and K. G. Zulak, *Opium poppy: a model system to investigate alkaloid biosynthesis in plants. Can. J. Bot.* **83**, 1189 (2005).
256. H. Fujiwara, N. Takeshita, Y. Terano, J. H. Fitchen, T. Tsujita, Y. Katagiri, F. Sato, and Y. Yamada, *Phytochemistry* **34**, 949 (1993).
257. P. J. Facchini, D. A. Bird, and B. St-Pierre, *Trends Plant Sci.* **9**, 116 (2004).
258. T. Hartmann and B. Dierich, *Planta* **206**, 443 (1998).
259. S. Moll, S. Anke, U. Kahmann, R. Hansch, T. Hartmann, and D. Ober, *Plant Physiol.* **130**, 47 (2002).
260. P. J. Facchini and B. St-Pierre, *Curr. Opin. Plant Biol.* **8**, 657 (2005).
261. V. De Luca and A. J. Cutler, *Plant Physiol.* **85**, 1099 (1987).
262. L. H. Stevens, T. J. M. Blom, and R. Verpoorte, *Plant Cell Rep.* **12**, 563 (1993).
263. T. D. McKnight, D. R. Bergey, R. J. Burnett, and C. L. Nessler, *Planta* **185**, 148 (1991).
264. K. M. Madyastha, J. E. Ridgway, J. G. Dwyer, and C. J. Coscia, *J. Cell Biol.* **72**, 303 (1977).
265. T. J. M. Blom, M. Sierra, T. B. van Vliet, M. E. I. Franke-van Dijk, P. de Koning, F. van Iren, R. Verpoorte, and K. R. Libbenga, *Planta* **183**, 170 (1991).
266. D. A. Bird and P. J. Facchini, *Planta* **213**, 888 (2001).
267. M. Rueffer, M. Amman, and M. H. Zenk, *Plant Cell Rep.* **3**, 182 (1986).
268. J. Alcantara, D. A. Bird, V. R. Franceschi, and P. J. Facchini, *Plant Physiol.* **138**, 173 (2005).
269. A. Bock, G. Wanner, and M. H. Zenk, *Planta* **216**, 57 (2002).
270. K. Sakai, N. Shitan, F. Sato, K. Ueda, and K. Yazaki, *J. Exp. Bot.* **53**, 1879 (2002).
271. N. Shitan, I. Bazin, K. Dan, K. Obata, K. Kigawa, K. Ueda, F. Sato, C. Forestier, and K. Yazaki, *Proc. Natl. Acad. Sci. USA* **100**, 751 (2003).
272. K. Terasaka, K. Sakai, F. Sato, H. Yamamoto, and K. Yazaki, *Phytochemistry* **62**, 483 (2003).
273. K. Terasaka, N. Shitan, F. Sato, F. Maniwa, K. Ueda, and K. Yazaki, *Plant Cell Physiol.* **44**, 198 (2003).
274. K. Yazaki, *Curr. Opin. Plant Biol.* **8**, 301 (2005).
275. A. Goossens, S. T. Hakkinen, I. Laakso, K. M. Oksman-Caldentey, and D. Inze, *Plant Physiol.* **131**, 1161 (2003).
276. T. M. Kutchan, S. Ayabe, R. J. Krueger, E. M. Coscia, and C. J. Coscia, *Plant Cell Rep.* **2**, 281 (1983).
277. M. Otani, N. Shitan, K. Sakai, E. Martinoia, F. Sato, and K. Yazaki, *Plant Physiol.* **138**, 1939 (2005).
278. G. Decker, G. Wanner, M. H. Zenk, and F. Lottspeich, *Electrophoresis* **21**, 3500 (2000).
279. A. G. Millgate, B. J. Pogson, I. W. Wilson, T. M. Kutchan, M. H. Zenk, W. L. Gerlach, A. J. Fist, and P. J. Larkin, *Nature* **431**, 413 (2004).

280. T. Morishige, E. Dubouzet, K. B. Choi, K. Yazaki, and F. Sato, *Eur. J. Biochem.* **269**, 5659 (2002).
281. D. I. Jacobs, M. Gaspari, J. van der Greef, R. van der Heijden, and R. Verpoorte, *Planta* **221**, 690 (2005).
282. A. Goossens, S. T. Hakkinen, I. Laakso, T. Seppanen-Laakso, S. Biondi, V. De Sutter, F. Lammertyn, A. M. Nuutila, H. Soderlund, M. Zabeau, D. Inze, and K. M. Oksman-Caldentey, *Proc. Natl. Acad. Sci. USA* **100**, 8595 (2003).
283. Y. H. Choi, E. C. Tapias, H. K. Kim, A. W. M. Lefeber, C. Erkelens, J. T. J. Verhoeven, J. Brzin, J. Zel, and R. Verpoorte, *Plant Physiol.* **135**, 2398 (2004).
284. M. Frédérich, Y. H. Choi, L. Angenot, G. Harnischfeger, A. W. M. Lefeber, and R. Verpoorte, *Phytochemistry* **65**, 1993 (2004).
285. C. Hinse, C. Richter, A. Provenzani, and J. Stöckigt, *Bioorg. Med. Chem.* **11**, 3913 (2003).
286. E. H. Hughes and J. V. Shanks, *Metab. Eng.* **4**, 41 (2002).
287. O. J. M. Goddijn, E. J. Pennings, P. van der Helm, R. A. Schilperoort, R. Verpoorte, and J. H. C. Hoge, *Transgenic Res.* **4**, 315 (1995).
288. A. H. Meijer, R. Verpoorte, and J. H. C. Hoge, *J. Plant Res.* **3**, 145 (1993).
289. C. Canel, M. I. Lopes-Cardoso, S. Whitmer, L. van der Fits, G. Pasquali, R. van der Heijden, J. H. Hoge, and R. Verpoorte, *Planta* **205**, 414 (1998).
290. S. Whitmer, C. Canel, D. Hallard, C. Goncalves, and R. Verpoorte, *Plant Physiol.* **116**, 853 (1998).
291. A. Geerlings, D. Hallard, A. Martinez Caballero, I. Lopes Cardoso, R. van der Heijden, and R. Verpoorte, *Plant Cell Rep.* **19**, 191 (1999).
292. D. Hallard, R. van der Heijden, R. Verpoorte, M. I. Lopes Cardoso, J. Memelink, and J. H. C. Hoge, *Plant Cell Rep.* **17**, 50 (1997).
293. M. J. Leech, K. May, D. Hallard, R. Verpoorte, V. De Luca, and P. Christou, *Plant Mol. Biol.* **38**, 765 (1998).
294. E. H. Hughes, S. B. Hong, S. I. Gibson, J. V. Shanks, and K. Y. San, *Metab. Eng.* **6**, 268 (2004).
295. E. H. Hughes, S. B. Hong, S. I. Gibson, J. V. Shanks, and K. Y. San, *Biotechnol. Bioeng.* **86**, 718 (2004).
296. S. B. Hong, C. A. Peebles, J. V. Shanks, K. Y. San, and S. I. Gibson, *J. Biotechnol.* **122**, 28 (2006).
297. C. A. Peebles, S. B. Hong, S. I. Gibson, J. V. Shanks, and K. Y. San, *Biotechnol. Prog.* **21**, 1572 (2005).
298. S. Whitmer, R. van der Heijden, and R. Verpoorte, *J. Biotechnol.* **96**, 193 (2002).
299. J. D. Hamill, R. J. Robins, A. J. Parr, D. M. Evans, J. M. Furze, and M. J. C. Rhodes, *Plant Mol. Biol.* **15**, 27 (1990).
300. D. Burtin and A. J. Michael, *Biochem. J.* **325**, 331 (1997).
301. T. Hashimoto and Y. Yamada, *Annu. Rev. Plant Physiol. Plant Mol. Biol.* **45**, 257 (1994).
302. E. Moyano, K. Jouhikainen, P. Tammela, J. Palazon, R. M. Cusido, M. T. Pinol, T. H. Teeri, and K. M. Oksman-Caldentey, *J. Exp. Bot.* **54**, 203 (2003).
303. E. Moyano, S. Fornale, J. Palazon, R. M. Cusido, N. Bagni, and M. T. Pinol, *Phytochemistry* **59**, 697 (2002).
304. Y. Chintapakorn and J. D. Hamill, *Plant Mol. Biol.* **53**, 87 (2003).
305. U. Richter, G. Rothe, A. K. Fabian, B. Rahfeld, and B. Drager, *J. Exp. Bot.* **56**, 645 (2005).
306. D. J. Yun, T. Hashimoto, and Y. Yamada, *Proc. Natl. Acad. Sci. USA* **89**, 11799 (1992).
307. T. Hashimoto, D. J. Yun, and Y. Yamada, *Phytochemistry* **32**, 713 (1993).
308. K. Jouhikainen, L. Lindgren, T. Jokelainen, R. Hiltunen, T. H. Teeri, and K. M. Oksman-Caldentey, *Planta* **208**, 545 (1999).
309. D. J. Yun, T. Hashimoto, and Y. Yamada, *Biosci. Biotech. Biochem.* **57**, 502 (1993).

310. S. T. Hakkinen, E. Moyano, R. M. Cusido, J. Palazon, M. T. Pinol, and K. M. Oksman-Caldentey, *J. Exp. Bot.* **56**, 2611 (2005).
311. L. Zhang, R. Ding, Y. Chai, M. Bonfill, E. Moyano, K. M. Oksman-Caldentey, T. Xu, Y. Pi, Z. Wang, H. Zhang, G. Kai, Z. Liao, X. Sun, and K. Tang, *Proc. Natl. Acad. Sci. USA* **101**, 6786 (2004).
312. V. Bonhomme, D. Laurain-Mattar, and M. A. Fliniaux, *J. Nat. Prod.* **63**, 1249 (2000).
313. V. Bonhomme, D. Laurain-Mattar, J. Lacoux, M. A. Fliniaux, and A. Jacquin-Dubreuil, *J. Biotechnol.* **81**, 151 (2000).
314. L. F. Fecker, C. Rügenhagen, and J. Berlin, *Plant Mol. Biol.* **23**, 11 (1993).
315. S. Herminghaus, D. Tholl, C. Rügenhagen, L. F. Fecker, C. Leuschner, and J. Berlin, *Transgenic Res.* **5**, 193 (1996).
316. L. Belbahri, L. Chevalier, L. Bensaddek, F. Gillet, M. A. Fliniaux, W. Boerjan, D. Inze, D. Thomas, and B. Thomasset, *Biotechnol. Bioeng.* **69**, 11 (2000).
317. S. U. Park and P. J. Facchini, *J. Plant Physiol.* **157**, 207 (2000).
318. J. A. Chitty, R. S. Allen, A. J. Fist, and P. J. Larkin, *Funct. Plant Biol.* **30**, 1045 (2003).
319. S. U. Park and P. J. Facchini, *J. Exp. Bot.* **51**, 1005 (2000).
320. V. Le Flem-Bonhomme, D. Laurain-Mattar, and M. A. Fliniaux, *Planta* **218**, 890 (2004).
321. M. Belny, D. Herouart, B. Thomasset, H. David, A. Jacquin-Dubreuil, and A. David, *Physiol. Plant* **99**, 233 (1997).
322. S. U. Park and P. J. Facchini, *Plant Cell Rep.* **19**, 421 (2000).
323. N. Samanani, S. U. Park, and P. J. Facchini, *Physiol. Plant.* **116**, 79 (2002).
324. S. Frick, J. A. Chitty, R. Kramell, J. Schmidt, R. S. Allen, P. J. Larkin, and T. M. Kutchan, *Transgenic Res.* **13**, 607 (2004).
325. F. Sato, T. Hashimoto, A. Hachiya, K. Tamura, K. B. Choi, T. Morishige, H. Fujimoto, and Y. Yamada, *Proc. Natl. Acad. Sci. USA* **98**, 367 (2001).
326. J. G. Dubouzet, T. Morishige, N. Fujii, C. I. An, E. Fukusaki, K. Ifuku, and F. Sato, *Biosci. Biotechnol. Biochem.* **69**, 63 (2005).
327. D. D. Songstad, V. De Luca, N. Brisson, W. G. W. Kurz, and C. L. Nessler, *Plant Physiol.* **94**, 1410 (1990).
328. C. Poulsen, O. J. M. Goddijn, J. H. C. Hoge, and R. Verpoorte, *Transgenic Res.* **3**, 43 (1994).
329. J. Berlin, C. Rügenhagen, P. Dietze, L. F. Frecker, O. J. M. Goddijn, and J. H. C. Hoge, *Transgenic Res.* **2**, 336 (1993).
330. D. D. Songstad, W. G. W. Kurz, and C. L. Nessler, *Phytochemistry* **30**, 3245 (1991).
331. S. Chavadej, N. Brisson, J. N. McNeil, and V. De Luca, *Proc. Natl. Acad. Sci. USA* **91**, 2166 (1994).
332. K. Yao, V. De Luca, and N. Brisson, *Plant Cell* **7**, 1787 (1995).
333. P. J. Facchini, M. Yu, and C. Penzes-Yost, *Plant Physiol.* **120**, 653 (1999).
334. G. Guillet and V. De Luca, *Plant Physiol.* **137**, 692 (2005).
335. G. Guillet, J. Poupart, J. Basurco, and V. De Luca, *Plant Physiol.* **122**, 933 (2000).
336. J. M. Hagel and P. J. Facchini, *Planta* **221**, 904 (2005).
337. P. Gantet and J. Memelink, *Trends Pharmacol. Sci.* **23**, 563 (2002).
338. A. Geerlings, F. J. Redondo, A. Contin, J. Memelink, R. Vander Heijden, and R. Verpoorte, *Appl. Microbiol. Biotechnol.* **56**, 420 (2001).

—CHAPTER 2—

# ERGOT ALKALOIDS – BIOLOGY AND MOLECULAR BIOLOGY☆

CHRISTOPHER L. SCHARDL[1], DANIEL G. PANACCIONE[2] AND PAUL TUDZYNSKI[3]

[1]*Department of Plant Pathology, University of Kentucky, Lexington, KY 40546-0312, USA;* [2]*Division of Plant and Soil Sciences, WV University, Morgantown, WV 26506-6108, USA;* [3]*Institut für Botanik, Westfälische Wilhelms Universität Münster, Münster D-48149, Germany*



## I. Introduction

The ergot alkaloids (EA) (Fig. 1) are among the most important natural pharmaceuticals and toxins in human history. Some EA have been used purposefully for medical benefit in many cultures over the ages (*2*). Some have been used illicitly for psychedelic recreation, again, throughout the ages into modern times (*3*,*4*). Some have been ingested unknowingly, leading to poisonings with dire physiological and social implications (*2*). In the past centuries, poor understanding of the sources of these toxins and their physiological effects exacerbated the problems arising from their intended and unintended uses, and the symptoms of ergotism in humans and livestock were often attributed to witchcraft (*5*). Even now, the EA pharmaceutical properties are not fully understood and are the subject of intense research. Gaps in our current understanding result from a combination of the shear complexity of the central and peripheral nervous systems on which EA act, the pleiotropic effects of some EA whereby multiple neuroreceptors may be affected simultaneously, and on structural variation among the EA with consequent vari ations in their specificities of action. However, as greater understanding of EA has developed with modern research, their utilization as pharmaceuticals has become

☆Dedicated to Prof. Dr. Heinz G. Floss in recognition of his 45 years of profound discoveries relating to the biosynthesis of ergot alkaloids and other microbial metabolites.

THE ALKALOIDS, Vol. 63
ISSN: 1099-4831
DOI: 10.1016/S1099-4831(06)63002-2

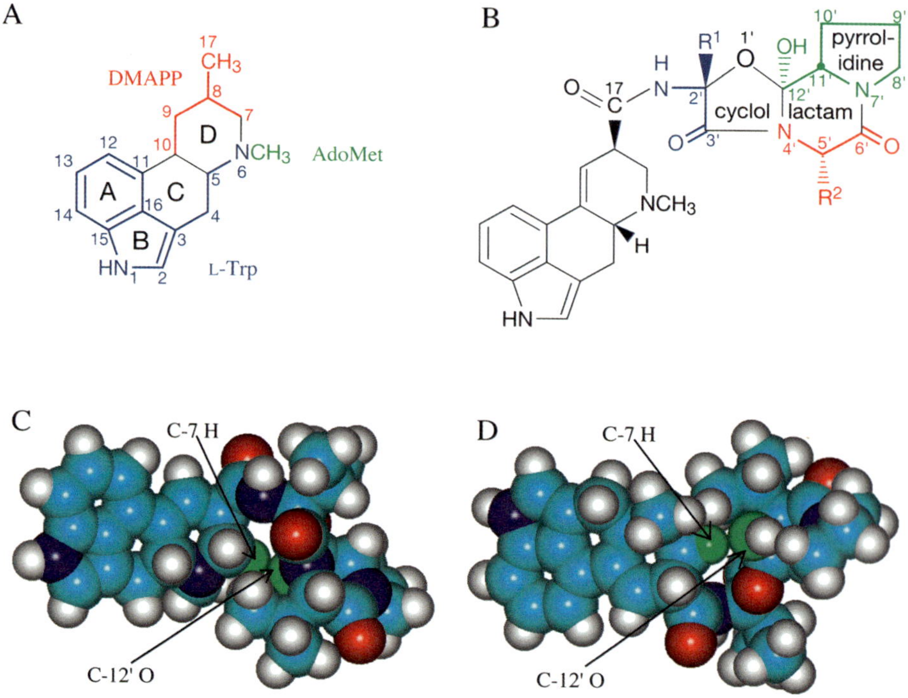


Fig. 1. Ergoline (A) and ergopeptine (B) structures showing atom numbering and ring designations. (A) Substituents are color-coded according to their origins: DMAPP, dimethylallyl-diphosphate (red); AdoMet, *S*-adenosyl-methionine (green), and L-Trp, L-tryptophan (blue). (B) Portions of the tripeptide substituent are color-coded according to the positions of L-amino acid precursors: A.A. I (blue), A.A. II (red), and A.A. III (green). (C, D) Two views of a space-filling model of ergocornine, demonstrating proximity of the lactam ring and the lysergic D-ring. Reproduced with permission from Lehner *et al.* (*1*).

more refined and effective, while human poisonings have become rare (*6*). Problems of livestock poisonings have also been reduced as some of the culprits – biological sources in the field and feed – have been identified. However, in many places, exposure of livestock to EA is unavoidable, and mitigation of EA toxicoses in livestock is another subject of major research efforts (*7–9*).

The EA are characterized by the tetracyclic ergoline ring system (Fig. 1), or by related tricyclic alkaloids open between N(6) and C(7) (ergoline numbering). They are categorized as clavines, lysergic acid (**1**) and its simple amides (Fig. 2), and ergopeptines (Fig. 1). Some clavines are biosynthetic intermediates in the lysergic acid pathway, although several fungal species make clavines, but not **1**. Other clavines are derivatives of lysergic acid pathway intermediates. The carboxylic acid group at position 17 characterizes lysergic acid (**1**), and most lysergic acid derivatives have substituents linked as amides to that function. Among the simpler lysergic acid amides are ergine (**2** = lysergic acid amide) and ergonovine (**3** = ergometrine = ergobasine) (Fig. 2). In the ergopeptines, which are the most complex

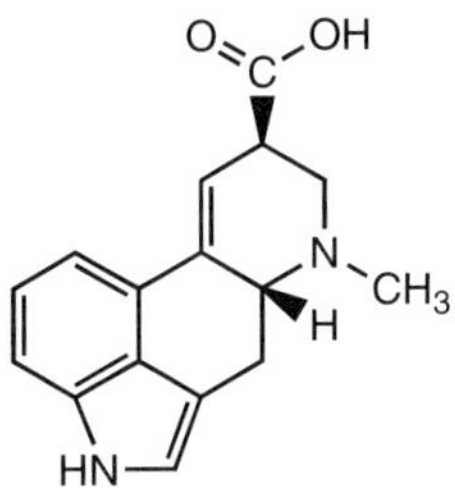

**1** D-lysergic acid

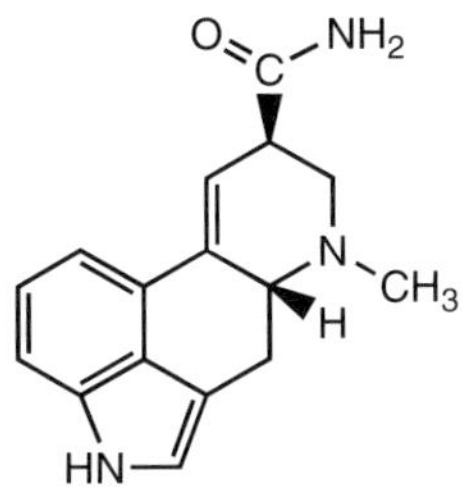

**2** ergine
= lysergic acid amide

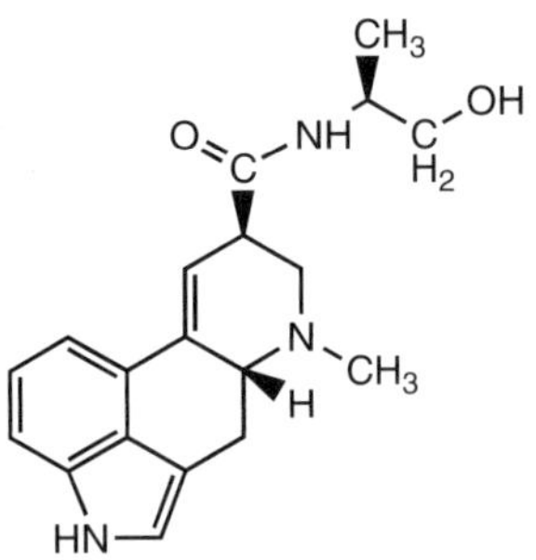

**3** ergonovine
= ergometrine
= ergobasine

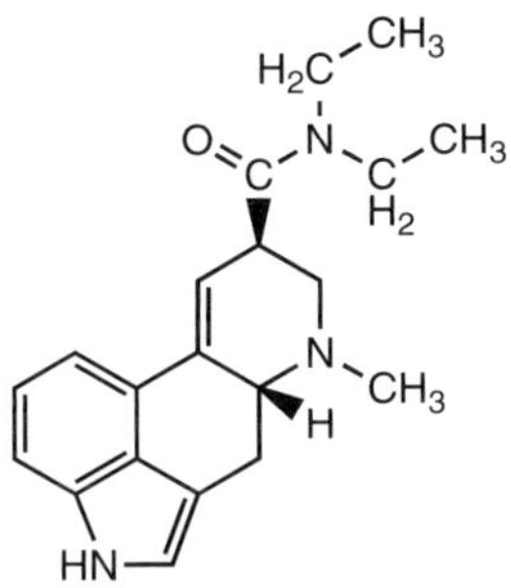

**5** LSD-25

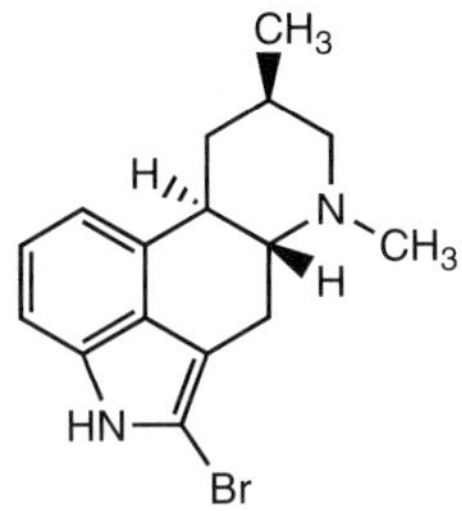

**11** pibocin

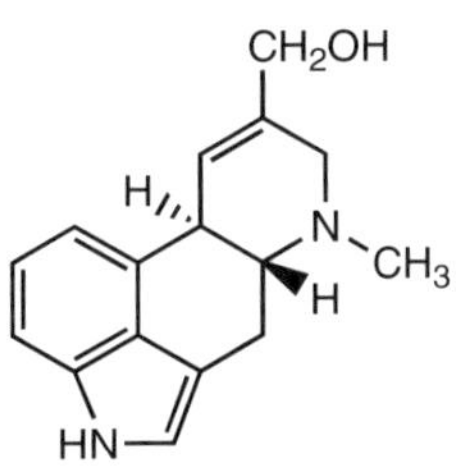

**6** elymoclavine

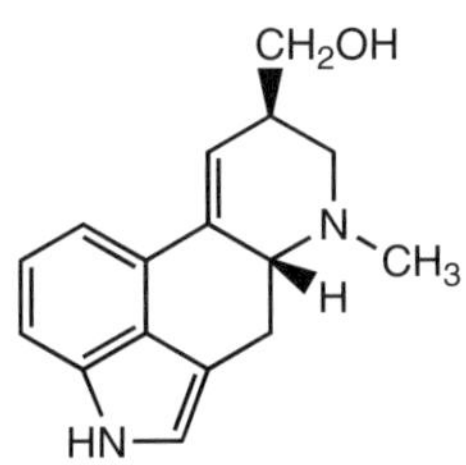

**6a** D-lysergol

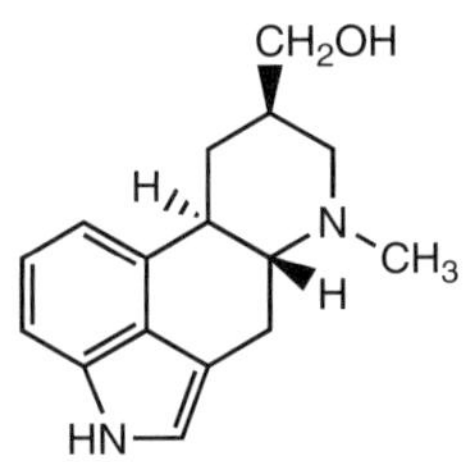

**6b** dihydrolysergol-I
= dihydroelymoclavine

Fig. 2. Structures of lysergic acid and some lysergyl amides and clavines.

of the natural EA, the C(17) amide substituent is a tripeptide-derived, cyclol-lactam structure (Fig. 1).

The biosynthetic pathway to EA in *Claviceps* species has been studied intensively since the 1950s, and was reviewed by Gröger and Floss in the 1998 volume of this series (*10*). That outstanding and thorough review historically punctuates the transition from mainly biochemical studies to investigations involving molecular genetics. Therefore, salient points of the previous review will be summarized here, and this review will emphasize contributions of recent molecular genetic studies to our understanding of the EA pathway. In addition, we will summarize the distribution and variation in EA, and activities of EA in humans and livestock, then discuss their possible ecological roles.

## II. Through the Ages: A History of Ergot Alkaloid Use, Abuse, and Poisoning

Ergots are hard, dense, and darkly pigmented resting structures (sclerotia) of certain fungi – the *Claviceps* species – that parasitize ears of grain, and are the most convenient and abundant source of EA. The medicinal use of ergots is quite ancient (*11*). Records extend from approximately 600 BCE., in Mesopotamia (now Iraq), for the use of ergots at specified doses to women in labor as an aid for parturition or to staunch hemorrhaging. This was a risky treatment, however, because the dosage of active ingredients could not be controlled well, and rupture due to hypercontraction of the uterus was a significant risk.

In the Americas, too, EA have been used in traditional medicine or for recreation. Seeds of *Rivea corymbosa* or *Ipomea* species were known as *ololiuqui* and used ceremonially by indigenous people of Mexico. Investigation of their chemical composition indicated **2** and other lysergic acid amides among the hallucinogenic constituents (*12*). Furthermore, indigenous people of South America have used *Cyperus* spp. (Cyperaceae; sedges), infected with the EA-producing fungus *Balansia cyperi,* to promote parturition (*13,14*).

With increased understanding of the chemistry and pharmacology of EA, careful dosing and use of chemically modified EA to treat an ever-greater array of conditions, this class of compounds emerged with a significant role in modern medicine.

In ergots produced by *Claviceps purpurea* on rye (*Secale cereale*), the concentration of ergotamine (**4**) and related ergopeptine alkaloids (Table I), plus simpler lysergyl amides, typically approximates 0.5–2% by dry mass (*15–20*). These alkaloids can readily be converted into **1** by alkaline or acid hydrolysis, providing a convenient starting material for legal pharmaceuticals or illicit recreational drugs. Semi-synthesis of lysergic acid diethylamide (LSD) (**5**; Fig. 2) is sufficiently straightforward to pose a major problem for law enforcement. This compound, originally named LSD-25, was 25th in a series of lysergyl amides produced by Albert Hofmann at Sandoz Research Laboratories, Basel, Switzerland (*21*). The acronym, LSD, which in English is a better fit to the song title "Lucy in the Sky with Diamonds," is an abbreviation of the German *lysergsäure diäthylamid.* This alkaloid is the most potent hallucinogen known (*3*). Originally tested as an experimental antidepressant drug and as a treatment for schizophrenia, its early tests turned up severe problems including paranoia, potentially fatal loss of judgment, and flashbacks. Nevertheless, "acid trips" are a feature of the cultural upheavals that gripped the Western World in the 1960s and 1970s.

Historically, infected ears of rye have been a common source of ergot-contaminated flour (*2*). Ergot contamination, particularly of rye, has been a major problem in Europe for millennia, and ergot poisoning is now widely believed to be the key trigger for a number of major historical events, such as the Salem Witch Trials in colonial Massachusetts (*5*). (It should be noted, however, that some strong arguments have been made against this "convenient" hypothesis (*22*). In the late 20th century, outbreaks were reported in Ethiopia and India. The Ethiopian ergot outbreak in 1977–1978 is thought to have arisen by ergot-infected wild oats harvested as a field contaminant of barley (*23*). Outbreaks of convulsive ergotism in

TABLE I.
Ergopeptine Alkaloids.

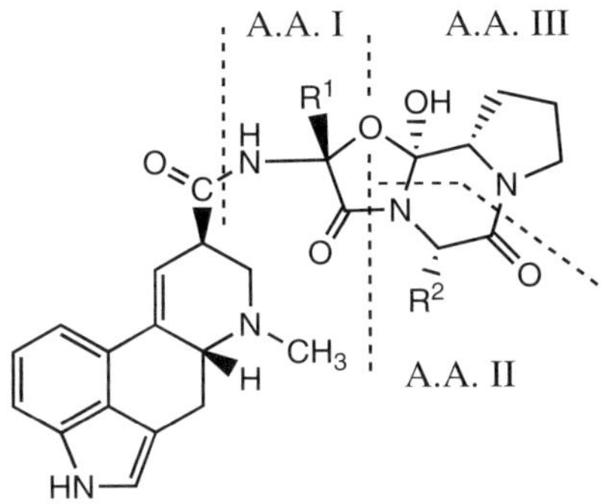


| | $R^1$ | Amino acid I | $R^2$ | Amino acid II |
|---|---|---|---|---|
| Ergotamine group | | | | |
| **4** Ergotamine | $CH_3$ | L-alanine | $CH_2Ph$ | L-phenylalanine |
| **8** Ergovaline | $CH_3$ | L-alanine | *i*-Pr | L-valine |
| Ergosine[a] | $CH_3$ | L-alanine | *i*-Bu | L-leucine |
| β-ergosine | $CH_3$ | L-alanine | *sec*-Bu | L-isoleucine |
| Ergobine | $CH_3$ | L-alanine | Et | L-2-aminobutyric acid |
| Ergotoxine group | | | | |
| Ergocristine | *i*-Pr | L-valine | $CH_2Ph$ | L-phenylalanine |
| Ergocornine | *i*-Pr | L-valine | *i*-Pr | L-valine |
| Ergocryptine[a,b] | *i*-Pr | L-valine | *i*-Bu | L-leucine |
| β-ergocryptine[b] | *i*-Pr | L-valine | *sec*-Bu | L-isoleucine |
| γ-ergocryptine[b,c] | *i*-Pr | L-valine | *n*-Bu | L-norleucine |
| Ergobutyrine | *i*-Pr | L-valine | Et | L-2-aminobutyric acid |
| Ergoladine[c] | *i*-Pr | L-valine | $EtSCH_3$ | L-methionine |
| Ergogaline | *i*-Pr | L-valine | 2-Me-*n*-Bu | L-homoisoleucine |
| Ergoxine group | | | | |
| Ergostine | Et | L-2-aminobutyric acid | $CH_2Ph$ | L-phenylalanine |
| Ergonine | Et | L-2-aminobutyric acid | *i*-Pr | L-valine |
| Ergoptine[a] | Et | L-2-aminobutyric acid | *i*-Bu | L-leucine |
| β-ergoptine | Et | L-2-aminobutyric acid | *sec*-Bu | L-isoleucine |
| ergobutine | Et | L-2-aminobutyric acid | Et | L-2-aminobutyric acid |
| Other (A.A. III = L-alanine) | | | | |
| **9** Ergobalansine | $CH_3$ | L-alanine | *i*-Bu | L-leucine |

[a]Synonyms: ergosine, α-ergosine; ergocryptine, α-ergocryptine; ergoptine, α-ergoptine.
[b]Synonyms: α-, β-, or γ-ergocryptine = α-, β-, or γ-ergokryptine.
[c]Only the isolysergyl isomers, ergoladinine, and γ-ergocryptinine, have been reported to date.

India were attributed to *Claviceps fusiformis* ergots on pearl millet (*Pennisetum glaucum*) (*24,25*).

The symptoms of ergot alkaloid poisoning vary, probably depending on the particular profiles of alkaloids present in the contaminated flour (*6*). Two syndromes have been described as convulsive and gangrenous ergotism. Gangrenous ergotism results from the extreme vasoconstrictive properties of certain EA (particularly **4** and other ergopeptines), resulting in ischaemia (restricted blood-flow to parts of the body). Limbs may become hypoxic, develop dry gangrene, and self-amputate or require amputation (*6*). The effect on the corotid artery is not as severe as effects on other arteries, thus sparing the brain of substantial damage.

Apparently, many outbreaks of convulsive ergotism occurred during the middle ages in Europe, where the malady was called *ignis sacer* (holy fire), or St. Anthony's fire (*2,5*). (The same terms were also applied to intense rashes resulting from bacterial infections.) Among the reported symptoms of convulsive ergotism are involuntary muscle contractions, painful flexion or extension of the fingers, wrists, and ankles, involuntary twisting (such as wryneck), paresthesia (skin-crawling and tingling), tinnitus, vertigo, headaches, double-vision, profuse sweating, fever, ravenous appetite, hallucinations, mania, melancholy, and delirium (*6*). Stiff, distorted postures (dyskinesias) could last for minutes or hours, with relatively normal intervals (except for a voracious appetite) lasting for hours to days. Dyskinesias are sometimes followed by fatal epileptic seizures.

Clavines are thought to contribute substantially to convulsive ergotism, since *C. fusiformis* ergots, which possess clavines, but no **1** or lysergyl amides, cause convulsive symptoms (*26*). However, the ergopeptines are known to produce similar symptoms, and are also thought to cause gangrenous ergotism (*6*). The occurrence of convulsive ergotism without dry gangrene suggests that other clavine or lysergyl alkaloids are involved, or that individual effects of specific ergopeptines may give clinically different syndromes (*6*).

## III. Ergot Alkaloid Producers

The distribution of organisms possessing EA appears disjointed, including two orders of fungi and three plant families. The EA-producing fungi are in the Eurotiales and Hypocreales, two distantly related orders within the phylum Ascomycota. Within the Hypocreales, EA are associated exclusively with plant-associated fungi of the family Clavicipitaceae, although not all members of the Clavicipitaceae produce EA. Some producers of EA lack a known sexual state, so according to the Botanical Code these are classified as Fungi Imperfecti. Nevertheless, the evolutionary derivation of EA producers from asexual Eurotiales and Hypocreales is clear.

The type species of the Clavicipitaceae is *C. purpurea*, otherwise known as the ergot fungus of rye and related grasses. Parasitism of host plants commences when ascospores (meiotically generated spores) are ejected from fungal fruiting structures and land on newly exerted stigmata of grass florets (*27*). The spores germinate, and hyphae grow down the style to the ovary. Should the floret be pollinated beforehand, the style will dry and become refractory to ergot parasitism, making the window of susceptibility very brief, and rendering self-pollinated wheat far less prone than rye to ergot infection. Should a successful infection of the floret occur, further proliferation of the fungus encases the ovary, preventing its maturation. In the ensuing "honeydew" stage the fungus produces abundant conidia (mitotically derived spores) in a sweet exudate, which attracts insects that facilitate secondary spread of the ergot fungus. The fungus then produces resting structures called sclerotia ("ergots" in the case of *Claviceps* species). An interesting and problematic characteristic of ergots is that their density prevents removal by winnowing, so that ergot contamination of rye flour was a major hazard for millennia. Modern screening techniques reduce contamination, but may be selecting for genotypes that produce smaller ergots similar in size to caryopses (*27*).

Ergots of *C. purpurea* tend to be rich in ergopeptines (*28*), and strains of *C. purpurea* have been developed for fermentation to produce ergopeptines and other EA (*29*). The production of EA in those cultures has been associated with a growth characteristic that resembles the early stage of sclerotium development on host plants (*4*).

Ergots and cultures of other *Claviceps* species often have different alkaloid profiles. For example, *C. fusiformis* from *Pennisetum typhoideum* (Kikuyu grass) and pearl millet produces elymoclavine (**6**; Fig. 2) (*30*). *Claviceps africana*, one of the ergot fungi that infects sorghum (*Sorghum bicolor*), produces dihydroergot alkaloids derived from festuclavine (**7**; Fig. 3), in which the D-ring is saturated (*31*). The *C. africana* EA include dihydroelymoclavine (**6b**), dihydrolysergic acid, and the ergopeptine dihydroergosine (see Table I for ergosine structure).

EA identified in the plant families Poaceae and Cyperaceae are products of systemic symbionts in the family Clavicipitaceae, particularly species of *Balansia, Epichloë*, and *Neotyphodium* (*13,32,33*). All of these fungi are systemic symbionts, growing intercellularly or subcuticularly throughout the aerial parts of host plants, and the symbioses are constitutive in that they persist through the life of the plant, or, in the cases of seed-transmissible fungi, through many generations of the host (*34,35*). These symbioses range from pathogenic to mutualistic (*36*). Indications of mutualism are especially evident in associations of Poöideae (a subfamily of C3 grasses) with seed-transmissible *Epichloë* and *Neotyphodium* species (*34*). (Since

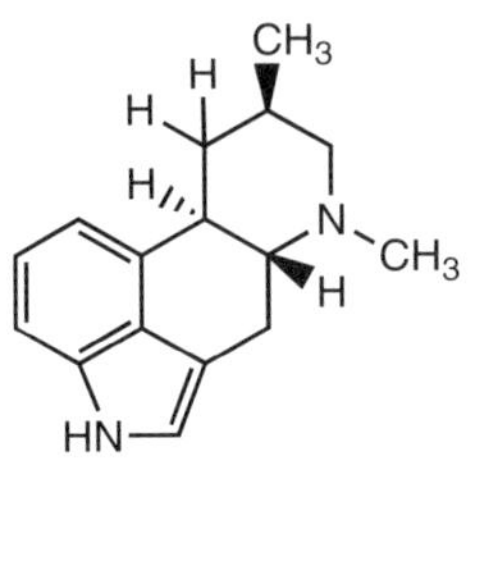

**7** festuclavine

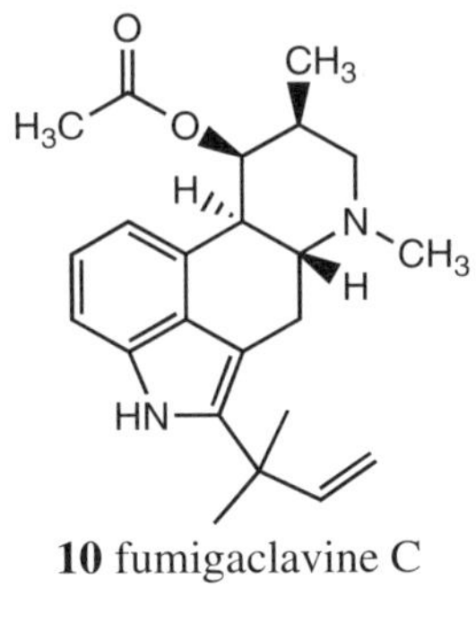

**10** fumigaclavine C

Fig. 3. Some dihydroclavines.

*Neotyphodium* species are asexual derivatives of *Epichloë* species, members of these two genera are collectively called "epichloë" hereafter.) Many, but not all, of these fungi produce EA when associated with their host plants, and a few were shown to produce EA in culture (*37*).

Like *C. purpurea,* other plant pathogenic Clavicipitaceae cause atypical diseases on their hosts. They form fruiting bodies or sclerotia on inflorescences, florets or at specific locations on vegetative leaves of the infected host. The affected inflorescences or florets are prevented from producing seeds, presumably benefiting the fungus by diverting resources to its fruiting bodies. The cause of this effect on the plant is probably hormonal, since even those *Balansia* species that fruit on leaves can reduce or eliminate host flowering (*38*). The EA are not involved in pathology, since many species or isolates fail to produce EA, yet exhibit the same effects on plants, and several *Neotyphodium* species produce EA, but cause no plant disease (*39,40*). Even so, EA, and the indolediterpenes (also produced by many of these Clavicipitaceae) have precursors in common with phytohormones, including auxin (from tryptophan), cytokinin (containing isoprene), gibberellin, and abscisic acid (from isoprene). To date, there has been rather little investigation of the potential for these fungi to produce these hormones or analogues, although some have been shown to produce auxin (*41*), and *Gibberella fujikuroi* (also in Hypocreales) is famous as a gibberellin producer (*42*). Thus, the metabolic flux of EA precursors into fungal products with phytohormone activity, and possible competition between these pathways, may be worthy of future investigations.

In many grass-epichloë symbiota the ergopeptine, ergovaline (**8**; Table I), is the principal EA. Generally **2,** and in some cases **3** or **6** and other clavines, can also be identified in these symbiota (*33,43–47*). In two symbiota involving *Achnatherum* species with epichloë endophytes, **2** and **3** are the principal EA, and no **8** or any other complex ergopeptine has been observed (*47,48*).

Ergobalansine (**9**; Table I) was first identified in *Cyperus* species that had symbiotic *Balansia* species, which also produce this alkaloid in culture (*13*). Later, ergobalansine was also identified in the plant, *Ipomoea asarifolia* (family Convolvulaceae) (*49*). Based on recent studies in which EA were eliminated from *I. asarifolia* by fungicide treatment (*50*), it seems reasonable to expect that **9** is produced by a fungal symbiont of this plant.

EA producers among the Eurotiales have ecological niches very distinct from those of the Clavicipitaceae. *Aspergillus fumigatus* is a notable example: a heat-resistant saprophyte, well adapted for survival and growth in compost, and currently the most common agent of invasive mycosis in humans (*51*). Spores of this fungus are ubiquitous, and inhaled spores can enter alveoli and induce aspergillosis. Very high levels of festuclavine (**7**) and fumigaclavines (e.g., **10**; Fig. 3) are associated with the conidia of *A. fumigatus* (*52*).

An interesting clavine structure, pibocin (**11**; Fig. 2), was isolated from a tunicate, *Eudistoma* species (*53*). This marine animal represents another taxonomic kingdom from which EA have been isolated. The preponderance of symbiotic fungi as EA sources in plants highlights the possibility that a fungal symbiont of the *Eudistoma* species might be the source of pibocin or a pibocin precursor, but this remains to be investigated. This alkaloid, which has antitumor activity, is a 2–bromo derivative of festuclavine. Interestingly, 2–brominated ergocryptine (bromocriptine) has been synthesized and used pharmacologically (*11*).

## IV. Biosynthetic Pathways and Genes

### A. ERGOT ALKALOID BIOSYNTHESIS GENE CLUSTERS

Nearly half a century of intensive investigation of the biosynthetic precursors, enzymes and pathways have been additionally informed by the recent identification of gene clusters that are likely to encode all or most of the enzymes for EA biosynthesis. The first pathway gene to be cloned, *dmaW* (*54*), encodes the determinant step in EA biosynthesis (Scheme 1). The gene was cloned from *C. fusiformis* SD58, which produces **6** as an end product. Subsequently, the orthologue was identified in *C. purpurea* P1 in a cluster of genes, of which many are predicted to encode biosynthetic enzymes (*55*). To date, 68 kb of the cluster has been sequenced (Fig. 4), and the likely or confirmed EA biosynthesis genes are listed in Table II. Here we designate the cluster *eas* (ergot alkaloid synthesis).

To facilitate discussion of the genes and their products, and comparisons among EA-producing fungi, we take this opportunity to adopt a systematic set of names for the *eas* cluster genes and their orthologues in other species (Fig. 4; Table II).

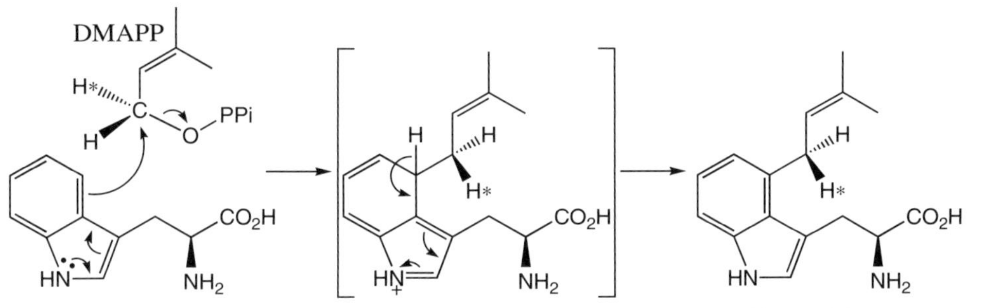


Scheme 1

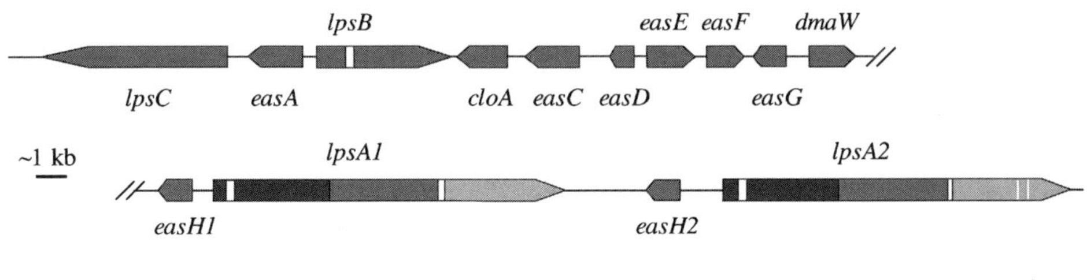


Fig. 4. Map of *eas* gene cluster in *Claviceps purpurea* P1. Arrows indicate direction of transcription. The corresponding modules in *lpsA1* and *lpsA2* are indicated with similar shading. White bars indicate intron positions in the *lpsA1*, *lpsA2*, and *lpsB* genes.

TABLE II.
Known and Predicted Genes in the Ergot Alkaloid Gene Cluster of *Claviceps purpurea* Strain P1.

| Gene | Synonym | Predicted product size (aa) | Function or putative function | Likely cofactors | Conserved domain | *E*-value |
|---|---|---|---|---|---|---|
| *easA* | *cpox3* | 369 | Reductase/dehydrogenase (OYE) | $FMNH_2$ | COG1902 | 4e-41 |
| *lpsB* | *cpps2* | 1308 | LPS subunit 2 | 4′-Phospho-pantetheine, ATP | COG1020 | 1e-54 |
| *cloA* | *cpP450-1* | 507 | Elymoclavine oxygenase | Heme-Fe | pfam00067 | 3e-18 |
| *easC* | *cpcat2* | 521 | Catalase | Heme-M | cd00328 | 3e-122 |
| *easD* | *cpox2* | 261 | Reductase/dehydrogenase | $NAD^+$ or NADH | pfam00106 | 3e-27 |
| *easE* | *cpox1* | 551 | Reductase/dehydrogenase | FAD | pfam01565 | 8e-14 |
| *easF* | *orfB* | 359 | Methyltransferase | AdoMet | COG4301 | 3e-24 |
| *easG* | *orfA* | 257 | Reductase/dehydrogenase | $NAD^+$ | COG0702 | 1e-11 |
| *dmaW* | *cpd1* | 448 | DMATrp synthase | $Ca^{2+}$? | n.d. | – |
| *easH1* | *orfC* | 314 | Oxygenase/hydroxylase | Fe(II) | pfam05721 | 8e-16 |
| *lpsA1* | *cpps1* | 3585 | LPS 1 (ergotamine?) | 4′-Phospho-pantetheine, ATP | COG1020 | 8e-69 |
| *easH2* | *orfE* | 154 | Hydroxylase or ψ | Fe(II) | n.d. | – |
| *lpsA2* | *cpps4* | 3524 | LPS 1 (ergocryptine?) | 4′-Phospho-pantetheine, ATP | COG1020 | 4e-74 |

*Note:* Abbreviations: OYE, old yellow enzyme; Heme-M, heme with metal ion; LPS, lysergyl peptide synthetase; DMATrp, dimethylallyltryptophan; n.d., none detected; ψ pseudogene.

Those *C. purpurea eas* genes that have not yet been fully characterized for function are designated *easA* through *easG*, plus the two closely related genes *easH1* and *easH2*. Those genes whose products have been characterized biochemically are designated according to the enzyme activities of the proteins they encode.

Four *eas* cluster genes encode enzymes that have been characterized in *C. purpurea* P1 or other EA-producing fungi. The *dmaW* gene product was well characterized in *C. fusiformis* SD58 (*54,56*), and its role in the determinant step of EA biosynthesis was established by disruption of the orthologue in the epichloë endophyte *Neotyphodium lolii* × *Epichloë typhina* strain Lp1 (*39*) (a natural hybrid of *N. lolii* and *E. typhina* (*57*)). Similarly, as will be discussed in detail later, the *lpsA1* and *lpsA2* genes were linked by protein sequences to the well-characterized lysergyl peptide synthetase (LPS) of *C. purpurea* P1 (*55,58,59*). Then, gene disruption demonstrated the role of a close homologue in synthesis of **8**, as well as **2** and lysergylalanine, by *N. lolii* × *E. typhina* Lp1 (*46,60*). Whereas the *lpsA* genes encode a large, multimodular subunit of LPS, the predicted protein sequence of the *lpsB* gene product indicates that it encodes the smaller subunit responsible for activation of the D-lysergic acid moiety. This role has been confirmed by *lpsB* knockout and heterologous expression (*61*), as detailed later. Finally, disruption of *cloA*, predicted to encode a cytochrome-P450 monooxygenase, has demonstrated that its product is involved in conversion of **6** into **1** (*62*).

A cluster of at least nine genes homologous to *C. purpurea eas* cluster genes has been identified in *C. fusiformis* SD58 (*54,63*). These genes are *lpsB, easA, cloA, easC, easD, easE, easF, easG,* and *dmaW*. All are transcribed, but the *lpsB* sequence suggests that it does not code for a functional product.

It is possible that the gene cluster so far characterized in *C. purpurea* encodes all of the enzymes for synthesis of **4**, yet some genes found in other fungal secondary metabolism gene clusters have no counterpart among those identified in *C. purpurea* to date. For example, no obvious transcriptional regulator is encoded in the cluster. However, in *A. fumigatus* several putative transcription factor genes are loosely linked to the ergot alkaloid gene cluster (*64*).

The *eas* gene cluster in *A. fumigatus* contains homologues for eight of the genes found in the *C. purpurea* cluster (Fig. 5; Table III). The genes common to the ergot alkaloid clusters of the two divergent fungi are hypothesized to encode enzymes that catalyze the early steps of the ergot alkaloid pathway, which are presumably shared between *A. fumigatus* and the clavicipitaceous fungi. The *A. fumigatus*

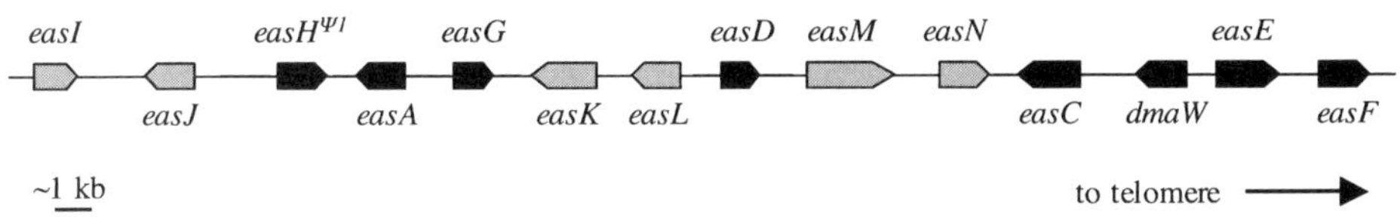


Fig. 5. Map of *eas* gene cluster identified on the long arm of chromosome 2 in the sequenced genome of *Aspergillus fumigatus*. Arrows indicate direction of transcription. Black arrows indicate orthologues of genes in the *C. purpurea eas* cluster. Gray arrows indicate genes present in *A. fumigatus*, but not identified in the *C. purpurea eas* cluster.

TABLE III.
Genes and Hypothetical Genes of the Ergot Alkaloid Cluster of *Aspergillus fumigatus*.

| *A. fumigatus* gene | | Closest matching protein (and other relevant descriptive match)[a] | | |
|---|---|---|---|---|
| (*eas* designation) (synonym) | Accession number | Deduced protein description (organism) | Accession number | %ID/ *E* value[b] |
| *easI* | XM_751037 | Hypothetical protein (*Aspergillus nidulans*) | EAA64052 | 61/1e-134 |
| | | Short chain alcohol dehydrogenase (*Trichodesmium* sp.) | ZP_00072819 | 27/2e-13 |
| *easJ* | XM_751038 | Hypothetical protein (*Magnaporthe grisea*) | EAA49815 | 32/1e-18 |
| | | FAD monoox., maackiain detoxification (*Nectria haematococca*) | ACC49410 | 31/3e-15 |
| *easHψ* | XM_751039 | Hypothetical protein, hydroxylase (*Pseudomonas aeruginos*) | AAK01512 | 33/3e-12 |
| | | fum3p dioxygenase (*Gibberella moniliformis*) | AAG27131 | 32/5e-07 |
| *easA* | XM_751040 | Hypothetical protein (*Neurospora crassa*) | XP_323805 | 50/5e-99 |
| | | Old yellow enzyme, NADPH dehydrogenase (*Saccharomyces cerivisiae*) | NP_015154 | 40/3e-70 |
| *easG* | XM_751041 | Hypothetical protein, ergot cluster (*Claviceps purpurea*) | AY836771 | 37/8e-41 |
| | | Hypothetical protein, epimerase (*Photorhabdus luminescens*) | NP_929597 | 37/2e-29 |
| *easK* | XM_751042 | Hypothetical protein (*A. nidulans*) | EAA58471 | 30/4e-47 |
| | | P450, pisatin demethylase activity (*N. haematococca*) | Q12645 | 31/4e-28 |
| *easL*(fgaPT1) | XM_751043 | DMATrp synthase, ergot pathway (*Neotyphodium lolii* × *E.pichloë typhina*) | AAP81206 | 27/1e-39 |
| *EasD* | XM_751044 | Ox/red, short chain dehydrogenase, ergot cluster (*C. purpurea*) | CAB39316 | 61/6e-81 |
| *easM* | XM_751045 | Cytochrome-P450 monooxygenase (*lovA*) (*N. crassa*) | XP_325231 | 46/8e-76 |
| *EasN* | XM_751046 | Hypothetical protein (*A. nidulans*) | EAA66317 | 30/3e-50 |
| | | Trichothecene 3-*O*-acetyltransferase (*Fusarium oxysporum*) | BAC65220 | 23/4e-11 |
| *EasC* | XM_751047 | Catalase, ergot cluster (*C. purpurea*) | AJ703808 | 55/1e-155 |
| *DmaW*(fgaPT2) | XM_751048 | DMATrp synthase, ergot pathway (*Neotyphodium coenophialum*) | AAP81207 | 63/1e-152 |
| *Ease* | XM_751049 | FAD-containing oxidoreductase, ergot cluster (*C. purpurea*) | CAB39328 | 40/1e-93 |
| *EasF* | XM_751050 | Hypothetical protein, ergot cluster (*C. purpurea*) | AY836772 | 50/1e-101 |

[a]If closest match was to a hypothetical protein, the closest descriptive match is also indicated.
[b]Percent amino acid sequence identity and number of similar matches expected in the database by chance.

cluster contains additional genes that are presumed to catalyze steps unique to the *A. fumigatus* pathway (*65*). These genes and the reactions with which they are likely associated are described below.

## B. ERGOT ALKALOID PRECURSORS

Although fermentation cultures had been established for the production of EA, the first experiments to determine the precursors of these molecules were conducted on rye ears infected with *C. purpurea* (*66*). The infected ears were injected with L-[$^{14}$C] tryptophan, which was incorporated into clavine alkaloids. The same result was obtained with a *Claviceps* species fermentation culture, which incorporated L-[$^{14}$C] tryptophan into **6** (*67*).

Incorporation of label from [2–$^{14}$C]mevalonate into the ergoline ring system supported the hypothesis that the pathway involved aromatic prenylation (*68*). This was further supported by the identification and purification of tryptophan dimethylallyltransferase (*69,70*), which transferred the dimethylallyl moiety from dimethylallyl-diphosphate (DMAPP) to the 4–position on the six-membered ring of L-tryptophan. Molecular genetic studies confirmed the role of this enzyme in EA biosynthesis (*39,54*).

The final component of the clavine intermediates is the N(6)-linked methyl group, derived from the *S*-methyl group of *S*-adenosylmethionine (AdoMet) (*71*).

The various ergopeptines that characterize many clavicipitaceous EA producers, are derived from **1** and three hydrophobic L-amino acids, the identities of which vary among the ergopeptines (see Table I).

The dihydroclavine alkaloids (e.g., **10**) that are produced by *A. fumigatus* have an unmodified C(17) methyl group, but may bear a hydroxyl or acetyl group at C(9), and a prenyl adduct at C(2).

## C. THE PATHWAY-DETERMINANT STEP

### 1. *dmaW and Aromatic Prenylation of Tryptophan*

Prenylation of L-tryptophan is the determinant step for the pathway to clavines. The enzyme that catalyzes this determinant step is EC 2.5.1.34, DMAPP: L-tryptophan dimethylallyltransferase (4–$\gamma$, $\gamma$-dimethylallyltryptophan synthase = DMATrp synthase; Scheme 1). The enzyme was purified to homogeneity from *C. fusiformis* SD58 (*70*), and shown to catalyze an electrophilic substitution, leading to addition of the 4–(3–methylbut-2–enyl) moiety in an all-*trans* configuration on C(4) of the L-tryptophan aromatic ring to form DMATrp (**12**) (*56*). Although no requirement was evident for divalent cations, it is conceivable that the purified enzyme might contain a tightly bound ion, or that divalent cations may be allosteric effectors, based on the observation that addition of 4 mM $CaCl_2$ to the reaction mixture approximately doubled the $V_{max}$. In the presence of 4 mM $CaCl_2$, $k_{cat} = 0.44$/s; $K_M = 8.0\,\mu$M for DMAPP, and $K_M = 17\,\mu$M for L-tryptophan were observed (*70*).

Native DMATrp synthase is a homodimer with an estimated molecular size of 105 kDa (*70*). The purified protein was cleaved on the carboxyl sides of the methionine residues by treatment with CNBr, and N-termini of the three peptides were sequenced. The data facilitated a plan for cloning the copy-DNA (cDNA) of its mRNA, and from this, the gene designated *dmaW* (where the W represents the single-letter code for L-tryptophan) (*54*). The structure of *dmaW* consists of three exons, interspersed with two short introns near the 3′-end of the coding sequence. The predicted polypeptide monomer size of 51,824 Da is in close agreement with the report of Gebler and Poulter (*70*), and contrasts with a previous estimate of 36 kDa (*69*). Possibly, the inclusion of phenylmethylsulfonyl fluoride, a serine-proteinase inhibitor, helped maintain the integrity of the polypeptide during its purification (*70*). Although comparison with sequences of farnesyl-diphosphate synthase and geranylgeranyl-diphosphate synthase suggested a short, but possibly conserved, motif (*54*), subsequent comparisons with *dmaW* from other fungi argued against any primary sequence relationship between these aliphatic prenyl transferases and DMATrp synthase (*55*).

Subsequent cloning of *dmaW* orthologues from several clavicipitaceous fungi indicated a conserved gene structure (*39,40,72*). Those from *C. purpurea, Balansia obtecta*, and three *Neotyphodium* species all encode proteins with greater than 60% sequence identity to that of *C. fusiformis*, and include introns at the same two positions near the 3′-ends of their open reading frames. The *A. fumigatus dmaW* (*65,73*) has two introns in positions likely to correspond to those in the Clavicipitaceae (but sequence similarity is low in the region, making this difficult to assess), plus a third intron between these.

The role of the *A. fumigatus* orthologue was confirmed by expression of the recombinant gene in the yeast, *Saccharomyces cerevisiae*, resulting in DMATrp synthase activity (*73*). When the gene was knocked out in the *A. fumigatus* genome, the mutant failed to produce the fumigaclavine alkaloids that characterized the wild-type parent (*65*).

As might be expected for the determinant step, DMATrp synthase is subject to regulation. L-tryptophan, supplied to *C. purpurea* fermentation cultures, induces expression of this enzyme, whereas medium with higher phosphate levels tends to inhibit expression (*74*). The phosphate inhibition operates partly or entirely through transcriptional control of *dmaW* (*72*).

### 2. *Genes related to dmaW in other systems*

Although the *dmaW* sequence indicates that its product represents a novel family of prenyltransferases, the family appears to be common in the fungal phylum Ascomycota. A BLAST search of the current Genbank database reveals related sequences in genomes of *Neurospora crassa, Magnaporthe oryzae, Fusarium heterosporum, Aspergillus nidulans, A. fumigatus, Penicillium roquefortii, Leptosphaeria maculans*, and *Sirodesmium diversum*. In several cases, the apparent *dmaW* homologues are associated with secondary metabolism gene clusters. For example, the *paxD* gene in *Penicillium paxilli* is located well within a cluster of genes known to encode enzymes for synthesis of the indolediterpene, paxilline (*75*), though the role of PaxD remains to be determined. Another *dmaW* homologue was identified in the sirodesmin synthesis cluster of the plant-pathogenic fungus, *L. maculans*, and was designated *sirD* (*76*). The most obvious role for the SirD protein is prenylation of the hydroxyl group of L-tyrosine either before or after its incorporation with L-serine into a cyclopeptide.

The EA gene cluster of *A. fumigatus* (Fig. 5; Table III) contains both the functional orthologue of *dmaW* and a second *dmaW* homologue (*easL*) (*65*). The *easL* gene product catalyzes "reverse prenylation," linking the clavine C(2) to the prenyl C(3) rather than C(1), to form fumigaclavine C (**10**) (*76*a). Numerous secondary metabolites are known with such reverse prenyl linkages. Thus, the *dmaW* family is surprisingly diverse in reaction specificities.

## D. SYNTHESIS OF CHANOCLAVINE-I

Following prenylation of L-tryptophan to yield **12**, the α-amine is methylated by an *N*-methyltransferase dependent on AdoMet (Scheme 2) (*71*). Haarmann *et al.* (*59*) suggest that the gene here designated *easF* (Fig. 4; Table II) might encode the *N*-methyltransferase for this step. The *N*-methylation must occur after L-tryptophan prenylation, but before the C-ring closure, since dual-labeled [6–$^{15}$N–C$^2$H$_3$]-**13** is incorporated intact into **6** (*77*). Furthermore, although norchanoclavines (which lack the *N*-methyl group) are detectable in fermentation cultures, labeled nor-chanoclavines are not incorporated into **6** (*77*). Also, when *C. purpurea* and *C. fusiformis dmaW* cDNAs were expressed in yeast, the resulting activity catalyzed prenylation of L-tryptophan, but not *N*-methyl-L-tryptophan (*40*).

Formation of the chanoclavine isomers from **13** requires two oxidation steps, and Scheme 2 shows the mechanism proposed by Gröger and Floss (*10*). The first

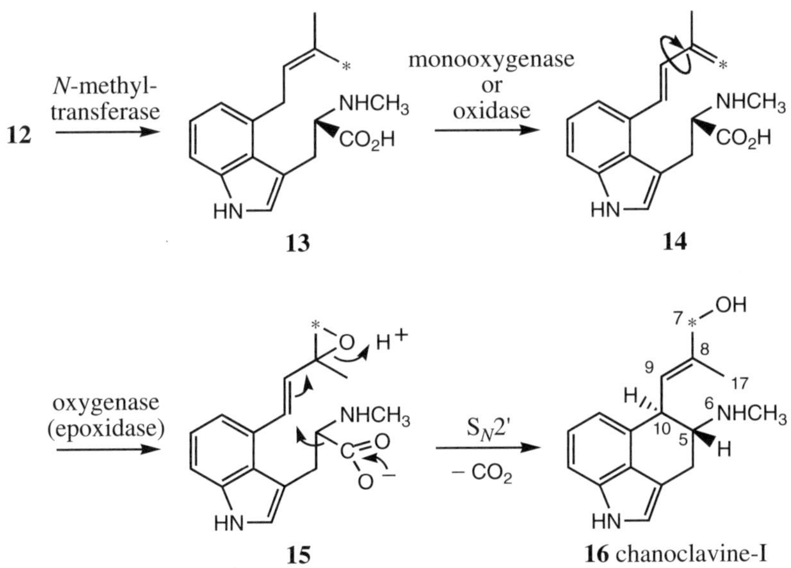


Scheme 2

oxidation results in diene **14**. Gröger and Floss (*10*) suggest that this reaction may proceed through an intermediate hydroxylated at the benzyl carbon, and note that 10–hydroxy-**13** is unstable and spontaneously dehydrates to form **14**. However, this instability precluded feeding 10–hydroxy-**13** to test the hypothesis. When cultures are fed deuterium-labeled **14**, the label is detected in **16**. Interestingly, if **12** is labeled in the methyl carbon *trans* to the vinyl hydrogen, the label that appears in **16** is *cis* to the vinyl hydrogen, evidencing the first of two epimerization steps in the pathway. This is further supported by the observation that label from [2–$^{14}$C]mevalonic acid incorporates into the methyl group *trans* to the vinyl hydrogen.

Diene **14** formation presents an obvious opportunity for the epimerization by rotation around the newly saturated C(8)–C(9) bond (Scheme 2). The orientation of this bond would likely be dictated by the active site of the next enzyme, which is predicted to be an epoxidase. The mechanism proposed for C-ring formation is epoxidation of **14**, followed by a possibly spontaneous $S_N2'$ reaction that couples C(5)–C(10) bond formation with decarboxylation at C(5) (Scheme 2) (*10*).

## E. SYNTHESIS OF AGROCLAVINE

Formation of the ergolene D-ring (Scheme 3) requires first the oxidation of **16** to chanoclavine-I-aldehyde (**17**), dehydration to the C(7)–N(6) iminium ion (**18**), and reduction of the iminium ion to form agroclavine (**19**) (*10*). During this process, the hydroxymethyl group, which is *cis* to the C(9)-hydrogen atom in **16**, ends up *trans* to the C(9)-hydrogen in **19**. Thus, a second epimerization occurs before or

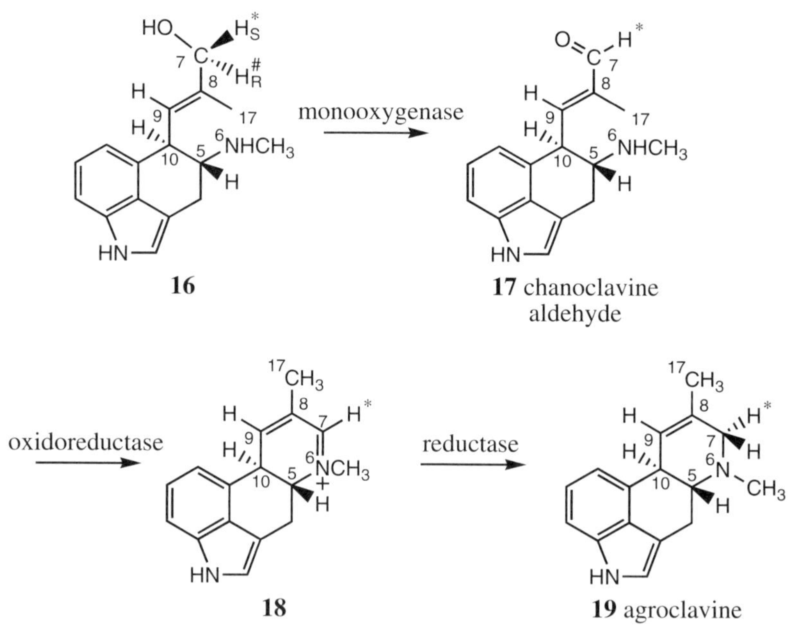


Scheme 3

during D-ring formation. Such an epimerization would be necessary to bring the aldehyde in proximity to N(6) to form the iminium ion **18**.

Chanoclavine-I-aldehyde (**17**) has been synthesized with a [17–$^3$H] label (ergoline numbering), and fed to cultures of *C. fusiformis* SD58, resulting in incorporation in **6** (*78,79*). Formation of **17** from **16** likely requires a monooxygenase. Although "chanoclavine-I cyclase" has been reported from crude extracts of fungal mycelium, it is likely that it represents the combined activities of two or three separate enzymes. Based on results reviewed by Floss *et al.* (*79*), we present a possible scheme for the path from **17** to **19** (Scheme 4). To close the D-ring, the aldehyde function must be brought into proximity with N(6) to permit formation of the iminium ion intermediate. This epimerization is likely to be catalyzed by an enzyme with two activities: a reductase that utilizes a reduced cofactor to saturate the C(8)–C(9) bond of **17**, and an oxidase that uses the oxidized form of the same cofactor to re-oxidize the same bond in either the reduction product **20** or its minium ion **21**. Furthermore, that cofactor should be tightly bound in the enzyme

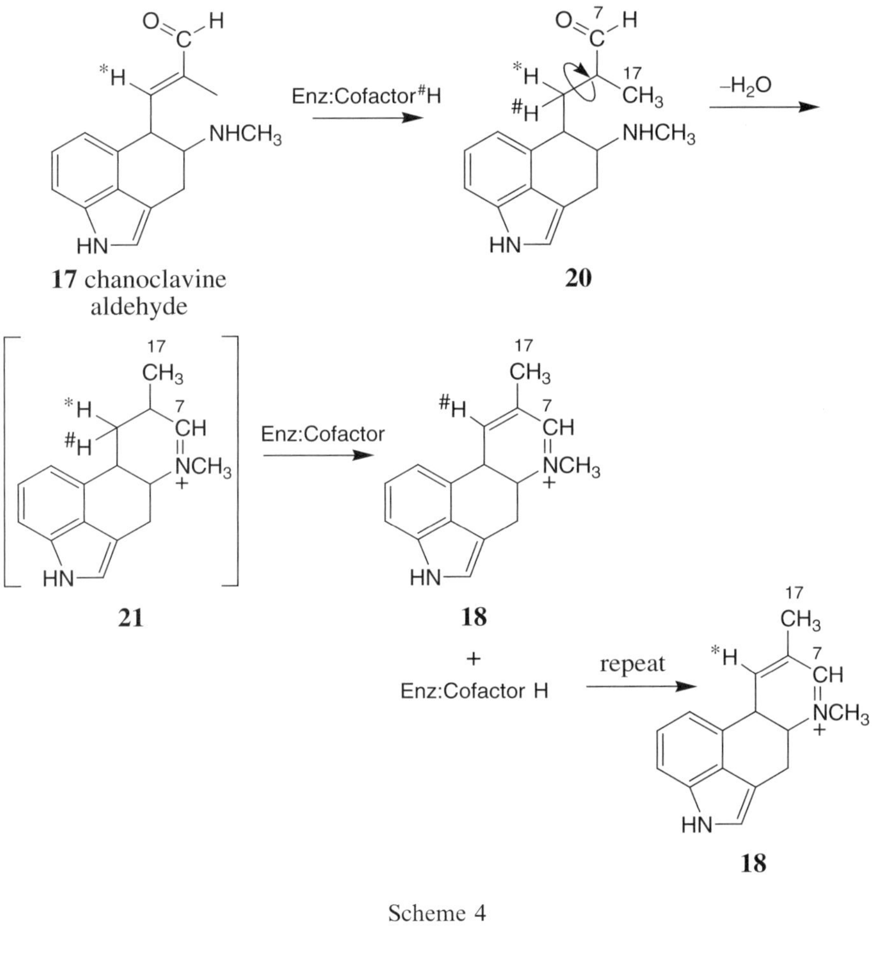


Scheme 4

active site. This would account for the observation that when fermentation cultures are fed **16** labeled with $^{3}H$ at C(9) (ergoline numbering), most, but not all, of the label is retained in the tetracyclic clavines (*78,79*). Furthermore, the reduction product **20** or **21** can be released from the active site during the catalytic cycle. This was demonstrated by an elegant experiment in which a mixture of [2–$^{13}$C]mevalonate and [4–$^{2}$H]mevalonate was fed to fermentation cultures, giving dual-labeled **9** and penniclavine, but not dual-labeled **16** (*79*). The implication is that the hydrogen at C(9) is extracted by the cofactor during re-oxidation of the C(8)–C(9) bond in **20** or **21**, which is then released without release of the reduced cofactor. This would allow a molecule of **17** to enter the active site and be reduced by that cofactor, thus acquiring the H-atom from the previous substrate molecule, and regenerating the oxidized cofactor in the active site of the enzyme.

A candidate enzyme for epimerization of **17** (Scheme 4) is the *easA* gene product, predicted to be similar to old yellow enzymes. These enzymes utilize tightly bound $FMNH_2$ as a cofactor and commonly reduce C–C double bonds conjugated with, and *trans* to, carbonyl groups (*80*). Different enzymes utilize different substrates to re-reduce the enzyme-bound FMN. During epimerization of **17** the reduced product **20** or **21** must, in turn, reduce the cofactor. It is noteworthy that retention of label from [9–$^{3}$H]-**16** (ergolene ring numbering) is incomplete, and the proportion of label retained in that position depends on the efficiency of **19** formation (*79*). As a possible explanation for this finding, we speculate that at each cycle the FMN cofactor may be re-reduced either by its reduction product (**20** or **21**) or by another reducing agent such as NADPH. Alternatively, the epimerization might be carried out by an enzyme that utilizes NAD(P)H for the reduction step, and the corresponding oxidized cofactor for product re-oxidation. If so, the oxidized cofactor should be exchanged slowly in the active site.

The product of the epimerization and loss of a water molecule would form iminium ion **18**, and cyclization would be completed by reduction of **18** (Scheme 3). This must require a separate reductase from the one involved in epimerization of **17**, or else the aforementioned pattern of H(9) atom retention would be highly unlikely. Thus, we predict that three distinct enzymes are likely to be involved in the "chanoclavine-I-cyclase" step from **16** to **19**.

The "chanoclavine cyclase" activity is dependent on NADH or NADPH, $Mg^{2+}$, and ATP (*10*). The scheme we propose here is consistent with an NAD(P)H requirement for the third enzyme (reductase). Also, we expect the oxidized form, NADP or $NAD^{+}$, to oxidize **16** to aldehyde **17**. Furthermore, and as stated above, it is possible that NAD(P)H is involved in the oxidation and reduction reactions to generate iminium ion **21**. However, it is not obvious why there is a requirement for ATP. Clearly, elucidation of the mechanisms underlying the steps in D-ring formation will continue to reveal some interesting biochemistry.

## F. SYNTHESIS OF ELYMOCLAVINE AND LYSERGIC ACID

Oxidation of C(17) of **19** generates **6**, and further oxidation steps lead to paspalic acid (**22**), the $\Delta^{8,9}$-isomer of **1** (Scheme 5). In cell-free extracts, the microsomal (membrane) fraction contains oxygenases for conversion of **19** into **6**, and

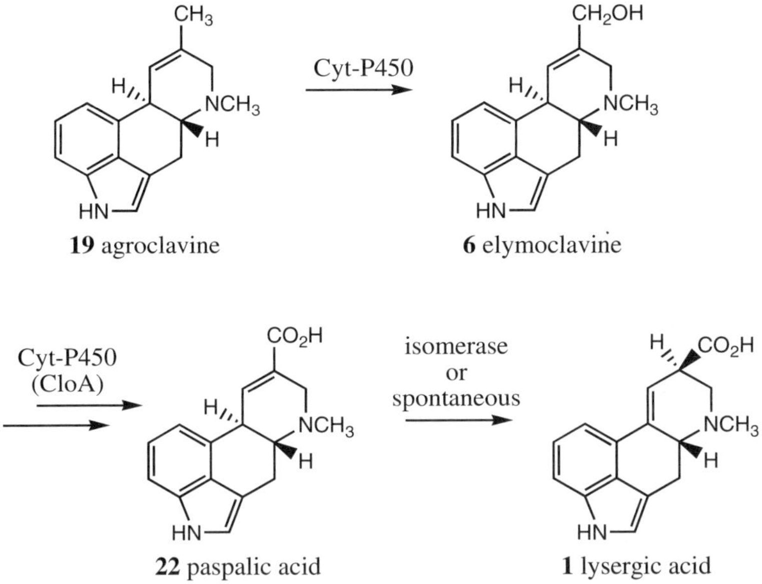


Scheme 5

conversion of **6** into **22** (*81*). These activities are NADPH-dependent, and inhibitor studies strongly suggest they are catalyzed by cytochrome-P450. Given that conversion of **6** into **22** is a total of a 4–electron oxidation, two sequential monooxygenase reactions by either the same enzyme or two different enzymes should be involved. It is also possible that a single enzyme catalyzes oxygenations of both **19** and **6**, but the observation that in some isolates of *Claviceps* species these two activities differ in the kinetics of expression suggests that distinct enzymes catalyze these conversions (*81*).

In the *eas* gene cluster of *C. purpurea*, one gene (*cloA*) has been identified with the heme-binding motif similar to cytochrome-P450. When this gene was knocked out in *C. purpurea* strain P1, the mutant produced **6** and **19**, but no **1**, lysergyl amides, or ergopeptines (*62*). Thus, CloA almost certainly catalyzes one or both monooxygenation reactions leading from **6** to **22** (Scheme 5). The observation that considerably more **19** than **6** accumulates in cultures of the mutant is in keeping with earlier findings that the agroclavine monooxygenase is inhibited by its product (*81*).

The enzyme responsible for the 2–electron oxidation of **19** to **6** remains to be identified. One possibility is that CloA catalyzes this reaction as well. The differences in induction profiles of agroclavine monooxygenase and elymoclavine monooxygenase activities (*81*), and the phenotype of the *cloA* knockout (*62*), suggest that there is a separate agroclavine monooxygenase, but do not exclude the possibility that CloA, as well as another monooxygenase, can act on **19** (as we consider below in discussing the *C. fusiformis eas* cluster). It is also conceivable that a second pathway to **6** is responsible for the accumulation of **6** in the mutant. Previous reports (reviewed by Gröger and Floss (*10*)) indicate that a microsomal fraction can hydroxylate **16**, and feeding experiments indicate that **16** hydroxylated on either allylic methyl group and labeled in the other allylic methyl carbon, can be converted into **6** by fermentation cultures. However, the labeled carbon becomes C(7) in **6**, in contrast to the specific labeling pattern in chanoclavine-I feeding experiments cited above. Therefore, the possibility seems remote that this alternative pathway to **6** operates *in vivo*. Rather, the accumulation of **6** in the Δ*cloA* mutant is probably due to an agroclavine monooxygenase for which a candidate gene is not apparent in the known cluster.

The final step in the synthesis of **1** from **22** is isomerization from $\Delta^{8,9}$ to $\Delta^{9,10}$. This may be catalyzed by a distinct enzyme, or occur spontaneously as observed in aqueous solutions of **22** (*10*). The prevalence of the 8(*β*)-diasteroisomer of **1** and its amides over the 8(α)-diastereoisomers suggests a directed, enzymatic conversion, but might alternatively be explained by selectivity of the lysergyl-peptide synthetase (LPS) discussed below.

## G. CLAVINES OF *ASPERGILLUS FUMIGATUS*

*A. fumigatus* produces **7** and three additional dihydroclavines called fumigaclavines (Scheme 6). Three catalytic steps would be required to complete the *A. fumigatus*-specific branch of the pathway from **7** to **10**. The products of each

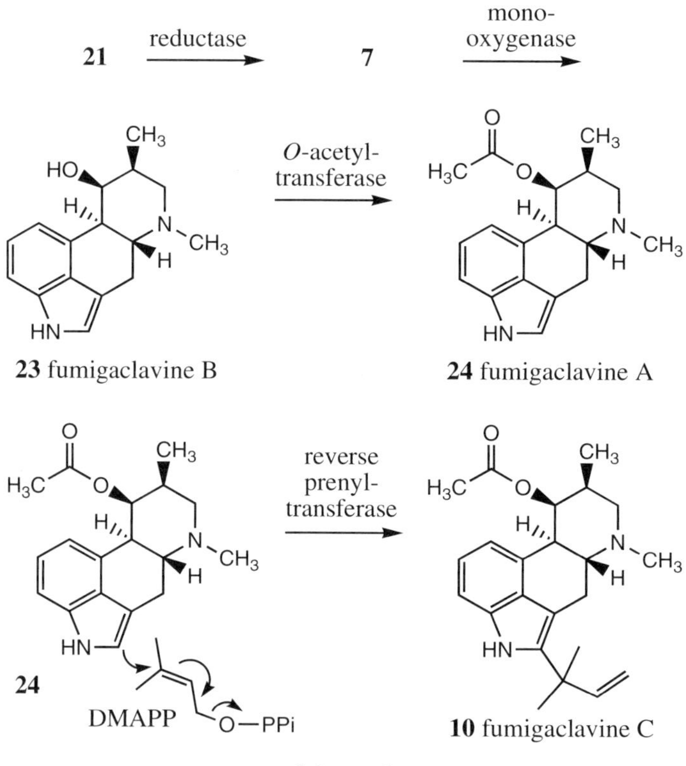


Scheme 6

of these steps accumulate to easily detectable levels in conidia of the fungus (*52*), indicating that the pathway is relatively inefficient in converting the product of the preceding step. Festuclavine (**7**) is hydroxylated at C(8) to yield fumigaclavine B (**23**). The *eas* gene cluster of *A. fumigatus* contains three genes that are candidates to control this step (*65*): *easJ* has the capacity to encode an FAD-containing monooxygenase, and *easK* and *easM* appear to encode cytochrome-P450 monooxygenases. All three of these oxidases lack orthologues in the *C. purpurea* and *C. fusiformis* clusters.

Acetylation of **23** to fumigaclavine A (**24**) is hypothesized to be catalyzed by the product of *easN*, which has high amino-acid-sequence identity with *O*-acetyl transferases (*65*). No orthologue of this gene is found in the *eas* clusters of the *Claviceps* spp. Moreover, there is no other acetylation step in any part of the ergot-alkaloid pathway or its shunts.

Fumigaclavine C (**10**) is the product of reverse prenylation of **24** (Scheme 6). In this ultimate product of the *A. fumigatus* branch, the prenyl group is attached to the ergoline ring system in a "reversed" orientation relative to the product of a "normal" prenylation. In a normal prenylation, such as the prenylation of tryptophan by DMATrp synthase (Scheme 1), the C(1) that is initially attached to the pyrophosphate group of DMAPP is presented to the indole co-substrate. In contrast, reverse prenylation is hypothesized to proceed with DMAPP presented in a "reverse" orientation relative to the indole cosubstrate promoting a facially non-selective $S_N'$ attack on the olefinic π electron system of C(3) of DMAPP (*82,83*). The mechanism of reverse prenylation has experimental support from the labeling studies of Stocking *et al.* (*82*). Interestingly, the *A. fumigatus* ergot alkaloid cluster contains a second *dmaW*-like gene, *easL*, encoding a product with 25% amino acid sequence identity to *dmaW* of *A. fumigatus* (*65*). The low, but significant, sequence identity with the normal prenyl transferase, coupled with its location in the ergot alkaloid gene cluster, makes this gene an excellent candidate to catalyze the reverse prenylation of fumigaclavine A to fumigaclavine C. This possibility has recently received experimental confirmation (*76*a).

The proposed sequence of events in which **23** is acetylated to **24**, followed by reverse prenylation to **10**, is supported by two observations. The first is the existence of an acetylated, but nonprenylated, fumigaclavine (**24**) and the lack of a prenylated, but not acetylated, fumigaclavine intermediate. Fumigaclavine C (**10**) can be experimentally deacetylated to yield such a compound, but this compound has not been detected in cultures of *A. fumigatus* (*52*). The second observation is simply that **10** is the most abundant ergot alkaloid in *A. fumigatus* (*52*), consistent with its position as the ultimate and inconvertible product of the pathway.

## H. COMMON STEPS AND GENES IN CLAVINE-ALKALOID PRODUCERS

Relationships among genes in different fungi capable of producing clavines provide clues to *eas* gene functions. C. *purpurea* P1 produces **6** as an intermediate to **1**, lysergic acid amides and ergopeptines, *C. fusiformis* SD58 accumulates **6** as the pathway end product, and *A. fumigatus* WVU1943 produces **7** (dihydro-**19**) and fumigaclavines. Functional genes shared between the *Claviceps* species are *easA,* *cloA, easC, easD, easE, easF, easG*, and *dmaW*. It is unknown whether *C. fusiformis* has an *easH* gene. Those genes shared between the *Claviceps* species and *A. fumigatus* are *easA, easC, easD, easE, easF, easG*, and *dmaW*. An *easH*-homologous sequence appears to be present in *A. fumigatus* as a pseudogene.

Given the *C. fusiformis* chemotype, and the recently established role of *cloA* (*62*) in conversion of **6** into **22**, the presence of a *C. fusiformis cloA* is surprising. Possibly, the *C. fusiformis cloA* product is nonfunctional. Alternatively, it is conceivable that the *C. fusiformis cloA* encodes a specific agroclavine monooxygenase. This remains an intriguing possibility considering that it has not been excluded that *C. purpurea* CloA might have agroclavine monooxygenase activity as well as elymoclavine monooxygenase activity.

The eight biosynthetic steps apparently shared between the *Claviceps* species are DMATrp (**12**) synthase (encoded by *dmaW*), methylation of **12** to **13**, oxidation of **13** to **14**, oxygenation (epoxidation) to give **16** *via* **15**, oxidation or oxygenation of **16** to **17**, the single-enzyme reduction and oxidation reactions to **18** (possibly by old yellow enzyme encoded by *easA*), reduction to **19**, and oxygenation to **6**. It is not clear that the eight-shared genes in the *eas* clusters correspond precisely to these eight steps. Two shared genes appear problematic in this regard: *cloA* (for reasons discussed above) and *easC*, predicted to encode a catalase.

Catalases are best known for disproportionating $H_2O_2$ to $H_2O$ and $O_2$; however, they can also use $H_2O_2$ to oxidize other substrates (*84*). Furthermore, flavine-utilizing enzymes can generate peroxides during oxidation reactions. So, we speculate that two successive oxidation/oxygenation reactions may be catalyzed by a FAD-containing enzyme (perhaps EasE) and the catalase, EasC. The obvious candidates would be the two steps from **13** to **16** (Scheme 2), or the two steps from **14** to **17** (*65*) (Scheme 3). Alternatively, the EasC catalase might detoxify peroxide produced during an FAD-dependent oxidation reaction (*64*).

Assuming that EasC catalyzes one of the oxygenations, the five genes encoding redox enzymes and known to be shared between the *Claviceps* species may be sufficient for all redox steps needed to produce **6**. However, if EasC does not act directly on a clavine intermediate, these five genes would appear to be insufficient. It is possible that a monooxygenase may catalyze multiple steps. For example, as argued earlier, CloA might catalyze oxygenation of both **19** and **6** (*62*). It must also be considered that some cluster genes may remain to be discovered, and that some steps might be carried out by enzymes encoded by genes outside of the cluster. Finally, EasH1, predicted to be a homologue of the fum3p hydroxylase in fumonisin biosynthesis (*85*), may catalyze an oxygenation in the clavine pathway. Identification of a homologue in *C. fusiformis* would provide evidence for this possibility. Signature sequences for a nonheme-iron oxygenase are evident in the predicted EasH1, but not EasH2 (which may be nonfunctional). The EasH step should occur after the point of divergence of the clavine pathway in *Claviceps* species from the festuclavine/fumigaclavine pathway in *A. fumigatus*, since the latter appears to lack a functional *easH/fum3* homologue. Therefore, we predict that EasH1 catalyzes either hydroxylation of **19** to **6**, or hydroxylation of ergopeptide lactams to form ergopeptines (discussed later).

Interestingly, except for *cloA*, the set of functional genes shared between *C. purpurea* and *C. fusiformis* is also shared with *A. fumigatus* (Fig. 5) (*65*). We speculate that the divergence point between the fumigaclavine pathway and the pathways in *Claviceps* species is marked either by **21** or **19**, intermediates that could be reduced to festuclavine (**7**). If the divergence point is at iminium ion **21**, this compound would not be re-oxidized in *A. fumigatus* as predicted to occur in the *Claviceps* species. Instead, it would be released from the enzyme in its reduced form, and its further reduction would yield **7**. If the divergence point is at agroclavine (**19**), then reduction of **19** would yield **7**. Further modification of **7** would produce fumigaclavines, and the genes likely to carry out those reactions are discussed above. *C. africana* might have the same pathway to **7**, which would then be used analogously to **19** in production of dihydrolysergic acid and subsequent dihydroergot alkaloids that characterize this species (*20,86*).

## I. ERGOPEPTINES AND OTHER LYSERGIC ACID AMIDES

### *1. Ergopeptines in Claviceps Species*

The amides of D-lysergic acid show a wide range of pharmacological effects, depending on the amide substituents, and different EA may act as agonists or antagonists of the various neurotransmitters in mammals. An extremely important group of D-lysergic acid amides are the ergopeptines, in which a tripeptide chain, modified to form a bicyclic cyclol-lactam structure, is joined with **1**. The two first amino acids in the tripeptide are variable, but always nonpolar, whereas the third position is taken by L-proline in all known ergopeptines except ergobalansine (see Table I). Two intramolecular reactions result in the unusual cyclol-lactam structure (Scheme 7). Amino acid III forms a lactam bond to the nitrogen atom of amino acid II. Subsequently, an α-hydroxy group is added to amino acid I and forms a cyclol bridge with the carbonyl carbon of amino acid III (*87*). Besides the ergopeptines, there are simpler derivatives of **1**, whose possible origins are discussed later.

Biochemical analyses of the assembly of the ergopeptines in *C. purpurea* have shown that ergopeptines are the products of an enzyme complex consisting of two nonribosomal peptide synthetase (NRPS) subunits (*58*). NRPSs generally exhibit modular structures, with each module responsible for the addition of an amino acid or other substituent. A typical module includes an adenylation (A-) domain, a thiolation (T-) domain (also known as a peptidyl carrier protein domain), and a condensation (C-) domain. The A-domain specifies the amino acid or other carboxylic acid substituent, and activates by it by an ATP-dependent adenylation reaction. The activated substituent then forms a thioester with the 4′-phosphopantetheine prosthetic group in the adjacent T-domain. Finally, the C-domain links the substituent to the next substituent in the chain. In a multimodular NRPS protein, the order in which substituents are added corresponds to the arrangement of modules from its N- to C terminus.

The two subunits of LPS—LPS 1 and LPS 2—have sizes of 370 and 140 kDa, respectively. Together they bind D-lysergic acid (**1**) and the three L-amino acids of the peptide portion of the alkaloid cyclopeptide as thioesters, catalyzing their successive condensation into the D-lysergyl mono-, di-, and tripeptide thioester intermediates,

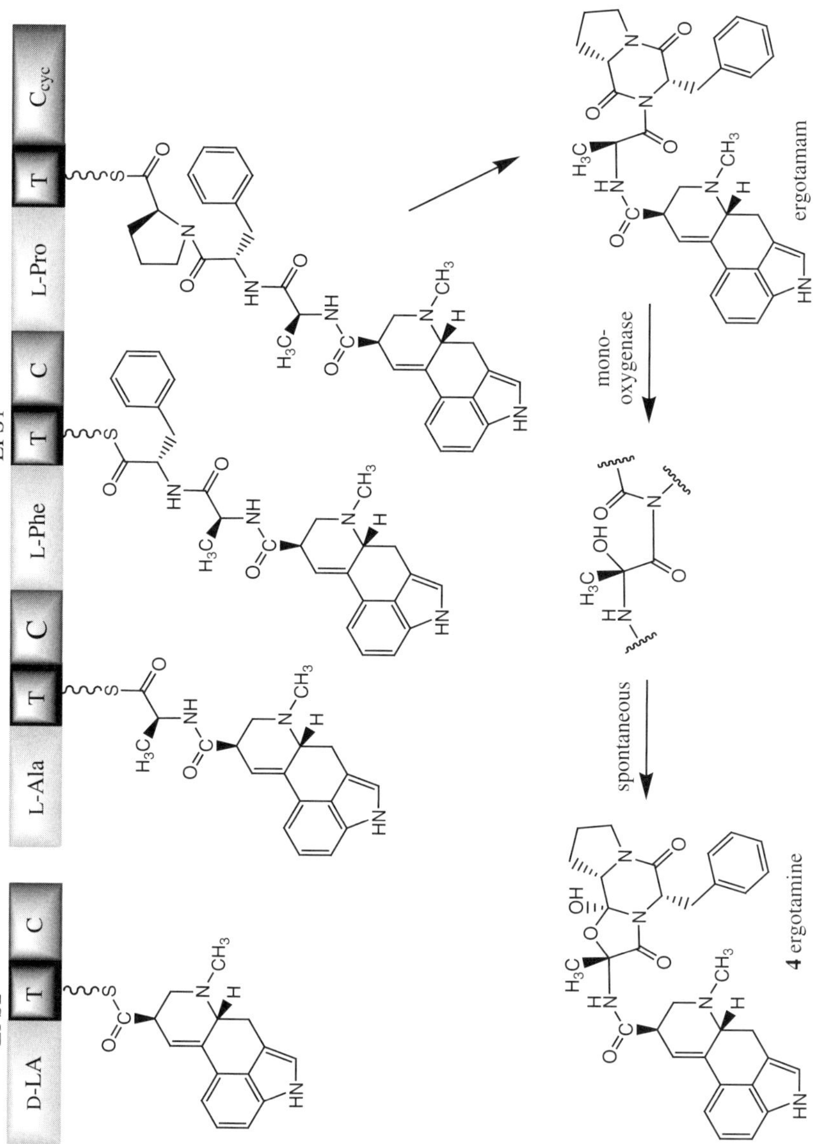

LPS2
D-LA
T
C
LPS1
L-Ala
L-Phe
L-Pro
Ccyc
mono-oxygenase
spontaneous
ergotamam
4 ergotamine


Scheme 7

and culminating in cyclization and release of the product D-lysergyltripeptide lactam (Scheme 7) (*58*). The monomodular subunit, LPS 2, activates **1**, and binds it as a thioester (*61*). LPS 2 also contains a C-domain near its C- terminus, which is presumed to catalyze the formation of the peptide bond between **1** and the amino acid in position I of what will be the tripeptide moiety of the ergopeptine. The remaining modules, with A-, T-, and C-domains for the three amino acids of the ergopeptine, are found in LPS 1 (*55,58,59*). The final C-domain catalyzes cyclization between amino acids at positions II and III, resulting in release of the lysergyl-peptide lactam from the enzyme complex.

The *eas* gene cluster of *C. purpurea* strain P1, stretching over ca. 68 kb, contains three characterized NRPS genes, designated *lpsA1* (formerly *cpps1*), *lpsA2* (formerly *cpps4*), and *lpsB* (formerly *cpps2*) (*55,61,62*). The cluster also contains a putative fourth NRPS gene tentatively designated *lpsC* (formerly *cpps3*). The *lpsA1* and *lpsA2* genes encode related trimodular proteins, each of which is a variant LPS 1 subunit. The *lpsB* gene encodes the monomodular LPS 2 subunit, and *lpsC* is predicted to encode another monomodular enzyme with a reduction domain in place of the condensation domain.

Comparison of the deduced amino acid sequence of *lpsA1* with a partial amino acid sequence of purified LPS 1 revealed matches (with a few exceptions discussed below) that indicated that *lpsA1* encodes the predominant LPS 1 subunit. Analysis of the deduced amino-acid sequence of LpsA1 confirmed that it harbors three modules, each responsible for the recruitment of an amino acid of the tripeptide substituent (Scheme 7) (*55*). Also, sequence data have demonstrated that both LPS 1 and LPS 2 are separately encoded enzymes and not breakdown products of a much larger NRPS enzyme (*55,61*). In contrast to the more common arrangement of catalytic domains in NRPS elongation modules, the first module of LPS 1 lacks an N-terminal C-domain. This suggests that the C-domain catalyzing formation of the peptide bond between **1** and the first amino acid of the tripeptide chain could be associated with the initiation module LPS 2 (a less likely possibility is that this is a stand-alone condensation domain yet to be identified). The size of the native LPS complex has been estimated to be between 500 and 550 kDa, roughly the sum of four modules contained within the two LPS subunits.

Analysis of the deduced amino-acid sequence of the *lpsB* product revealed that the protein comprises an A-domain, a T-domain, and a C-terminal C-domain (*61*). The predicted molecular mass of the protein is 140 kDa (1308 amino acids), in accordance with the size previously estimated for the LPS 2 subunit (*58*). Most importantly, the *lpsB* gene product carries at its C-terminus a typical C-domain that is almost certainly responsible for peptide bond formation between D-lysergic acid and the first amino acid of the tripeptide moiety of the ergopeptine. Such a C-domain was surprisingly not found at the N-terminus of LPS 1. This domain arrangement (A–T–C) is unusual in NRPS systems because C-domains usually lie on the N-terminal side of their corresponding A- and T-domains. This placement may add to their substrate specificity, which has been reported to be high for the substrate bound to the downstream T-domain, and low for the substrate bound to the upstream module (*88*). Thus, many elongation subunits in NRPSs start at their N-terminal ends with a C-domain. In the case of LPS 2, the location of the

C-domain at the C-terminal end of the protein may add to the extremely high specificity of the A-domain for D-lysergic acid (*58*), as measured with LPS 2.

Functional analysis in *C. purpurea* P1 has provided unequivocal evidence that *lpsB* encodes the LPS 2 subunit (*61*). Gene replacement of *lpsB* created an ergopeptine-nonproducing phenotype, which resulted in accumulation of **1** instead of ergotamine (**4**) and other ergopeptines. This biochemical phenotype is consistent with a block in the assembly of these complex alkaloid peptides. Comparative enzymatic studies of protein extracts derived from the Δ*lpsB* mutant and its parent strain showed that the parent contained both LPS 1 and LPS 2, whereas extracts from the mutant were devoid of any LPS 2–related activity. Furthermore, recombinant *lpsB* product expressed in the bacterium, *Escherichia coli*, catalyzed D-lysergic acid-dependent ATP-pyrophosphate exchange, demonstrating the specificity of the module for D-lysergic acid.

The putative D-lysergic acid-binding pocket of LPS 2 is similar to pockets from various A-domains that activate hydrophobic amino acids (*61*). The best match is with the *p*-hydroxyphenylglycine-activating module of chloroeremomycin synthetase. Interestingly, no similarity is seen with known tryptophan-activating A-domains, suggesting that the indole portion of D-lysergic acid is not responsible for its recognition by LPS 2.

The distribution of four modules for ergopeptine assembly between two polypeptides is unique among the known eukaryotic NRPSs, which otherwise contain all modules on one polypeptide chain. Although the significance of their independence is as yet unclear, the stand-alone D-lysergic acid module may provide an advantage to the fungus to use that module in combination with other NRPSs for other lysergyl-peptides such as ergonovine (**3**) (Scheme 8). In fact, a number of *C. purpurea* strains produce **3** as well as ergopeptines. Precursors of **3** are **1** and L-alanine (*89*), which would require an NRPS with a D-lysergic acid module and an

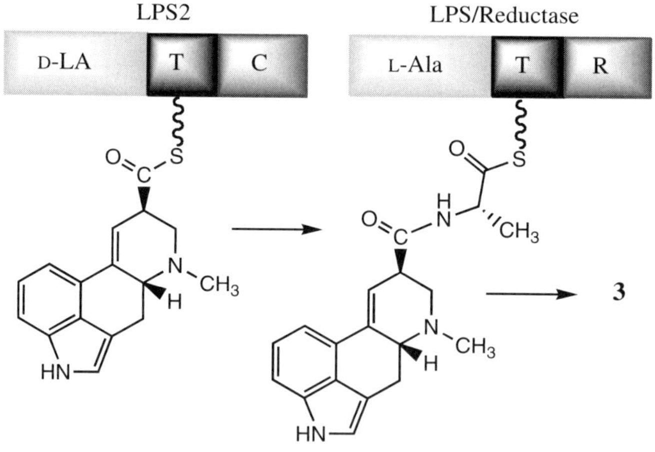


Scheme 8

L-alanine module, along with additional reductase and releasing domains. Possibly, the LPS 2 subunit sometimes associates with LPS 1 and sometimes with a different NRPS subunit containing an L-alanine module, thereby serving both synthetic processes in a natural combinatorial biosynthesis (Scheme 8). One candidate for the L-alanine-activating module would be the product of *lpsC*, for which functional analysis is in progress.

The theme of combinatorial biosynthesis is also apparent in the presence of another peptide-synthetase gene in the *eas* cluster (*59*). The *lpsA2* gene lies nearby, and is highly similar to *lpsA1*, implying a duplication in the origin of these two genes. The function of the trimodular enzyme encoded by *lpsA2* has not yet been demonstrated. However, comparison of the A-domain substrate-binding pockets strongly suggests that *lpsA2* encodes a version of LPS 1 with a variant specificity. The *lpsA2* product, in complex with LPS 2, could catalyze formation of ergocryptine, the second ergopeptine present in fermentation cultures of *C. purpurea* P1. Evaluation of the partial amino acid sequences from LPS 1 preparations showed that the few sequences not matching with the *lpsA1*-derived sequence matched the *lpsA2*-derived sequence, indicating that the enzyme preparations contained both LPS 1 variants. Analysis of the putative *lpsA1* and *lpsA2* orthologues in a different chemical race (producing ergocristine as a major ergopeptine) indicated that the chemical races differ in the specificity of the LPS 1 modules (*59*).

### 2. *Ergopeptines and Lysergic Acid Amides of Epichloë Endophytes*

Several epichloë endophytes of *Lolium* spp. grasses, as well as endophytes of several other cool-season grasses, produce **8** (*90,91*), a member of the ergotamine group composed of D-lysergic acid, L-alanine, L-valine, and L-proline (Table I). *Neotyphodium coenophialum* also produces small amounts of dehydroergovaline, which has a mass that is 2 atomic mass units lower than that of ergovaline, presumably as a result of desaturation in either the valine or proline substituents (*43*). Reports of other ergopeptines from epichloë endophytes may be due to extractions from endophyte-infected grass seeds that were contaminated with sclerotia or honeydew of *C. purpurea* (*43*).

Ergovaline (**8**) is produced in *N. lolii* × *E. typhina* Lp1 by an LPS complex very similar to the *C. purpurea* LPS described above. Functional analysis of the epichloë *lpsA* gene has been conducted by gene knockout, using gene sequences from *N. lolii* to direct the recombination event (*60*). Perennial ryegrass plants symbiotic with the Δ*lpsA* mutant lacked **8**, as well as two simpler amides of lysergic acid, **2** and lysergylalanine (**25**; Scheme 9), alkaloids that were all present in plants with the wild-type strain (*46*). The structural characterization of *N. lolii lpsB* also indicates that, as in *C. purpurea*, the D-lysergic acid-activating LPS 2 subunit is encoded separately. In keeping with this model, *Epichloë festucae* (the sexual ancestor of *N. lolii*) contains a gene homologous to the LPS 2–encoding gene from *C. purpurea* (D. Fleetwood, A. Tanaka, B. Scott, and R. Johnson, personal communication).

The chemical profile of the Δ*lpsA* mutant (*46*) indicates that accumulation of **2** and **25** requires the activity or products of the same LPS complex that generates the ergopeptine. However, it is unlikely that **2** and **25** are direct products of LPS, since NRPSs require specific domains to catalyze release of covalently bound intermediates

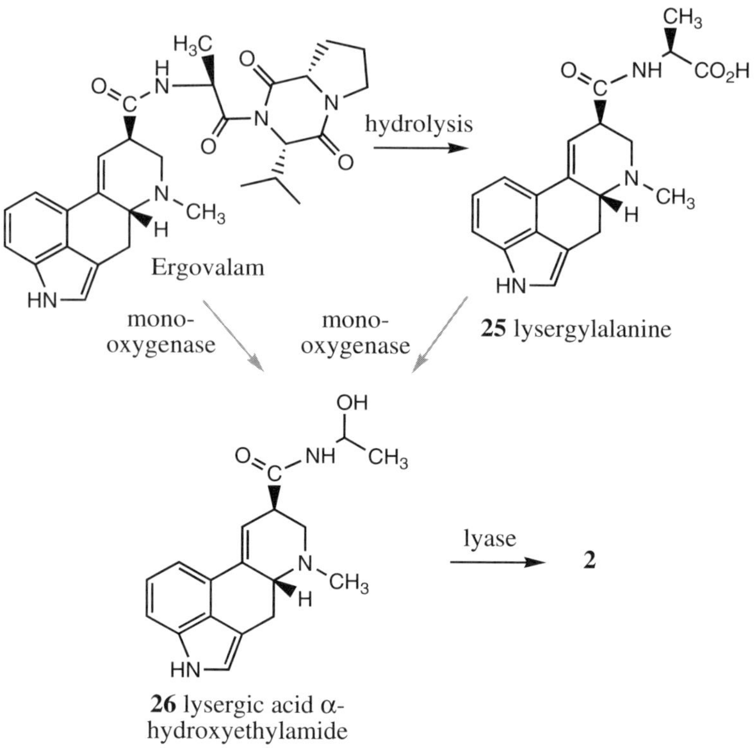


Scheme 9

(*92*). Perhaps occasional hydrolysis of the lysergyl-peptide lactam generates **25**. Generation of **2** from the lactam may proceed *via* lysergyic acid α-hydroxyethylamide (**26**; Scheme 9). After oxygenation at the α-carbon of the alanine residue, some of the intermediate may cleave at the peptide linkage to the lactam ring, to give **26**, whereas most of this intermediate would cyclize to form the ergopeptine. A lyase reaction of **26** could then yield **2** (*30*). This scheme, though still speculative, is also in keeping with *Claviceps paspali* feeding experiments, in which label from L-[2–$^{14}$C]alanine and L-[$^{15}$N]alanine, but not labeled **25**, incorporates into **3** (*10*).

The LPS 1–encoding genes from *N. lolii* and *C. purpurea* are clearly homologous (*93*). Their domain sequences throughout the three homologous modules range from 49% to 62% identical, and each contains two introns at precisely the same positions. However, the N-terminal regions of the two deduced polypeptides are much more divergent (32% identity). The N-terminal region of *N. lolii* LpsA contains a partial C-domain, though it appears to lack a critical active site (*93*). The N-terminal region of the deduced LPS 1 of *C. purpurea* contains no evidence of a C-domain (*59*).

The signature sequences that define the amino acid-binding pocket, and thus substrate specificity of the individual A-domains, reflect the differences in ergopeptines produced by *C. purpurea* P1 (primarily **4**) and *N. lolii* × *E. typhina* Lp1 (exclusively **8**) (*93*). The signature sequences for the first and third A-domains of the deduced LPS 1 products are nearly identical between these two fungi. Among the 10 amino acids lining the amino acid-binding pocket, nine are identical in each of module I (specifying L-alanine) and module III (specifying L-proline). However, the signature sequences for the second A-domain differ at five of the 10 sites between the corresponding sequences from the two fungi. These data, together with the aforementioned comparison of *lpsA1* and *lpsA2* in *C. purpurea* P1, indicate a strong genetic component to the determination of ergopeptine composition.

## J. PATHWAY INTERMEDIATES, SHUNTS AND SPURS

Typically, EA-producing fungi build up substantial levels of intermediates in addition to pathway end products. For example, quantitative assays indicated that *A. fumigatus* conidia possessed **7**, **23** and **24** at a combined molar concentration that was 25% that of **10** (20), and sclerotia of *C. africana* had intermediates **7** and dihydrolysergol (**6b**) at a combined molar concentration 49% that of dihydroergosine (*86*). Most dramatically, in leaf blades of ryegrass symbiotic with *N. lolii* × *E. typhina* Lp1, chanoclavines were 2.5–fold more abundant than **8** on a molar basis (20).

An EA pathway intermediate in one fungus may be an endpoint in another. For example, **6** is an intermediate in *C. purpurea* P1, but an end product in *C. fusiformis* SD58. Also, shunts provide multiple end products, such as the production of **2**, **3**, **25**, and 6,7–secolysergine (**28**; Fig. 6), in addition to **8**, by the epichloë endophyte (*46*). Two tricyclic clavines, **28** and its isomer, 6,7–secoagroclavine (**29**) appear to be reduced forms of **16** or **19**.

Interestingly, knockout of epichloë *lpsA*, which controls the penultimate step in the pathway to **8**, affected the profile of alkaloids from earlier stages of the pathway and a shunt pathway (*20,46*). Perennial ryegrass with the Δ*lpsA* mutant accumulated elevated concentrations of **1**, which is the pathway intermediate prior to the LPS step. However, the amount observed to accumulate was only 13% of the amount incorporated into D-lysergic acid derivatives in perennial ryegrass containing the wild-type endophyte. The amounts of **16** and setoclavine (**27**; Scheme 10) produced by the mutant did not differ significantly from those produced by the wild-type strain. However, in grass containing the Δ*lpsA* mutant the amount of shunt

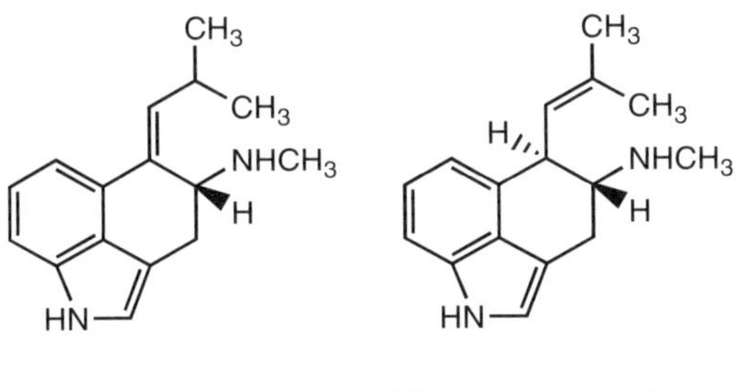


**28** 6,7-secolysergine **29** 6,7-secoagroclavine

Fig. 6. Secoclavines.

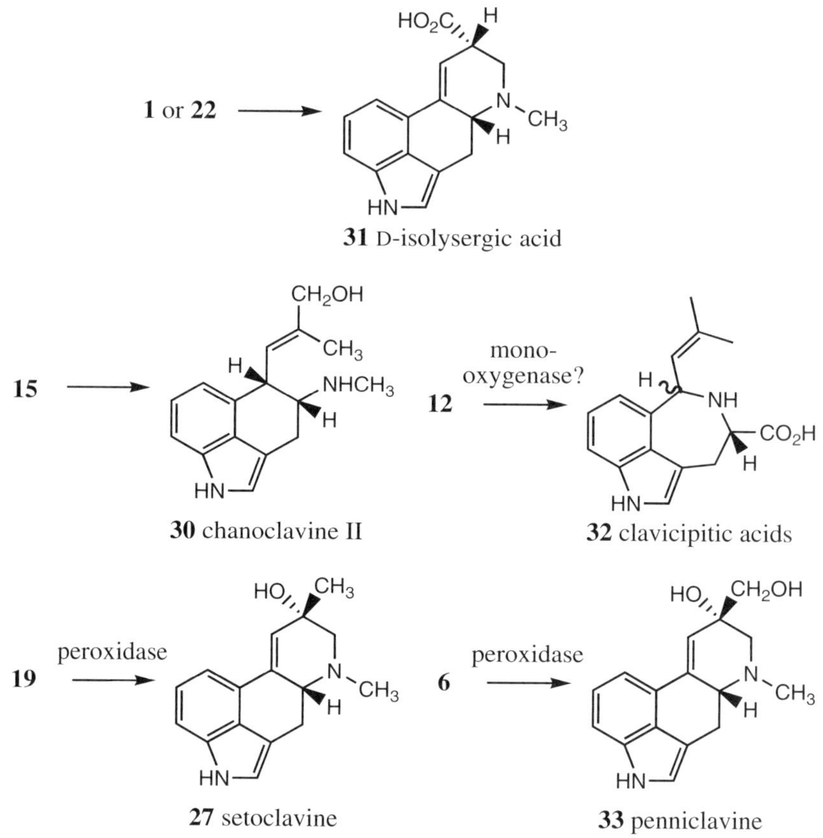


Scheme 10

product **28** increased approximately two-fold to levels that actually exceeded the amount of **1**, the substrate for LPS (*46*). These data indicate that expression or activity of several upstream enzymes in the ergot-alkaloid pathway were affected by the Δ*lpsA* knockout, perhaps by negative feedback from the accumulated **1**.

Pathway shunts and the apparently well regulated, but inefficient, flux through intermediates to end products may be selectively favorable because of the various biological activities of the different clavine, lysergyl, and ergopeptine alkaloids. Evidence pertinent to this hypothesis is reviewed by Panaccione (*20*).

Incomplete substrate specificity or stereoselectivity of enzyme reactions appears to be responsible for several spur products, examples of which are shown in Scheme 10. Although different chanoclavine isomers are produced (*10*), the stereochemistry of the tetracyclic clavines and **1** implies that chanoclavine I (**16**), but not chanoclavine II (**30**), is an intermediate. Also, **22** can isomerize spontaneously to **1** or isolysergic acid (**31**). Typically, both C(8) diastereomers of ergopeptines are obtained in preparations, and those with an isolysergic acid moiety – which are pharmacologically inactive (*6*) – are specified by the "inine" ending (e.g., ergotaminine, ergovalinine, etc.). Isomerization at C(8), can occur in aqueous solutions, so it is generally thought that **1**, and not **31**, predominates or is selectively incorporated into ergopeptides by the LPS, and that the "inine" isomers are purification artifacts (*94*).

Oxidation or hydroxylation of C(10) of **13** is involved in the synthesis of **16**. If the enzyme for this reaction has a low level of activity on **12**, the product may rearrange to form clavicipitic acids (**32**), which are present in small amounts (Scheme 10) (*95,96*). More abundant spur products, setoclavine (**33**) and penniclavine (**33**) (plus their C(8) stereoisomers), arise by peroxidase action on **19** and **6**, respectively. In the case of grass-epichloë symbiota, both plant and fungal peroxidases can catalyze those conversions (*46*).

## V. Routine Analytical Methods for Ergot Alkaloids

Analytical methods for studying EA have changed over the years in response to changes in technology. This topic was reviewed comprehensively by Flieger *et al.* (*97*), so here we will only briefly provide our perspective on routine analyses of EA, with particular emphasis on notable changes over the past several years.

Historically, EA have been isolated *via* repeated alternating extractions with an alkaline organic solvent followed by an acidic aqueous solvent. This classical procedure exploits the solubility of alkaloids in both the solvents, whereas most other molecules lack solubility in one of the solvents and are thereby separated from the alkaloids.

One of the earliest methods for separating EA from each other was to separate water-soluble forms (e.g., **3** and several clavine alkaloids) from water-insoluble forms (primarily ergopeptines). However, the solubility differences are not absolute. Alternatively, ergopeptine alkaloids have been separated from other EA by their preferential retention on flash chromatography columns packed with silica containing HL binder (e.g., Ergosil) (*98*). Surprisingly, Ergosil binds ergopeptines, but not their 8($\alpha$) stereoisomers (ergopeptinines).

A more recent approach has been to simplify the extraction procedure and rely on extended chromatographic separation to obtain a complete EA profile in a single step. An extraction solvent composed of 50% 2–propanol + 1% lactic acid can recover EA of a wide spectrum of polarities, from the extremely polar lysergic acid, to moderately polar clavines or simple amides of lysergic acid, to the nonpolar ergopeptines (*46,52,99*). Alkaloids of varying polarity may then be separated by reverse phase HPLC by application of a gradual gradient beginning with a mostly aqueous solvent (such as 5% acetonitrile in 50 mM ammonium acetate) to a nonpolar solvent (such as 75% acetonitrile in 50 mM ammonium acetate).

Following chromatographic separation, EA are routinely detected with great sensitivity by fluorescence. Those alkaloids with a $\Delta^{9,10}$-double bond can be effectively detected with an excitation wavelength of 310 nm and an emission wavelength of 410 nm. EA with a $\Delta^{8,9}$-double bond and those with a saturated D-ring fluoresce maximally with excitation and emission wavelengths of 272 and 372 nm, respectively. Thus extracts that contain EA with a combination of these fluorescence properties must be analyzed twice to detect and quantify a complete profile. Alternatively, two fluorescence detectors connected in series allow for analysis of the full ergot-alkaloid profile in a single HPLC run (*20*).

Inclusion of a mass spectrum (MS) detector in the analytical system provides mass information for the eluted alkaloids to confirm their identities. In many cases, EA can be fragmented in such detectors and the masses of the fragments facilitate identification of alkaloids (*1,43,46*). Moreover, MS detectors operating in single ion mode are exceptionally sensitive. However, unless appropriately labeled standards are available for each alkaloid, data from MS detectors do not provide information on the quantities of EA in the sample.

The primary alternative to the instrumental methods summarized above is immunodetection of EA with specific antibodies. Antiserum raised in response to immunization with lysergol (**6a**) coupled to human serum albumin recognizes several EA (*100*). Used for competitive enzyme-linked immunosorbent assay (ELISA), this broad-spectrum antiserum allows rapid detection of total EA in many samples concurrently. However, it does not separately quantify the different EA that contribute to the profile, and different EA show different reactivities to the antiserum. Monoclonal antibodies specific for individual EA species, such as ergotamine, also have been raised and provide an effective means for detection and quantification of the particular EA, as well as a valuable tool for physiological studies (*101*).

## VI. Ergot Alkaloid Activities and Roles

### A. PHARMACOLOGICAL ACTIVITIES

Three important neurotransmitters are derived from aromatic L-amino acids by decarboxylation followed by hydroxylation. Likewise, the ergoline ring system is derived from L-tryptophan, which has been decorated (prenylated and *N*-methylated), and decarboxylated in the process. The implication of this similarity in origins is that structures of noradrenaline, dopamine and 5–hydroxytryptamine (serotonin) can be mapped almost entirely onto the ergoline ring structure. This, it is believed, is why many clavine and EA can interact with receptors for all three of these neurotransmitters (*102*).

The clinical utility of natural EA like **3** and **4** was a prelude to broader and more precise clinical applications of semisynthetic derivatives of **1** and ergopeptines. Mukherjee and Menge (*103*) provide a detailed review of ergot-alkaloid modifications and the effects on their pharmacological activities.

Simple derivatives of ergopeptines include dihydroergotamine (with a saturated D-ring) for the treatment of migraines and dihydroergotoxine for the treatment of hypertension (*11*). Also, bromocriptine, derived from ergocryptine by bromination at C(2), is used in treatment of Parkinsonism. Bromocriptine is also effective in cases of hyperprolactinaemia (*104*), a condition that can result in reproductive disorders such as galactorrhoea and anovulation. Interaction of ergopeptines with $D_2$ dopamine receptors is key to the depression of prolactin levels.

Several EA and dihydro-EA also exhibit agonistic activity on serotonin receptors. Potency of **4** and **8** as agonists of arterial $5\text{–}HT_{2A}$ and $5\text{–}HT_{1B/1D}$ is comparable to that of serotonin (*105*), and seems to be partially responsible for vasoconstrictive activity. In treatments of migraine, which is thought to be due to vasodilatation of carotid arteriovenous anastomoses, **4** and dihydroergotamine can affect $\alpha_1$- and $\alpha_2$-adrenergic receptors (*106,107*).

Serotonin agonism in the brain is also thought to be a key factor in hallucinogenic activities. Inhibition of the serotonergic activity by ketanserin, believed to be a selective antagonist of $5\text{–}HT_{2A}$ receptors, supports the contention that **5**, as well as other hallucinogens such as psilocybin, act on 5–HT receptors (*108*). However, the bases for overall effects of EA are still not fully understood. For example, the potency of **5** as a $5\text{–}HT_2$-receptor agonist is less than that of amphetamine hallucinogens, yet the human potency of **5** as a hallucinogen is about 10–fold higher. In fact, **5** is only a partial agonist of brain $5\text{–}HT_{2A}$ (*108*), and shows high affinity for $5\text{–}HT_{1A}$, as well as the $5\text{–}HT_{5a}$, $5\text{–}HT_6$, $5\text{–}HT_7$, and dopamine $D_2$ receptors, so activity at these receptors might account for the high potency of **5** as a hallucinogen (*3*).

## B. LIVESTOCK POISONING

The possibility that ergot poisoning was the key trigger of the Salem witch trials is supported by the recorded symptoms exhibited both by the citizens of the village and their livestock, for which rye was also a major component of the diet (*2*). Although human ergotism remains much more widely recognized by the public, in modern developed countries livestock poisoning is far more common and economically important. There are two major sources of EA that may be ingested by livestock. One source is the ergots produced by *Claviceps* species on ears of pasture grasses and feed grain crops (such as *C. africana* on sorghum (*109*)). To minimize exposure to ergots, it is important to time grazing and mowing to minimize availability of maturing seed heads in pasture grasses, and to avoid feed with substantial ergot contamination. However, the second source of EA is much less conspicuous, and often much more difficult to avoid: epichloë endophytes in pasture grasses. In particular, two *Lolium* species have become extremely popular for hay and perennial pastures in temperate zones of North America, Australia, and New Zealand, even though they often host EA-producing endophytes: tall fescue (*Lolium arundinaceum* = *Festuca arundinacea*) is the host of *N. coenophialum*, and perennial ryegrass (*Lolium perenne*) is the host of *N. lolii*.

Toxicosis due to EA is prevalent in animals that are grazed on, or fed the hay of, tall fescue with *N. coenophialum* (*7,110*), whereas "staggers" symptoms suffered by animals fed perennial ryegrass with *N. lolii* have usually been attributed to indolediterpenes, such as the lolitrems (*110,111*). A possibility that has not been tested is that EA might exacerbate staggers. Another endophyte of perennial ryegrass, *N. lolii* × *E. typhina*, is rare in nature, but has been very useful in some of the aforementioned molecular genetic studies of EA biosynthesis genes (*39,46,60*). This natural hybrid was initially incorporated into perennial ryegrass cultivars in New

Zealand, but they were withdrawn because of their high EA levels (*112*). Various *Festuca* species, particularly the fine fescues, harbor *E. festucae*, and many strains produce EA (*113,114*).

To date, little difficulty has been observed in livestock grazing fine fescues, though circumstantial evidence suggests that their EA may have dramatic effects on populations of feral Soay sheep in the Outer Hebrides (*115*). On the island of Hirta, the Soay sheep undergo population crashes of 40–65% every 3–5 years, and red fescue (*Festuca rubra*) infection by *E. festucae* correlates with grazing pressure. Furthermore, mock herbivory (clipping) induced production of **8** in plants from the island, whereas **8** was undetectable in control plants. Thus, **8** appears to be an inducible defense in red fescue – *E. festucae* symbiota.

Livestock grazing or eating hay of tall fescue with *N. coenophialum* can show toxicosis symptoms similar to animals poisoned by ingesting ergots (*7,116*). When associated with the tall fescue endophytes, the syndromes are known collectively as fescue toxicosis (*117*). Symptoms depend heavily on temperature stresses (*118*). The syndrome associated with heat stress is called "summer slump." The animals "go off feed" (lose appetite), show poor weight gain or even lose weight, and retain winter coats. In cold conditions, vasoconstrictive activity of the EA can cause extremities to become hypoxic. As a result, the animals may exhibit dry gangrene, with loss of switches (tails) and the tips of ears, and in more extreme cases, loss of hooves, a syndrome called "fescue foot" (*7,110*). Physiological indications of fescue toxicosis include elevated core body temperatures, very low prolactin levels, and fat necrosis (*7,116*). In cows, a major reduction in conception rates, as well as agalactia (low or no milk production), are probably linked to low prolactin (*119*). In cows, and especially mares, grazing EA-containing tall fescue can cause long gestations culminating in calves or foals that are dysmature, and weak or still-borne (*120,121*).

Given the symptoms of tall fescue toxicosis, there is little doubt that EA are etiological agents. Traditionally, ergovaline (**8**) has been measured as a proxy to EA levels, based in part on the presumption that **8** is the most important toxin. However, most toxicological studies have employed, not **8**, but **4** and other ergopeptines that could be obtained in abundance from *C. purpurea* ergots and are commercially available. Also, in most labs, clavines and simpler lysergic acid amides are not as easy to measure as ergopeptines. Since epichloë-symbiotic plants can have substantial amounts of the simpler EA (*20,33,46*), it is reasonable to consider that the clavines or simple derivatives of **1** can be as or more important in fescue toxicosis compared to ergopeptines. An indication of the importance of EA in addition to **8** comes from the work of Gadberry *et al.* (*116*) with *N. lolii*-infected perennial ryegrass and heat-stressed lambs. In this study, the effect of endophyte-infected grass seed (containing **8**, other EA, and additional *N. lolii* metabolites), was compared to the effect of an equivalent amount of **8** fed as an additive to endophyte-free ryegrass seed, and to controls without endophyte or **8**. Addition of **8** alone to the diet had no significant effect on feed intake, and a marginally significant effect on weight gain ($P = 0.10$), but significantly lowered the thermocirculation index ($P = 0.05$) and serum prolactin levels ($P < 0.01$). Lowered thermocirculation index implies vasoconstriction and reduced blood flow to extremities. Interestingly, prolactin levels were reduced to a significantly greater extent in animals fed endophyte-infected seed (94%) than in those with endophyte-free seed supplemented with an equivalent amount of **8** (34%), suggesting that other toxins (perhaps other EA) produced by *N. lolii* contributed to the prolactin reduction. Also, *N. lolii* infection significantly reduced feed intake and daily weight gain. Thus, the effect of endophyte on appetite seems to be due primarily to toxins other than **8**.

The effect of **8** on serum prolactin levels is thought to be due to D2 receptor agonism. Treatment of cultured rat pituitary cells with 10 nM **8** inhibited prolactin release at least 40%, and the effect was reversed by the D2 receptor antagonist domperidone (at 1000 nM) (*122*).

The relative importance of different EA depends on their kinetics of interaction with various receptors, as well as absorption, distribution, metabolism, and excretion. Uptake of different EA has been investigated for bovine reticular, ruminal, and omasal tissues (*123*). Ergopeptines (**4** and ergocryptine), **3**, lysergol (**6a**; Fig. 2), and **1** were actively transported across these tissues, and **6a** and **1** were transported at significantly higher rates than were the ergopeptines or **3**.

The distribution of EA in infected plant tissues is an important consideration when assessing risk to livestock of epichloë-infected pasture grass. In perennial ryegrass with *N. lolii*, a strong gradient of **8** was evident in the leaves (*124*), with highest levels in the lower half of the leaf sheath (pseudostem), less in the upper half of the leaf sheath, and very little in leaf blades. High concentrations (approx. 0.4–3 μg/g dry mass) were also measured in the true stems at the bases of the plants. If pasture grasses are overgrazed, the animals are more likely to consume material from the base of the plant, ingesting the lower leaf sheathes and stems that have the greatest concentration of **8**. When plants flower, levels of **8** in the inflorescences and seeds can be very high. In studies of tall fescue with *N. coenophialum*, panicles contained **8** at over 2 μg/g dry mass (*125*), and seeds had 1–8 μg/g (*126*). Therefore, proper management of livestock on pasture should minimize overgrazing and exposure to developing or ripe seed heads.

Despite their production of toxic alkaloids, it is often necessary to retain epichloë endophytes in cultivars and pastures because they confer increased stress tolerances and protection to the grasses (*34,127*). For this reason, surveys of natural *Lolium* species populations in Europe and North Africa have been conducted to identify endophytes lacking EA in general, or **8** in particular as well as lacking the indolediterpenes. Some such EA-nonproducers have been used to replace the toxin-producing endophytes in popular forage cultivars (*91,128,129*). As hoped, animals that grazed on pastures with these novel endophytes showed none of the symptoms of toxicosis associated with EA or indolediterpenes, and exhibited weight gains similar to those that grazed on endophyte-free grasses (*129–133*). However, field performance of tall fescue with the novel endophytes is not always equal to that of tall fescue with the endemic endophyte (*128*).

## C. ECOLOGICAL ROLES

Available literature is fairly devoid of experimental studies that would address directly the possible ecological roles of EA in natural circumstances without the confounding issues of other toxins and the effects of the pathogens or symbionts that produce them. The phenomenon of livestock toxicosis due to alkaloids produced by epichloës suggests a key role in defense against grazing mammals. However, in most cases the animals must consume such a large amount of infected grass to suffer toxicity, that this protective effect would rarely be significant except in large, nearly monocultural stands of grass with EA-producing endophytes, as would typify some modern pastures. However, the widespread distribution of EA within the Clavicipitaceae clearly indicates that their origins predate modern pastoral agriculture by many millions of years. There are some situations that approximate natural ecosystems, in which roles are apparent for EA produced by epichloë endophytes. Drunken-horse grass (*Achnatherum inebrians*) in desert regions of northern China often harbors an epichloë endophyte that produces **2** and **3** in abundance (*47*). Likewise, these toxins, **16**, and other lysergic acid derivatives (*48*) are produced by the endophyte(s) in a population of sleepygrass (*Achnatherum robustum* = *Stipa robusta*) near Cloudcroft, New Mexico. In both cases, naive horses (and presumably other animals) may consume the grass, and suffer several days of stupor as a result. Once exposed, the animals will avoid these grasses. While illustrative of the protective effects of EA, these two grasses are characterized by extremely high levels of these alkaloids compared to those that typify other grass species with EA-producing endophytes.

Conceivably, EA produced by *Claviceps* species also discourage large grazing mammals. However, to our knowledge there is no evidence that heavily ergotized ears are detected and avoided by large grazers. A deterrent effect against small mammals and birds seems more plausible. The ergots might otherwise be a significant source of nutrition for small vertebrates unless they find them distasteful. Indeed, the bad taste of ergots has led to the tradition of spicing rye breads with caraway seeds. Activity of ergopeptines and other EA against invertebrates, particularly insects, may be as or more important than activities against vertebrates. EA in concentrations typical of epichloë-symbiotic tall fescue or perennial ryegrass significantly slow growth and increase mortality of fall army worm (*Spodoptera frugiperda*) (*134*). Resistance to insects may be an important selection on the *Claviceps* species to devote considerable resources to produce high levels of EA in ergots. However, not all *Claviceps* species and strains produce EA, leaving an interesting puzzle: Which selection conditions favor, and which do not favor, EA production?

In considering the roles of EA, it is important to appreciate the diversity of the clavine and lysergyl alkaloids. Such diversity may have been selected by multiple challenges on the producing organisms (*20*). In fact, current literature suggests that clavines tend to be much more active against bacteria than are ergopeptines (*20,135,136*), whereas ergopeptines are much more effective than clavines against insects and nematodes (*20,134,137*). Fumigaclavine levels in *A. fumigatus* conidia are very high, totaling approximately 1% of the dry mass (*52*). Perhaps the fumigaclavines help to protect these fungal propagules from degradation by microbes, and deter their consumption by invertebrates.

## VII. Summary and Conclusions

EA have been a major benefit, and a major detriment, to humans since early in recorded history. Their medicinal properties have been used, and continue to be used, to aid in childbirth, with new uses being found in the treatment of neurological and cardiovascular disorders. The surprisingly broad range of pharmaceutical uses for EA stems from their affinities for multiple receptors for three distinct neurotransmitters (serotonin, dopamine, and adrenaline), from the great structural diversity of natural EA, and from the application of chemical techniques that further expand that structural diversity. The dangers posed by EA to humans and their livestock stem from the ubiquity of ergot fungi (*Claviceps* species) as parasites of cereals, and of related grass endophytes (*Epichloë, Neotyphodium*, and *Balansia* species) that may inhabit pasture grasses and produce toxic levels of EA. Further concerns stem from saprophytic EA producers in the genera *Aspergillus* and *Penicillium*, especially *A. fumigatus*, an opportunistic pathogen of humans.

Numerous fungal species produce EA with a wide variety of structures and properties. These alkaloids are associated with plants in the families Poaceae, Cyperaceae, and Convolvulaceae, apparently because these plants can have symbiotic fungi that produce EA. Pharmacological activities of EA relate to their specific structures. Known as potent vasoconstrictors, the ergopeptines include a lysergic acid substituent with an amide linkage to a complex cyclol-lactam ring structure generated from three amino acids. Simpler lysergyl amides and clavines are more apt to have oxytonic or psychotropic activities. One of the lysergyl amides is LSD (**5**), the most potent hallucinogen known.

The EA biosynthetic pathway in *Claviceps* species has been studied extensively for many decades, and recent studies have also employed epichloës and *A. fumigatus*. The early pathway, shared among these fungi, begins with the action of an aromatic prenyl transferase, DMATrp synthase, which links a dimethylallyl chain to L-tryptophan. When the *dmaW* gene encoding DMATrp synthase was cloned and sequenced, the predicted product bore no identifiable resemblance to other known prenyl transferases. The *dmaW* genes of *Claviceps* species are present in clusters of genes, several of which also have demonstrated roles in EA biosynthesis. In many other fungi, *dmaW* homologues are identifiable in otherwise very different gene clusters. The roles of DMATrp synthase homologues in these other fungi are probably quite variable. One of them is thought to prenylate the phenolic oxygen of L-tyrosine, and another catalyzes the unusual reverse prenylation reaction in the biosynthesis of fumigaclavine C(**10**), an EA characteristic of *A. fumigatus*.

The second step of the EA pathway is *N*-methylation of DMATrp (**12**) to form **13**, which is then subjected to a series of oxidation/oxygenation and reduction reactions to generate, in order, chanoclavine-I (**16**), agroclavine (**19**), and elymoclavine (**6**). Shunt reactions generate a wide variety of other clavines. Two epimerizations occur in this pathway: one from **12** to **16**, the other from **16** to **19**. Further oxidation of **6**, catalyzed by the cytochrome-P450 CloA, generates lysergic acid (**1**).

An unusual NRPS complex, lysergyl peptide synthetase (LPS), is responsible for linking **1** to three hydrophobic L-amino acids to generate the ergopeptide lactams. The LPS complex includes two polypeptides, one (LPS 2) possessing a single module for activation of **1**, and the other (LPS 1) possessing three modules, each specifying one of the L-amino acids. Variations in LPS 1 sequences are associated with variations in the incorporated amino acids, leading to differences between strain chemotypes, and even multiple ergopeptines within strains. For example,

*C. purpurea* P1 produces two distinct ergopeptines (ergotamine (**4**) and ergocryptine (Table I)), each of which is believed to be generated by multiple LPS 1 subunits encoded by separate, but related, genes (*lpsA1* and *lpsA2*).

The main ecological roles of EA in nature are probably to protect the fungi from consumption by vertebrate and invertebrate animals. The EA produced by plant-symbiotic fungi (such as epichloë endophytes) may protect the fungus by protecting the health and productivity of the host, which may otherwise suffer excessive grazing by animals. The EA, at levels typical of plants bearing these symbionts, can negatively affect the health of large mammals as well herbivorous insects. Some clavines have substantial anti-bacterial properties, which might protect the fungus and, in some cases, their host plants from infection. However, the fact that a large number of epichloë, and even several *Claviceps* species, produce no detectable EA indicates that the selection for their production is not universal. An unfortunate fact for many livestock producers is that some of the most popular forage grasses tend to possess EA-producing epichloë endophytes. Such endophytes are easily eliminated, but confer such fitness enhancements to their hosts that their presence is often preferred, despite the toxic EA.

The future looks promising for continued interest in EA. Research continues into their pharmacological properties, medicinal uses, and structure–function relationships. New clavines and lysergic acid derivatives are identified regularly from new sources, such as marine animals. Also, programs are well underway to modify or replace epichloë endophytes of forage grasses in order to produce new grass cultivars that lack these toxins.

## Acknowledgments

Research of the authors cited here was supported by United States Department of Agriculture (USDA) National Research Initiative (grants 2001-35319-10930 and 2005-35318-16184 to D.P. and C.S.), by a USDA Specific Cooperative Agreement (grant 200403171013 to C.S.), and by the Deutsche Forschungsgemeinschaft (Tu 50/13; SPP 1152: Evolution of metabolic diversity to P.T.). We thank Thomas Haarmann for assistance with artwork.

## References

1. A. F. Lehner, M. Craig, N. Fannin, L. Bush, and T. Tobin, *J. Mass Spectrom.* **39**, 1275 (2004).
2. M. K. Matossian, "Poisons of the Past: Molds, Epidemics, and History", Yale University Press, New Haven, CT, 1989.
3. D. E. Nichols, *in* "The Heffter Review of Psychedelic Research", vol. 2, p. 80. Heffter Research Institute, Santa Fe, NM, 2001.
4. L. C. Vining, *in* "Genetics of Industrial Microorganisms II: Actinomycetes and Fungi", (Z. Vanek, Z. Hostálek, and J. Cudlín, eds.), vol. II, p. 405. Elsevier, Amsterdam, 1973.
5. L. R. Caporael, *Science* **192**, 21 (1976).
6. M. J. Eadie, *Lancet Neurol.* **2**, 429 (2003).
7. F. N. Thompson and J. A. Stuedemann, *Agric. Ecosyst. Environ.* **44**, 263 (1993).
8. J. R. Strickland, J. W. Oliver, and D. L. Cross, *Vet. Hum. Toxicol.* **35**, 454 (1993).

9. J. K. Porter, *J. Anim. Sci.* **73**, 871 (1995).
10. D. Gröger and H. G. Floss, *in* "Alkaloids", (G. A. Cordell, ed.), vol. 50, p. 171. Academic Press, New York, 1998.
11. A. N. de Groot, P. W. van Dongen, T. B. Vree, Y. A. Hekster, and J. van Roosmalen, *Drugs* **56**, 523 (1998).
12. A. Hoffmann and H. Tscherter, *Experientia* **16**, 414 (1960).
13. R. G. Powell, R. D. Plattner, S. G. Yates, K. Clay, and A. Leuchtmann, *J. Nat. Prod.* **53**, 1272 (1990).
14. T. C. Plowman, A. Leuchtmann, C. Blaney, and K. Clay, *Econ. Bot.* **44**, 452 (1990).
15. L. Cvák, J. Minár, S. Pakhomova, J. Ondrácek, B. Kratochvíl, P. Sedmera, V. Havlícek, and A. Jegorov, *Phytochemistry* **42**, 231 (1996).
16. E. Németh, *in* "Ergot: The Genus *Claviceps*", (V. Kren and L. Cvák, eds.), p. 303. Harwood Academic Publishers, Amsterdam, The Netherlands, 1999.
17. L. Cvák, *in* "Ergot: The Genus *Claviceps*", (V. Kren and L. Cvák, eds.), p. 373. Harwood Academic Publishers, Amsterdam, The Netherlands, 1999.
18. S. Mainka, S. Danicke, H. Bohme, J. Wolff, S. Matthes, and G. Flachowsky, *Arch. Anim. Nutr.* **59**, 81 (2005).
19. P. Tudzynski, T. Correia, and U. Keller, *Appl. Microbiol. Biotechnol.* **57**, 593 (2001).
20. D. G. Panaccione, *FEMS Microbiol. Lett.* **251**, 9 (2005).
21. A. Hofmann, *Pharmacology* **16**, 1 (1978).
22. A. Woolf, *J. Toxicol. Clin. Toxicol.* **38**, 457 (2000).
23. K. Urga, A. Debella, Y. W'Medihn, N. Agata, A. Bayu, and W. Zewdie, *Ethiop. J. Health Dev.* **16**, 317 (2002).
24. K. A. V. R. Krishnamachari and R. V. Bhat, *Ind. J. Med. Res.* **64**, 1624 (1976).
25. P. G. Tulpule and R. V. Bhat, *Ind. J. Med. Res.* **68**, 99 (1978).
26. M. Peraica, B. Radic, A. Lucic, and M. Pavlovic, *Bull. World Health Org.* **77**, 754 (1999).
27. P. Tudzynski and J. A. N. Scheffer, *Mol. Plant Pathol.* **5**, 377 (2004).
28. S. Pazoutova, J. Olsovska, M. Linka, R. Kolinska, and M. Flieger, *Appl. Env. Microbiol.* **66**, 5419 (2000).
29. C. Spalla, *in* "Genetics of Industrial Microorganisms II: Actinomycetes and Fungi", (Z. Vanek, Z. Hostálek, and J. Cudlín, eds.), vol. II, p. 393. Elsevier, Amsterdam, 1973.
30. H. G. Floss, *Tetrahedron* **32**, 873 (1976).
31. K. D. Barrow, P. G. Mantle, and F. R. Quigley, *Tetrahedron Lett.* **16**, 1557 (1974).
32. A. E. Glenn and C. W. Bacon, *in* "*Neotyphodium*/Grass Interactions", (N. S. Hill and C. W. Bacon, eds.), p. 53. Plenum Press, New York, 1997.
33. P. C. Lyons, R. D. Plattner, and C. W. Bacon, *Science* **232**, 487 (1986).
34. C. L. Schardl, A. Leuchtmann, and M. J. Spiering, *Annu. Rev. Plant Biol.* **55**, 315 (2004).
35. J. F. White, Jr., G. Morgan-Jones, and A. C. Morrow, *Agric. Ecosyst. Environ.* **44**, 13 (1993).
36. C. L. Schardl and K. Clay, *in* "The Mycota V: Plant Relationships", (G. C. Carroll and P. Tudzynski, eds.), vol. V, part B, p. 221. Springer, Berlin, 1997.
37. C. W. Bacon, *Appl. Environ. Microbiol.* **54**, 2615 (1988).
38. K. Clay, G. P. Cheplick, and S. Marks, *Oecologia* **80**, 374 (1989).
39. J. Wang, C. Machado, D. G. Panaccione, H.-F. Tsai, and C. L. Schardl, *Fung. Genet. Biol.* **41**, 189 (2004).
40. J. Wang, "*dmaW* Encoding Tryptophan Dimethylallyltransferase in Ergot Alkaloid Biosynthesis from Clavicipitaceous Fungi." Ph.D. thesis, University of Kentucky, Lexington, 2000.
41. J. P. de Battista, C. W. Bacon, R. Severson, R. D. Plattner, and J. H. Bouton, *Agron. J.* **82**, 878 (1990).
42. K. Mende, V. Homann, and B. Tudzynski, *Mol. Gen. Genet.* **255**, 96 (1997).

43. R. A. Shelby, J. Olsovska, V. Havlicek, and M. Flieger, *J. Agric. Food Chem.* **45**, 4674 (1997).
44. J. K. Porter, C. W. Bacon, and J. D. Robbins, *J. Agric. Food Chem.* **27**, 595 (1979).
45. J. K. Porter, C. W. Bacon, J. D. Robbins, and D. Betowski, *J. Agric. Food Chem.* **29**, 653 (1981).
46. D. G. Panaccione, B. A. Tapper, G. A. Lane, E. Davies, and K. Fraser, *J. Agric. Food Chem.* **51**, 6429 (2003).
47. C. O. Miles, G. A. Lane, M. E. Di Menna, I. Garthwaite, E. L. Piper, O. J. P. Ball, G. C. M. Latch, J. M. Allen, M. B. Hunt, L. P. Bush, F. K. Min, I. Fletcher, and P. S. Harris, *J. Agric. Food Chem.* **44**, 1285 (1996).
48. R. Petroski, R. G. Powell, and K. Clay, *Nat. Tox.* **1**, 84 (1992).
49. K. Jenett-Siems, M. Kaloga, and E. Eich, *J. Nat. Prod.* **57**, 1304 (1994).
50. S. Kucht, J. Gross, Y. Hussein, T. Grothe, U. Keller, S. Basar, W. A. Konig, U. Steiner, and E. Leistner, *Planta* **219**, 619 (2004).
51. J. P. Latgé, *Clin. Microbiol. Rev.* **12**, 310 (1999).
52. D. G. Panaccione and C. M. Coyle, *Appl. Environ. Microbiol.* **71**, 3106 (2005).
53. T. N. Makarieva, S. G. Ilyin, V. A. Stonik, K. A. Lyssenko, and V. A. Denisenko, *Tetrahedron Lett.* **40**, 1591 (1999).
54. H. -F. Tsai, H. Wang, J. C. Gebler, C. D. Poulter, and C. L. Schardl, *Biochem. Biophys. Res. Commun.* **216**, 119 (1995).
55. P. Tudzynski, K. Hölter, T. Correia, C. Arntz, N. Grammel, and U. Keller, *Mol. Gen. Genet.* **261**, 133 (1999).
56. J. C. Gebler, A. B. Woodside, and C. D. Poulter, *J. Amer. Chem. Soc.* **114**, 7354 (1992).
57. C. L. Schardl, A. Leuchtmann, H. -F. Tsai, M. A. Collett, D. M. Watt, and D. B. Scott, *Genetics* **136**, 1307 (1994).
58. B. Riederer, M. Han, and U. Keller, *J. Biol. Chem.* **271**, 27524 (1996).
59. T. Haarmann, C. Machado, Y. Lübbe, T. Correia, C. L. Schardl, D. G. Panaccione, and P. Tudzynski, *Phytochemistry* **66**, 1312 (2005).
60. D. G. Panaccione, R. D. Johnson, J. H. Wang, C. A. Young, P. Damrongkool, B. Scott, and C. L. Schardl, *Proc. Natl. Acad. Sci. USA* **98**, 12820 (2001).
61. T. Correia, N. Grammel, I. Ortel, U. Keller, and P. Tudzynski, *Chem. Biol.* **10**, 1281 (2003).
62. T. Haarmann, I. Ortel, P. Tudzynski, U. Keller, *Chembiochem* 7, 645(2006).
63. C. Machado, "Studies of Ergot Alkaloid Biosynthesis Genes in Clavicipitaceous Fungi." Ph.D. thesis, University of Kentucky, Lexington, 2004.
64. D. G. Panaccione, C. L. Schardl, and C. M. Coyle, *Rec. Adv. Phytochem.* **40** in press (2006), in press.
65. C. M. Coyle and D. G. Panaccione, *Appl. Environ. Microbiol.* **71**, 3112 (2005).
66. K. V. Mothes, F. Weygand, D. Gröger, and H. Grisebach, *Z. Naturforsch. Teil* **13b**, 41 (1958).
67. D. Gröger, H. J. Wendt, K. Mothes, and F. Weygand, *Z. Naturforsch. Teil* **14b**, 355 (1959).
68. D. Gröger, K. Mothes, H. Simon, H. G. Floss, and F. Weygand, *Z. Naturforsch. Teil* **15b**, 141 (1960).
69. P. F. Heinstein, S.-L. Lee, and H. G. Floss, *Biochem. Biophys. Res. Comm.* **44**, 1244 (1971).
70. J. C. Gebler and C. D. Poulter, *Arch. Biochem. Biophys.* **296**, 308 (1992).
71. H. Otsuka, F. R. Quigley, D. Gröger, J. A. Anderson, and H. G. Floss, *Planta Med.* **40**, 109 (1980).
72. C. Arntz and P. Tudzynski, *Curr. Genet.* **31**, 357 (1997).
73. I. A. Unsöld and S. M. Li, *Microbiology (Reading, Engl.)* **151**, 1499 (2005).
74. V. M. Krupinski, J. E. Robbers, and H. G. Floss, *J. Bacteriol.* **125**, 158 (1976).
75. C. Young, L. McMillan, E. Telfer, and B. Scott, *Mol. Microbiol.* **39**, 754 (2001).

76. D. M. Gardiner, A. J. Cozijnsen, L. M. Wilson, M. S. Pedras, and B. J. Howlett, *Mol. Microbiol.* **53**, 1307 (2004).
76a. I. A. Unsold, and S. M. Li, Chembiochem **7, 158** (2006).
77. T. Otsuka, J. A. Anderson, and H. G. Floss, *Chem. Comm.* **15**, 660 (1979).
78. B. Naidoo, J. M. Cassady, G. E. Blair, and H. G. Floss, *Chem. Comm.* **258**, 471–472 (1970).
79. H. G. Floss, M. Tcheng-Lin, C.-j. Chang, T. Bala Naidoo, G. E. Blair, C. I. Abou-Chaar, and J. M. Cassady, *J. Amer. Chem. Soc.* **96**, 1898 (1974).
80. R. E. Williams and N. C. Bruce, *Microbiology (Reading, Engl.)* **148**, 1607 (2002).
81. W. Maier, B. Schumann, and D. Gröger, *J. Basic Microbiol.* **28**, 83 (1988).
82. E. M. Stocking, J. F. Sanz-Cervera, and R. M. Williams, *Angew. Chem. Int. Ed.* **38**, 786 (1999).
83. R. M. Williams, E. M. Stocking, and J. F. Sanz-Cervera, *Top. Curr. Chem.* **209**, 97 (2000).
84. M. Zamocky and F. Koller, *Prog. Biophys. Mol. Biol.* **72**, 19 (1999).
85. Y. Ding, R. S. Bojja, and L. Du, *Appl. Environ. Microbiol.* **70**, 1931 (2004).
86. B. J. Blaney, R. Maryam, S. -A. Murray, and M. J. Ryley, *Austral. J. Agric. Res.* **54**, 167 (2003).
87. F. R. Quigley and H. G. Floss, *J. Org. Chem.* **46**, 464 (1981).
88. P. Belshaw, C. T. Walsh, and T. Stachelhaus, *Science* **284**, 486 (1999).
89. N. Castagnoli, Jr., K. Corbett, E. B. Chain, and R. Thomas, *Biochem. J.* **117**, 451 (1970).
90. M. R. Siegel, G. C. M. Latch, L. P. Bush, F. F. Fannin, D. D. Rowan, B. A. Tapper, C. W. Bacon, and M. C. Johnson, *J. Chem. Ecol.* **16**, 3301 (1990).
91. M. J. Christensen, A. Leuchtmann, D. D. Rowan, and B. A. Tapper, *Mycol. Res.* **97**, 1083 (1993).
92. T. A. Keating, D. E. Ehmann, R. M. Kohli, C. G. Marshall, J. W. Trauger, and C. T. Walsh, *Chembiochem* **2**, 99 (2001).
93. P. Damrongkool, A. B. Sedlock, C. A. Young, R. D. Johnson, K. E. Goetz, B. Scott, C. L. Schardl, and D. G. Panaccione, *DNA Seq.* **16**, 379 (2005).
94. J. K. Porter, *in* "Biotechnology of Endophytic Fungi of Grasses", (C. W. Bacon and J. F. White, Jr., eds.), p. 103. CRC Press, Boca Raton, FL, 1994.
95. J. E. Robbers and H. G. Floss, *Tetrahedron Lett.* **23**, 1857 (1969).
96. Y. Yokoyama, H. Hikawa, M. Mitsuhashi, A. Uyama, Y. Hiroki, and Y. Murakami, *Eur. J. Org. Chem.* **2004**, 1244 (2004).
97. M. Flieger, M. Wurst, and R. Shelby, *Folia Microbiol* **42**, 3 (1997).
98. G. E. Rottinghaus, G. B. Garner, C. N. Cornel, and J. L. Ellis, *J. Agric. Food Chem.* **39**, 112 (1991).
99. M. J. Spiering, E. Davies, B. A. Tapper, J. Schmid, and G. A. Lane, *J. Agric. Food Chem.* **50**, 5856 (2002).
100. N. S. Hill and C. S. Agee, *Crop Sci.* **34**, 530 (1994).
101. R. L. Rice, G. G. Schurig, and D. J. Blodgett, *Amer. J. Vet. Res.* **59**, 1258 (1998).
102. B. Berde and E. Stürmer, *in* "Ergot Alkaloids and Related Compounds", (B. Berde and H. O. Schild, eds.), p. 1. Springer, Berlin, 1978.
103. J. Mukherjee and M. Menge, *Adv. Biochem. Eng. Biotechnol.* **68**, 1 (2000).
104. P. G. Crosignani, *Eur. J. Obstet. Gynecol. Reprod. Biol.* **125**, 152 (2006).
105. C. Schöning, M. Flieger, and H. H. Pertz, *J. Anim. Sci.* **79**, 2202 (2001).
106. C. M. Villalon, P. De Vries, G. Rabelo, D. Centurion, A. Sanchez-Lopez, and P. Saxena, *Br. J. Pharmacol.* **126**, 585 (1999).
107. E. W. Willems, M. Trion, P. De Vries, J. P. C. Heiligers, C. M. Villalon, and P. R. Saxena, *Br. J. Pharmacol.* **127**, 1263 (1999).
108. E. Sanders-Bush, K. D. Burris, and K. Knoth, *J. Pharmacol. Exp. Ther.* **246**, 924 (1988).

109. B. J. Blaney, J. S. Kopinski, M. H. Magee, R. A. McKenzie, G. W. Blight, R. Maryam, and J. A. Downing, *Austral. J. Agric. Res.* **51**, 785 (2000).
110. J. Tor-Agbidye, L. L. Blythe, and A. M. Craig, *Vet. Hum. Toxicol.* **43**, 140 (2001).
111. R. T. Gallagher, A. D. Hawkes, P. S. Steyn, and R. Vleggaar, *J. Chem. Soc., Chem. Comm.* **1984**, 614 (1984).
112. Anonymous, "Manawatu Evening Standard", Palmerston North, New Zealand, p. 1, October 16, 1992.
113. A. Leuchtmann, D. Schmidt, and L. P. Bush, *J. Chem. Ecol.* **26**, 1025 (2000).
114. C. L. Schardl, *Fung. Genet. Biol.* **33**, 69 (2001).
115. D. R. Bazely, M. Vicari, S. Emmerich, L. Filip, D. Lin, and A. Inman, *J. Appl. Ecol.* **34**, 847 (1997).
116. M. S. Gadberry, T. M. Denard, D. E. Spiers, and E. L. Piper, *J. Anim. Sci.* **81**, 1538 (2003).
117. C. W. Bacon and M. R. Siegel, *J. Prod. Agric.* **1**, 45 (1988).
118. D. E. Spiers, Q. Zhang, P. A. Eichen, G. E. Rottinghaus, G. B. Garner, and M. R. Ellersieck, *J. Anim. Sci.* **73**, 1954 (1995).
119. J. M. Burke, D. E. Spiers, F. N. Kojima, G. A. Perry, B. E. Salfen, S. L. Wood, D. J. Patterson, M. F. Smith, M. C. Lucy, W. G. Jackson, and E. L. Piper, *Biol. Reprod.* **65**, 260 (2001).
120. D. L. Cross, L. M. Redmond, and J. R. Strickland, *J. Anim. Sci.* **73**, 899 (1995).
121. D. L. Cross, *in* "Neotyphodium/Grass Interactions", (N. S. Hill and C. W. Bacon, eds.), p. 289. Plenum Press, New York, 1997.
122. J. R. Strickland, D. L. Cross, G. P. Birrenkott, and L. W. Grimes, *Am. J. Vet. Res.* **55**, 716 (1994).
123. N. S. Hill, F. N. Thompson, J. A. Stuedemann, G. W. Rottinghaus, H. J. Ju, D. L. Dawe, and E. E. Hiatt, *J. Anim. Sci.* **79**, 542 (2001).
124. M. J. Spiering, G. A. Lane, M. J. Christensen, and J. Schmid, *Phytochemistry* **66**, 195 (2005).
125. J. T. Roylance, N. S. Hill, and C. S. Agee, *J. Chem. Ecol.* **20**, 2171 (1994).
126. R. E. Welty, A. M. Craig, and M. D. Azevedo, *Plant Dis.* **78**, 845 (1994).
127. D. P. Malinowski and D. P. Belesky, *Crop Sci.* **40**, 923 (2000).
128. J. H. Bouton, G. C. M. Latch, N. S. Hill, C. S. Hoveland, M. A. McCann, R. H. Watson, J. A. Parish, L. L. Hawkins, and F. N. Thompson, *Agron. J.* **94**, 567 (2002).
129. M. E. Nihsen, E. L. Piper, C. P. West, R. J. Crawford, T. M. Denard, Z. B. Johnson, C. A. Roberts, D. A. Spiers, and C. F. Rosenkrans, *J. Anim. Sci.* **82**, 878 (2004).
130. S. J. Bluett, E. R. Thom, D. A. Clark, and C. D. Waugh, *Aust. J. Exp. Agric.* **45**, 11 (2005).
131. S. J. Bluett, E. S. Kolver, M. J. Auldist, E. R. Thom, S. R. Davis, V. C. Farr, and B. A. Tapper, *N. Zeal. J. Agric. Res.* **46**, 9 (2003).
132. J. A. Parish, M. A. McCann, R. H. Watson, N. N. Paiva, C. S. Hoveland, A. H. Parks, B. L. Upchurch, N. S. Hill, and J. H. Bouton, *J. Anim. Sci.* **81**, 2856 (2003).
133. J. A. Parish, M. A. McCann, R. H. Watson, C. S. Hoveland, L. L. Hawkins, N. S. Hill, and J. H. Bouton, *J. Anim. Sci.* **81**, 1316 (2003).
134. K. Clay and G. P. Cheplick, *J. Chem. Ecol.* **15**, 169 (1989).
135. E. Eich and H. Pertz, *in* "Ergot: The Genus *Claviceps*", (V. Kren and L. Cvák, eds.), p. 441. Harwood Academic Publishers, Amsterdam, The Netherlands, 1999.
136. G. Schwarz and E. Eich, *Planta Med.* **47**, 212 (1983).
137. O. J. P. Ball, C. O. Miles, and R. A. Prestidge, *J. Econ. Entomol.* **90**, 1382 (1997).

—Chapter 3—

# CHEMICAL AND BIOLOGICAL ASPECTS OF *NARCISSUS* ALKALOIDS

Jaume Bastida[1], Rodolfo Lavilla[2] and Francesc Viladomat[1]

[1]*Departament de Productes Naturals, Facultat de Farmàcia, Universitat de Barcelona, 08028 Barcelona, Spain* [2]*Parc Científic de Barcelona, Universitat de Barcelona, 08028 Barcelona, Spain*



## I. Introduction

### A. GEOGRAPHICAL DISTRIBUTION AND TAXONOMICAL ASPECTS

The genus *Narcissus* L. belongs to the Narcisseae, one of the 15 tribes of the Amaryllidaceae, a widely distributed monocotyledonous family of 59 genera and about 850 species (*1*). The Amaryllidaceae are richly represented in the tropics and have pronounced centers of diversity in South Africa and the Andean region. Some genera are also found in the Mediterranean area and the temperate regions of Asia. The family's phylogenetic relationships closely follow geographic distribution, with much regional endemism, which adds credence to a Gondwana origin for the family at a time when the continents were much closer together (*1*). The hypothesis that the family evolved in Africa and subsequently spread to other continents, further suggesting that South America is the center of secondary diversification is supported by the *matK* sequence data (*2*).

*Narcissus* comprises approximately 80–100 wild species of perennial geophytes, geographically concentrated in southwestern Europe, with a center of diversity in the Iberian Peninsula and North Africa, which substantiates the hypotheses that landmasses of Europe and Africa were once joined. A few species extend into France and Italy, and even fewer are found in the Balkans and the eastern Mediterranean. Records outside this area, such as *N. tazetta* variants in China and Japan, are almost certainly ancient introductions (*3,4*). The native habitats of *Narcissus* species are very varied, ranging from lowland to mountain sites, including grassland, scrub, woods, river banks, and rocky crevices (*3*). Most of the species

THE ALKALOIDS, Vol. 63
ISSN: 1099-4831
DOI: 10.1016/S1099-4831(06)63003-4

flower in late winter and spring, although a few flower in the autumn (*5*). All species are insect pollinated, with the majority possessing showy flowers, some of which are highly scented (*6*). The major pollinators are bees, butterflies, flies, and hawkmoths (*5,7,8*).

The taxonomy of the genus *Narcissus* is complex and unsettled because of its very varied wild populations, the ease with which hybridization occurs naturally, accompanied by extensive cultivation, breeding, selection, escape, and naturalization, and also because many descriptions of taxons have been based on garden specimens, several of which were probably of hybrid origin. However, although the status bestowed upon the individual groups may vary somewhat from author to author, the basic content of each group is similar in the various classifications (*3–5*). The most recent classification (Scheme 4), which incorporates elements of those of Fernandes and Webb (*9–11*), also notes that substantial work of revision is needed for some taxa in the genus. The inferred plastid-based phylogeny, on the other hand, reveals that floral morphology has been remarkably dynamic during the history of *Narcissus*, indicating extensive and repeated diversification and convergent evolution. Polymorphic sexual systems in *Narcissus* have evolved independently from the ancestral monomorphic state on at least six occasions (*5,12*). As these morphologies are associated with particular suites of pollinators, the recurring transitions from one floral design to another have probably accompanied pollinator transitions. Additionally, the remarkable resemblance of flower features between unrelated species indicates that pollinator activity might have driven flower convergence (*13*).

Most of the species of *Narcissus* will hybridize but, significantly, there is great variation in the fertility of the offspring, depending upon the degree of relationship between the parents (*4*). Hybridization has become very popular, resulting in thousands of commercial *Narcissus* cultivars that are in most cases larger and more robust than their wild parents (*3*). Moreover, large garden varieties of daffodils have recently been crossed with many of the small wild species to produce delightfully graceful blossoms. Unfortunately, the ability to produce new kinds of miniature daffodils is hampered by the disappearance of many of the tiny wild species (*14*). Thus, the survival of a number of *Narcissus* species is under threat because of over-collection and habitat destruction. There is a need to maintain vigilance in the conservation of wild species and of their many variants (*15*).

*Narcissus* bulbs have been an important floricultural crop in Western Europe since the late nineteenth century, although the bulbs have been grown in the Netherlands since the sixteenth century. At the start of the twenty-first century, the daffodil remains one of the major ornamental spring-flowering bulb crops grown in temperate regions, with large areas of field-grown crops providing both bulbs and flowers (*15,16*). Rees (*17*) estimated that the area of *Narcissus* grown in gardens, parks, cemeteries, etc., is five times the area grown commercially.

The development of the horticultural classification of *Narcissus* cultivars in 13 divisions was described by Kington (*18*) and, in order to avoid confusion due to the re-use of earlier names, since 1998, the International Daffodil Register and Classified List is updated annually by supplements of newly registered names (*4*).

## B. THE AMARYLLIDACEAE ALKALOIDS

A particular characteristic of the Amaryllidaceae is a consistent presence of an exclusive group of alkaloids, which have been isolated from the plants of all the genera of this family. The Amaryllidaceae alkaloids represent a large, and still expanding, group of isoquinoline alkaloids, the majority of which are not known to occur in any other family of plants. Since the isolation of the first alkaloid, lycorine, from *Narcissus pseudonarcissus* in 1877, substantial progress has been made in examining the Amaryllidaceae plants, although they still remain a relatively untapped phytochemical source (*1*). At present, over 300 alkaloids have been isolated from plants of this family (*19*) and, although their structures vary considerably, these alkaloids are considered to be biogenetically related.

The large number of structurally diverse Amaryllidaceae alkaloids are classified mainly into nine skeleton types, for which the representative alkaloids are norbelladine, lycorine, homolycorine, crinine, hemanthamine, narciclasine, tazettine, montanine, and galanthamine (Fig. 1). With the aim of unifying the numbering system of the different skeleton types, Ghosal's model will be used in this review (*20*).

As the alkaloids of the Amaryllidaceae family species fall mainly into one of these subgroups, they can serve as a classifying tool for including genera and species in this family. Thus, the genus *Behria*, in spite of having been classified as

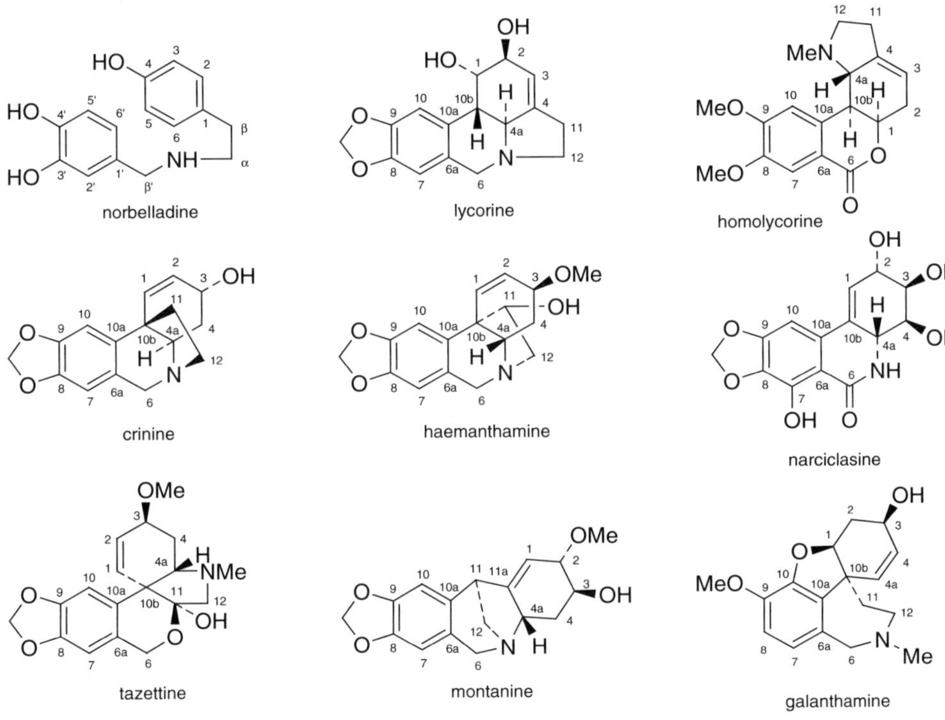


Figure 1. Amaryllidaceae alkaloid types.

Amaryllidaceae (*21*), does not have any Amaryllidaceae alkaloids (*22*) and should therefore be included in the Alliaceae family. Furthermore, although it is unusual to find other types of alkaloids in this family, if present, they are always accompanied by typical Amaryllidaceae alkaloids. The classical example is the reported presence of the mesembrane (*Sceletium*) alkaloids, generally found in the Aizoaceae family (*23,24*), in a few species of Amaryllidaceae such as *Hymenocallis arenicola, Crinum oliganthum, N. pallidulus*, and *N. triandrus* (*25–27*). In turn, the unexpected isolation of (−)-capnoidine and (+)-bulbocapnine from *Galanthus nivalis* subsp. *cilicicus* is the first report of the occurrence of classical isoquinoline alkaloids in a typical member of the Amaryllidaceae (*28*). On the contrary, the isolation of the Amaryllidaceae alkaloid crinamine from the wild yam *Dioscorea dregeana* (Dioscoreaceae) has also proven that Amaryllidaceae alkaloids may well be encountered in other plant families (*29*). Both of these results suggest that the definition of the Amaryllidaceae alkaloids should be reconsidered, and also the alkaloidal profile of this plant family.

The presence of alkaloids in plants is believed to be a protective adaptation, which in Amaryllidaceae is connected with the seasonal cycle of development, many species grow in early spring when other genera are only just starting to grow (*30*). Thus, Amaryllidaceae plants present an ontogenic variation of alkaloids, and the effects of stress, such as incisional injuries or insect attacks, on alkaloid metabolism causes almost complete hydrolysis of the alkaloidal conjugates and also produces oxidized metabolites (*31*). Regarding co-evolutionary adaptation, a notable example is the elaboration of a new conjugate alkaloid (telastasine) by the insect *Polytela gloriosa* Fab. (Noctuidae), a smokey-gray moth that is adapted to toxic Amaryllidaceae plants (*32,33*). Plants and insects use the same detoxification mechanism. These alkaloids, essential for plant survival, have a wide range of interesting physiological effects, including antitumor, antiviral, acetylcholinesterase inhibitory, immunostimulatory, and antimalarial activities, some of them being of particular interest because of their potential use in clinical therapy (*34*).

Plants of the Amaryllidaceae family have been used for thousands of years as herbal remedies. The alkaloids from their extracts have been the object of active chemical investigation for nearly 200 years. Over the past two decades many have been isolated, screened for different biological activities, and synthesized by a number of research groups. An important handicap is the availability of these alkaloids, which are obtained only in minute quantities from natural sources. Since isolation in larger quantities is not practical, there is a strong case for the development of syntheses or semisyntheses of these alkaloids and their derivatives as potential prodrugs (*35*).

The history of the Amaryllidaceae alkaloids, their structural elucidation, and their biological profiles, as well as their synthesis, have been summarized on several occasions (*20,33,36–48*) which, together with the regular publications in the journal *Natural Product Reports* (*34,49–66*), represent a valuable source of information.

The present review provides coverage of the isolation, spectroscopy, biological activity, and chemical synthesis of the Amaryllidaceae alkaloids present in the genus *Narcissus* up to July 2005.

## II. *Narcissus* Alkaloids and their Occurrence

Numerous alkaloids have been isolated from *Narcissus* species as a result of the continuing search for novel alkaloids with pharmacological activity in the Amaryllidaceae family. The alkaloids isolated from this genus, classified in relation to the different skeleton types, are shown in Tables I–VII. Table VIII lists the different *Narcissus* wild species and intersectional hybrids, grouped into subgenera and sections, with their corresponding alkaloids, arranged according to their ring system. The occurrence of alkaloids in *Narcissus* cultivars is shown in Table IX.

### A. ALKALOID TYPES AND DISTRIBUTION

The presence of almost 100 alkaloids in this genus (Tables I–VII) means that approximately a third of the total number of alkaloids isolated from Amaryllidaceae have been found in the genus *Narcissus*. Amaryllidaceae alkaloids are present in all the species and varieties of *Narcissus* studied. In general, a series of related alkaloids is found in each plant: often a few major metabolites and several minor components, which differ in the position of their substitutents. Up to now, about 40 wild species and around 100 cultivars have been studied in relation to the presence of alkaloids (Tables VIII and IX), which means that more than half of the *Narcissus* species or varieties have still to be explored in this aspect.

#### *1. Lycorine and Homolycorine Types*

The alkaloids of the lycorine and homolycorine groups are, on the whole, the most common alkaloids in this genus. Lycorine (**1**), galanthine (**7**), and pluviine (**11**) (lycorine type) and homolycorine (**26**) and lycorenine (**35**) (homolycorine type) are particularly frequent, lycorine being the most abundant. The presence of these alkaloids is very significant in the sections Narcissus (mainly lycorine type), Pseudonarcissi (mainly homolycorine type), and Tazettae of the wild species, and in the Divisions 1, 2, and 4 of cultivars.

Almost all of the lycorine-type alkaloids isolated from *Narcissus* present a *trans*-union between B/C rings, making them especially vulnerable to oxidative processes. It is interesting to observe that four of these alkaloids, namely vasconine (**22**), tortuosine (**23**), ungeremine (**24**), and roserine (**25**) possess an unusual quaternary nitrogen (*67–71*). The species *N. pallidulus* (section Ganymedes), unusual because it contains mesembrane-type alkaloids, is even more exceptional due to the presence of roserine.

The alkaloids of the homolycorine series, formed by a restructuring of lycorine-type alkaloids, are absent from some tribes of the Amaryllidaceae, such as the Amaryllideae or Hemantheae (*44*). For that reason, the presence of these alkaloids is a distinctive feature of the Narcisseae tribe. Moreover, all the *Narcissus* alkaloids of the homolycorine series display a B/C ring junction with a *cis* stereochemistry. An exceptional homolycorine-type alkaloid is dubiusine (**33**), which has an unusual hydroxybutyryl substituent (*72*).

TABLE I.
*Narcissus* Alkaloid Structures. Lycorine Type.

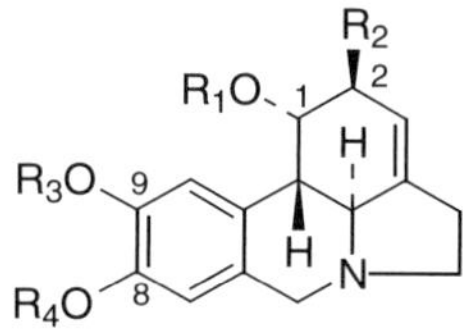


| | | |
|---|---|---|
| **1** | $R_1$=H, $R_2$=OH, $R_3$+$R_4$=$CH_2$ | lycorine |
| **2** | $R_1$=Ac, $R_2$=OH, $R_3$+$R_4$=$CH_2$ | poetaminine |
| **3** | $R_1$=H, $R_2$=OH, $R_3$=H, $R_4$=Me | pseudolycorine |
| **4** | $R_1$=Ac, $R_2$=OH, $R_3$=H, $R_4$=Me | 1-*O*-acetylpseudolycorine |
| **5** | $R_1$=H, $R_2$=OAc, $R_3$=H, $R_4$=Me | 2-*O*-acetylpseudolycorine |
| **6** | $R_1$=H, $R_2$=OH, $R_3$=Me, $R_4$=Me | 9-*O*-methylpseudolycorine |
| **7** | $R_1$=H, $R_2$=OMe, $R_3$=Me, $R_4$=Me | galanthine |
| **8** | $R_1$=H, $R_2$=OMe, $R_3$=Me, $R_4$=H | goleptine |
| **9** | $R_1$=Ac, $R_2$=O, $R_3$+$R_4$= $CH_2$ | jonquilline |
| **10** | $R_1$=H, $R_2$=H, $R_3$+$R_4$= $CH_2$ | caranine |
| **11** | $R_1$=H, $R_2$=H, $R_3$=Me, $R_4$=Me | pluviine |
| **12** | $R_1$=H, $R_2$=H, $R_3$=Me, $R_4$=H | norpluviine (= 8-*O*-demethylpluviine) |
| **13** | $R_1$=H, $R_2$=H, $R_3$=H, $R_4$=Me | 9-*O*-demethylpluviine |
| **14** | $R_1$=Ac, $R_2$=H, $R_3$=H, $R_4$=Me | 1-*O*-acetyl-9-*O*-demethylpluviine |
| **15** | $R_1$=Ac, $R_2$=H, $R_3$=Ac, $R_4$=Me | 1,9-*O*-diacetyl-9-*O*-demethylpluviine |

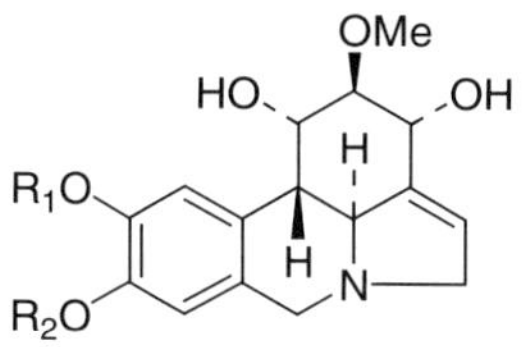


**16** $R_1$=Me, $R_2$=Me narcissidine
**17** $R_1$+$R_2$=$CH_2$ ungiminorine (= ungminorine)

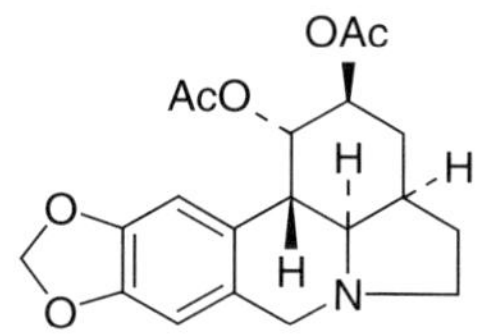


**18** nartazine

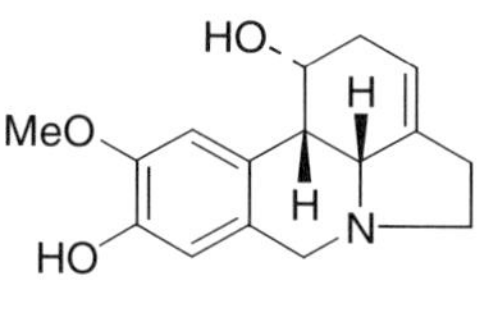


**19** kirkine

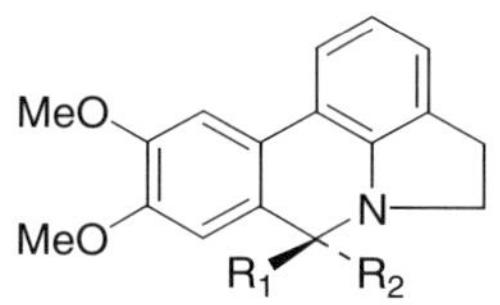


**20** $R_1$=H, $R_2$=H assoanine
**21** $R_1$+$R_2$=O oxoassoanine

(*continued*)

TABLE I.
Continued.

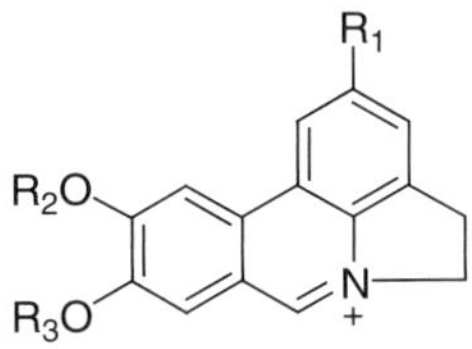

**22** $R_1$=H, $R_2$=Me, $R_3$=Me vasconine
**23** $R_1$=OMe, $R_2$=Me, $R_3$=Me tortuosine
**24** $R_1$=$O^-$, $R_2$+$R_3$=$CH_2$ ungeremine

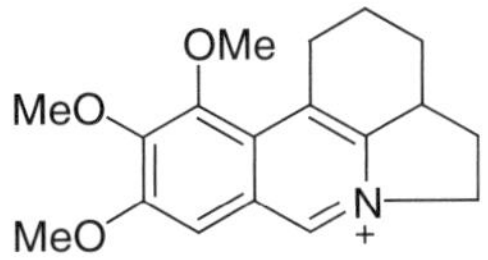

**25** roserine

### 2. *Hemanthamine, Tazettine, Narciclasine, and Montanine Types*

The main groups of alkaloids originating from a *para–para′* oxidative phenolic coupling of *O*-methylnorbelladine (**87**) (hemanthamine, tazettine, and narciclasine types) are frequently represented in this genus by the model alkaloid of each group, hemanthamine (**53**), tazettine (**62**), and narciclasine (**68**), respectively. Their presence is significant in the Pseudonarcissi, Tazettae, and Bulbocodii sections of wild species, and in the Divisions 1, 2, and 8 of cultivars. The hemanthamine-type alkaloids are the most abundant, being the biogenetic source of the other types.

The lack of crinine-type alkaloids (β-5,10b-ethano bridge configuration) in this genus, which only has representatives of the α-5,10b-ethano bridge series (hemanthamine type), is a very significant taxonomic feature. Additionally, in *Narcissus*, the hemanthamine-type alkaloids never show any substitution in the aromatic ring at position 7, which is usual in crinine-type alkaloids from tribes such as the Amaryllideae or Hemantheae (*44*). The pairs of alkaloids with a hydroxyl substituent at C-6, like papyramine/6-epipapiramine (**48/49**) or hemanthidine/6-epihemanthidine (**55/56**), always appear as a mixture of epimers in solution (*73*). It is also interesting to mention the unusual structure of bujeine (**61**), an alkaloid of the hemanthamine type that has a modified bridge with a heteroatom between C-11 and C-12 and an acetoxymethyl substituent at the 11- *endo* position (*74*). Ismine (**72**) is considered to be a catabolic product from the hemanthamine series.

Tazettine (**62**) is a widely reported alkaloid in the Amaryllidaceae family, particularly in *Narcissus* (see Tables VIII and IX), although it is known to be an extraction artifact of pretazettine (**64**) (*75*). All the alkaloids of the tazettine type that are isolated from *Narcissus* species show the typical methylenedioxy group between the C-8 and C-9 positions. Obesine (**67**) is an exceptional tazettine-type alkaloid with a seven-membered ring (*76*).

Narciclasine (**68**) is reasonably abundant in some *Narcissus* spp. and has served as a very useful intermediate for synthetic conversion into (+)-pancratistatin (see Section IV) and to conduct a series of structure–activity relationship studies (*77,78*). Bicolorine (**71**), another member of the narciclasine series, is an unusual, completely aromatized quaternary alkaloid with an *N*-methyl group (*79*).

TABLE II.
*Narcissus* Alkaloid Structures. Homolycorine Type.

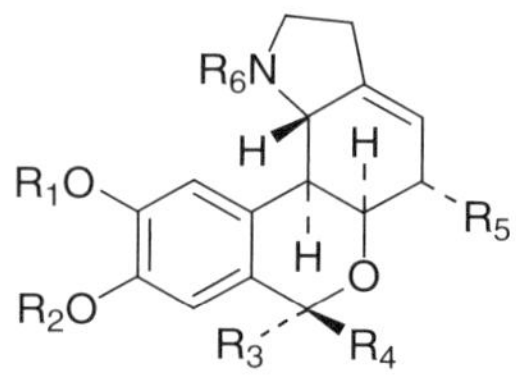

| | | |
|---|---|---|
| **26** | $R_1$=Me, $R_2$=Me, $R_3+R_4$=O, $R_5$=H, $R_6$=Me | homolycorine |
| **27** | $R_1$=Me, $R_2$=H, $R_3+R_4$=O, $R_5$=H, $R_6$=Me | 8-*O*-demethylhomolycorine |
| **28** | $R_1$=Me, $R_2$=Ac, $R_3+R_4$=O, $R_5$=H, $R_6$=Me | 8-*O*-demethyl-8-*O*-acetylhomolycorine |
| **29** | $R_1$=H, $R_2$=Me, $R_3+R_4$=O, $R_5$=H, $R_6$=Me | 9-*O*-demethylhomolycorine |
| **30** | $R_1+R_2$=$CH_2$, $R_3+R_4$=O, $R_5$=H, $R_6$=Me | masonine |
| **31** | $R_1+R_2$=$CH_2$, $R_3+R_4$=O, $R_5$=H, $R_6$=H | normasonine (= *N*-demethylmasonine) |
| **32** | $R_1$=H, $R_2$=Me, $R_3+R_4$=O, $R_5$=OH, $R_6$=Me | 9-*O*-demethyl-2α-hydroxyhomolycorine |
| **33** | $R_1$=3OHbut, $R_2$=Me, $R_3+R_4$=O, $R_5$=OAc, $R_6$=Me | dubiusine |
| **34** | $R_1+R_2$=$CH_2$, $R_3+R_4$=O, $R_5$=OH, $R_6$=Me | hippeastrine |
| **35** | $R_1$=Me, $R_2$=Me, $R_3$=OH, $R_4$=H, $R_5$=H, $R_6$=Me | lycorenine |
| **36** | $R_1$=Me, $R_2$=Me, $R_3$=OMe, $R_4$=H, $R_5$=H, $R_6$=Me | *O*-methyllycorenine |
| **37** | $R_1+R_2$= $CH_2$, $R_3$=OH, $R_4$=H, $R_5$=H, $R_6$=Me | oduline |
| **38** | $R_1+R_2$= $CH_2$, $R_3$=OMe, $R_4$=H, $R_5$=H, $R_6$=Me | 6-*O*-methyloduline |
| **39** | $R_1+R_2$= $CH_2$, $R_3$=OMe, $R_4$=H, $R_5$=OH, $R_6$=Me | 2α-hydroxy-6-*O*-methyloduline |

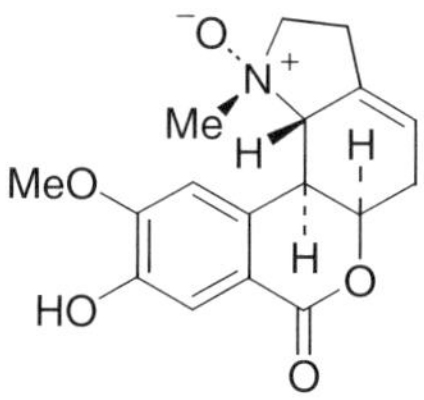

**40** 8-*O*-demethylhomolycorine-*N*-oxide

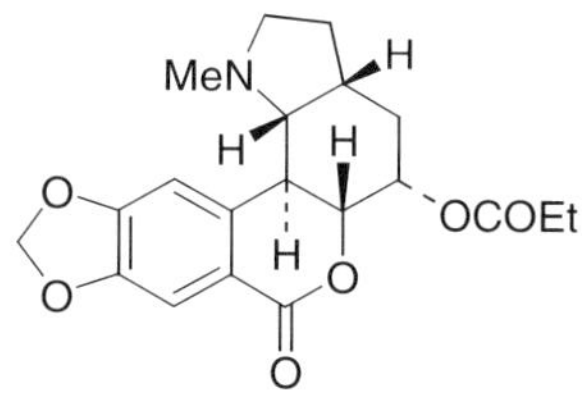

**41** poetinatine

The montanine-type alkaloids, also deriving from the hemanthamine series, are very unusual. Only two, namely pancracine (**73**) and nangustine (**74**), have been isolated from two species, both belonging to the Narcissus section of this genus. Nangustine has a unique substitution pattern (*80*), being the first 5,11-methanomorphanthridine alkaloid with a C-3/C-4 substitution, instead of a typical C-2/C-3.

### 3. *Galanthamine Type*

The galanthamine-type alkaloids are the only group among the Amaryllidaceae alkaloids showing two *ortho* aromatic protons in ring A. This type of alkaloid often shows an *N*-methyl group, or occasionally *N*-formyl or *N*-acetyl

TABLE III.
*Narcissus* Alkaloid Structures. Hemanthamine Type.

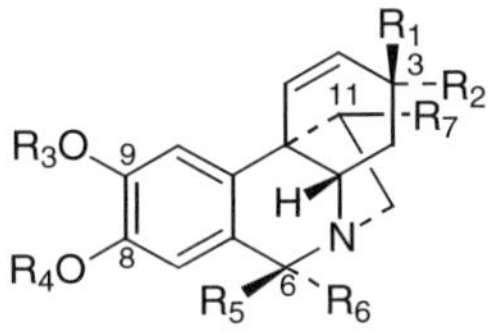


| | | |
|---|---|---|
| **42** | $R_1$=OH, $R_2$=H, $R_3$+$R_4$=$CH_2$, $R_5$=H, $R_6$=H, $R_7$=H | vittatine |
| **43** | $R_1$=OH, $R_2$=H, $R_3$+$R_4$=$CH_2$, $R_5$=H, $R_6$=H, $R_7$=OH | 11-hydroxyvittatine |
| **44** | $R_1$=OH, $R_2$=H, $R_3$=Me, $R_4$=Me, $R_5$=H, $R_6$=H, $R_7$=H | maritidine |
| **45** | $R_1$=OH, $R_2$=H, $R_3$=Me, $R_4$=H, $R_5$=H, $R_6$=H, $R_7$=H | 8-*O*-demethylmaritidine |
| **46** | $R_1$=OH, $R_2$=H, $R_3$=H, $R_4$=Me, $R_5$=H, $R_6$=H, $R_7$=H | 9-*O*-demethylmaritidine |
| **47** | $R_1$=OMe, $R_2$=H, $R_3$=Me, $R_4$=Me, $R_5$=H, $R_6$=H, $R_7$=H | *O*-methylmaritidine |
| **48** | $R_1$=OMe, $R_2$=H, $R_3$=Me, $R_4$=Me, $R_5$=H, $R_6$=OH, $R_7$=H | papyramine |
| **49** | $R_1$=OMe, $R_2$=H, $R_3$=Me, $R_4$=Me, $R_5$=OH, $R_6$=H, $R_7$=H | 6-epipapyramine |
| **50** | $R_1$=OMe, $R_2$=H, $R_3$=Me, $R_4$=Me, $R_5$=OMe, $R_6$=H, $R_7$=H | *O*-methyl-6-epipapyramine |
| **51** | $R_1$=H, $R_2$=OMe, $R_3$=Me, $R_4$=Me, $R_5$=H, $R_6$=OH, $R_7$=H | 6α-hydroxy-3-*O*-methylepimaritidine |
| **52** | $R_1$=H, $R_2$=OMe, $R_3$=Me, $R_4$=Me, $R_5$=OH, $R_6$=H, $R_7$=H | 6β-hydroxy-3-*O*-methylepimaritidine |
| **53** | $R_1$=OMe, $R_2$=H, $R_3$+$R_4$=$CH_2$, $R_5$=H, $R_6$=H, $R_7$=OH | haemanthamine |
| **54** | $R_1$=OMe, $R_2$=H, $R_3$+$R_4$=$CH_2$, $R_5$=H, $R_6$=H, $R_7$=OAc | 11-*O*-acetylhaemanthamine |
| **55** | $R_1$=OMe, $R_2$=H, $R_3$+$R_4$=$CH_2$, $R_5$=H, $R_6$=OH, $R_7$=OH | haemanthidine |
| **56** | $R_1$=OMe, $R_2$=H, $R_3$+$R_4$=$CH_2$, $R_5$=OH, $R_6$=H, $R_7$=OH | 6-epihaemanthidine |
| **57** | $R_1$=H, $R_2$=OMe, $R_3$+$R_4$=$CH_2$, $R_5$=H, $R_6$=H, $R_7$=OH | crinamine |
| **58** | $R_1$=OMe, $R_2$=H, $R_3$=Me, $R_4$=H, $R_5$=H, $R_6$=H, $R_7$=OH | narcidine |

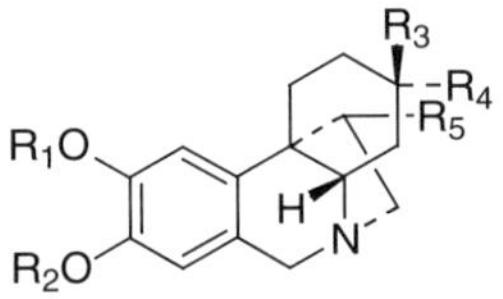


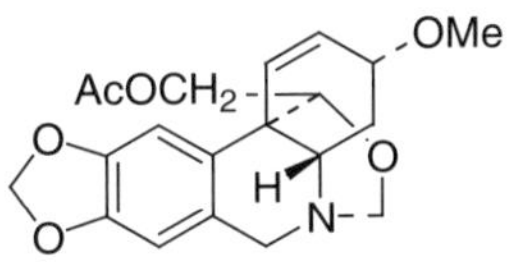


**59** $R_1$=H, $R_2$=Me, $R_3$=H, $R_4$=OAc, $R_5$=H cantabricine

**60** $R_1$+$R_2$=$CH_2$, $R_3$=OMe, $R_4$=H, $R_5$=$OCH_2CH(OH)Et$ narcimarkine

**61** bujeine

derivatives. They are frequent in *Narcissus*, galanthamine (**75**) being the most usual and representative of them (*30,81*). The concentration of galanthamine in *Narcissus* has been found to vary widely between species and varieties, from trace amounts to 2.5% (*30*). The potential of some *Narcissus* cultivars ('Carlton', 'Gigantic Star', Ice Follies', and 'Fortune') and wild species (*N. confusus*) as sources of galanthamine and related alkaloids has been recognized (*81–84*), and research has been initiated into the agronomic factors that affect their content. *Narcissus* has two important

TABLE IV.
*Narcissus* Alkaloid Structures. Tazettine Type.

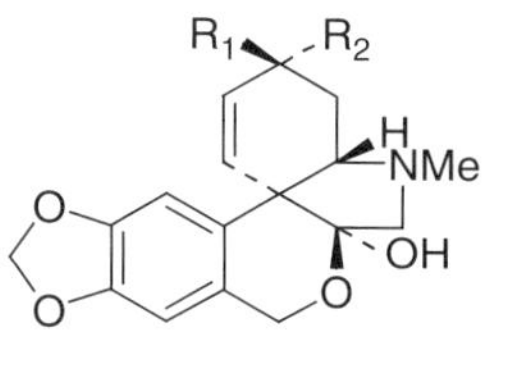

**62** $R_1$=OMe, $R_2$=H tazettine
**63** $R_1$=H, $R_2$=OMe criwelline

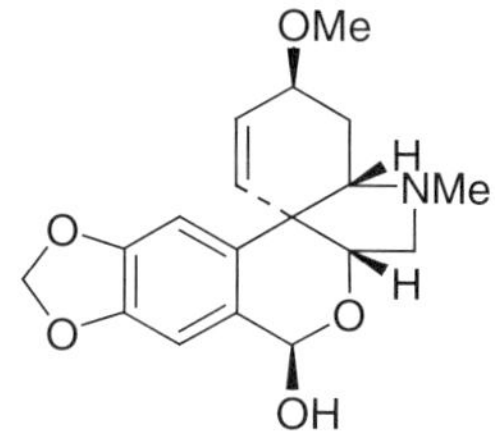

**64** pretazettine

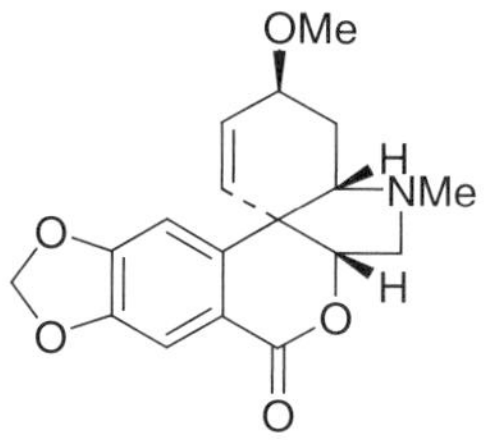

**65** 3-epimacronine

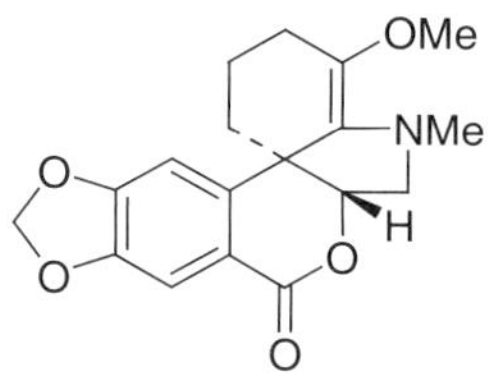

**66** 3-epimacronine isomer
(structure under revision)

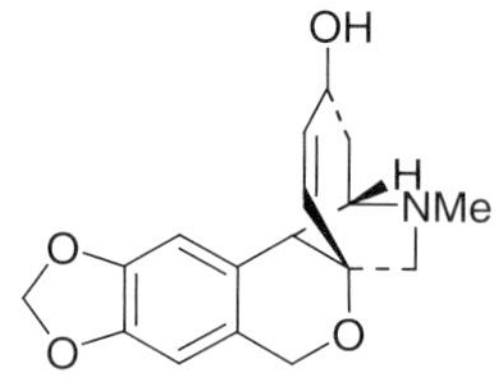

**67** obesine

advantages over *Leucojum aestivum*: first, bulbs of many *Narcissus* cultivars are available in commercial quantities, offering the possibility of establishing large-scale cultivation in a short time; second, a comprehensive body of information already exists regarding narcissus propagation, physiology, breeding, and cultivation (*85*).

### 4. Other Alkaloids

The mesembrane (*Sceletium*)-type alkaloids are also present in *Narcissus* plants. It is noteworthy that the alkaloids of this group (mesembrenone (**90**), mesembrenol (**91**), and mesembrine (**92**)) isolated from *Narcissus* are restricted to the species of the Ganymedes section, which is therefore of chemotaxonomic interest because they are generally found in the Aizoaceae, a dicotyledonous family (*23,24*). Both the Amaryllidaceae and *Sceletium*-type alkaloids have common biogenetic precursors, although their biosyntheses are fundamentally different (*33*).

There are also two unclassified alkaloids, namely cherylline (**88**) and pallidiflorine (**89**). Cherylline, a 4-substituted tetrahydroisoquinoline derivative, is a unique representative of rare phenolic Amaryllidaceae alkaloids. Pallidiflorine, a heterobis alkaloid isolated from *N. pallidiflorus* (*86*), is the only bis-alkaloid present in this genus isolated up to now. This alkaloid is formed from two directly joined moieties, one of them being the alkaloid lycoramine (galanthamine type), and the other being the reverse form of the hemiacetal at C-11 of tazettine.

TABLE V.
*Narcissus* Alkaloid Structures. Narciclasine and Montanine Types.

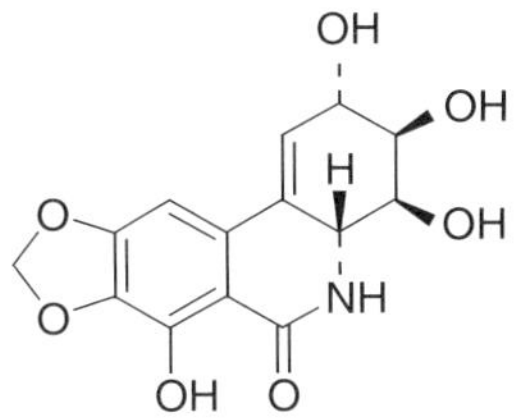

**68** narciclasine

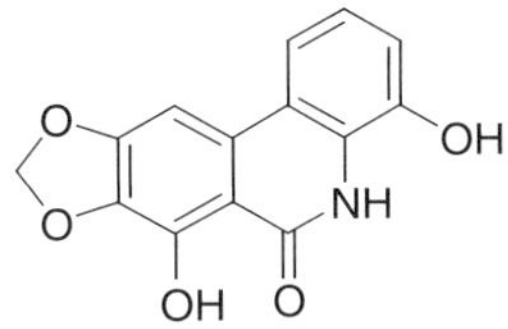

**69** narciprimine

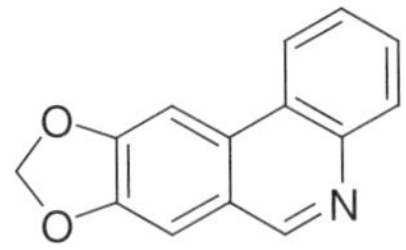

**70** trisphaeridine

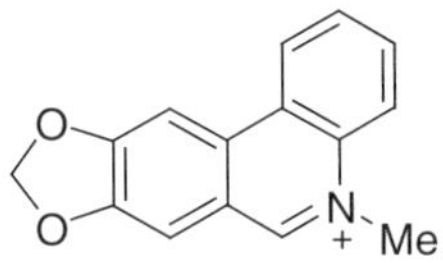

**71** bicolorine

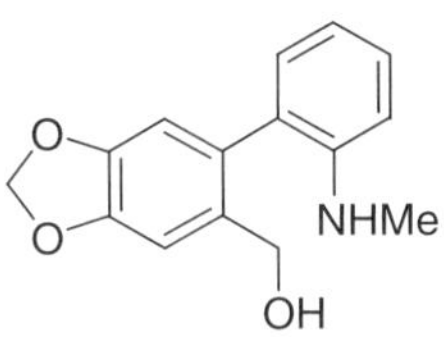

**72** ismine

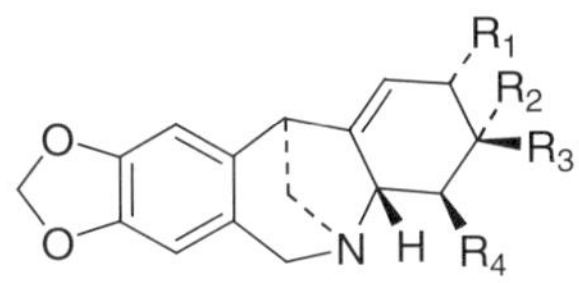

**73** $R_1$=OH, $R_2$=H, $R_3$=OH, $R_4$=H pancracine
**74** $R_1$=H, $R_2$=OH, $R_3$=H, $R_4$=OH nangustine

## B. ONTOGENIC VARIATIONS

It is well established that profiles of alkaloids vary with time, location, and developmental stage. In many instances, the site of biosynthesis is restricted to a single organ, but accumulation of the corresponding products can be detected in several other plant tissues. Long-distance transport must take place in these instances. There are only a few data on the ontogenic variations and distribution of alkaloids in species of the Amaryllidaceae family, and some results have been obtained in *Narcissus* species, such as *N. assoanus* (with only lycorine-type alkaloids) or *N. confusus* (with alkaloids of the homolycorine, hemanthamine, tazettine, and galanthamine types) (*84,87*).

During the vegetative period, the bulb of *N. assoanus* has a lower concentration of alkaloids than the aerial part. Pseudolycorine (**3**), the major alkaloid of both the aerial parts and the bulb, represents around 50% of the total alkaloids. The synthesis of alkaloids might take place mainly in the aerial part, where they accumulate. Once flowers have been fecundated, they are transported to the bulb. It seems that, in order to avoid autotoxicity, pseudolycorine is stored in the bulbs as

TABLE VI.
*Narcissus* Alkaloid Structures. Galanthamine Type.

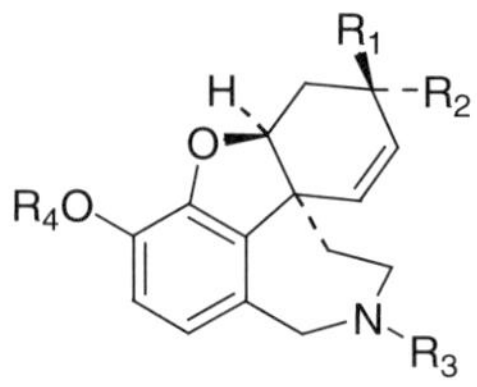

| | | |
|---|---|---|
| **75** | $R_1$=OH, $R_2$=H, $R_3$=Me, $R_4$=Me | galanthamine |
| **76** | $R_1$=H, $R_2$=OH, $R_3$=Me, $R_4$=Me | epigalanthamine |
| **77** | $R_1$=OAc, $R_2$=H, $R_3$=Me, $R_4$=Me | *O*-acetylgalanthamine |
| **78** | $R_1$=OH, $R_2$=H, $R_3$=H, $R_4$=Me | norgalanthamine (= *N*-demethylgalanthamine) |
| **79** | $R_1$=H, $R_2$=OH, $R_3$=H, $R_4$=Me | epinorgalanthamine |
| **80** | $R_1$=OH, $R_2$=H, $R_3$=CHO, $R_4$=Me | *N*-formylnorgalanthamine |
| **81** | $R_1$=OH, $R_2$=H, $R_3$=Me, $R_4$=H | sanguinine |
| **82** | $R_1$=OH, $R_2$=H, $R_3$=Ac, $R_4$=Me | narcisine |
| **83** | $R_1$+$R_2$=O, $R_3$=Me, $R_4$=Me | narwedine |

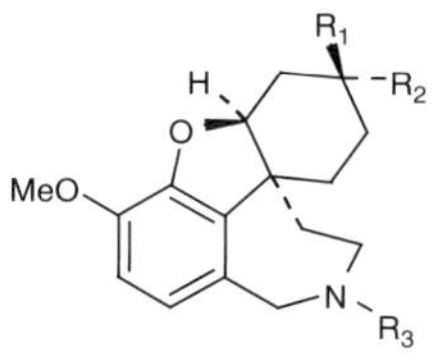

| | | |
|---|---|---|
| **84** | $R_1$=OH, $R_2$=H, $R_3$=Me | lycoramine |
| **85** | $R_1$=OH, $R_2$=H, $R_3$=H | norlycoramine (=*N*-demethyllycoramine) |
| **86** | $R_1$=H, $R_2$=OH, $R_3$=H | epinorlycoramine |

the 1-*O*-Ac (**4**) and 2-*O*-Ac (**5**) derivatives (*87*). The specific localization of alkaloids make sense if their role as defense and/or signal compounds is accepted. As a rule, vulnerable tissues are defended more than old, senescent tissues.

Galanthamine (**75**) and four other alkaloids of *N. confusus* were found to be present in all of the organs at every stage, with the exception of hemanthamine (**53**), which does not occur in senescent flowers. The main alkaloid is galanthamine, followed by *N*-formylnorgalanthamine (**80**) or tazettine (**62**), depending on the sample, while homolycorine (**26**) is in general the least common. The organ with the highest total alkaloid accumulation throughout the ontogenic cycle is the bulb, except at the end of the cycle, when the alkaloids accumulate mainly in the flower stem, reaching a concentration of up to 2.5% by dry weight (*84*).

TABLE VII.
*Narcissus* Alkaloid Structures. Other Types.

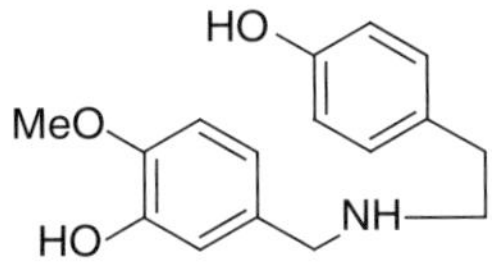


**87** *O*-methylnorbelladine

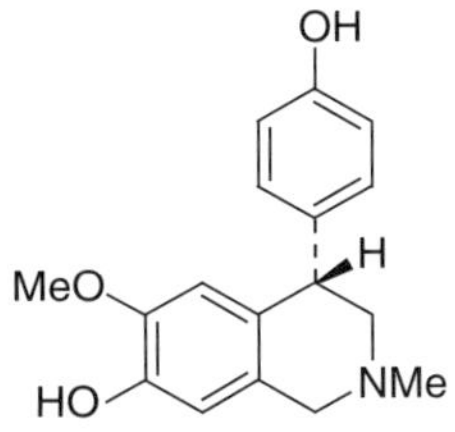


**88** cherylline

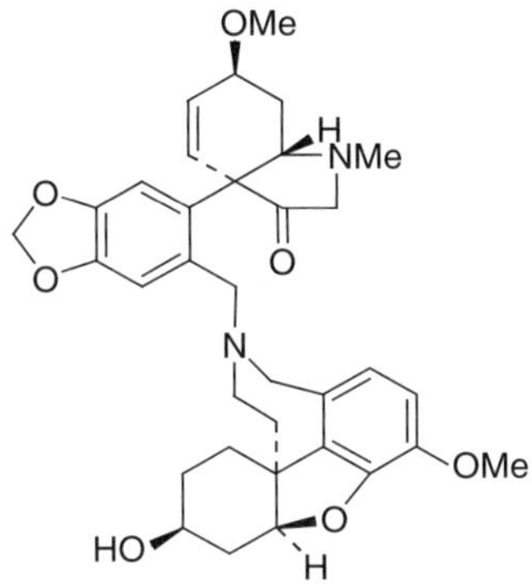


**89** pallidiflorine

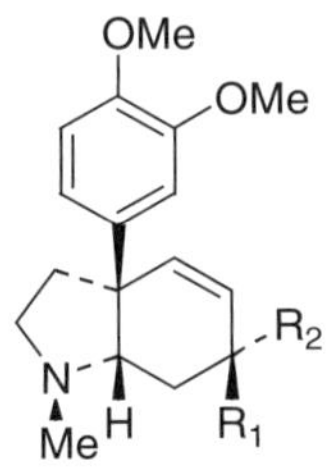


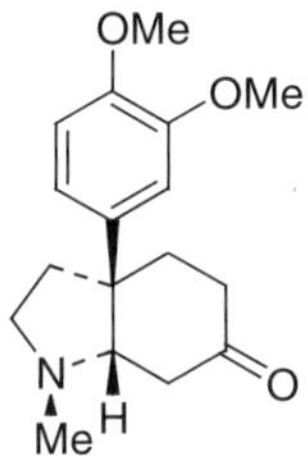


| | | |
|---|---|---|
| **90** | $R_1$=$R_2$=O | mesembrenone |
| **91** | $R_1$=OH, $R_2$=H | mesembrenol |

**92** mesembrine

## III. Biosynthetic Pathways

Most of the biosynthetic research done on Amaryllidaceae alkaloids was carried out in the 1960s and early 1970s. Since then, the only noteworthy study has been the biosynthesis of galanthamine (**75**) and related alkaloids (*181*). As in most alkaloid biosyntheses, that of the Amaryllidaceae follows a sequential pattern.

TABLE VIII.
Occurrence of *Narcissus* Alkaloids in Wild Species and Intersectional Hybrids.

| Species[a] | Skeleton types: (alkaloids)[b] | References |
|---|---|---|
| 1. Subgenus *Narcissus* | | |
| a- *Section* Narcissus L. | | |
| *N. angustifolius* Curtis ex Haw. | GA: (**75**) | (*88*) |
| *N. angustifolius* Curtis subsp. *transcarpathicus* Kricsfalusy | LY: (**3, 22, 24**); HL: (**27**); MN: (**73, 74**); OA: (**88**) | (*80*) |
| *N. poeticus* L. | LY: (**1, 2, 6, 7, 11, 16, 18**); HL: (**26, 35**); MN: (**73**); GA: (**75, 79**) | (*89–95*) |
| *N. poeticus* L. var. *ornatus* Hort. | LY: (**1, 2, 7**); HL: (**26, 35, 41**); HT: (**53**); TZ: (**62**); GA: (**75**) | (*89,96–98*) |
| b- *Section* Pseudonarcissi DC. | | |
| Group A [Plant small, usually less than 15 cm] | | |
| *N. asturiensis* (Jordan) Pugsley | HT: (**53, 55**) ; TZ: (**62, 65**); NC: (**70, 72**) | (*99*) |
| *N. cyclamineus* DC | NC: (**68**); GA: (**83**) | (*100,101*) |
| *N. jacetanus* Fdez. Casas | LY: (**1, 3, 20, 21**) | (*102*) |
| *N. lobularis* Hort. | HT: (**53**); GA: (**75**) | (*103*) |
| *N. muñozii-garmendiae* Fdez. Casas | HL: (**26, 35, 36**) | (*104*) |
| *N. vasconicus* Fdez. Casas | LY: (**1, 22**); HL: (**26, 28**) | (*70*) |
| Group B [Plant often 15–60 cm or more and/or with large flowers] | | |
| *N. bicolor* L. | LY: (**22**); HL: (**29**); TZ: (**64, 65**); NC: (**71, 72**) | (*79*) |
| *N. bujei* (Fdez. Casas) Fdez. Casas | HL: (**26, 27, 30, 35, 36, 38**); HT: (**53, 54, 57, 61**); TZ: (**62**); GA: (**75**) | (*74,93*) |
| *N. confusus* Pugsley | HL: (**26, 29**); HT: (**53**); TZ: (**64**); GA: (**75, 79, 80**) | (*82,83,93*) |
| *N. eugeniae* Fdez. Casas | HL: (**26, 35**); GA: (**75**) | (*93,105,106*) |
| *N. leonensis* Pugsley | LY: (**1**); GA: (**78, 79, 86**) | (*93,107*) |
| *N. pallidiflorus* Pugsley | HL: (**26, 27**); HT: (**53**); TZ: (**64**); NC: (**72**); OA: (**89**) | (*86*) |
| *N. perez-chiscanoi* Fdez. Casas | GA: (**75, 79**) | (*93*) |
| *N. primigenius* (Laínz) Fdez. Casas & Laínz | HL: (**26, 27**); HT: (**45, 53**) | (*108*) |
| *N. pseudonarcissus* L. | LY: (**1**); HL: (**26, 30**); HT: (**53**); OA: (**87**) | (*95,109,110*) |
| *N. radinganorum* Fdez. Casas | HL: (**26, 27**); HT: (**46**) | (*111*) |
| *N. tortuosus* Haworth | LY: (**1, 23**) | (*68*) |
| c- *Section* Ganymedes Salisbury ex Schultes and Schultes | | |
| *N. pallidulus* Graells | LY: (**25**); OA: (**90**) | (*26,67*) |
| *N. triandrus* L. | OA: (**90, 91, 92**) | (*112*) |

(*continued*)

TABLE VIII.
Continued.

| Species[a] | Skeleton types: (alkaloids)[b] | References |
|---|---|---|
| d- *Section* Jonquillae De Candolle | | |
| *N. assoanus* Léon-Dufour = *N. requienii* Roemer | LY: (**3, 4, 5, 20, 21**) | (*87,113,114*) |
| *N. jonquilla* L. | NC: (**68**); GA: (**75, 84**) | (*115*) |
| 2. Subgenus *Hermione* (Salisbury) Spach | | |
| a- *Section* Hermione | | |
| i. Subsection *Hermione* [syn. Section *Tazettae* De Candolle] | | |
| A. Series *Hermione* | | |
| *N. aureus* Loiseleur = *N. tazetta* L. ssp. *aureus* (Loisel.) Baker | LY: (**1**) | (*116*) |
| *N. canaliculatus* Guss. | HT: (**53**); TZ: (**62**); NC: (**68**) | (*89,101*) |
| *N. tazetta* L. | LY: (**1, 3, 7, 16, 18**); HL: (**26, 27, 30, 34**); HT: (**47, 53, 55, 56**); TZ: (**62, 63, 64**); NC: (**68, 71**) GA: (**75, 82, 83**) | (*89,117–129*) |
| *N. tazetta* L. ssp. *tazetta* | LY: (**1**); HT: (**43, 53**); TZ: (**62**); GA: (**78**) | (*130*) |
| *N. tazetta* L. var. *chinensis* Roem | LY: (**1, 3, 11**); HL: (**26, 35**); HT: (**44, 47, 48, 49, 51, 52, 55, 56**); TZ: (**62, 64**); GA: (**75, 76, 84**) | (*131–133*) |
| B. Series *Albiflorae* | | |
| *N. dubius* Gouan | LY: (**3**); HL: (**32, 33**) | (*72,134*) |
| *N. panizzianus* Parlatore | LY: (**7**); HL: (**26**); HT: (**48, 49**); TZ: (**64**) | (*135*) |
| *N. papyraceus* Ker-Gawler. | LY: (**1, 3**); HL: (**26, 27, 40**); HT: (**44, 47, 48, 49, 50, 60**); TZ: (**62**); GA: (**75, 84**) | (*91,136,137*) |
| *N. tortifolius* Fdez. Casas | HL: (**26, 27, 32, 33**); GA: (**75**) | (*138*) |
| ii. Subsection *Serotini* [syn. Section *Serotini* Parlatore] | | |
| *N. serotinus* Löfl. ex L. | TZ: (**66**); NC: (**68**) | (*101,139*) |
| 3. Subgenus *Corbularia* (Salisb.) Pax [syn. Section *Bulbocodii* De Candolle] | | |
| *N. bulbocodium* L. | HT : (**53**); TZ : (**62, 64**); NC: (**72**); GA: (**75, 81**) | (*112*) |
| *N. cantabricus* DC. | HT: (**42, 57, 59**); TZ: (**62**) | (*140*) |
| *N. nivalis* Graells | LY: (**6**); GA: (**75, 78**) | (*141*) |
| *N. obesus* Salisbury | HT: (**53**); TZ: (**64, 65, 67**); NC: (**71, 72**); GA: (**75**) | (*76*) |
| Intersectional hibrids | | |
| Intersection [*Jonquillae* DC. *x Narcissus* L.] | | |
| *N. x gracilis* Sabine | LY: (**1**); TZ: (**62**); GA: (**75**) | (*142*) |

(*continued*)

TABLE VIII.
Continued.

| Species[a] | Skeleton types: (alkaloids)[b] | References |
|---|---|---|
| Intersection [*Narcissus* L. *x Tazettae* DC.] | | |
| *N. x biflorus* Curtis | LY: (**1**) | (*116*) |
| Intersection [*Jonquillae* DC. *x Pseudonarcissii* DC.] | | |
| *N. x odorus* L. var. *rugulosus* | LY: (**1**); HL: (**26, 34, 37**); TZ: (**62**); NC: (**68**); GA: (**75**) | (*101,142*) |
| Other *Narcissus* | | |
| *Narcissus* sp. | LY: (**2**); NC: (**68, 69**); GA: (**75**) | (*95,125, 143,144*) |

[a]Wild *Narcissus* species and hybrids are grouped into Subgenera and Sections according to the *Narcissus* classification of Mathew (*4*). Some taxonomical aspects are also based on the works of Barra and López-González (*145,146*), Dorda and Fernández-Casas (*147–152*), Fernandes (*10,153*), and Fernández-Casas (*154–157*).

[b]Skeleton types: LY, lycorine; HL, homolycorine; HT, hemanthamine; TZ, tazettine; NC, narciclasine; MN, montanine; GA, galanthamine; OA, other alkaloids. Alkaloid names: see Tables I–VII.

## A. ENZYMATIC PREPARATION OF THE PRECURSORS

Although L-phenylalanine (L-phe) and L-tyrosine (L-tyr) are closely related in their chemical structure, they are not interchangeable in plants. In the Amaryllidaceae alkaloids, L-phe serves as a primary precursor of the $C_6$–$C_1$ fragment, corresponding to ring A and the benzylic position (C-6), and L-tyr is the precursor of ring C, the two-carbon side chain (C-11 and C-12) and nitrogen, $C_6$–$C_2$–N. The conversion of L-phe to the $C_6$–$C_1$ unit requires the loss of two carbon atoms from the side chain as well as the introduction of at least two oxygenated substituents into the aromatic ring, which is performed *via* cinnamic acids. The presence of the enzyme phenylalanine ammonia lyase (PAL) has been demonstrated in Amaryllidaceae plants (*182*), and the elimination of ammonia mediated by this enzyme is known to occur in an antiperiplanar manner to give *trans*-cinnamic acid, with loss of the $\beta$-*pro*-S hydrogen (*183*). Thus, it may be expected that L-phe would be incorporated into Amaryllidaceae alkaloids with retention of the $\beta$-*pro*-R hydrogen. However, feeding experiments in *Narcissus* ‘King Alfred’ showed that tritium originally present at C-$\beta$ of L-phe, whatever the configuration, was lost in the formation of several hemanthamine- and homolycorine-type alkaloids, which led to the conclusion that fragmentation of the cinnamic acids involves the oxidation of C-$\beta$ to the ketone or acid level, the final product being protocatechuic aldehyde or its derivatives (Fig. 2). On the other hand, L-tyr is degraded no further than tyramine before incorporation into the Amaryllidaceae alkaloids.

## B. PRIMARY CYCLIZATION MECHANISMS

Tyramine and protocatechuic aldehyde or its derivatives are logical components for the biosynthesis of the precursor norbelladine (**93**). This pivotal reaction

TABLE IX.
Occurrence of Alkaloids in *Narcissus* Cultivars.

| *Narcissus* cultivars[a] | Skeleton types: (alkaloids)[b] | References |
|---|---|---|
| Division 1 – Trumpet Daffodils | | |
| *Narcissus* 'Celebrity' | NC: (**68**) | (*101*) |
| *Narcissus* 'Covent Garden' | LY: (**1, 7, 11**); HL: (**35**); HT: (**53**); GA: (**75, 78, 85**) | (*103,158*) |
| *Narcissus* 'Dutch Master' | GA: (**75**) | (*81*) |
| *Narcissus* 'Early Glory' | LY: (**1, 7**); HT: (**53**); GA: (**75**) | (*103*) |
| *Narcissus* 'Flower Carpet' | NC: (**68**) | (*101*) |
| *Narcissus* 'Golden Harvest' | NC: (**68**); GA: (**75**) | (*81,101*) |
| *Narcissus* 'Grand Maître' | LY: (**1, 7**); HL: (**26**); HT: (**53**); TZ: (**62**); GA: (**75**) | (*103*) |
| *Narcissus* 'Imperator' | LY: (**1, 7, 11**); HL: (**26, 35**); HT: (**53**); GA: (**75**) | (*103*) |
| *Narcissus* 'King Alfred' | LY: (**1, 6, 7, 11, 16, 20**); HL: (**26, 27, 34, 35, 37**); HT: (**53, 58**); NC: (**68**); GA: (**75, 78, 85**) | (*92,101,103,125, 158–161*) |
| *Narcissus* 'Magnet' | LY: (**7, 11**); HT: (**53**) | (*103*) |
| *Narcissus* 'Magnificience' | LY: (**1, 7, 11**); HT: (**53**); GA: (**75, 78, 85**) | (*103,158*) |
| *Narcissus* 'Mount Hood' | HL: (**26, 34**); NC: (**68**); GA: (**75, 84, 85**) | (*101,162–164*) |
| *Narcissus* 'Mrs. Ernst H. Krelage' | LY: (**1, 7, 11**); HT: (**53**); NC: (**68**); GA: (**75**) | (*103,163*) |
| *Narcissus* 'Music Hall' | LY: (**1, 7**); HT: (**53**) | (*103*) |
| *Narcissus* 'Oliver Cromwell' | LY: (**7, 11**); HT: (**53**); GA: (**75**) | (*103*) |
| *Narcissus* 'President Lebrun' | NC: (**68**) | (*101,163*) |
| *Narcissus* 'Queen of Bicolors' | LY: (**1, 7, 11**); HT: (**53**) | (*103*) |
| *Narcissus* 'Rembrandt' | LY: (**1**); HT: (**53**); NC: (**68**); GA: (**75, 78, 85**) | (*101,103*) |
| *Narcissus* 'Rockery Beauty' | LY: (**7, 16**); HT: (**53**); GA: (**78, 85**) | (*103*) |
| *Narcissus* 'Romaine' | LY: (**7, 11**); HL: (**35**); HT: (**53**) | (*103*) |
| *Narcissus* 'Spring Glory' | LY: (**1, 7, 11**); HT: (**53**); GA: (**78, 85**) | (*103*) |
| *Narcissus* 'Unsurpassable' | LY: (**11**); HL: (**35**); HT: (**53**); GA: (**75**) | (*81,103*) |
| *Narcissus* 'Victoria' | LY : (**1, 7, 11**); HT: (**53**); GA: (**78, 85**) | (*103*) |
| *Narcissus* 'Wrestler' | LY: (**7, 11**); HT: (**53**); GA: (**75**) | (*103*) |
| Division 2 – Large-Cupped Daffodils | | |
| *Narcissus* 'Carabiniere' | NC: (**68**) | (*101*) |
| *Narcissus* 'Carlton' | LY: (**13, 14, 15**); HL: (**26, 30, 31, 34, 35, 36, 37, 38**); HT: (**42, 53**); NC: (**68**); GA: (**75, 77, 78, 83, 84, 86**) | (*101,165–168*) |

(*continued*)

TABLE IX.
Continued.

| *Narcissus* cultivars[a] | Skeleton types: (alkaloids)[b] | References |
|---|---|---|
| *Narcissus* 'Clamor' | NC: (**68**) | (*101,163*) |
| *Narcissus* 'Daisy Schäffer' | LY: (**7, 11**); GA: (**75**) | (*103*) |
| *Narcissus* 'Deanna Durbin' | LY: (**1, 7, 11, 16**); HT: (**53**) | (*103,158*) |
| *Narcissus* 'Flower Record' | LY: (**1, 7, 11**); HT: (**53**); GA: (**75**) | (*81,103*) |
| *Narcissus* 'Folly' | LY: (**1**); TZ : (**62**) | (*118*) |
| *Narcissus* 'Fortune' | LY: (**19**); HL: (**34, 37**); HT: (**53**); GA: (**75**) | (*81,103,169,170*) |
| *Narcissus* 'Gigantic Star' | GA: (**75**) | (*81*) |
| *Narcissus* 'Helios' | LY: (**7, 11**); HL: (**26, 35**); HT: (**53**); NC: (**68**); GA: (**75**) | (*101,103,125*) |
| *Narcissus* 'Ice Follies' | LY: (**1, 7, 10, 13, 14**); HL: (**34**); HT: (**53**); GA: (**75, 84, 85**) | (*81,164,171,172*) |
| *Narcissus* 'John Evelyn' | LY: (**1, 7,11**); HT: (**53**); GA: (**75**) | (*103*) |
| *Narcissus* 'Marion Cran' | LY: (**1, 7**); HT: (**53**); GA: (**75**) | (*103*) |
| *Narcissus* 'Mercato' | NC: (**68**) | (*101*) |
| *Narcissus* 'Mrs. R.O. Backhouse' | NC: (**68**) | (*101*) |
| *Narcissus* 'Nova Scotia' | LY: (**1, 7**); HT: (**53**); GA: (**75**) | (*103*) |
| *Narcissus* 'Oranje Bruid' | NC: (**68**) | (*101*) |
| *Narcissus* 'Pluvius' | LY: (**1, 7, 11, 16**); HT: (**53**); GA: (**75**) | (*103,158*) |
| *Narcissus* 'Salome' | LY: (**3, 22, 23**); HL: (**34, 39**); HT: (**57**); GA: (**75, 78**) | (*71,81*) |
| *Narcissus* 'Scarlet Elegance' | NC: (**68**) | (*101,163*) |
| *Narcissus* 'Sempre Avanti' | LY: (**1, 7, 11, 16**); HT: (**53**); NC: (**68**) | (*101,103,125,163*) |
| *Narcissus* 'Suda' | LY: (**7, 11**); HL: (**35**); HT: (**53**); GA: (**75**) | (*103*) |
| *Narcissus* 'Toronto' | LY: (**1, 7**); HT: (**53**) | (*103*) |
| *Narcissus* 'Tunis' | NC: (**68**) | (*101,163*) |
| *Narcissus* 'Walt Disney' | NC: (**68**) | (*101,163*) |
| *Narcissus* 'Yellow Sun' | GA: (**75**) | (*81*) |
| Division 3 – Small-Cupped Daffodils | | |
| *Narcissus* 'Barrett Browning' | GA: (**75**) | (*81*) |
| *Narcissus* 'Daphne' | LY: (**1**); HL: (**26, 35**); GA: (**75**) | (*103*) |
| *Narcissus* 'Verger' | NC: (**68**); GA: (**75**) | (*81,101,163*) |
| Division 4 – Double Daffodils | | |
| *Narcissus* 'Cheerfulness' | NC: (**68**); GA: (**75**) | (*101,164*) |
| *Narcissus* 'Dick Wilden' | GA: (**75**) | (*81*) |
| *Narcissus* 'Flower Drift' | GA: (**75**) | (*81*) |
| *Narcissus* 'Inglescombe' | LY: (**1, 11**); HL: (**26, 34, 35**); HT: (**53**); GA: (**75, 84, 85**) | (*103,162*) |
| *Narcissus* 'Insulinde' | LY: (**1, 7, 11**); HT: (**53**) | (*103*) |

(*continued*)

TABLE IX.
Continued.

| *Narcissus* cultivars[a] | Skeleton types: (alkaloids)[b] | References |
|---|---|---|
| *Narcissus* 'Irene Copeland' | LY: (**1**); GA: (**83**) | (*103*) |
| *Narcissus* 'Livia' | LY: (**1, 7, 10, 11**); HT: (**53**) | (*103*) |
| *Narcissus* 'Sir Winston Churchill' | LY: (**17**) | (*173*) |
| *Narcissus* 'Texas' | LY: (**1, 11, 12**); HT: (**53**); NC: (**68**); GA: (**75, 83**) | (*101,103,163, 174–176*) |
| *Narcissus* 'Twink' | LY: (**1, 7, 11**); HT: (**53**); GA: (**75**) | (*103,158*) |
| *Narcissus* 'Van Sion' | LY: (**1, 7, 11**); HL: (**26, 35**); HT: (**53**); TZ: (**62**); GA: (**75, 78, 85**) | (*81,103,158*) |
| Division 5 – Triandrus Daffodils | | |
| *Narcissus* 'Thalia' | LY: (**1**); HL: (**26, 35**); HT: (**53**); NC: (**68**) | (*101,125,142*) |
| *Narcissus* 'Tresamble' | HL: (**35**); HT: (**53**); NC: (**68**); GA: (**75**) | (*101,125,142*) |
| Division 6 – Cyclamineus Daffodils | | |
| *Narcissus* 'Beryl' | LY: (**1, 7, 16**) | (*142*) |
| *Narcissus* 'Cairhays' | LY: (**7, 11**); HT: (**55, 56**) | (*142*) |
| *Narcissus* 'February Gold' | HL: (**26, 35**); GA: (**75, 84**) | (*81,142*) |
| *Narcissus* 'Peeping Tom' | LY: (**11**); HL: (**35**); TZ: (**62**) | (*142*) |
| *Narcissus* 'Tête-a-Tête' | GA: (**75**) | (*81*) |
| Division 7 – Jonquilla and Apodanthus Daffodils | | |
| *Narcissus* 'Golden Sceptre' | LY: (**1, 7, 8, 9**); HL : (**26, 30, 34, 35, 37**); HT: (**53**); TZ: (**62**); GA: (**75**) | (*142,177–180*) |
| *Narcissus* 'Pipit' | GA: (**75**) | (*168*) |
| *Narcissus* 'Trevithian' | LY: (**1**); HL: (**35**); TZ: (**62**); NC: (**68**); GA: (**75**) | (*101,142*) |
| Division 8 – Tazetta Daffodils | | |
| *Narcissus* 'Cragford' | LY: (**1**); HL: (**26**); HT: (**53**); TZ: (**62**) | (*89*) |
| *Narcissus* 'Early Perfection' | LY: (**1, 11**); HL: (**26, 34**); HT: (**53**); TZ: (**62**) | (*89*) |
| *Narcissus* 'Geranium' | LY: (**1, 16**); HL: (**26**); HT: (**53**); TZ: (**62**); NC: (**68**); GA: (**75**) | (*89,122,143,164*) |
| *Narcissus* 'La Fiancée' | LY: (**1, 7**); HL: (**26**); HT: (**53**); TZ: (**62**) | (*89*) |
| *Narcissus* 'Laurens Koster' | LY: (**1**); HL: (**26**); HT: (**53**); TZ: (**62**) | (*89*) |
| *Narcissus* 'L'innocence' | LY: (**1, 16**); HL: (**26**); HT: (**53**); TZ: (**62**) | (*89*) |
| *Narcissus* 'Minnow' | GA: (**75**) | (*81*) |

(*continued*)

TABLE IX.
Continued.

| *Narcissus* cultivars[a] | Skeleton types: (alkaloids)[b] | References |
|---|---|---|
| *Narcissus* 'Scarlet Gem' | LY: (**1, 7**); HL: (**26**); HT: (**53**); TZ: (**62**) | (*89*) |
| *Narcissus* 'Silver Chimes' | LY: (**1**); HT: (**53**); TZ: (**62**) | (*142*) |
| *Narcissus* 'St. Agnes' | LY: (**1**); HL: (**26**); HT: (**53**); TZ: (**62**) | (*89*) |
| *Narcissus* 'Totus Albus' | NC: (**68**) | (*101,163*) |
| Division 9 – Poeticus Daffodils | | |
| *Narcissus* 'Actaea' | LY: (**1, 7, 16**); HL: (**35**); NC: (**68**); GA: (**75**) | (*89,101*) |
| *Narcissus* 'Sarchedon' | LY: (**1, 7, 16**); HL: (**35**); GA: (**75**) | (*89*) |
| Division 12 – Other Daffodils | | |
| *Narcissus* 'Brougshane' | GA: (**75**) | (*81*) |
| *Narcissus* 'Kristalli' | LY: (**1**); TZ: (**62**); GA: (**75, 83**) | (*118*) |

[a]Horticultural classification of *Narcissus* cultivars according to the International Daffodil Register and Classified List of the Royal Horticultural Society (*18*).
[b]Skeleton types: LY, lycorine; HL, homolycorine; HT, hemanthamine; TZ, tazettine; NC, narciclasine; MN, montanine; GA, galanthamine. Alkaloid names: see Tables I–VII.

represents the entry of primary metabolites into a secondary metabolic pathway. The junction of the amine and the aldehyde results in a Schiff's base, two of which have been isolated up to now from several *Crinum* species: craugsodine (*184*) and isocraugsodine (*185*). The existence of Schiff's bases in nature, as well as their easy conversion into the different ring systems of the Amaryllidaceae alkaloids, suggest that the initial hypothesis about this biosynthetic pathway was correct.

## C. ENZYMATIC PREPARATION OF INTERMEDIATES

In 1957, Barton and Cohen (*186*) proposed that norbelladine (**93**) or related alkaloids could undergo oxidative coupling in Amaryllidaceae plants, once ring A had been suitably protected by methylation, resulting in the different skeletons of the Amaryllidaceae alkaloids (Fig. 3). The key intermediate in most of cases is *O*-methylnorbelladine (**87**).

## D. SECONDARY CYCLIZATION, DIVERSIFICATION, AND RESTRUCTURING

Secondary cyclization is produced by an oxidative coupling of *O*-methylnorbelladine.

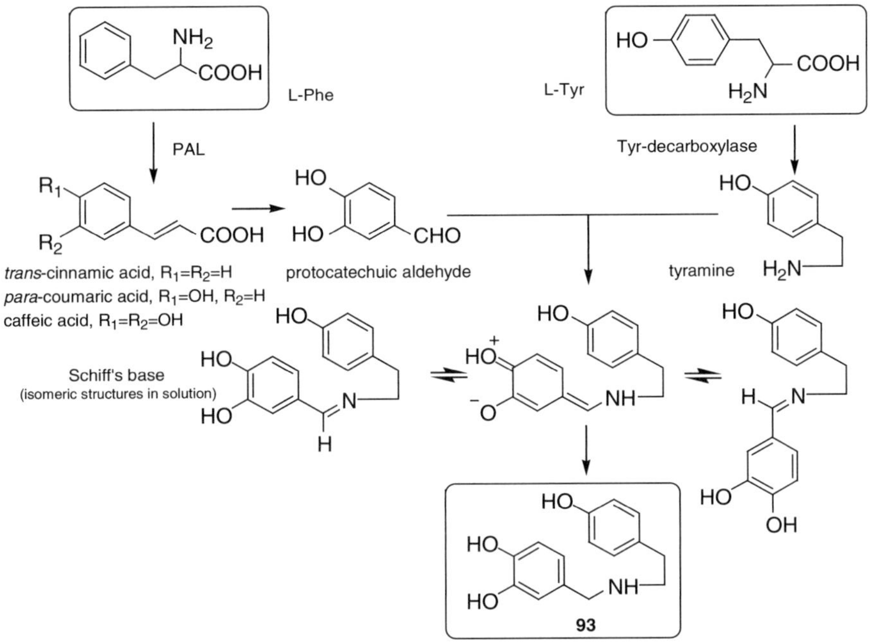


Figure 2. Biosynthetic pathway to norbelladine (**93**).

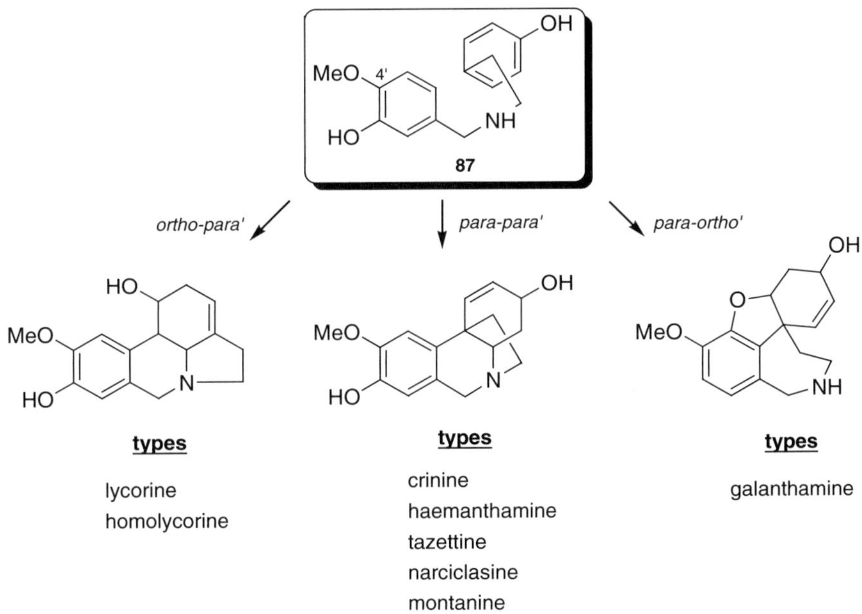


Figure 3. Phenol oxidative coupling in Amaryllidaceae alkaloids.

### *1. Lycorine and Homolycorine Types*

The alkaloids of this group are derivatives of the pyrrolo[de]phenanthridine (lycorine type) and the 2-benzopirano-[3,4-*g*]indole (homolycorine type) skeletons, and both types originate from an *ortho–para′* phenol oxidative coupling (Fig. 4).

The biological conversion of cinnamic acid *via* hydroxylated cinnamic acids into the $C_6$–$C_1$ unit of norpluviine (**12**) has been used in a study of hydroxylation mechanisms in higher plants (*187*). When [3-$^3$H, β-$^{14}$C] cinnamic acid was fed to *Narcissus* 'Texas' a tritium retention in norpluviine (**12**) of 28% was observed, which is very close to the predicted value resulting from *para*-hydroxylation with hydrogen migration and retention.

In the conversion of *O*-methylnorbelladine (**87**) into lycorine (**1**), the labeling position [3-$^3$H] on the aromatic ring of L-tyr afterward appears at C-2 of norpluviine (**12**), which is formed as an intermediate, the configuration of the tritium apparently being β (*176*). This tritium is retained in subsequently formed lycorine (**1**), which means that hydroxylation at C-2 proceeds with an inversion of configuration (*188*) by a mechanism involving an epoxide, with ring opening followed by allylic rearrangement of the resulting alcohol (Fig. 5). Supporting evidence comes from the incorporation of [2β-$^3$H]caranine (**10**) into lycorine (**1**) in *Zephyranthes candida* (*189*). However, hydroxylation of caranine (**10**) in *Clivia miniata* occuring with retention of configuration was also observed (*190*). Further, [2α-$^3$H,11-$^{14}$C]caranine (**10**) was incorporated into lycorine (**1**) with high retention of tritium at C-2, indicating that no 2-oxo-compound can be implicated as an intermediate.

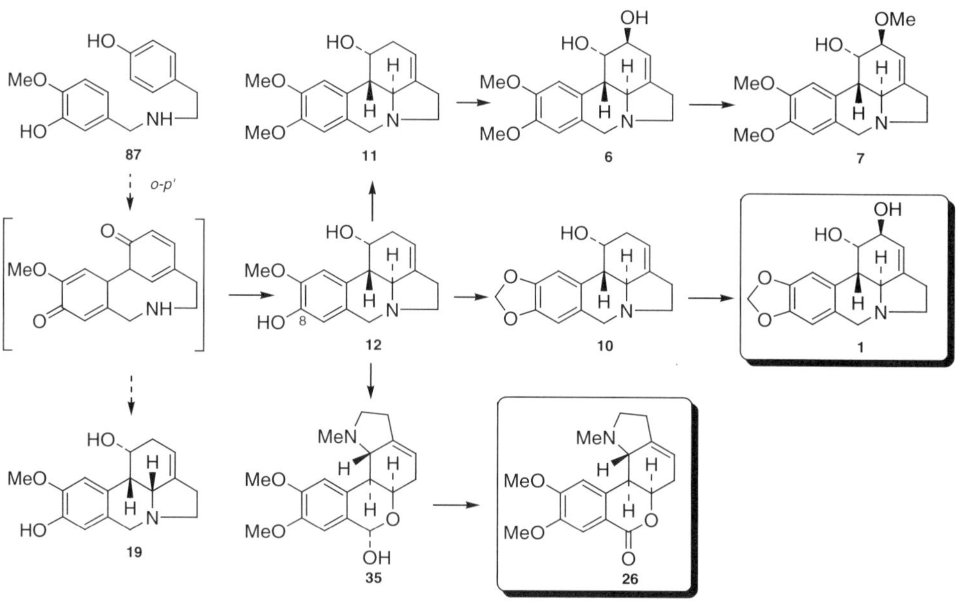


Figure 4. Alkaloids proceeding from an *ortho–para′* coupling.

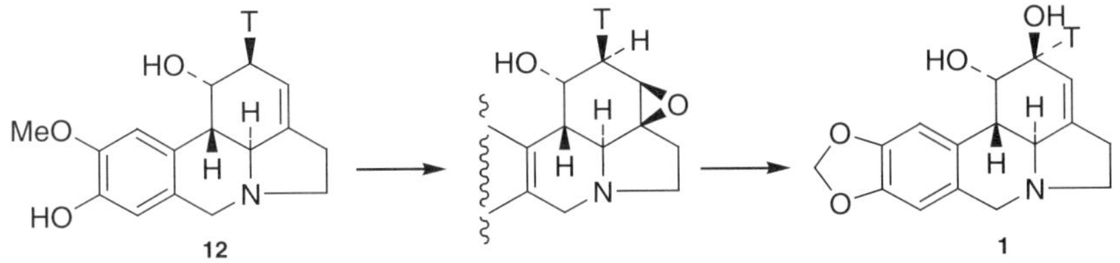


Figure 5. Biosynthesis of lycorine (**1**) with inversion of the configuration.

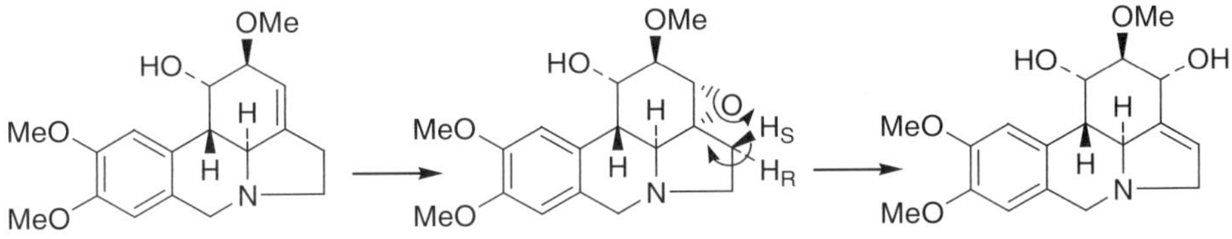


Figure 6. Conversion of galanthine (**7**) to narcissidine (**16**) *via* epoxide (**94**).

The conversion of the *O*-methoxyphenol to the methylenedioxy group may occur late in the biosynthetic pathway. Tritiated norpluviine (**12**) is converted to tritiated lycorine (**1**) by *Narcissus* 'Deanna Durbin', which demonstrates the previously mentioned conversion and also indicates that the C-2 hydroxyl group of lycorine (**1**) is derived by allylic oxidation of either norpluviine (**12**) or caranine (**10**) (*191*).

Regarding the conversion of [2$\beta$-$^3$H,8-OMe-$^{14}$C]pluviine (**11**) into galanthine (**7**), in *Narcissus* 'King Alfred', the retention of 79% of the tritium label confirms that hydroxylation of C-2 may occur with inversion of configuration (*92*).

It was considered (*192*) that another analogous epoxide **94** could give narcissidine (**16**) in the way shown by loss of the *pro*-S hydrogen from C-11, galanthine (**7**) being a suitable substrate for epoxidation. Labeled [$\alpha$-$^{14}$C, $\beta$-$^3$H]-*O*-methylnorbelladine (**87**), when fed to *Narcissus* 'Sempre Avanti', afforded galanthine (**7**) (98% tritium retention) and narcissidine (**16**) (46% tritium retention). Loss of hydrogen from C-11 of galanthine (**7**) was therefore stereospecific. In the 1990s, Kihara *et al.* (*193*) isolated a new alkaloid, incartine (**94**), from the flowers of *Lycoris incarnata*, which could be considered as the biosynthetic intermediate of this pathway (Fig. 6).

The biological conversion of protocatechuic aldehyde into lycorenine (**35**), which proceeds *via* *O*-methylnorbelladine (**87**) and norpluviine (**12**), first involves a reduction of the aldehyde carbonyl, and afterward, in the generation of lycorenine (**35**), oxidation of this same carbon atom. The absolute stereochemistry of these processes has been elucidated in subsequent experiments (*194*), and the results show that hydrogen addition and removal take place on the *re*-face of the molecules concerned (*195*), the initially introduced hydrogen being the one later removed (*196*). It is noteworthy that norpluviine (**12**), unlike pluviine (**11**), is converted in *Narcissus* 'King Alfred' primarily to alkaloids of the homolycorine type. Benzylic

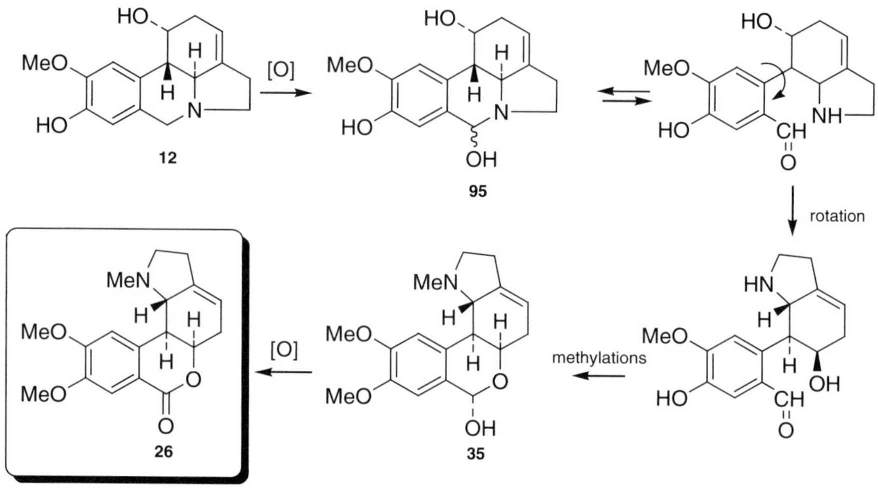


Figure 7. Conversion of norpluviine (**12**) to homolycorine-type alkaloids.

oxidation of position 6 would afford **95**, followed by ring opening to form an amino aldehyde. Hemiacetal formation and methylation could provide lycorenine (**35**) (*92*), and, on subsequent oxidation, could afford homolycorine (**26**), as shown in Fig. 7.

*2. Crinine, Hemanthamine, Tazettine, Narciclasine, and Montanine Types*

This group includes the alkaloids derived from 5,10b-ethanophenanthridine (crinine and hemanthamine types), 2-benzopyrano[3,4-*c*]indole (tazettine type), phenanthridine (narciclasine type), and 5,11-methanomorphanthridine (montanine type) skeletons, originating from a *para–para′* phenol oxidative coupling (Fig. 8).

Results of experiments with labeled crinine (**96**), and less conclusively with oxovittatine, indicate that the two, naturally occurring enantiomeric series, represented in Fig. 8 by crinine (**96**) and vittatine (**42**), are not interchangeable in *Nerine bowdenii* (*197*).

Incorporation of *O*-methylnorbelladine (**87**), labeled in the methoxy carbon and also in positions [3,5-$^3$H], into the alkaloid hemanthamine (**53**) was without loss of tritium, half of which was sited at C-2 of **53**. Consideration of the possible mechanisms involved in relation to tritium retention led to the suggestion that the tritium which is expected at C-4 of **53** might not be stereospecific (*198*). The conversion of *O*-methylnorbelladine (**87**) into hemanthamine (**53**) involves loss of the *pro*-R hydrogen from the C-$\beta$ of the tyramine moiety as well as the further entry of a hydroxyl group at this site (*199*). The subsequent benzylic oxidation results in a **55/56** epimeric mixture that even HPLC cannot separate. The epimeric forms were proposed to be interchangeable through **97**. The biosynthetic conversion of the 5,10b-ethanophenanthridine alkaloids to the 2-benzopyrano[3,4-*c*]indole was demonstrated by feeding tritium-labeled alkaloids to *Sprekelia formosissima*. It was shown that this plant converts hemanthamine (**53**) to hemanthidine/epihemanthamine (**55/56**), and subsequently to pretazettine (**64**), in an essentially irreversible

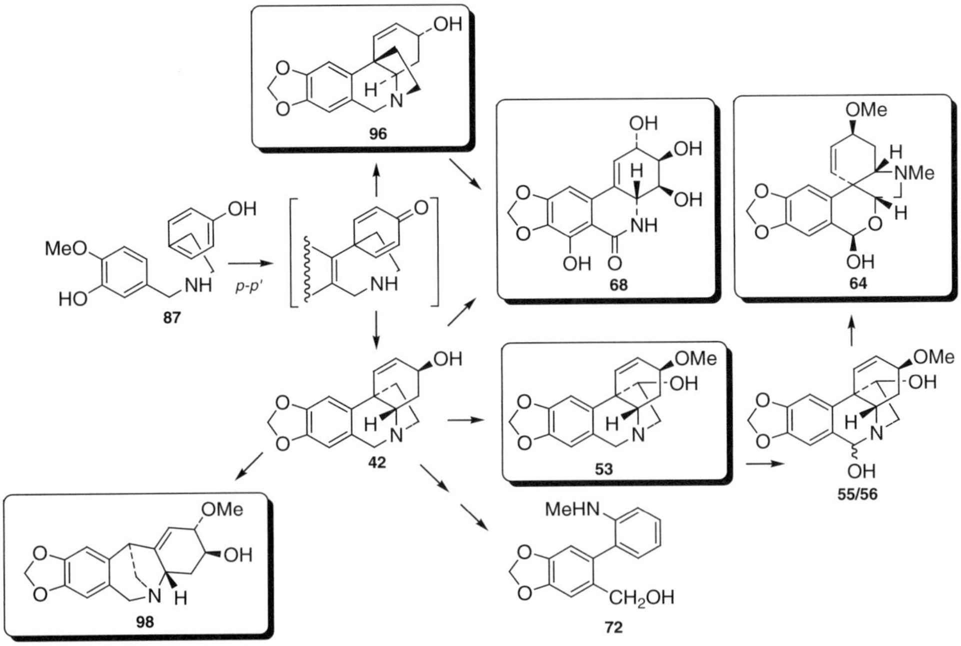


Figure 8. Alkaloids proceeding from a *para–para'* coupling.

manner (*200*). This transformation was considered to proceed through **97** or the related alkoxide anion, although intermediate **97** have never been detected by spectral methods (*201*) (Fig. 9).

It has also been proved that formation of the alkaloid narciclasine (**68**) proceeds from the pathway for crinine- and hemanthamine-type alkaloids, and not through norpluviine (**12**) and lycorine (**1**) derivatives. In fact, in view of its structural affinity to both hemanthamine (**53**) and lycorine (**1**), narciclasine (**68**) could be derived by either of the pathways. When *O*-methylnorbelladine (**87**), labeled in the methoxy carbon and in both protons of positions 3 and 5 of the tyramine aromatic ring, was administered to *Narcissus* plants, all four alkaloids incorporated activity. The isotopic ratio [$^3$H:$^{14}$C] for norpluviine (**12**) and lycorine (**1**) was, as expected, 50% that of the precursor, because of its *ortho–para'* coupling. On the contrary, in hemanthamine (**53**) the ratio was unchanged. These results show clearly that the methoxy group of **87** is completely retained in the alkaloids mentioned, providing a satisfactory internal standard and showing that the degree of tritium retention is a reliable guide to the direction of phenol coupling. Narciclasine (**68**) showed an isotopic ratio (75%) higher than that of lycorine (**1**) or norpluviine (**12**), though lower than that of hemanthamine (**53**). However, the fact that more than 50% of tritium is retained suggests that *O*-methylnorbelladine (**87**) is incorporated into narciclasine (**68**) *via para–para'* phenol oxidative coupling.

*O*-methylnorbelladine (**87**) and vittatine (**42**) are implicated as intermediates in the biosynthesis of narciclasine (**68**) (*202–204*), and the loss of the ethane bridge

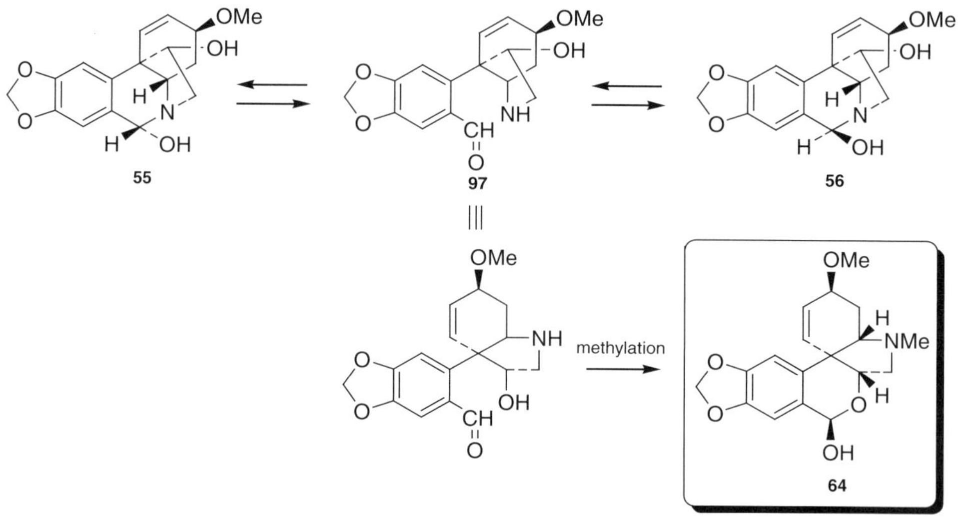

Figure 9. Biosynthesis of pretazettine (**64**).

from the latter could occur by a retro-Prins reaction on 11-hydroxyvittatine (**43**). Strong support for this pathway was obtained by labeling studies. 11-Hydroxyvittatine (**43**) has also been proposed as an intermediate in the biosynthesis of hemanthamine (**53**) and montanine (**98**) (a 5,11-methanomorphanthridine alkaloid) following the observed specific incorporation of vittatine (**42**) into the two alkaloids in *Rhodophiala bifida* (*197*) (Fig. 10).

Fuganti and Mazza (*203,204*) concluded that in the late stages of narciclasine (**68**) biosynthesis, the two-carbon bridge is lost from the oxocrinine skeleton, passing through intermediates bearing a pseudoaxial hydroxy-group at the C-3 position and further hydrogen removal from this position does not occur. Noroxomaritidine was also implicated in the biosynthesis of narciclasine (**68**) and further experiments (*205*) showed that it is also a precursor for ismine (**72**).

The alkaloid ismine (**72**) has also been shown (*206*) to be a transformation product of the crinine-hemanthamine series. The precursor, oxocrinine labeled with tritium in the positions 2 and 4, was administered to *Sprekelia formosissima* plants and the radioactive ismine (**72**) isolated was shown to be specifically labeled at the expected positions.

### *3. Galanthamine Type*

These alkaloids have a dibenzofuran nucleus (galanthamine type) and are obtained from a *para–ortho′* phenol oxidative coupling.

The initial studies of this pathway suggested that the *para–ortho′* coupling does not proceed from *O*-methylnorbelladine (**87**), but from *N,O*-dimethylnorbelladine (**99**) to finally give galanthamine (**75**) (*207*). *O,N*-dimethylnorbella-dine (**99**) was

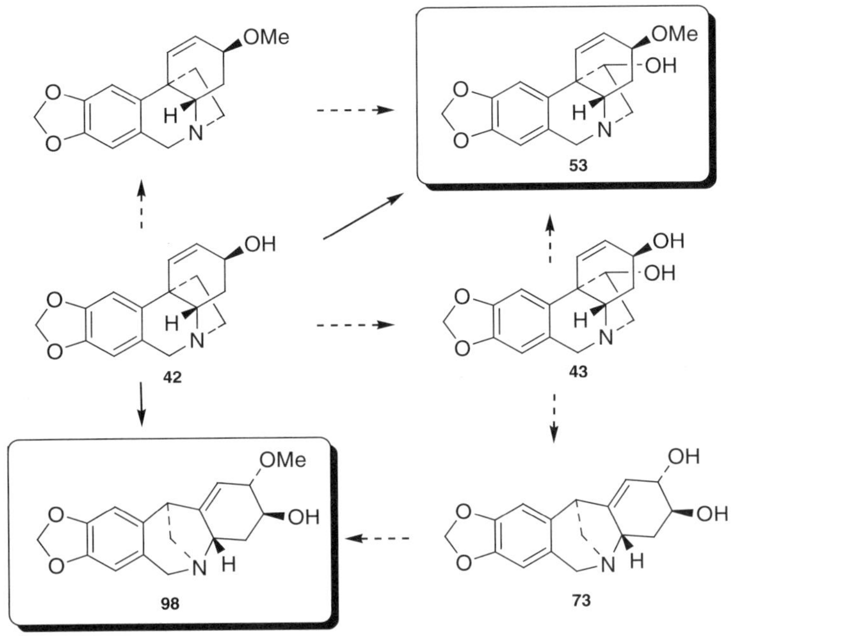


Figure 10. Proposed biosynthetic pathways to hemanthamine (**53**) and montanine (**98**).

first isolated from *Pancratium maritimum* (*208*), a species that also contains galanthamine (**75**).

However, the most recent study seems to contradict the evidence set forth above. Experiments carried out with application of $^{13}$C-labeled *O*-methylnorbelladine (**87**) to organs of field-grown *L. aestivum* have shown that the biosynthesis of galanthamine (**75**) involves the phenol oxidative coupling of *O*-methylnorbelladine (**87**) to a postulated dienone, which undergoes spontaneous closure of the ether bridge to yield *N*-demethylnarwedine (**100**), giving norgalanthamine (**78**) after stereoselective reduction. Furthermore, it was shown that norgalanthamine (**78**) is *N*-methylated to galanthamine (**75**) in the final step of biosynthesis (*181*) (Fig. 11). In contrast to the literature, *N,O*-dimethylnorbelladine (**99**) was metabolized to a lesser extent in *L. aestivum* and incorporated into galanthamine (**75**) as well as norgalanthamine (**78**) at about one-third of the rate of *O*-methylnorbelladine (**87**).

According to Eichhorn *et al.* (*181*), narwedine (**83**) is not the direct precursor of galanthamine (**75**), and could possibly exist in equilibrium with galanthamine (**75**), a reaction catalyzed by a hypothetically reversible oxido-reductase.

Chlidanthine (**101**), by analogy with the known conversion of codeine to morphine, might be expected to arise from galanthamine (**75**) by *O*-demethylation. This was shown to be true when both galanthamine (**75**) and narwedine (**83**), with tritium labels, were incorporated into chlidanthine (**101**) (*209*).

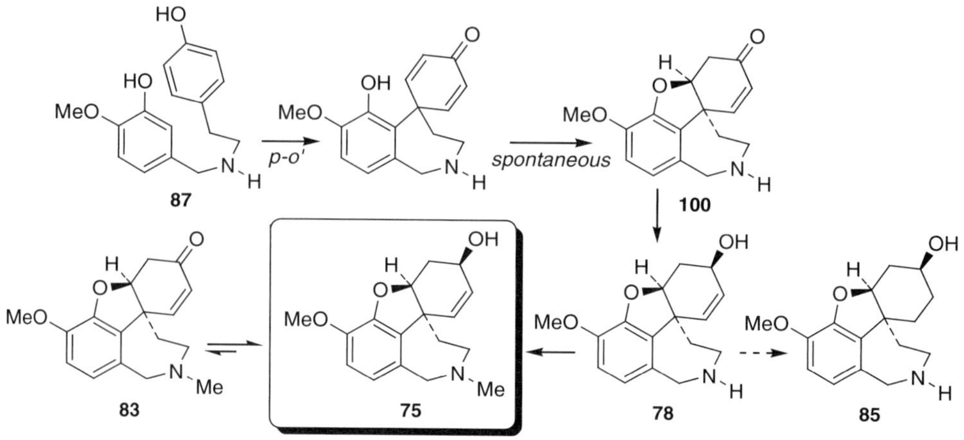


Figure 11. Biosynthesis of galanthamine (**75**) and derivatives.

## IV. Synthetic Studies

The synthesis of *Narcissus* alkaloids has been the subject of intense efforts over the past few years. An exhaustive report is beyond the scope of this chapter, and the reader is referred to recent and excellent reviews in the field for a more thorough account (*34,60–66,210,211*). Several approaches to the total synthesis of these alkaloids have demonstrated their efficiency. We describe herein some representative examples, classified according to mechanistic criteria.

### A. Oxidative Couplings

Biomimetic oxidative phenolic coupling still stands as a practical method for the preparation of these alkaloids. New syntheses of (−)-narwedine (**83**) and (−)-galanthamine (**75**) have been disclosed, the former having been prepared in its enantiopure form from the corresponding racemate through a crystallization-induced chiral transformation (*212*). Derivatives of 1-methylgalanthamine (*213*), as well as galanthamine analogues having the N atom at altered positions in the azepine ring (*214*), have also been reported. A significant improvement to the oxidation protocol is the use of PIFA as the oxidant on a symmetrical substrate, allowing efficient synthesis of racemic narwedine and galanthamine (*215*) (Scheme 1).

The same group later reported an efficient route to (−)-galanthamine (**75**) by a variation of the method whereby remote asymmetric induction was used (*216*) as well as the total syntheses of siculine, oxocrinine, epicrinine, and buflavine (*217*).

A solid-phase version of this biomimetic method has been developed in the context of diversity-oriented synthesis to generate a library of galanthamine-like molecules, which allowed the identification of compound **A**, a potent new blocker of protein trafficking from the Golgi apparatus to the plasma membrane. Interestingly, galanthamine (**75**) does not exhibit this biological activity (*218*) (Scheme 2).

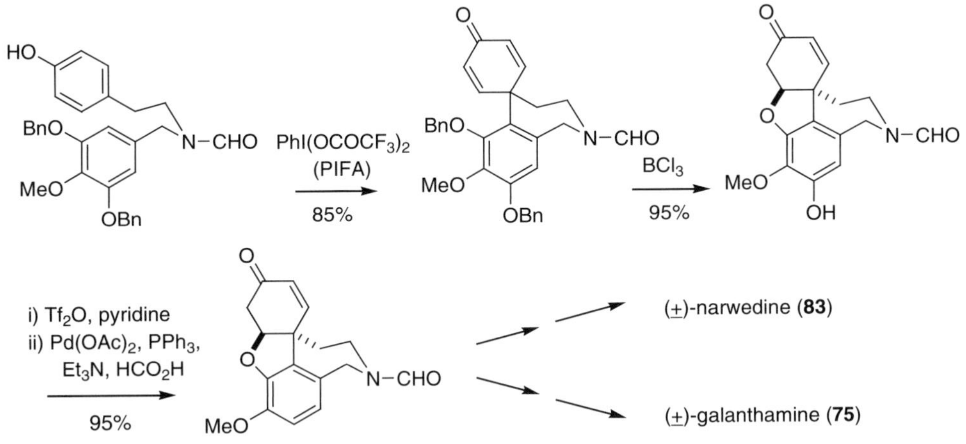


Scheme 1. Synthesis of racemic narwedine (**83**) and galanthamine (**75**).

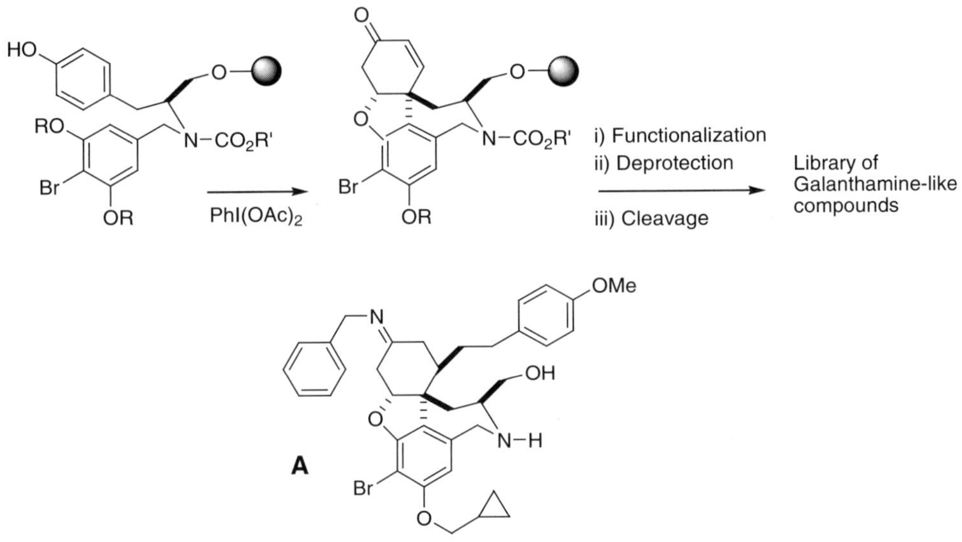


Scheme 2. Diversity-oriented synthesis of galanthamine-like compounds.

A new oxidative rearrangement has led to the synthesis of (±)-lycoramine (*219*), and a related Lewis-acid-promoted transformation has allowed the preparation of mesembrine (**92**) and crinane in racemic form (*220*).

An impressive demonstration of the use of solid-supported reagents and scavengers in the total synthesis of natural products was provided by the group of Ley (for previous work, see ref. *221*). As an application of this new methodology (intended to expedite synthetic routes, as the only work up required in these protocols involves filtration, followed by evaporation of the solvent), (+)-plicamine was synthesized without the need for chromatography. The sequence starts

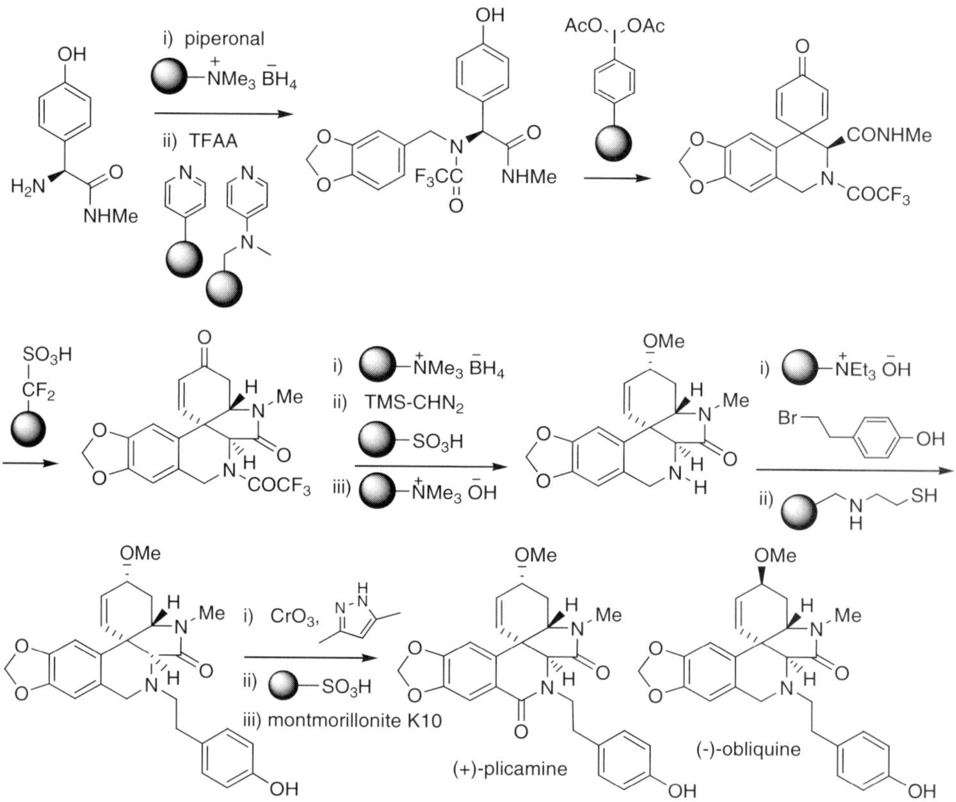

Scheme 3. Synthesis of plicamine and obliquine using solid-supported reagents.

with *p*-hydroxyphenylglycine which, upon functionalization, undergoes a reductive amination with piperonal. The resulting product is subjected to biomimetic phenol oxidative coupling with solid-supported iodonium diacetate. An acid-promoted conjugate addition closed the lactam ring, and subsequent functional group transformations led to the desired product. Every single step used solid-supported reagents or solid-supported scavengers (Scheme 3) (*222,223*). The same group later disclosed the syntheses of (−)-obliquine and (+)-plicane (*224*).

## B. ORGANOMETALLIC-PROMOTED REACTIONS

Transition metals, especially palladium, have been used for the stereoselective synthesis of these alkaloids with remarkable success. Since Overman's approach to (±)-tazettine (**62**) using an intramolecular Heck reaction (*225*), this methodology for the construction of the critical quaternary stereogenic center has been implemented for the preparation of many alkaloids. Grigg later reported access to (*R,R*)-crinane *via* regiospecific palladium-mediated cyclization (*226*). An efficient total synthesis of racemic galanthamine (**75**) following this approach has been recently disclosed (*227*). After the key intramolecular Heck reaction, deprotection of the

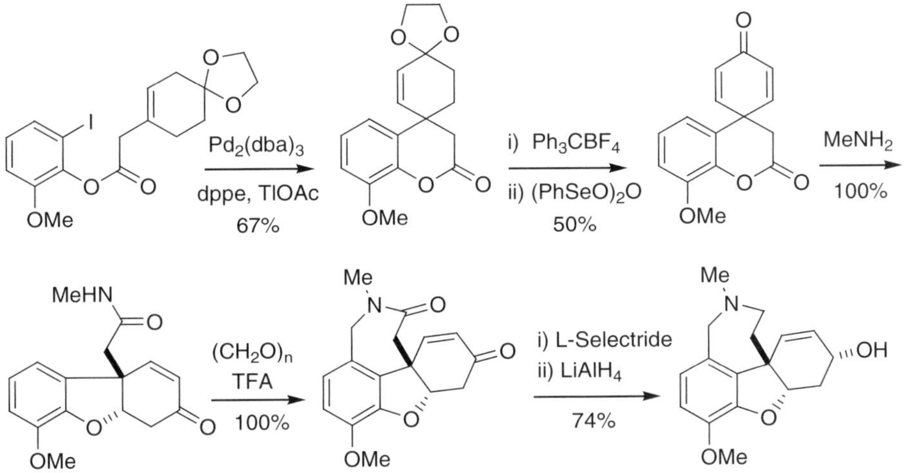


(±)-galanthamine (**75**)

Scheme 4. Palladium-mediated synthesis of galanthamine (**75**).

acetal moiety, followed by oxidation and aminolysis of the lactone, promoted the spontaneous conjugate addition of the phenolic hydroxyl, to afford the tricyclic intermediate. A Pictet–Spengler reaction was then used to close the azepine ring. Subsequent functional group transformations allowed the completion of the total synthesis (Scheme 4).

Related strategies were used to prepare racemic maritidine (**44**) (*228*) and lycoramine (**84**) (*229,230*). An elegant and unified approach to synthesize (−)-pancracine (**73**), (−)-montanine (**98**), (−)-coccinine, and (−)-brunsvigine used an intramolecular Heck reaction for the ring closure of the central azepine ring, connecting the bromoarene and terminal alkene moieties. The key precursor was prepared *via* stereospecific intramolecular allenylsilane addition to an imine (*231*). Enders recently reported an asymmetric synthesis of the 1-*epi* aglycon of cripowellins A and B, based on an intramolecular Heck reaction of a functionalized bromoaryl-azacyclononene, which in turn was prepared *via* ring closing metathesis of an open-chain precursor (*232*). Buflavine was prepared in a very short route by assembling a biphenyl intermediate through a Suzuki coupling (*233*).

Recently, palladium-catalyzed asymmetric allylic substitution of an activated cyclohexenol derivative has allowed two enantioselective syntheses of (−)-galanthamine (**75**) (*234,235*). Both approaches rely on the enantioselective preparation of the same tricyclic intermediate, which is subsequently converted to the alkaloid *via* stereocontrolled transformations; the most efficient of which comprised stereoselective allylic oxidation of the cyclohexene moiety (Scheme 5). The same methodology allowed the synthesis of (−)-codeine and (−)-morphine (*236*). The same group had earlier reported the synthesis of (+)-pancratistatin following a related strategy (*237*). Use of a tosylamide as the nucleophile in the displacement of an activated aryl-cyclohexenol derivative enabled the preparation of a chiral intermediate which

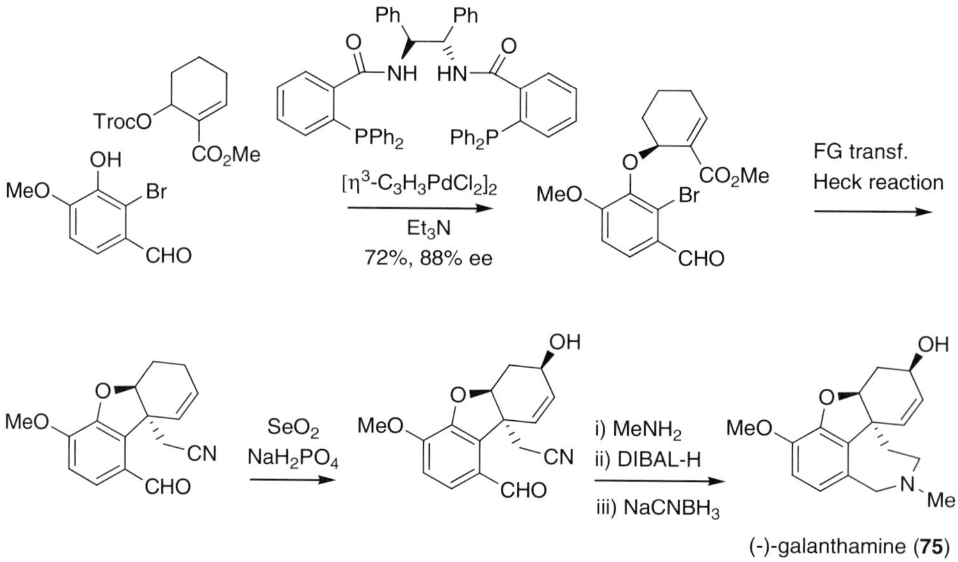


Scheme 5. Enantioselective synthesis of (−)-galanthamine (**75**).

was sequentially converted into (+)-crinamine (**57**), (−)-hemanthidine (**55**), and (+)-pretazettine (**64**) (*238*).

Regioselective C–H activation is often used for intramolecular biaryl preparations in Pd-mediated transformations. Thus concise syntheses of a large collection of alkaloids have been reported using variations of this approach (*239–241*), which differ mainly in terms of the functionalization of the precursor, palladium catalysts, and ligands. An efficient synthesis of racemic γ-lycorane was recently disclosed whereby the key step is the intramolecular α-arylation of a cyclohexanone derivative (*242*). Some heterocyclic synthetic precursors (e.g. substituted indoles, quinolines, quinolones, phenanthridines, and phenanthridinones) have also been prepared *en route* to the alkaloids [hippadine, trisphaeridine (**70**), and crinasiadine] *via* Pd- and Cu-mediated reactions (*243,244*).

The use of main group metals has also facilitated alkaloid synthesis, exemplified in the preparation of (−)-brunsvigine by intramolecular anionic cyclization (the addition of an organolithium derivative to an amide) (*245*). Also, a Meyers biaryl coupling has been used as the key step in a total synthesis of buflavine (*246*). The connection of directed *ortho* metalations with cross-coupling reactions is the basis of very elegant syntheses of buflavine, 8-*O*-demethylbuflavine, pratosine, and hippadine (*247,248*). Synthesis of (−)-mesembrine has been achieved through a route involving stereoselective ring opening of an aryl-substituted epoxide with a Grignard reagent (*249*).

## C. ENZYMATIC DIHYDROXYLATION

The development of enzymatic dihydroxylation of aromatics has enabled synthetic access to a large collection of cyclohexadiene diols in enantiopure form.

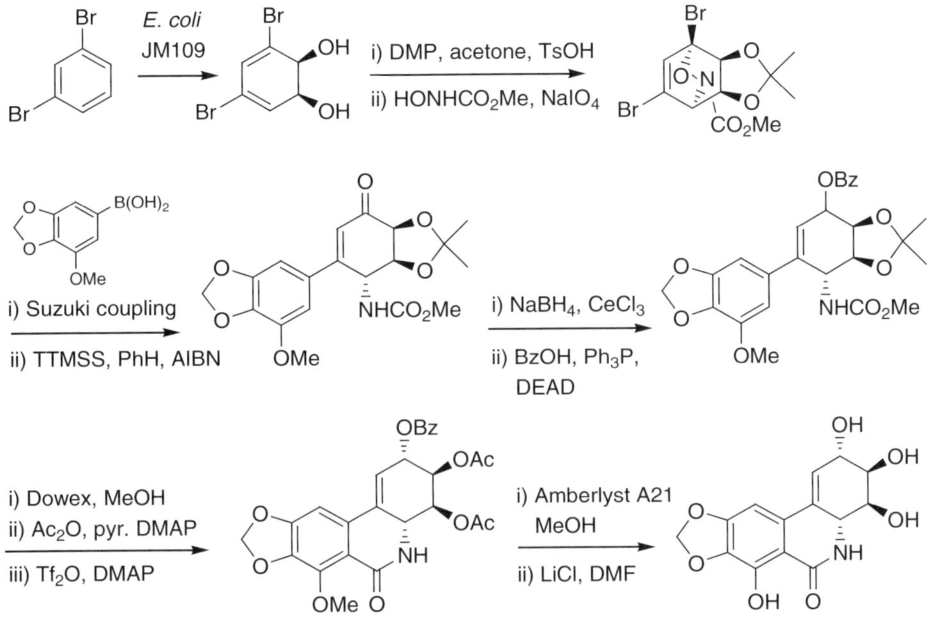


narciclasine (**68**)

Scheme 6. Synthesis of narciclasine (**68**).

These compounds have become invaluable synthetic materials and are pivotal intermediates in the total syntheses of structurally diverse natural products (*250*). In particular, the group of Hudlicky which developed the methodology has devised short, efficient, and enantioselective total syntheses of alkaloids and related compounds to meet quantity demands for materials used in biological studies. Lycoricidine was prepared in only nine steps from bromobenzene (*251*), and the same starting material was used in straightforward syntheses of pancratistatin and its 7-deoxy derivative (*252*). Recently, the Hudlicky group published the synthesis of narciclasine (**68**), based on the whole-cell fermentation of 1,3-dibromobenzene by a recombinant *Escherichia coli* (Scheme 6). The resulting dienediol was subjected to a hetero-Diels–Alder reaction with a nitroso derivative and the corresponding adduct underwent Suzuki coupling followed by reductive opening of the oxazine N–O bond. Stereoselective reduction of the resulting ketone, exchange of the acetonide moiety with acetate groups, a Bischler–Napieralski-type ring closure and the removal of the ester and phenolic methyl protecting groups led to the desired product (*253*).

The same group also disclosed the synthesis of *epi*-7-deoxypancratistatin *via* an aza-Payne rearrangement (*254*) (Scheme 7). Analogues of narciclasine (**68**), pancrastistatin, and 7-deoxypancratistatin have been synthesized using modifications of the reported procedures as well as new methodologies (e.g. addition of indoles to oxiranes and aziridines derived from cyclohexadiene diols) (*255–258*).

Pancratistatin analogues having a carbohydrate-derived structure in place of the polyhydroxycyclohexane moiety have been prepared for studying the minimum pharmacophore of the parent molecule (*259*).

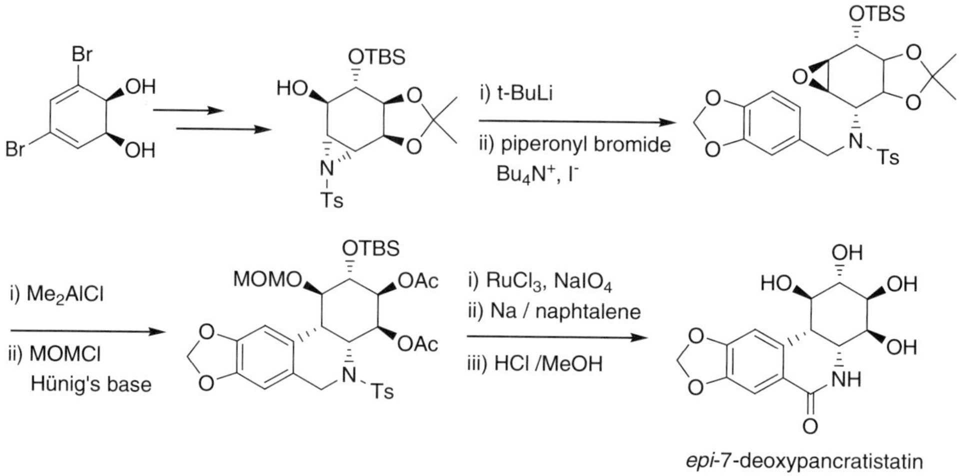


Scheme 7. Synthesis of epi-7-deoxypancratistatin.

## D. RADICAL REACTIONS

The well-established radical reactions have been used to drive selective connectivity, often without the need of protecting groups normally required in polar approaches. Also the 'living' nature of some radical additions has triggered elegant and efficient reaction cascades for the formation of multiple C–C bonds.

The group of Magnus has continued to exploit the $\beta$-azidonation reaction of silyl enol ethers, which they used in the synthesis of pancratistatin (*260*), and has applied the same type of approach for the synthesis of a lycorane derivative. The pyrrolidine ring was closed *via* a cobalt-promoted intramolecular radical addition-dehydrogenation reaction (*261*). Parsons has reported model studies on the cyclization of diversely substituted 2-amino-cyclohex-2-enone derivatives to yield octahydroindole derivatives related to the Amaryllidaceae alkaloids. In these processes, radical cyclizations were promoted with tributyltin hydride, samarium (II) iodide, or manganese (III) acetate (*262*). Curran has reported a radical cyclization leading to spirocyclohexadienones, which includes the formal synthesis of aza-galanthamine (*263*). Keck reported an impressive radical cascade process as the crucial step in his total synthesis of 7-deoxypancratistatin (*264*). A modified approach led to the synthesis of (+)-lycoricidine, *via* 6-*exo*-cyclizations of substituted alkenyl radicals with oxime ethers (*265*) (Scheme 8). The precursor was prepared from D-gulonolactone, which was protected, selectively transformed, and coupled through a Sonogashira reaction with a substituted iodoarene to yield the key alkyne intermediate. PhSH was added under photochemical activation to generate a vinyl radical, which was then added to the oxime ether moiety to close the six-membered ring with high stereoselectivity. Treatment of the resulting compound with $SmI_2$ allowed the cleavage of the N–O bond, the removal of the thioether, and the formation of the lactam. The final step was the deprotection of the acetonide group. *Ent*-lycoricidine, and (+)-narciclasine (**68**) have been similarly prepared.

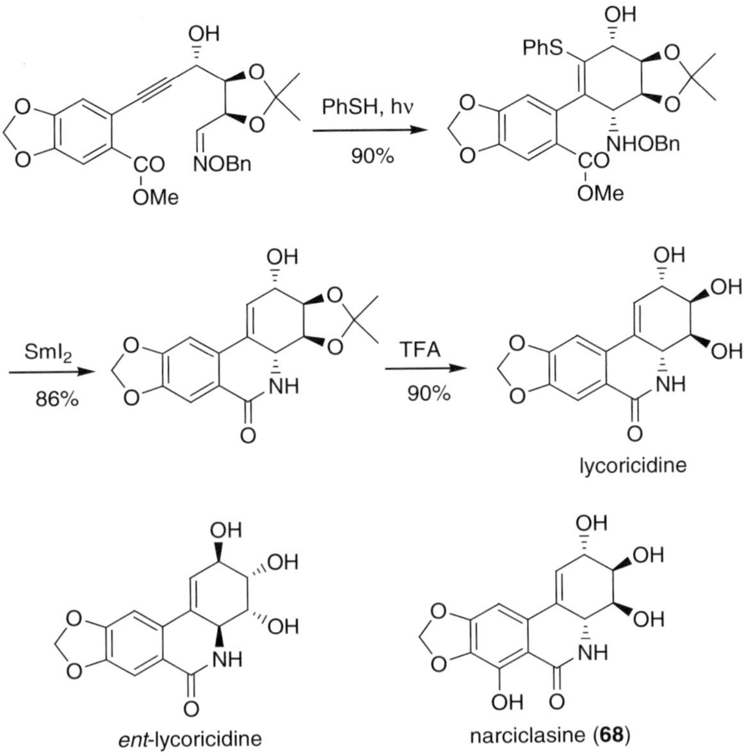


Scheme 8. Radical synthesis of lycoricidine.

The formal synthesis of racemic pancracine (**73**) *via* radical cyclization of a *N*-(2-cyclohexenyl)-α-phenylthioacetamide has been reported (*266*). Racemic montanine (**98**), coccinine, and pancracine (**73**) have been prepared using a phenylthiotetrahydroisoquinoline derivative as the radical precursor (*267*). Recently, Zard has described the construction of the carbon skeleton of kirkine (**19**). In these model studies, thiosemicarbazide proradicals interact with lauroyl peroxide (a tin-free initiator) to trigger a reaction cascade that ultimately yields the desired tetracyclic system of kirkine (**19**) (*268*).

An interesting $Csp^2$–$Csp^2$ coupling *via* an aryl radical cyclization has been used as the key step in a unified strategy to synthesize vasconine (**22**), assoanine (**20**), oxoassoanine (**21**), and pratosine. Tin hydride cyclization of an *N*-(*o*-bromobenzyl)-aniline yielded a substituted phenanthridine which was subsequently transformed into vasconine (**22**), from which the remaining natural products were prepared (*269*) (Scheme 9).

## E. PERICYCLIC PROCESSES

The aza-Cope–Mannich cyclization protocol, developed by Overman has been strategically applied in the synthesis of racemic and enantiopure pancracine (**73**) (*270*). Later, Kim developed a synthetic approach based on a [3,3]-sigmatropic

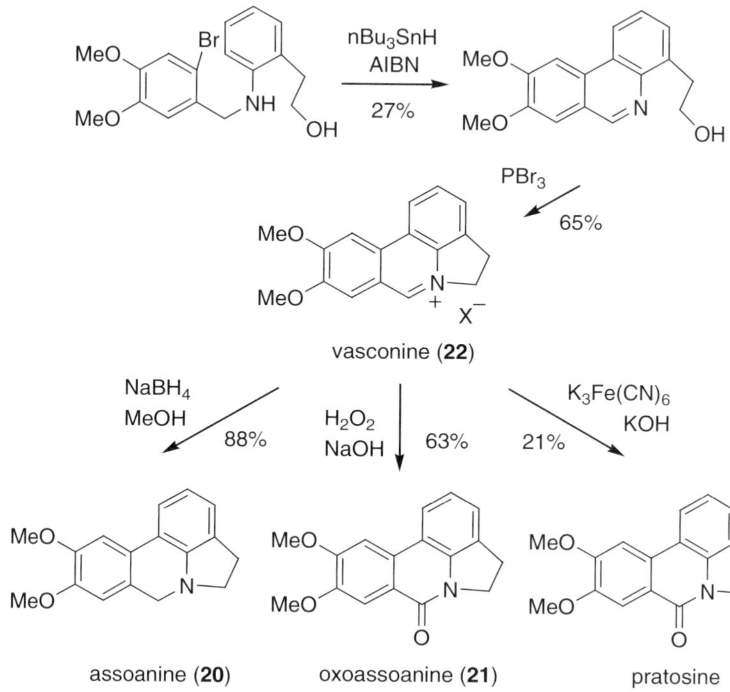


Scheme 9. Radical synthesis of vasconine (**22**), assoanine (**20**), oxoassoanine (**21**), and pratosine.

rearrangement as the key step in the stereoselective synthesis of racemic pancratistatin (*271*). Horner–Wasdsworth–Emmons reaction provided a reactive olefin that upon heating afforded the rearranged (aryl)cyclohexenylcarbaldehyde which, in turn, was further oxidized to the corresponding acid and subjected to a halolactonization and elimination protocol to yield the unsaturated lactone. This intermediate was converted into the hydroxyester, and subsequent functional-group transformations allowed the stereocontrolled installation of the remaining stereogenic centers and the final ring closure. The sequence ended with deprotection of the esters and methyl phenolic groups (Scheme 10).

Rigby's studies on the synthesis of alkenylisocyanates fostered the preparation of a suitable substituted aryl enamide, which on photocyclization yielded the polysubstituted pentacyclic system. Key to the success of this process is the hydrogen bond between the phenolic OH and the carbonyl group, which restricts the rotation around the aryl–amide bond and directs the cyclization. Further functionalization allowed the total synthesis of pancratistatin (*272*) and narciclasine (**68**) (*273*) (Scheme 11). The [4+1] cycloaddition of bis(alkylthio)carbenes with vinyl isocyanates was the key process in a recent synthesis of (±)-mesembrine (**92**) (*274*).

The Diels–Alder cycloaddition has played an important role in the synthesis of *Narcissus* alkaloids and related structures. Boger has prepared anhydrolycorinone using an intramolecular hetero-Diels–Alder reaction in which a 1,3,4-oxadiazole

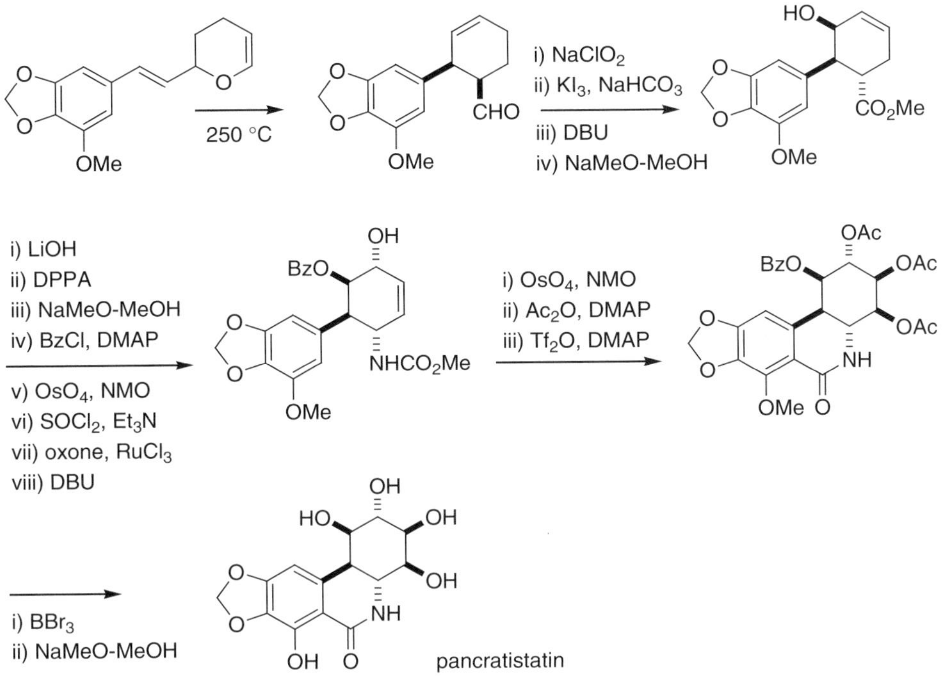


Scheme 10. Synthesis of racemic pancratistatin.

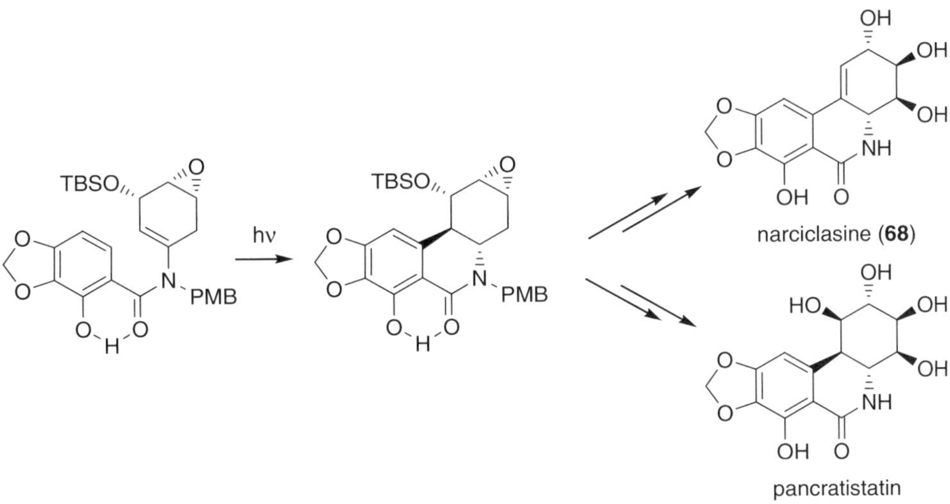


Scheme 11. Synthesis of narciclasine (**68**) and pancratistain.

unit interacted with an electron-rich double bond to yield an intermediate furan adduct that underwent another intramolecular [4 + 2] cycloaddition with a second olefin moiety (*275*). Padwa has reported a conceptually attractive entry to this family of alkaloids based on intramolecular cycloadditions on substituted furanyl

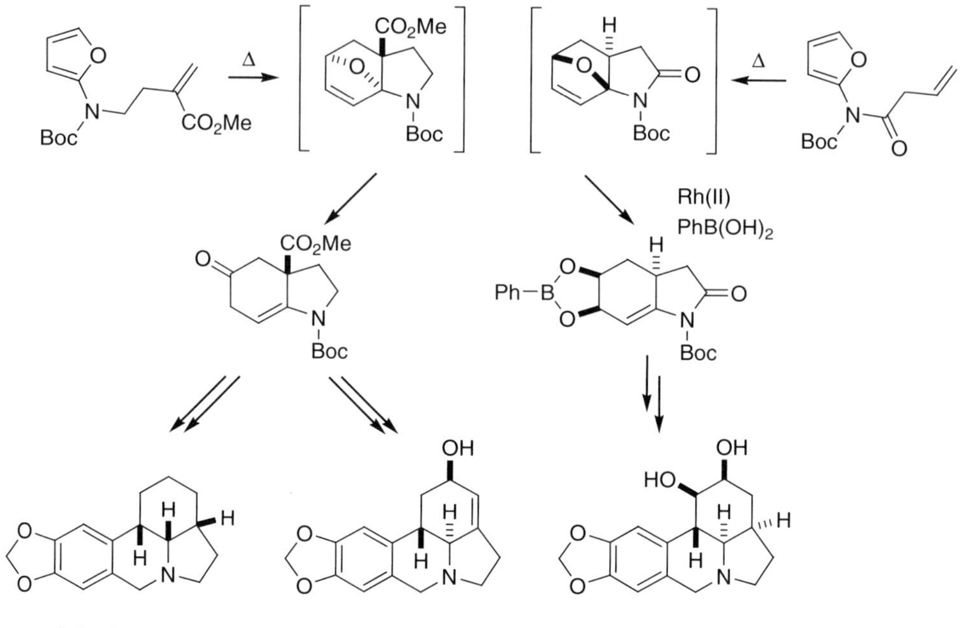


Scheme 12. Synthesis of γ-lycorane, 1-deoxylycorine, and epi-zephyranthine.

carbamates. The resulting adduct can be elaborated into diversely substituted hydroindolinones *via* nitrogen-assisted ring opening (*276*) or Rh(II)-promoted nucleophilic reaction (*277*). The former has provided compounds which on subsequent transformations yielded γ-lycorane and 1-deoxylycorine as racemates, whereas the latter afforded materials for the preparation of *epi*-zephyranthine (Scheme 12).

## F. POLAR REACTIONS

Haseltine has described an enantioselective formal synthesis of pancratistatin in which the stereocontrol is driven by the acetonide of conduritol A. The enantioselective hydrolysis (desymmetrization) of this compound was achieved with a lipase, and the aryl–cyclohexane ring bond was formed through an intramolecular cyclization of the activated benzene ring with an allylic triflate (*278*). Plumet reported a total synthesis of (+)-7-deoxypancratistatin based on the conjugate addition of an aryl-lithium species to a bicyclic conjugated sulfone derived from furan, which enabled the efficient installation of the six stereogenic centers of the cyclohexane ring (*279*) (Scheme 13).

The next series of syntheses is based on conjugate additions. A 2-arylcyclohexanone was regio- and stereoselectively added to nitroethylene to access the octahydroindole core present in the alkaloids. This has enabled the total synthesis of (±)-γ-lycorane and (±)-crinane (*280*). Tomioka described a chemoselective conjugate addition – nitro Michael reaction sequence to prepare α- and β-lycoranes in their racemic form (*281*). The addition of an arylcuprate to a D-mannitol-derived

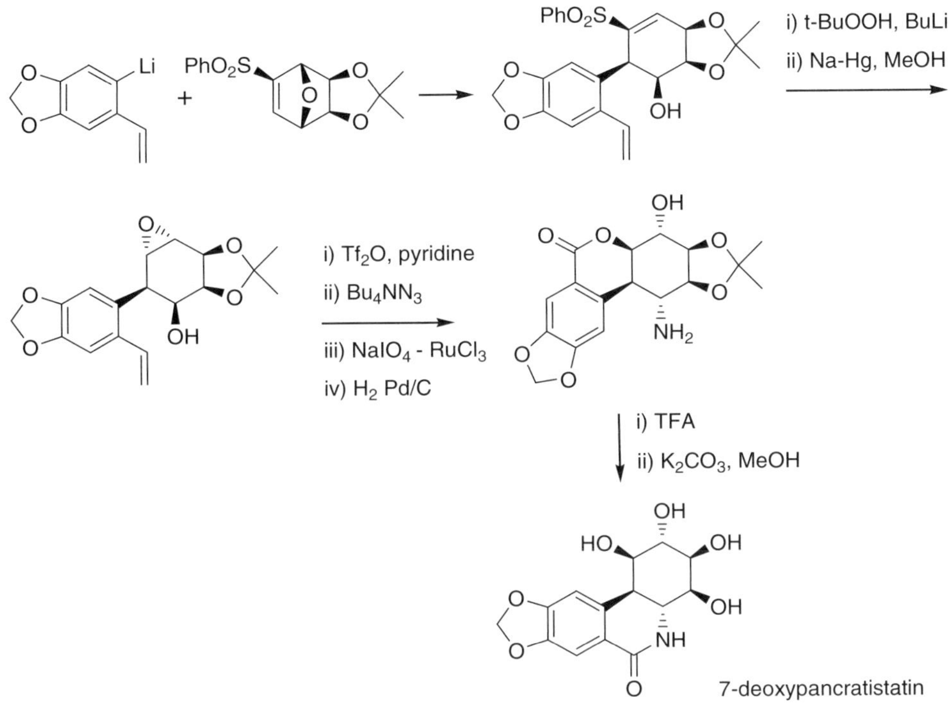


Scheme 13. Synthesis of 7-deoxypancratistatin.

conjugate ester provided access to synthetic precursors of the Amaryllidaceae alkaloids (*282*). Similarly, the addition of arylcuprates to enantiomerically pure conjugate esters derived from D-xylose, allowed, after pertinent FG transformations and ring-closing metathesis, the preparation of novel 1-aryl-1-deoxyconduritols F. These compounds are advanced intermediates *en route* to pancratistatin derivatives (*283*). A 15-step synthesis of racemic mesembrine (**92**) featuring the intramolecular addition of an amine to a cyclohexenone moiety to close the octahydroindolone ring system was recently disclosed (*284*). Taber has described an elegant approach to (−)-mesembrine relying on a conjugate addition and an intramolecular alkylidene C–H insertion (*285*).

Intramolecular additions to *N*-acyliminium ions (generated by Pummerer reaction) were used to prepare highly functionalized tricyclic intermediates for the synthesis of the putative alkaloid jamtine (*286*). Synthesis of cherylline (**88**) in both of its enantiopure forms was achieved using a chiral auxiliary through a sequence involving reductive amination-acid-promoted cyclization (*287*).

Short syntheses of enantiomerically pure narciclasine (**68**) and lycoricidine based on the intramolecular acid-catalyzed arene-epoxide coupling have been described (*288,289*). Bromohydroxylation of a protected aminocyclohexenol afforded the corresponding bromohydrin as a mixture of two *trans* stereoisomers, which was subsequently transformed to link an arylmethyl moiety in basic medium with the concomitant formation of the epoxide ring, thereby setting the stage for the

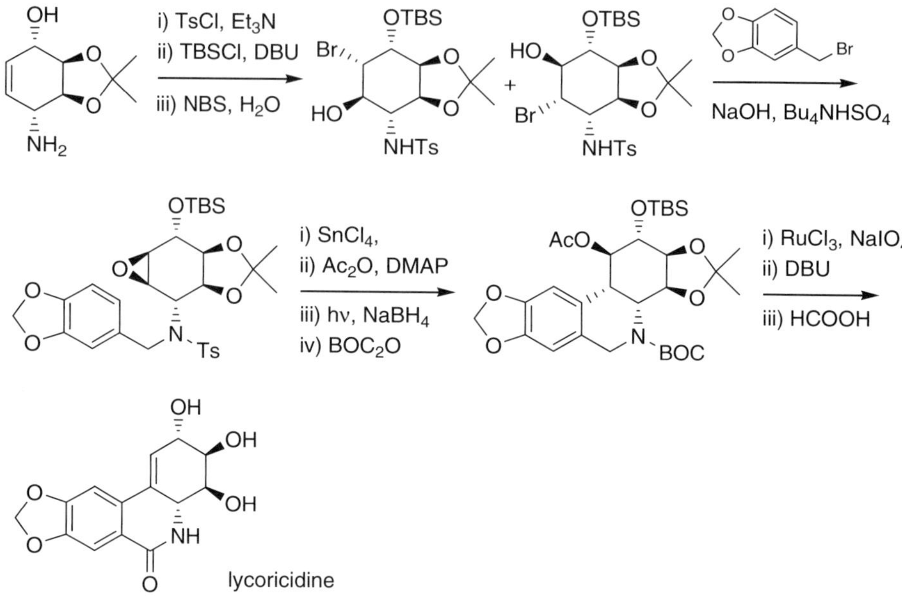


Scheme 14. Total synthesis of lycoricidine.

crucial carbon–carbon bond formation required for cyclization. Subsequent functional group transformations, exchange of protecting groups, oxidation of the benzylic position, and finally deprotection, afforded lycoricidine (Scheme 14). Narciclasine (**68**) has been prepared following a related pathway. The same group has also reported a convenient approach to chiral *O*-isopropylidene-protected 4-aminocyclohexenol (*290*).

## G. SEMISYNTHESIS AND CHIRAL POOL APPROACHES

Taking advantage of the availability of narciclasine (**68**) from plant extracts, Pettit used the compound as a starting material in an efficient synthesis of pancratistatin (*77*). The same group has also described related approaches for the preparation of a pancratistatin phosphate prodrug (*291*), as well as for the natural product 7-deoxy-*trans*-dihydronarciclasine and related derivatives (*292*). In another context, an improved protocol for the synthesis of (−)-galanthamine (**75**), based on the spontaneous resolution of either of the enantiomers of narwedine (**83**), has been reported (*293*).

A short route to enantiomerically pure lactone analogues of narciclasine and lycoricidine uses D-gulonolactone as the chiral pool source, as it contains all of the stereogenic centers of the products in their correct configuration. Connectivity with the functionalized aryl moiety arises from the addition of an organolithium reagent to the carbonyl group of the gulonolactone (*294*). The total syntheses of (−)-hemanthidine (**55**), (+)-pretazettine (**64**), and (+)-tazettine (**62**) were successfully achieved

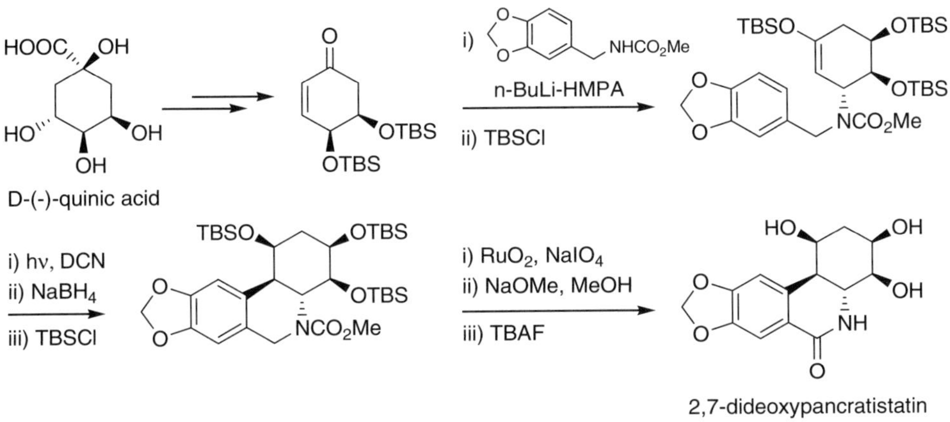


Scheme 15. Synthesis of 2,7-dideoxypancratistatin from D-(−)-quinic acid.

starting from the α-methyl-D-mannopyranoside (*295*). Vittatine (**42**) has been prepared from D-glucose. The sequence involved a Ferrier carbocyclization to yield a cyclohexenone derivative that, after functionalization and a Claisen rearrangement, gave an intermediate with the quaternary carbon atom and the required stereochemistry for the closure of the aryloctahydroindole ring system to provide the desired product (*296*).

D-(−)-Quinic acid has been efficiently used as the synthetic precursor for the incorporation of the aminocyclohexanetriol moiety present in 2,7-dideoxypancratistatin (*297*) (Scheme 15).

## V. Spectroscopy and Alkaloid Data

There follows a discussion of proton nuclear magnetic resonance ($^1$H NMR), carbon nuclear magnetic resonance ($^{13}$C NMR), and mass spectrometry (MS) of the *Narcissus* alkaloids. A list of the different *Narcissus* alkaloids, their spectroscopic properties, and literature with the most recent spectroscopic data is given in Table X.

### A. PROTON NUCLEAR MAGNETIC RESONANCE

$^1$H NMR spectroscopy gives important information about the different types of Amaryllidaceae alkaloids. In the last two decades, the routine use of 2D NMR techniques has facilitated the structural assignments and the establishment of their stereochemistry. A compilation of the different $^1$H NMR spectra arranged according to the different skeleton types is shown in Tables XI–XVII.

#### 1. *Lycorine Type*

This group has been subject to several $^1$H NMR studies and lycorine (**1**), as well as its main derivatives has been completely assigned. The general characteristics of the $^1$H NMR spectra are

TABLE X.
*Narcissus* Alkaloid Data.

| Alkaloid[a] | MF (MW)[b] | Spectroscopic data | References |
|---|---|---|---|
| **1** Lycorine | $C_{16}H_{17}NO_4$ (287) | UV, IR, MS, $^1$H NMR, $^{13}$C NMR, CD, X-ray | *(298–301)* |
| **2** Poetaminine | $C_{18}H_{19}NO_5$ (329) | UV, IR | *(302)* |
| **3** Pseudolycorine | $C_{16}H_{19}NO_4$ (289) | UV, IR, MS, $^1$H NMR, $^{13}$C NMR | *(113)* |
| **4** 1-*O*-acetyl-pseudolycorine | $C_{18}H_{21}NO_5$ (331) | UV, IR, MS, $^1$H NMR | *(113)* |
| **5** 2-*O*-acetyl-pseudolycorine | $C_{18}H_{21}NO_5$ (331) | UV, IR, MS, $^1$H NMR, $^{13}$C NMR | *(113)* |
| **6** 9-*O*-methyl-pseudolycorine | $C_{17}H_{21}NO_4$ (303) | UV, IR, MS, $^1$H NMR | *(159,303)* |
| **7** Galanthine | $C_{18}H_{23}NO_4$ (317) | UV, MS, $^1$H NMR, $^{13}$C NMR | *(135,304)* |
| **8** Goleptine | $C_{17}H_{21}NO_4$ (303) | IR | *(179)* |
| **9** Jonquilline | $C_{18}H_{17}NO_5$ (327) | UV, IR | *(180)* |
| **10** Caranine | $C_{16}H_{17}NO_3$ (271) | UV, IR, MS, $^1$H NMR, CD | *(305,306)* |
| **11** Pluviine | $C_{17}H_{21}NO_3$ (287) | UV, IR | *(176,307)* |
| **12** Norpluviine | $C_{16}H_{19}NO_3$ (273) | UV, IR | *(176,308)* |
| **13** 9-*O*-demethyl- pluviine | $C_{16}H_{19}NO_3$ (273) | UV, MS, $^1$H NMR, $^{13}$C NMR | *(167)* |
| **14** 1-*O*-acetyl-9-*O*-demethylpluviine | $C_{18}H_{21}NO_4$ (315) | UV, MS, $^1$H NMR, $^{13}$C NMR | *(167)* |
| **15** 1,9-*O*-diacetyl-9-*O*-demethylpluviine | $C_{20}H_{23}NO_5$ (357) | UV, MS, $^1$H NMR, $^{13}$C NMR | *(167)* |
| **16** Narcissidine | $C_{18}H_{23}NO_5$ (333) | UV, IR, MS, $^1$H NMR, X-ray | *(309–313)* |
| **17** Ungiminorine | $C_{17}H_{19}NO_5$ (317) | UV, MS, $^1$H NMR, $^{13}$C NMR | *(173,314,315)* |
| **18** Nartazine | $C_{20}H_{23}NO_6$ (373) | IR | *(89)* |
| **19** Kirkine | $C_{16}H_{19}NO_3$ (273) | IR, MS, $^1$H NMR, $^{13}$C NMR | *(316)* |
| **20** Assoanine | $C_{17}H_{17}NO_2$ (267) | UV, IR, MS, $^1$H NMR, $^{13}$C NMR | *(114)* |
| **21** Oxoassoanine | $C_{17}H_{15}NO_3$ (281) | UV, IR, MS, $^1$H NMR, $^{13}$C NMR | *(114)* |
| **22** Vasconine | $C_{17}H_{16}NO_2$ (266) | IR, MS, $^1$H NMR, $^{13}$C NMR | *(70,71)* |
| **23** Tortuosine | $C_{18}H_{18}NO_3$ (296) | IR, MS, $^1$H NMR, $^{13}$C NMR | *(68,69)* |
| **24** Ungeremine | $C_{16}H_{11}NO_3$ (265) | UV, IR, MS, $^1$H NMR, $^{13}$C NMR | *(317–320)* |
| **25** Roserine | $C_{18}H_{22}NO_3$ (300) | MS, $^1$H NMR, $^{13}$C NMR | *(67)* |
| **26** Homolycorine | $C_{18}H_{21}NO_4$ (315) | UV, IR, MS, $^1$H NMR, $^{13}$C NMR, CD, X-ray | *(82,300,321,322)* |

| | | | |
|---|---|---|---|
| **27** 8-*O*-demethyl-homolycorine | $C_{17}H_{19}NO_4$ (301) | UV, IR, MS, $^1$H NMR, $^{13}$C NMR, CD, X-ray | (*111,300,323,324*) |
| **28** 8-*O*-demethyl-8-*O*-acetylhomolycorine | $C_{19}H_{21}NO_5$ (343) | IR, MS, $^1$H NMR, $^{13}$C NMR | (*70*) |
| **29** 9-*O*-demethyl-homolycorine | $C_{17}H_{19}NO_4$ (301) | UV, IR, MS, $^1$H NMR, $^{13}$C NMR | (*82*) |
| **30** Masonine | $C_{17}H_{17}NO_4$ (299) | UV, IR, MS, $^1$H NMR, $^{13}$C NMR, CD | (*165,321*) |
| **31** Normasonine | $C_{16}H_{15}NO_4$ (285) | UV, IR, MS, $^1$H NMR, $^{13}$C NMR | (*165*) |
| **32** 9-*O*-demethyl-2α-hydroxy-homolycorine | $C_{17}H_{19}NO_5$ (317) | IR, MS, $^1$H NMR, $^{13}$C NMR | (*138*) |
| **33** Dubiusine | $C_{23}H_{27}NO_8$ (445) | UV, IR, MS, $^1$H NMR, $^{13}$C NMR | (*72*) |
| **34** Hippeastrine | $C_{17}H_{17}NO_5$ (315) | UV, IR, MS, $^1$H NMR, $^{13}$C NMR, CD | (*71,300,325,326*) |
| **35** Lycorenine | $C_{18}H_{23}NO_4$ (317) | UV, MS, $^1$H NMR, $^{13}$C NMR, X-ray | (*106,327–329*) |
| **36** *O*-methyllycorenine | $C_{19}H_{25}NO_4$ (331) | IR, MS, $^1$H NMR, $^{13}$C NMR, X-ray | (*74,104*) |
| **37** Oduline | $C_{17}H_{19}NO_4$ (301) | UV, IR, MS, $^1$H NMR, $^{13}$C NMR | (*165*) |
| **38** 6-*O*-methyloduline | $C_{18}H_{21}NO_4$ (315) | UV, IR, MS, $^1$H NMR, $^{13}$C NMR | (*165*) |
| **39** 2α-Hydroxy-6-*O*-methyloduline | $C_{18}H_{21}NO_5$ (331) | IR, MS, $^1$H NMR, $^{13}$C NMR | (*71*) |
| **40** 8-*O*-demethyl-homolycorine-*N*-oxide | $C_{17}H_{19}NO_5$ (317) | UV, IR, MS, $^1$H NMR, $^{13}$C NMR | (*137*) |
| **41** Poetinatine | $C_{20}H_{23}NO_6$ (373) | IR, MS, $^1$H NMR | (*98*) |
| **42** Vittatine | $C_{16}H_{17}NO_3$ (271) | UV, IR, MS, $^1$H NMR, $^{13}$C NMR, CD | (*208,300,330,331*) |
| **43** 11-Hydroxyvittatine | $C_{16}H_{17}NO_4$ (287) | UV, IR, MS, $^1$H NMR, $^{13}$C NMR, CD | (*208,300*) |
| **44** Maritidine | $C_{17}H_{21}NO_3$ (287) | UV, IR, MS, $^1$H NMR, $^{13}$C NMR, CD, X-ray | (*136,332–336*) |
| **45** 8-*O*-demethyl-maritidine | $C_{16}H_{19}NO_3$ (273) | IR, MS, $^1$H NMR, $^{13}$C NMR | (*108,331,337*) |
| **46** 9-*O*-demethyl-maritidine | $C_{16}H_{19}NO_3$ (273) | IR, MS, $^1$H NMR | (*111*) |
| **47** *O*-methylmaritidine | $C_{18}H_{23}NO_3$ (301) | UV, IR, MS, $^1$H NMR, $^{13}$C NMR, CD | (*133,137*) |
| **48** Papyramine | $C_{18}H_{23}NO_4$ (317) | UV, IR, MS, $^1$H NMR, $^{13}$C NMR | (*135,137*) |
| **49** 6-Epipapyramine | $C_{18}H_{23}NO_4$ (317) | UV, IR, MS, $^1$H NMR, $^{13}$C NMR | (*135,137*) |

TABLE X.
Continued.

| Alkaloid[a] | MF (MW)[b] | Spectroscopic data | References |
|---|---|---|---|
| **50** *O*-methyl-6-epipapyramine | $C_{19}H_{25}NO_4$ (331) | UV, IR, MS, $^1$H NMR, $^{13}$C NMR | (*137*) |
| **51** 6α-Hydroxy-3-*O*-methylepimaritidine | $C_{18}H_{23}NO_4$ (317) | UV, IR, MS, $^1$H NMR, CD | (*133*) |
| **52** 6β-Hydroxy-3-*O*-methylepimaritidine | $C_{18}H_{23}NO_4$ (317) | UV, IR, MS, $^1$H NMR, CD | (*133*) |
| **53** Hemanthamine | $C_{17}H_{19}NO_4$ (301) | UV, IR, MS, $^1$H NMR, $^{13}$C NMR, CD, X-ray | (*83,331, 338,339*) |
| **54** 11-*O*-acetyl- hemanthamine | $C_{19}H_{21}NO_5$ (343) | IR, MS, $^1$H NMR, $^{13}$C NMR, CD | (*74*) |
| **55** Hemanthidine | $C_{17}H_{19}NO_5$ (317) | UV, IR, MS, $^1$H NMR, $^{13}$C NMR, CD | (*73,300,325, 331,340,341*) |
| **56** 6-Epihemanthidine | $C_{17}H_{19}NO_5$ (317) | UV, IR, MS, $^1$H NMR, $^{13}$C NMR, CD | (*73,300,325, 331,340,341*) |
| **57** Crinamine | $C_{17}H_{19}NO_4$ (301) | UV, IR, MS, $^1$H NMR, $^{13}$C NMR, CD, X-ray | (*298,342–344*) |
| **58** Narcidine | $C_{17}H_{21}NO_4$ (303) | UV, IR, MS, $^1$H NMR | (*161*) |
| **59** Cantabricine | $C_{18}H_{23}NO_4$ (317) | IR, MS, $^1$H NMR, $^{13}$C NMR | (*140*) |
| **60** Narcimarkine | $C_{21}H_{29}NO_5$ (375) | IR, MS | (*91*) |
| **61** Bujeine | $C_{20}H_{23}NO_6$ (373) | IR, MS, $^1$H NMR, $^{13}$C NMR, CD | (*74*) |
| **62** Tazettine | $C_{18}H_{21}NO_5$ (331) | UV, IR, MS, $^1$H NMR, $^{13}$C NMR, CD, X-ray | (*300,345–351*) |
| **63** Criwelline | $C_{18}H_{21}NO_5$ (331) | UV, MS, $^1$H NMR, $^{13}$C NMR, CD | (*352–355*) |
| **64** Pretazettine | $C_{18}H_{21}NO_5$ (331) | UV, IR, MS, $^1$H NMR, CD | (*300,347*) |
| **65** 3-Epimacronine | $C_{18}H_{19}NO_5$ (329) | IR, MS, $^1$H NMR, $^{13}$C NMR, CD, X-ray | (*79,300,337,356*) |
| **66** 3-Epimacronine isomer | $C_{18}H_{19}NO_5$ (329) | IR, MS, $^1$H NMR, $^{13}$C NMR | (*139*) |
| **67** Obesine | $C_{16}H_{17}NO_4$ (287) | MS, $^1$H NMR, $^{13}$C NMR | (*76*) |
| **68** Narciclasine | $C_{14}H_{13}NO_7$ (307) | UV, IR, MS, $^1$H NMR, $^{13}$C NMR, X-ray | (*120,122,125*) |
| **69** Narciprimine | $C_{14}H_9NO_5$ (271) | UV, IR, MS, $^1$H NMR | (*125,357*) |
| **70** Trisphaeridine | $C_{14}H_9NO_2$ (223) | UV, IR, MS, $^1$H NMR, $^{13}$C NMR | (*99,358*) |

| | | | |
|---|---|---|---|
| **71** Bicolorine | $C_{15}H_{12}NO_2$ (238) | IR, MS, $^1$H NMR, $^{13}$C NMR | (*79*) |
| **72** Ismine | $C_{15}H_{15}NO_3$ (257) | UV, IR, MS, $^1$H NMR, $^{13}$C NMR, X-ray | (*99,358–360*) |
| **73** Pancracine | $C_{16}H_{17}NO_4$ (287) | UV, MS, $^1$H NMR, $^{13}$C NMR, CD | (*80,361*) |
| **74** Nangustine | $C_{16}H_{17}NO_4$ (287) | IR, MS, $^1$H NMR, $^{13}$C NMR | (*80*) |
| **75** Galanthamine | $C_{17}H_{21}NO_3$ (287) | UV, IR, MS, $^1$H NMR, $^{13}$C NMR, CD, X-ray | (*83,354,362,363*) |
| **76** Epigalanthamine | $C_{17}H_{21}NO_3$ (287) | UV, IR, MS, $^1$H NMR, CD | (*354,364,365*) |
| **77** *O*-acetyl-galanthamine | $C_{19}H_{23}NO_4$ (329) | UV, MS, $^1$H NMR, $^{13}$C NMR | (*167*) |
| **78** Norgalanthamine | $C_{16}H_{19}NO_3$ (273) | UV, IR, MS, $^1$H NMR, $^{13}$C NMR, CD, X-ray | (*141,366–368*) |
| **79** Epinorgalanthamine | $C_{16}H_{19}NO_3$ (273) | IR, MS, $^1$H NMR, $^{13}$C NMR | (*107*) |
| **80** *N*-formyl-norgalanthamine | $C_{17}H_{19}NO_4$ (301) | UV, IR, MS, $^1$H NMR, $^{13}$C NMR | (*83*) |
| **81** Sanguinine | $C_{16}H_{19}NO_3$ (273) | UV, IR, MS, $^1$H NMR, $^{13}$C NMR | (*369*) |
| **82** Narcisine | $C_{18}H_{21}NO_4$ (315) | UV, IR, MS, $^1$H NMR, $^{13}$C NMR | (*117*) |
| **83** Narwedine | $C_{17}H_{19}NO_3$ (285) | UV, IR, MS, $^1$H NMR | (*364,367*) |
| **84** Lycoramine | $C_{17}H_{23}NO_3$ (289) | IR, MS, $^1$H NMR, $^{13}$C NMR | (*336,367,369*) |
| **85** Norlycoramine | $C_{16}H_{21}NO_3$ (275) | IR, MS, $^1$H NMR | (*337*) |
| **86** Epinorlycoramine | $C_{16}H_{21}NO_3$ (275) | IR, MS, $^1$H NMR, $^{13}$C NMR | (*107*) |
| **87** *O*-methyl-norbelladine | $C_{16}H_{19}NO_3$ (273) | $^1$H NMR, $^{13}$C NMR | (*181*) |
| **88** Cherylline | $C_{17}H_{19}NO_3$ (285) | IR, MS, $^1$H NMR, $^{13}$C NMR, X-ray | (*370*) |
| **89** Pallidiflorine | $C_{34}H_{40}N_2O_7$ (588) | IR, MS, $^1$H NMR, $^{13}$C NMR | (*86*) |
| **90** Mesembrenone | $C_{17}H_{21}NO_3$ (287) | IR, MS, $^1$H NMR, $^{13}$C NMR | (*26,371*) |
| **91** Mesembrenol | $C_{17}H_{23}NO_3$ (289) | UV, IR, MS, $^1$H NMR, CD | (*372*) |
| **92** Mesembrine | $C_{17}H_{23}NO_3$ (289) | IR, MS, $^1$H NMR, $^{13}$C NMR | (*284,285,373*) |

[a]Alkaloids are listed in numerical order and grouped according to their ring system (see Tables I–VII).
[b]MF (MW) = Molecular formula (Molecular weight).

TABLE XI.
$^{1}$H NMR Data of Lycorine-Type Alkaloids.

| **Alkaloid** | **1** | **3** | **4** | **5** | **6** | **7** | **10** | **13** | **14** | **15** | **19** |
|---|---|---|---|---|---|---|---|---|---|---|---|
| H-1 | 4.27 br s | 4.85 br s | 5.58 m | 4.43 t | 4.53 dd | 4.55 s | 4.70 m | 4.26 dd | 5.54 dd | 5.51 m | 4.37 q |
| H-2$\alpha$ | 3.97 br s | 4.50 t | 4.15 dd | 5.29 dd | 4.18 m | 3.72 m | 2.59 m | 2.3–2.6 m | 2.3–2.5 m | 2.36 m | 2.46 ddd |
| H-2$\beta$ | — | — | — | — | — | — | 2.59 m | 2.3–2.4 m | 2.3–2.5 m | 2.36 m | 2.34 dddd |
| H-3 | 5.37 br s | 5.60 t | 5.69 dd | 5.45 t | 5.55 m | 5.55 br s | 5.41 m | 5.54 br d | 5.39 br s | 5.32 br t | 5.87 br dt |
| H-4a | 2.60 d | 3.02 br d | 2.90 br s | 2.85 d | 2.87 dd | 2.65 s | 2.78 dd | 2.70 br d | 2.3–2.5 m | 2.2–2.3 m | 4.24 br d |
| H-6$\alpha$ | 3.32 d | 3.68 br d | 3.61 dd | 3.54 d | 3.55 dd | 3.40 br d | 3.52 dd | 3.31 d | 3.54 br d | 3.5–3.6 br s | 4.40 d |
| H-6$\beta$ | 4.02 d | 4.16 d | 4.17 d | 4.14 d | 4.16 d | 4.05 d | 4.13 d | 3.93 d | 3.59 br d | 3.5–3.6 br s | 4.61 dt |
| H-7 | 6.68 s | 6.71 s | 6.70 s | 6.61 s | 6.75 s | 6.52 s | 6.58 s | 6.59 s | 6.64 s | 6.71 s | 6.60 s |
| H-10 | 6.81 s | 6.89 s | 6.74 s | 6.83 s | 6.96 s | 6.78 s | 6.82 s | 6.82 s | 7.08 s | 7.14 br s | 7.08 s |
| H-10b | 2.50 m | 2.74 br d | 2.90 br s | 2.69 d | 2.76 ddd | 2.65 s | 2.41 ddd | 2.98 dd | 3.33 dd | 3.27 dd | 3.52 dd |
| H-11 (2H) | 2.44 m | 2.6–2.7 m | 2.66 t | 2.63 m | 2.64 m | 2.4–2.6 m | 2.59 m | 2.5–2.6 m | 2.3–2.5 m | 2.2–2.4 m | 2.8–2.9 m |
| H-12$\alpha$ | 2.19 ddd | 2.6–2.7 m | 2.59 br t | 2.42 dt | 2.42 br q | 2.25 dd | 2.33 br q | 2.3–2.4 m | 2.79 m | 2.6–2.8 m | 3.71 ddd |
| H-12$\beta$ | 3.19 dd | 3.37 dd | 3.33 m | 3.35 dt | 3.35 ddd | 3.25 ddd | 3.32 ddd | 3.32 m | 2.79 m | 2.6–2.8 m | 3.83 ddd |
| $OCH_2O$ | 5.95 s | — | — | — | — | — | 5.91 (2d) | — | — | — | — |
| OMe | — | 3.84 s | 3.84 s | 3.82 s | 3.85 s | 3.78 s | — | 3.89 s | 3.86 s | 3.74 s | 3.87 s |
| OMe | — | — | — | — | 3.80 s | 3.74 s | — | — | — | — | — |
| OMe | — | — | — | — | — | 3.40 s | — | — | — | — | — |
| OAc | — | — | 1.92 s | 2.06 s | — | — | — | — | 1.98 s | 2.23 s | — |
| OAc | — | — | — | — | — | — | — | — | — | 1.92 s | — |
| Solvent | a | b | b | b | c | d | d | d | d | d | c |
| MHz | 300 | 200 | 200 | 200 | 270 | 200 | 270 | 400 | 400 | 400 | 500 |
| Reference | (*298*) | (*113*) | (*113*) | (*113*) | (*303*) | (*135*) | (*305*) | (*167*) | (*167*) | (*167*) | (*316*) |

Solvent a: DMSO-$d_6$, b: $CDC_{13}$-$CDC_3OD$, c: $CD_3OD$, d: $CDCl_3$

TABLE XI.
Continued.

| **Alkaloid** | **20** | **21** | **22** | **23** | **24** |
|---|---|---|---|---|---|
| H-1 | 7.31 dd | 7.85 br d | 8.32 d | 8.13 d | 7.54 s |
| H-2 | 6.75 t | 7.27 t | 7.84 t | — | — |
| H-3 | 6.99 dt | 7.33 br d | 7.70 d | 7.59 br s | 7.34 s |
| H-6 | 4.09 s | — | 10.42 s | 9.55 s | 9.19 s |
| H-7 | 6.64 s | 7.57 s | 8.09 s | 7.92 s | 7.59 s |
| H-10 | 7.17 s | 7.81 s | 7.86 s | 8.34 s | 7.88 s |
| H-11 (2H) | 3.00 br t | 3.45 br t | 3.76 t | 3.88 t | 3.70 t |
| H-12 (2H) | 3.32 t | 4.45 br t | 5.42 t | 5.35 t | 5.12 t |
| OMe | 3.93 s | 4.09 s | 4.19 s | 4.33 s | — |
| OMe | 3.87 s | 4.03 s | 4.08 s | 4.20 s | — |
| OMe | — | — | — | 4.17 s | — |
| $OCH_2O$ | — | — | — | — | 6.32 s |
| Solvent | d | b | b | c | e |
| MHz | 200 | 200 | 500 | 500 | 600 |
| Reference | (*114*) | (*114*) | (*71*) | (*68*) | (*320*) |

Solvent a: DMSO-$d_6$, b: $CDC_{13}$-$CDC_3OD$, c: $CD_3OD$, d: $CDCl_3$, e: $D_2O + 0.01\%$ TFA-$d_1$

TABLE XI.
Continued.

| **Alkaloid** | **16** | **17** | **25** |
|---|---|---|---|
| H-1α | — | — | 2.85 ddd |
| H-1β | 4.66 m | 4.69 br s | 3.13 dd |
| H-2α | 3.4–4.2 m | 3.69 m | 1.8–1.95 m |
| H-2β | — | — | 2.25–2.4 m |
| H-3α | — | — | 1.38 qd |
| H-3β | 4.66 m | 4.60 br s | 2.25–2.4 m |
| H-4 | — | — | 3.4–3.5 m |
| H-4a | 3.4–4.2 m | 4.09 m | — |
| H-6α | 3.54 d | 3.69 m | 9.50 s (1H) |
| H-6β | 4.09 d | 4.15 br d | |
| H-7 | 6.68 s | 6.78 s | 6.95 s |
| H-10 | 6.88 s | 6.93 s | — |
| H-10b | 2.70 d | 2.84 br d | — |
| H-11α | 5.56 br s (1H) | 5.75 br d (1H) | 2.05 ddd |
| H-11β | | | 2.70 dt |
| H-12α | 3.4–4.2 m (2H) | 3.69 m | 4.75–4.95 m (2H) |
| H-12β | | 4.15 br d | |
| OMe | 3.86 s | — | 4.15 s |
| OMe | 3.82 s | — | 4.05 s |
| OMe | 3.44 s | 3.44 s | 3.90 s |
| $OCH_2O$ | — | 5.92 br s | — |
| Solvent | d | c | d |
| MHz | 400 | 600 | 250 |
| Reference | *(313)* | *(173)* | *(67)* |

Solvent a: DMSO-$d_6$, b: $CDCl_3$-$CD_3OD$, c: $CD_3OD$, d: $CDCl_3$, e: $D_2O + 0.01\%$ TFA-$d_1$.

TABLE XII.
$^1$H NMR Data of Homolycorine-Type Compounds: Lactone Alkaloids.

| **Alkaloid** | **26** | **27** | **28** | **29** | **30** | **31** | **32** | **33** | **34** | **40** |
|---|---|---|---|---|---|---|---|---|---|---|
| H-1 | 4.81 ddd | 4.78 m | 4.86 br d | 4.80 ddd | 4.75 m | 4.85 s | 4.59 br d | 4.63 dd | 4.58 br s | 4.83 m |
| H-2 (2H) | 2.49 m | 2.59 m | 2.5–2.7 m | 2.51 m | 2.60 m | 2.62 m | 4.28 br t (H$\beta$) | 5.46 dd (H$\beta$) | 4.38 br s (H$\beta$) | 2.64 m |
| H-3 | 5.50 m | 5.50 d | 5.63 br d | 5.55 m | 5.50 br d | 5.64 m | 5.68 br s | 5.62 m | 5.63 br s | 5.76 br d |
| H-4a | 2.72 dd | 2.73 dd | 2.96 br d | 2.71 br d | 2.74 m | 5.50 br d | 2.66 br d | 2.74 br d | 2.62 d | 3.87 br d |
| H-7 | 7.57 s | 7.60 s | 7.47 s | 7.54 s | 7.49 s | 7.52 s | 7.47 s | 7.52 s | 7.45 s | 7.44 s |
| H-10 | 6.99 s | 6.98 s | 7.13 s | 6.91 s | 6.96 s | 7.11 s | 6.95 s | 6.97 s | 6.92 s | 7.24 s |
| H-10b | 2.64 dd | 2.63 dd | 2.86 dd | 2.60 dd | 2.74 m | 2.66 d | 2.83 dd | 2.79 br d | 2.85 dd | 3.50 dd |
| H-11 (2H) | 2.6–2.7 m | 2.50 m | 2.5–2.7 m | 2.5–2.6 m | 2.50 m | 2.52 m | 2.5–2.6 m | 2.5–2.7 m | 2.48 m | 3.57 m |
| H-12$\alpha$ | 3.14 ddd | 3.14 ddd | 3.31 ddd | 3.15 ddd | 3.18 ddd | 3.15 ddd | 3.17 ddd | 3.19 ddd | 3.13 ddd | 2.88 m |
| H-12$\beta$ | 2.24 ddd | 2.25 dd | 2.45 dd | 2.30 ddd | 2.27 dd | 2.89 dd | 2.31 dd | 2.37 dd | 2.23 ddd | 2.72 m |
| $OCH_2O$ | — | — | — | — | 6.07 (2d) | 6.05 d | — | — | 6.04 (2d) | — |
| OMe | 3.96 s | 3.95 s | 3.97 s | 3.94 s | — | — | 3.95 s | 3.94 s | — | 3.96 s |
| OMe | 3.95 s | — | — | — | — | — | — | — | — | — |
| NMe | 2.00 s | 2.00 s | 2.16 s | 2.01 s | 2.06 s | — | 2.08 s | 2.06 s | 2.03 s | 2.94 s |
| OAc | — | — | 2.00 s | — | — | — | — | 2.00 s | — | — |
| $CH_A$ | — | — | — | — | — | — | — | 2.53 dd | — | — |
| $CH_B$ | — | — | — | — | — | — | — | 2.65 dd | — | — |
| CHOH | — | — | — | — | — | — | — | 5.24 m | — | — |
| Me | — | — | — | — | — | — | — | 1.29 d | — | — |
| Solvent | d | d | b | b | d | d | b | b | d | b |
| MHz | 200 | 400 | 250 | 200 | 400 | 400 | 200 | 400 | 500 | 200 |
| Reference | *(82)* | *(324)* | *(70)* | *(82)* | *(165)* | *(165)* | *(138)* | *(72)* | *(71)* | *(137)* |

Solvent a: DMSO-$d_6$, b: $CDCl_3$-$CD_3OD$, c: $CD_3OD$, d: $CDCl_3$

TABLE XII.
Continued.

| **Alkaloid** | **35** | **36** | **37** | **38** | **39** |
|---|---|---|---|---|---|
| H-1α | 4.35 br d | 4.29 br d | 4.35 d | 4.17 d | 4.15 br s |
| H-2α | 2.35 m | 2.35 m | 2.31 dm | 2.26 dd | — |
| H-2β | 2.65 m | 2.65 m | 2.62 dm | 2.53 dm | 4.21 br s |
| H-3 | 5.50 br s | 5.50 br d | 5.46 br d | 5.39 br s | 5.69 br s |
| H-4a | 2.78 br d | 2.77 dd | 2.72 br d | 2.69 br d | 2.87 d |
| H-6β | 5.93 s | 5.54 s | 5.99 s | 5.39 br s | 5.43 s |
| H-7 | 6.93 s | 6.80 s | 6.85 s | 6.67 s | 6.73 s |
| H-10 | 6.99 s | 6.93 s | 6.90 s | 6.77 s | 6.94 s |
| H-10b | 2.44 dd | 2.44 dd | 2.4–2.5 m | 2.35 dd | 2.85 d |
| H-11 (2H) | 2.4–2.6 m | 2.4–2.6 m | 2.4–2.5 m | 2.3–2.4 m | 2.54 m |
| H-12α | 3.15 ddd | 3.15 ddd | 3.14 ddd | 3.08 dd | 3.33 m |
| H-12β | 2.27 dd | 2.23 dd | 2.25 dd | 2.17 dd | 2.38 dt |
| $OCH_2O$ | — | — | 5.97 d | 5.83 d | 5.91 (2d) |
| OMe | 3.88 s | 3.89 s | — | 3.43 s | 3.51 s |
| OMe | 3.87 s | 3.88 s | — | — | — |
| OMe | — | 3.55 s | — | — | — |
| NMe | 2.08 s | 2.10 s | 2.11 s | 2.05 s | 2.22 s |
| Solvent | c | d | d | d | d |
| MHz | 250 | 200 | 400 | 400 | 500 |
| Reference | (*106*) | (*104*) | (*165*) | (*165*) | (*71*) |

Solvent a: DMSO-$d_6$; b: $CDCl_3$-$CD_3OD$; c: $CD_3OD$; d: $CDCl_3$.

TABLE XIII.
$^1$H NMR Data of Hemanthamine-Type Alkaloids.

| **Alkaloid** | **42** | **43** | **44** | **45** | **46** | **47** | **48/49** | **53** | **54** | **55/56** | **57** | **58** |
|---|---|---|---|---|---|---|---|---|---|---|---|---|
| H-1 | 6.44 d | 6.41 d | 6.71 d | 6.65 d | 6.58 d | 6.49 d | 6.65 d/6.66 d | 6.36 d | 6.40 d | 6.33 d | 6.22 br s | 6.48 d |
| H-2 | 6.06 dd | 6.19 dd | 5.93 dd | 5.99 dd | 5.91 ddd | 6.08 dd | 6.00 dd | 6.25 dd | 6.21 dd | 6.27 dd | 6.22 br s | 6.35 dd |
| H-3 | 4.46 m | 4.28 ddd | 4.27 ddd | 4.36 m | 4.31 td | 3.85 m | 3.88 m/3.85 m | 3.82 m | 3.91 m | 3.85 m | 3.98 m | 3.86 m |
| H-4$\alpha$ | 1.85 ddd | 2.28 ddd | 1.81 ddd | 1.75 ddd | 1.73 td | 1.73 ddd | 1.77 ddd/1.60 ddd | 2.11 ddd | 2.07 ddd | 2.36 ddd/2.21 ddd | 2.08 m | 2.11 ddd |
| H-4$\beta$ | 2.58 ddd | 1.85 ddd | 1.95 m | 2.03 ddd | 1.90 dddd | 2.95 ddd | 2.21 br d/2.09 m | 1.96 ddd | 1.98 ddd | 2.12 ddd/2.00 ddd | 2.08 m | 2.04 ddd |
| H-4a | 3.87 m | 3.47 dd | 3.48 dd | 3.37 m | 3.43 dd | 3.90 dd | 3.59 dd/3.90 m | 3.25 dd | 3.41 m | 3.56 dd/3.20 m | 3.22 dd | 3.38 m |
| H-6$\alpha$ | 4.07 d | 3.84 d | 3.84 d | 3.79 d | 3.81 d | 4.25 d | —/5.16 s | 3.72 d | 3.77 d | —/5.02 s | 3.66 d | 3.70 d |
| H-6$\beta$ | 4.71 d | 4.35 d | 4.39 d | 4.39 d | 4.41 d | 4.80 d | 5.86 s/— | 4.25 d | 4.40 d | 5.69 s/— | 4.28 d | 4.31 d |
| H-7 | 6.53 s | 6.56 s | 6.66 s | 6.56 s | 6.53 s | 6.57 s | 7.03 s/6.89 s | 6.41 s | 6.52 s | 6.94 s/6.79 s | 6.47 s | 6.53 s |
| H-10 | 6.87 s | 6.94 s | 6.97 s | 6.82 s | 6.87 s | 6.81 s | 6.80 s/6.83 s | 6.74 s | 6.95 s | 6.70 s/6.73 s | 6.79 s | 6.78 s |
| H-11 *endo* | 2.15 m | 3.98 dd | 1.95 m | 1.95 m | 1.8–2.0 m | 2.22 ddd | 2.00 m | 3.96 dd | 5.05 dd | 3.92 m | 3.92 m | 3.99 ddd |
| H-11 *exo* | 2.32 m | — | 2.17 m | 2.15 m | 2.19 ddd | 2.37 ddd | 2.00 m | — | — | — | — | — |
| H-12 *endo* | 3.17 m | 3.20 dd | 2.93 ddd | 2.95 m | 2.91 ddd | 3.27 ddd | 3.02 ddd/2.84 ddd | 3.30 dd | 3.41 m | 4.20 dd/3.30 m | 3.40 m | 3.39 m |
| H-12 *exo* | 3.87 m | 3.50 dd | 3.34 m | 3.37 m | 3.31 ddd | 4.08 ddd | 3.73 ddd/3.39 m | 3.19 dd | 3.41 m | 2.96 dd/3.20 m | 3.40 m | 3.26 dd |
| $OCH_2O$ | 5.95 s | 5.89 s | — | — | — | — | — | 5.81 (2d) | 5.95 s | 5.83 (2d)/5.86 (2d) | 5.86 s | — |
| OMe | — | — | 3.81 s | 3.89 s | 3.82 s | 3.86 s | 3.88 s | — | — | — | — | 3.86 s |
| OMe | — | — | 3.77 s | — | — | 3.82 s | 3.88 s | — | — | — | — | — |
| OMe | — | — | — | — | — | 3.34 s | 3.37 s/3.31 s | 3.36 s | 3.41 s | 3.32 s/3.28 s | 3.40 s | 3.35 s |
| OAc | — | — | — | — | — | — | — | — | 2.03 s | — | — | — |
| Solvent | d | c | c | d | d | d | d | d | d | d | d | d |
| MHz | 360 | 270 | 300 | 360 | 200 | 200 | 200 | 360 | 500 | 360 | 300 | 360 |
| Reference | *(331)* | *(376)* | *(336)* | *(331)* | *(111)* | *(137)* | *(135)* | *(331)* | *(74)* | *(331)* | *(298)* | *(161)* |

Solvent a: DMSO-$d_6$, b: $CDCl_3$-$CD_3OD$, c: $CD_3OD$, d: $CDCl_3$

TABLE XIII.
Continued.

| **Alkaloid** | **59** | **61** |
|---|---|---|
| H-1ax | 1.77 ddd | 6.38 d (1H) |
| H-1eq | 2.39 dt | |
| H-2ax | 1.61 dddd | 6.08 dd (1H) |
| H-2eq | 2.06 m | |
| H-3 | 4.67 tt ($\beta$) | 3.89 ddd ($\alpha$) |
| H-4ax | 1.44 ddd | 2.43 ddd |
| H-4eq | 2.23 m | 2.06 ddd |
| H-4a | 3.18 dd | 3.29 dd |
| H-6$\alpha$ | 3.91 d | 4.04 d |
| H-6$\beta$ | 4.50 d | 4.45 d |
| H-7 | 6.48 s | 6.57 s |
| H-10 | 6.74 s | 6.81 s |
| H-11 *exo* | 2.30 ddd | 3.93 dd |
| H-11 *endo* | 1.83 ddd | — |
| H-12 *exo* | 3.60 m | 4.82 d |
| H-12 *endo* | 2.94 m | 4.51 d |
| OMe | 3.82 s | 3.36 s |
| $OCH_2O$ | — | 5.91 s |
| $CH_A$ | — | 4.27 dd |
| $CH_B$ | — | 3.58 dd |
| OAc | 2.02 s | 2.00 s |
| Solvent | d | d |
| MHz | 500 | 500 |
| Reference | (*140*) | (*74*) |

Solvent a: DMSO-$d_6$, b: $CDCl_3$-$CD_3OD$, c: $CD_3OD$, d: $CDCl_3$.

TABLE XIV.
$^1$H NMR Data of Tazettine-Type Alkaloids.

| **Alkaloid** | **62** | **63** | **65** | **67** |
|---|---|---|---|---|
| H-1 | 5.60 ddd | 5.78 d | 5.44 ddd | 5.86 d |
| H-2 | 6.15 ddd | 6.20 dd | 5.97 d | 6.13 d |
| H-3 | 4.13 m | 3.89 ddd | 4.12 m | 4.3–4.4 m |
| H-4α | 2.20 m | 1.93 ddd | 1.70 m | 2.39 br d |
| H-4β | 1.60 m | 2.09 ddd | 2.55 m | 1.68 ddd |
| H-4a | 2.83 m | 2.95 t | 3.10 m | 3.15 br s |
| H-6 | 4.65 d | 4.68 d | — | 4.02 d |
| H-6′ | 4.95 dd | 4.94 d | — | 4.38 d |
| H-7 | 6.50 br s | 6.55 s | 7.52 s | 6.66 s |
| H-10 | 6.85 s | 6.52 s | 6.73 s | 6.70 s |
| H-10b | — | — | — | 2.81 br d |
| H-11 | — | — | 4.43 dd | — |
| H-12 | 2.65 d | 2.83 d | 2.76 dd | 3.01 d |
| H-12′ | 3.30 d | 3.30 d | 3.16 dd | 3.10 d |
| $OCH_2O$ | 5.90 s | 5.92 s | 6.01 s | 5.99 s |
| OMe | 3.45 s | 3.45 s | 3.40 s | — |
| NMe | 2.40 s | 2.38 s | 2.50 s | — |
| Solvent | d | d | d | c |
| MHz | 90 | 300 | 100 | 250 |
| Reference | (*347*) | (*355*) | (*337*) | (*76*) |

Solvent a: DMSO-$d_6$, b: $CDCl_3$-$CD_3OD$, c: $CD_3OD$, d: $CDCl_3$.

TABLE XV.
$^1$H NMR Data of Narciclasine-Type Alkaloids.

| **Alkaloid** | **68** | **70** | **71** | **72** |
|---|---|---|---|---|
| H-1 | 6.17 ddd | 8.36 dddd | 8.73 dd | 6.98 dd |
| H-2 | 4.23 ddd | 7.61 ddd | 8.02 td | 6.81 ddd |
| H-3 | 3.92 ddd | 7.67 ddd | 8.09 td | 7.28 ddd |
| H-4 | 3.90 dd | 8.11 ddd | 8.31 dd | 6.73 dd |
| H-4a | 4.35 ddd | — | — | — |
| H-6 | — | 9.06 s | 10.42 s | 4.26 d |
| H-6′ | — | — | — | 4.20 d |
| H-7 | — | 7.32 s | 7.97 s | 7.00 s |
| H-10 | 6.75 s | 7.89 s | 8.16 s | 6.67 s |
| $OCH_2O$ | 6.01 (2d) | 6.15 s | 6.41 s | 5.99 s |
| NMe | — | — | 4.73 s | 2.73 s |
| Solvent | c | d | b | d |
| MHz | 500 | 200 | 200 | 200 |
| Reference | (*122*) | (*358*) | (*79*) | (*358*) |

Solvent a: DMSO-$d_6$, b: $CDCl_3$-$CD_3OD$, c: $CD_3OD$, d: $CDCl_3$.

TABLE XVI.
$^1$H NMR Data of Montanine-Type Alkaloids.

| **Alkaloid** | **73** | **74** |
|---|---|---|
| H-1 | 5.37 br dddd | 5.52 dt |
| H-2α | — | 2.05 ddt |
| H-2β | 3.73 dddd | 2.57 dddd |
| H-3 | 3.65 ddddd (β) | 3.62 ddd (α) |
| H-4α | 1.36 ddd | 3.31 t |
| H-4β | 1.83 ddd | — |
| H-4a | 3.20 ddd | 3.16 br d |
| H-6α | 4.16 br d | 4.32 d |
| H-6β | 3.63 br d | 3.83 d |
| H-7 | 6.60 br s | 6.51 s |
| H-10 | 6.67 br s | 6.56 s |
| H-11 | 3.25 br s | 3.33 br d |
| H-12ax | 2.86 br s | 3.03 d |
| H-12eq | 2.86 br s | 2.94 dd |
| $OCH_2O$ | 5.90 (2d) | 5.86 (2d) |
| Solvent | a | c |
| MHz | 400 | 500 |
| Reference | (*361*) | (*80*) |

Solvent a: DMSO-$d_6$, b: $CDCl_3$-$CD_3OD$, c: $CD_3OD$, d: $CDCl_3$.

a. Two singlets for the *para*-oriented aromatic protons, together with a unique olefinic proton.
b. Two doublets as an AB system corresponding to the benzylic protons of C-6. The deshielding observed in the β-protons of positions 6 and 12 in relation to their α-homologues is due to the effect of the *cis*-lone pair of the nitrogen atom.
c. Like almost all other lycorine-type examples, the alkaloids isolated from the *Narcissus* genus show a *trans* B/C ring junction, the coupling constant being $J_{4a,10b}$ ~11 Hz. Only kirkine (**19**) shows a *cis* B/C ring junction, with a smaller coupling constant $J_{4a,10b}$ 8 Hz.

In the plant, the alkaloid lycorine (**1**) is particularly vulnerable to oxidation processes, giving several ring-C aromatized products.

## 2. *Homolycorine Type*

This group includes lactone, hemiacetal, or the more unusual cyclic ether alkaloids. The general traits for this type of alkaloids can be summarized as follows:

a. Two singlets for the *para*-oriented aromatic protons. In lactone alkaloids, the deshielding of H-7 is caused by the *peri*-carbonyl group.
b. The hemiacetal alkaloids always show the substituent at C-6 in an α-disposition.
c. The majority of alkaloids belong to a single enantiomeric series containing a *cis* B/C ring junction, which is congruent with the small size of the coupling constant $J_{1,10b}$. In the *Narcissus* genus, no exception to this rule has been observed.

TABLE XVII.
$^{1}$H NMR Data of Galanthamine-Type Alkaloids.

| **Alkaloid** | **75** | **76** | **77** | **78** | **79** | **80** | **81** | **83** | **84** | **86** |
|---|---|---|---|---|---|---|---|---|---|---|
| H-1 | 4.73 m | 4.58 m | 4.58 m | 4.67 m | 4.62 br s | 4.50 m | 4.52 br s | 4.72 m | 4.28 m | 4.37 t |
| H-2$\alpha$ | 2.12 ddd | 1.69 ddd | 2.10 ddd | 2.24 ddd | 2.04 ddd | 1.95 ddd | 2.10 ddd | 2.73 dd | 1.4–1.9 m | 1.88 m |
| H-2$\beta$ | 2.80 ddt | 2.77 dddd | 2.69 ddd | 2.60 m | 2.68 dd | 2.61 ddt | 2.49 dm | 3.14 m | 2.40 br d | 2.50 ddd |
| H-3 | 4.25 m | 4.61 dddd | 5.34 dd | 4.28 m | 4.15 m | 4.07 m | 4.15 br t | — | 3.98 m | 4.09 m |
| H-4 | 6.10 ddd | 6.05 dt | 5.92 d | 6.06 ddd | 5.98 d | 6.00 ddd | 5.91 dd | 6.02 d | 1.4–1.9 m (2H) | 1.5–1.7 m (2H) |
| H-4a | 6.21 dd | 5.79 dt | 6.28 d | 6.23 dt | 6.05 d | 5.88 dd | 6.11 d | 6.92 d | 1.4–1.9 m (2H) | 1.7–1.9 m (2H) |
| H-6 | 3.79 dd | 3.61 d | 3.77 d | 4.21 d | 3.93 d | 3.94 dd | 3.64 d | 3.76 d | 3.54 d | 3.98 s |
| H-6′ | 4.21 d | 4.06 d | 4.22 d | 4.44 d | 4.04 d | 4.80 d E | 4.06 d | 4.12 d | 3.92 d | 3.98 s |
| | | | | | | 5.10 d Z | | | | |
| H-7 | 6.74 d | 6.55 d | 6.61 d | 6.84 d | 6.62 d | 6.60 d | 6.57 d | 6.64 d | 6.51 d | 6.65 d |
| H-8 | 6.78 d | 6.61 d | 6.68 d | 6.90 d | 6.68 d | 6.76 d | 6.52 d | 6.68 d | 6.57 d | 6.62 d |
| H-11$\alpha$ | 2.20 ddd | 2.16 dt | 2.18 dd | 2.08 m | 1.88 dt | 1.71 ddd | 1.65 dm | 2.26 dt | 1.4–1.9 m | 1.7–1.9 m |
| H-11$\beta$ | 1.69 ddd | 1.63 ddd | 1.65 dd | 2.08 m | 1.80 td | 1.81 ddd | 2.05 m | 1.87 d | 1.4–1.9 m | 1.7–1.9 m |
| H-12$\alpha$ | 3.16 dt | 3.04 br d | 3.14 br d | 2.42 m | 3.22 m | 3.82 ddd E | 3.01 dm | 3.14 m | 2.96 t | 3.19 dt |
| | | | | | | 3.15 ddd Z | | | | |
| H-12$\beta$ | 3.39 ddd | 3.25 dt | 3.40 br t | 3.57 m | 3.38 dt | 3.61 ddd | 3.22 br t | 3.27 t | 3.12 t | 3.43 m |
| OMe | 3.95 s | 3.82 s | 3.85 s | 3.91 s | 3.84 s | 3.84 s | — | 3.82 s | 3.76 s | 3.85 s |
| NMe | 2.52 s | 2.56 s | 2.45 s | — | — | — | 2.38 s | 2.45 s | 2.29 s | — |
| NCHO | — | — | — | — | — | 8.02 s | — | — | — | — |
| | | — | — | — | | 8.07 s | — | — | — | — |
| OAc | — | — | 2.04 s | — | — | — | | — | — | — |
| Solvent | d | d | d | c | d | d | c | d | d | d |
| MHz | 200 | 400 | 400 | 200 | 200 | 200 | 250 | 400 | 400 | 200 |
| Reference | (*83*) | (*365*) | (*167*) | (*141*) | (*107*) | (*83*) | (*377*) | (*367*) | (*367*) | (*107*) |

Solvent a: DMSO-$d_6$, b: $CDCl_3$-$CD_3OD$, c: $CD_3OD$, d: $CDCl_3$.

d. The large coupling constant between H-4a and H-10b ($J_{4a,10b}$~10 Hz) is only consistent with a *trans*-diaxial relationship.
e. In general, ring C presents a vinylic proton. If position 2 is substituted by an OH, OMe, or OAc group, it always displays an $\alpha$-disposition.
f. The singlet corresponding to the *N*-methyl group is in the range of $\delta$ 2.0–2.2 ppm, its absence being very unusual.
g. The H-12$\alpha$ is more deshielded than H-12$\beta$ as a consequence of the *cis*-lone pair of the nitrogen atom.

Homolycorine-type alkaloids with a saturated ring C have been studied by Jeff and coworkers (*374*). They describe empirical correlations of *N*-methyl chemical shifts with stereochemical assignments of the B/C and C/D ring junction.

### 3. *Hemanthamine Type*

The absolute configuration of these alkaloids is determined through the circular dichroism spectrum. The alkaloids of the *Narcissus* genus are exclusively of the hemanthamine type, while in genera such as *Brunsvigia, Boophane*, etc., the crinine-type alkaloids are predominant. It is also noteworthy that the alkaloids isolated from the *Narcissus* genus do not show additional substitutions in the aromatic ring, apart from those of C-8 and C-9. On the contrary, in the genera where crinine-type alkaloids predominate, the presence of alkaloids with a methoxy substituent at C-7 is quite common. Thus, hemanthamine-type alkaloids show the following characteristics:

a. Two singlets for the *para*-oriented aromatic protons.
b. Using $CDCl_3$ as the solvent, the magnitude of the coupling constants between each olefinic proton (H-1 and H-2) and H-3 gives information about the configuration of the C-3 substituent. Thus, in those alkaloids in which the two-carbon bridge (C-11 and C-12) is *cis* to the substituent at C-3, H-1 shows an allylic coupling with H-3 ($J_{1,3}$~1–2 Hz) and H-2 shows a smaller coupling with H-3 ($J_{2,3}$~0–1.5 Hz), as it occurs in crinamine (**57**). On the contrary, in the corresponding C-3 epimeric series, e.g. hemanthamine (**53**), a larger coupling between H-2 and H-3 ($J_{2,3}$ 5 Hz) is shown, the coupling between H-1 and H-3 not being detectable.
c. There is frequently an additional coupling of H-2 with the equatorial H-4$\beta$ in a W-mechanism, while the proton H-4$\alpha$ shows a large coupling with H-4a ($J_{4\alpha,4a}$~13 Hz) due to their *trans*-diaxial disposition, characteristic of the hemanthamine series.
d. Two doublets for an AB system corresponding to the benzylic protons of position C-6.
e. The pairs of alkaloids with a hydroxy substituent at C-6, like papyramine/6-epipapyramine (**48/49**), hemanthidine/6-epihemanthidine (**55/56**), etc., appear as a mixture of epimers not separable even by HPLC.
f. Also, in relation to position C-6, it is interesting to note that ismine (**72**), a catabolic product from the hemanthamine series, shows a restricted rotation around the biarylic bond, which makes the methylenic protons at the benzylic position magnetically non-equivalent.

### 4. *Tazettine Type*

Although tazettine (**62**) is one of the most widely reported alkaloids in the Amaryllidaceae family, it was found to be an extraction artifact from pretazettine (**64**) (*75*).

The presence of an *N*-methyl group (2.4–2.5 ppm) in tazettine-type alkaloids immediately distinguishes them from the hemanthamine type, from which they proceed biosynthetically. Moreover, the $^{1}$H NMR spectrum always shows the signal corresponding to the methylenedioxy group.

We have also included the alkaloid obesine (**67**) in this group, although it exhibits some structural differences with the skeleton type.

*5. Galanthamine Type*

Among the Amaryllidaceae alkaloids, only the galanthamine type shows an *ortho*-coupling constant between both the aromatic protons of ring A. The general characteristics of their $^{1}$H NMR spectra are

a. Two doublets for the two *ortho*-oriented aromatic protons with a coupling constant of $J_{7,8}$ ~8 Hz.
b. The assignment of the substituent stereochemistry at C-3 is made in relation with the coupling constants of the olefinic protons H-4 and H-4a. When the coupling constant $J_{3,4}$ is about 5 Hz, the substituent is pseudoaxial, while if it is ~0 Hz this indicates that the substituent at C-3 is pseudoequatorial.
c. Two doublets as an AB system corresponding to the benzylic protons of C-6.
d. The existence of the furan ring results in a deshielding effect in H-1.
e. This type of alkaloid often shows an *N*-methyl group, but occasionally an *N*-formyl group has been reported.

## B. CARBON$^{13}$ NUCLEAR MAGNETIC RESONANCE

$^{13}$C NMR spectroscopy has been extensively used for determining the carbon framework of Amaryllidaceae alkaloids, and there are several major contributions (*330,346,375*). The preliminary assignments are made on the basis of chemical shifts and multiplicities of the signals (by DEPT experiment). The use of 2D NMR techniques, such as HMQC and HMBC, allow the assignments to be corroborated. Tables XVIII–XXIV show a compilation of the different $^{13}$C NMR spectra classified according to the different types.

The $^{13}$C NMR spectra of *Narcissus* alkaloids can be divided in two regions. The low-field region (>90 ppm) contains signals of the carbonyl group, the olefinic and aromatic carbons, as well as that of the methylenedioxy group. The other signals corresponding to the saturated carbon resonances are found in the high-field region, the *N*-methyl being the only characteristic group, easily recognizable by a quartet signal between 40 and 46 ppm.

The effect of a substituent (OH, OMe, OAc) on the carbon resonances is of considerable importance in localizing the position of the functional groups.

The analysis of the spectra allows conclusions to be drawn about the following aspects:

- The number of methine olefinic carbons.
- The presence and nature of the nitrogen substituent.

TABLE XVIII.
$^{13}C$ NMR Data of Lycorine-Type Alkaloids.

| Alkaloid | 1 | 3 | 5 | 7 | 13 | 14 | 15 | 19 | 21 | 22 | 23 | 24 | 25 |
|---|---|---|---|---|---|---|---|---|---|---|---|---|---|
| C-1 | 70.2 | 70.7 | 68.5 | 68.3 | 69.8 | 70.3 | 69.9 | 65.4 | 119.4 | 120.0 | 102.2 | 105.4 | 31.7 |
| C-2 | 71.7 | 71.8 | 73.9 | 81.0 | 33.5 | 28.8 | 28.8 | 34.7 | 123.9 | 131.2 | 164.5 | 163.8 | 21.8 |
| C-3 | 118.5 | 118.6 | 114.0 | 115.1 | 116.2 | 115.0 | 114.6 | 123.2 | 123.9 | 125.6 | 118.0 | 119.1 | 23.0 |
| C-4 | 141.7 | 141.7 | 144.6 | 143.9 | 139.6 | 139.2 | 139.4 | 133.9 | 124.3 | 123.2 | 139.7 | 140.9 | 40.0 |
| C-4a | 60.8 | 61.4 | 60.9 | 60.9 | 59.5 | 60.4 | 60.2 | 71.2 | 131.3 | 136.0 | 133.0 | 133.2 | 162.4 |
| C-6 | 56.7 | 56.4 | 56.7 | 56.6 | 55.3 | 51.9 | 51.8 | 66.2 | 157.5 | 144.7 | 142.6 | 143.1 | 138.5 |
| C-6a | 129.7 | 127.3 | 127.9 | 129.3 | 126.0 | 125.0 | 129.2 | 124.8 | 129.0 | 121.3 | 122.9 | 125.3 | 119.3 |
| C-7 | 107.0 | 110.5 | 110.4 | 110.8 | 112.4 | 113.4 | 111.5 | 115.3 | 108.6 | 110.5 | 111.0 | 109.2 | 96.8 |
| C-8 | 145.2 | 146.2 | 146.0 | 147.8 | 146.0 | 145.4 | 149.4 | 149.1 | 149.8 | 151.7 | 153.1 | 153.1 | 150.5 |
| C-9 | 145.6 | 145.1 | 146.0 | 147.6 | 144.4 | 144.0 | 137.5 | 150.2 | 153.3 | 157.9 | 159.1 | 158.6 | 146.8 |
| C-10 | 105.1 | 111.3 | 111.3 | 108.0 | 110.2 | 110.2 | 121.2 | 111.9 | 103.3 | 102.4 | 104.6 | 103.1 | 142.0 |
| C-10a | 129.6 | 126.5 | 127.2 | 126.6 | 128.8 | 129.2 | 132.5 | 123.7 | 120.3 | 130.6 | 131.4 | 134.3 | 127.1 |
| C-10b | 40.2 | 39.4 | 41.1 | 41.5 | 41.6 | 38.0 | 38.2 | 42.1 | 117.0 | 136.0 | 126.1 | 127.6 | 135.5 |
| C-11 | 28.1 | 28.3 | 28.8 | 28.5 | 27.3 | 28.4 | 28.6 | 26.2 | 27.5 | 27.4 | 27.8 | 29.4 | 26.8 |
| C-12 | 53.3 | 53.9 | 53.9 | 53.8 | 52.1 | 52.4 | 52.3 | 67.5 | 46.8 | 55.4 | 57.0 | 58.0 | 57.4 |
| $OCH_2O$ | 100.6 | — | — | — | — | — | — | — | — | — | — | 106.4 | — |
| OMe | — | 56.1 | 56.0 | 57.3 | 55.9 | 56.0 | 55.9 | 56.2 | 56.3 | 56.7 | 57.8 | — | 62.5 |
| OMe | — | — | — | 56.0 | — | — | — | — | 56.2 | 56.4 | 57.4 | — | 62.4 |
| OMe | — | — | — | 55.9 | — | — | — | — | — | — | 57.0 | — | 56.8 |
| O$\underline{\text{OC}}$Me | — | — | 170.9 | — | — | 170.6 | 170.6 | — | — | — | — | — | — |
| OOC$\underline{\text{Me}}$ | — | — | 21.3 | — | — | 21.4 | 20.6 | — | — | — | — | — | — |
| O$\underline{\text{OC}}$Me | — | — | — | — | — | — | 169.2 | — | — | — | — | — | — |
| OOC$\underline{\text{Me}}$ | — | — | — | — | — | — | 21.3 | — | — | — | — | — | — |
| Solvent | a | b | b | d | d | d | d | c | b | b | c | c | d |
| MHz | 75 | 50 | 50 | 50 | 100 | 100 | 100 | 50 | 50 | 50 | 50 | 50 | 62 |
| Reference | *(298)* | *(113)* | *(113)* | *(135)* | *(167)* | *(167)* | *(167)* | *(316)* | *(114)* | *(71)* | *(68)* | *(319)* | *(67)* |

Solvent a: DMSO-$d_6$, b: $CDCl_3$-$CD_3OD$, c: $CD_3OD$, d: $CDCl_3$.

$^{13}C$ NMR Data of Homolycorine-Type Alkaloids.

| Alkaloid | 26 | 27 | 28 | 29 | 30 | 31 | 32 | 33 | 34 | 35 | 36 | 37 | 38 | 39 | 40 |
|---|---|---|---|---|---|---|---|---|---|---|---|---|---|---|---|
| C-1 | 77.7 | 79.2 | 77.5 | 77.9 | 77.3 | 75.6 | 82.5 | 79.7 | 82.2 | 67.6 | 66.6 | 66.7 | 66.4 | 72.8 | 78.8 |
| C-2 | 31.3 | 32.1 | 30.9 | 31.3 | 31.1 | 31.2 | 66.2 | 69.2 | 66.2 | 32.6 | 31.3 | 31.7 | 31.3 | 68.3 | 31.4 |
| C-3 | 115.2 | 117.1 | 116.5 | 116.1 | 115.6 | 115.1 | 118.5 | 115.1 | 119.4 | 117.2 | 115.7 | 115.7 | 115.6 | 119.4 | 121.0 |
| C-4 | 140.9 | 141.3 | 138.7 | 140.0 | 140.2 | 139.4 | 143.3 | 147.4 | 142.4 | 140.8 | 140.7 | 140.6 | 140.3 | 144.0 | 135.6 |
| C-4a | 66.6 | 68.0 | 66.8 | 66.9 | 66.8 | 59.1 | 66.4 | 67.5 | 67.1 | 68.5 | 67.3 | 67.5 | 67.5 | 68.3 | 79.3 |
| C-6 | 165.9 | 167.9 | 166.3 | 167.0 | 165.4 | 165.1 | 165.7 | 165.4 | 165.0 | 92.3 | 98.2 | 91.8 | 98.2 | 98.5 | 167.1 |
| C-6a | 116.9 | 117.9 | 116.5 | 115.6 | 118.5 | 118.0 | 118.5 | 115.4 | 118.0 | 131.3 | 130.5 | 132.0 | 131.8 | 126.9 | 117.6 |
| C-7 | 111.9 | 116.9 | 115.9 | 112.7 | 109.7 | 109.9 | 112.0 | 115.0 | 109.8 | 111.9 | 109.7 | 107.4 | 107.3 | 107.6 | 117.1 |
| C-8 | 148.9 | 147.9 | 146.3 | 148.1 | 151.8 | 152.2 | 147.6 | 148.2 | 148.0 | 149.6 | 148.3 | 147.0 | 146.8 | 147.1 | 147.9 |
| C-9 | 153.1 | 153.8 | 152.2 | 152.0 | 147.8 | 147.8 | 151.8 | 152.0 | 151.9 | 149.6 | 148.3 | 147.0 | 146.7 | 147.1 | 153.4 |
| C-10 | 110.8 | 112.3 | 110.9 | 115.0 | 108.6 | 107.4 | 114.7 | 112.7 | 108.5 | 114.0 | 112.4 | 109.8 | 109.5 | 109.8 | 112.6 |
| C-10a | 137.8 | 137.6 | 135.7 | 138.0 | 139.7 | 138.7 | 137.2 | 136.7 | 138.8 | 128.6 | 125.6 | 128.2 | 126.9 | 131.2 | 134.9 |
| C-10b | 43.8 | 43.7 | 42.2 | 43.0 | 43.8 | 43.0 | 38.2 | 39.9 | 38.4 | 44.7 | 44.0 | 44.2 | 43.9 | 39.2 | 37.4 |
| C-11 | 28.1 | 28.6 | 27.4 | 27.9 | 27.9 | 29.6 | 27.0 | 28.0 | 27.3 | 28.7 | 27.8 | 28.1 | 27.9 | 27.7 | 26.2 |
| C-12 | 56.6 | 57.3 | 56.0 | 56.5 | 56.3 | 44.1 | 55.8 | 56.3 | 55.9 | 57.6 | 56.6 | 56.7 | 56.5 | 56.3 | 70.7 |
| $OCH_2O$ | — | — | — | — | 102.0 | 102.0 | — | — | 102.1 | — | — | 101.0 | 100.8 | 101.1 | — |
| OMe | 56.4 | 56.7 | 56.0 | 56.2 | — | — | 55.2 | 56.2 | — | 56.5 | 55.8 | — | 55.2 | 55.5 | 56.8 |
| OMe | 56.2 | — | — | — | — | — | — | — | — | 56.5 | 55.5 | — | — | — | — |
| OMe | — | — | — | — | — | — | — | — | — | — | 55.1 | — | — | — | — |
| NMe | 44.2 | 44.1 | 42.9 | 43.3 | 43.5 | — | 42.2 | 43.3 | 42.9 | 44.3 | 43.8 | 44.3 | 44.0 | 43.8 | 56.2 |
| OO$\underline{C}$Me | — | — | 174.9 | — | — | — | — | 169.4 | — | — | — | — | — | — | — |
| OOC$\underline{Me}$ | — | — | 21.0 | — | — | — | — | 21.1 | — | — | — | — | — | — | — |
| OO$\underline{C}$CH$_2$CHOHMe | — | — | — | — | — | — | — | 170.8 | — | — | — | — | — | — | — |
| OOC$\underline{CH_2}$CHOHMe | — | — | — | — | — | — | — | 40.8 | — | — | — | — | — | — | — |
| OOCCH$_2$$\underline{CH}$OHMe | — | — | — | — | — | — | — | 66.7 | — | — | — | — | — | — | — |
| OOCCH$_2$CHOH$\underline{Me}$ | — | — | — | — | — | — | — | 19.9 | — | — | — | — | — | — | — |
| Solvent | d | c | b | b | d | d | b | b | d | c | d | d | d | d | b |
| MHz | 63 | 50 | 62 | 50 | 100 | 100 | 50 | 50 | 50 | 62 | 50 | 100 | 100 | 50 | 50 |
| Reference | *(321)* | *(324)* | *(70)* | *(82)* | *(165)* | *(165)* | *(138)* | *(72)* | *(71)* | *(106)* | *(104)* | *(165)* | *(165)* | *(71)* | *(137)* |

Solvent a: DMSO-$d_6$, b: $CDCl_3$-$CD_3OD$, c: $CD_3OD$, d: $CDCl_3$.

TABLE XX.
$^{13}C$ NMR Data of Hemanthamine-Type Alkaloids.

| **Alkaloid** | **42** | **43** | **44** | **45** | **47** | **48** | **49** | **50** | **53** | **54** | **55** | **56** | **57** | **59** | **61** |
|---|---|---|---|---|---|---|---|---|---|---|---|---|---|---|---|
| C-1 | 132.1 | 133.2 | 132.3 | 130.3 | 128.2 | 132.4 | 132.0 | 132.4 | 128.0 | 127.5 | 126.7 | 126.3 | 123.6 | 25.4 | 131.4 |
| C-2 | 127.4 | 127.6 | 128.6 | 127.6 | 127.3 | 125.6 | 125.9 | 125.2 | 127.2 | 129.5 | 132.3 | 132.8 | 136.0 | 26.4 | 129.1 |
| C-3 | 64.0 | 64.6 | 64.5 | 63.3 | 70.7 | 72.4 | 72.4 | 72.4 | 73.0 | 72.4 | 72.5 | 72.5 | 76.0 | 69.9 | 72.5 |
| C-4 | 32.7 | 32.8 | 33.5 | 31.6 | 25.6 | 28.7 | 28.1 | 28.5 | 29.5 | 28.2 | 27.8 | 27.6 | 30.1 | 30.3 | 26.7 |
| C-4a | 62.3 | 64.0 | 64.1 | 62.8 | 64.8 | 62.2 | 56.3 | 57.1 | 62.7 | 62.8 | 61.6 | 56.2 | 66.1 | 66.0 | 53.4 |
| C-6 | 62.8 | 63.7 | 62.6 | 60.9 | 57.1 | 87.0 | 88.9 | 96.4 | 63.3 | 61.0 | 85.8 | 88.4 | 63.4 | 59.1 | 57.6 |
| C-6a | 126.3 | 126.3 | 125.8 | 121.1 | 116.5 | 127.4 | 126.5 | 125.5 | 126.9 | 126.2 | 129.2 | 127.8 | 126.5 | 118.4 | 131.8 |
| C-7 | 106.9 | 107.9 | 111.9 | 109.8 | 110.0 | 111.0 | 112.5 | 112.4 | 106.9 | 106.6 | 108.2 | 109.5 | 106.9 | 109.0 | 105.7 |
| C-8 | 145.6 | 147.9 | 149.2 | 144.8 | 148.9 | 147.8 | 147.8 | 147.3 | 146.5 | 146.5 | 146.5 | 146.4 | 146.2 | 146.3 | 146.2 |
| C-9 | 146.0 | 148.4 | 149.1 | 146.0 | 149.0 | 148.5 | 148.7 | 148.2 | 147.0 | 146.5 | 147.4 | 147.7 | 146.5 | 145.4 | 145.5 |
| C-10 | 102.7 | 104.4 | 107.8 | 109.2 | 106.2 | 105.4 | 105.4 | 105.0 | 103.3 | 103.4 | 102.7 | 102.9 | 103.2 | 109.6 | 105.4 |
| C-10a | 138.3 | 136.9 | 138.7 | 136.5 | 138.0 | 136.4 | 137.4 | 138.1 | 135.0 | 134.2 | 134.7 | 135.8 | 135.3 | 137.1 | 129.1 |
| C-10b | 44.2 | 51.4 | 45.4 | 44.0 | 44.9 | 44.7 | 44.1 | 43.7 | 50.0 | 49.2 | 50.7 | 50.3 | 50.3 | 42.4 | 36.9 |
| C-11 | 44.2 | 80.8 | 45.0 | 42.8 | 40.6 | 42.3 | 40.9 | 41.1 | 80.0 | 80.2 | 79.2 | 78.3 | 80.0 | 35.5 | 75.3 |
| C-12 | 53.5 | 61.5 | 54.0 | 52.8 | 52.3 | 42.0 | 47.9 | 48.4 | 61.5 | 60.4 | 52.0 | 57.8 | 61.2 | 50.8 | 81.0 |
| $OCH_2O$ | 100.6 | 102.4 | — | — | — | — | — | — | 101.0 | 100.8 | 101.0 | 101.0 | 100.9 | — | 100.8 |
| OMe | — | — | 56.8 | 55.8 | 56.1 | 56.6 | 56.6 | 56.8 | 56.0 | 56.5 | 56.8 | 56.5 | 55.8 | 55.8 | 56.5 |
| OMe | — | — | 56.5 | — | 56.1 | 56.6 | 56.6 | 56.0 | — | — | — | — | — | — | — |
| OMe | — | — | — | — | 56.1 | 56.1 | 56.0 | 55.8 | — | — | — | — | — | — | — |
| OMe | — | — | — | — | — | — | — | 55.7 | — | — | — | — | — | — | — |
| $OO\underline{C}Me$ | — | — | — | — | — | — | — | — | — | 170.0 | — | — | — | 170.5 | — |
| $OOC\underline{Me}$ | — | — | — | — | — | — | — | — | — | 21.2 | — | — | — | 21.0 | — |
| $\underline{CH_2}OOCMe$ | — | — | — | — | — | — | — | — | — | — | — | — | — | — | 64.6 |
| $CH_2OO\underline{C}Me$ | — | — | — | — | — | — | — | — | — | — | — | — | — | — | 170.8 |
| $CH_2OOC\underline{Me}$ | — | — | — | — | — | — | — | — | — | — | — | — | — | — | 20.9 |
| Solvent | d | c | c | b | d | d | d | d | d | d | d | d | d | d | d |
| MHz | 20 | 67.5 | 75 | 50 | 50 | 50 | 50 | 50 | 20 | 50 | 125 | 125 | 75 | 50 | 50 |
| Reference | *(330)* | *(376)* | *(336)* | *(108)* | *(137)* | *(135)* | *(135)* | *(137)* | *(330)* | *(74)* | *(73)* | *(73)* | *(298)* | *(140)* | *(74)* |

Solvent a: DMSO-$d_6$, b: $CDCl_3$-$CD_3OD$, c: $CD_3OD$, d: $CDCl_3$.

TABLE XXI.
$^{13}$C NMR Data of Tazettine-Type Alkaloids.

| Alkaloid | 62 | 63 | 65 | 67 |
|---|---|---|---|---|
| C-1 | 130.5 | 130.1 | 131.3 | 132.4 |
| C-2 | 128.4 | 128.9 | 126.0 | 136.5 |
| C-3 | 72.4 | 72.1 | 72.7 | 63.6 |
| C-4 | 26.6 | 25.4 | 29.8 | 34.4 |
| C-4a | 69.9 | 68.2 | 63.3 | 68.5 |
| C-6 | 65.0 | 62.6 | 168.5 | 62.2 |
| C-6a | 127.8 | 126.2 | 118.6 | 131.0 |
| C-7 | 103.7 | 108.5 | 103.8 | 107.3 |
| C-8 | 146.3 | 146.6 | 147.1 | 148.4 |
| C-9 | 146.3 | 146.2 | 152.3 | 147.3 |
| C-10 | 109.1 | 104.2 | 111.0 | 111.0 |
| C-10a | 125.4 | 130.9 | 142.2 | 125.0 |
| C-10b | 50.1 | 50.0 | 46.2 | 50.3 |
| C-11 | 101.7 | 102.6 | 80.1 | 82.7 |
| C-12 | 61.7 | 64.5 | 53.5 | 55.7 |
| $OCH_2O$ | 100.6 | 100.9 | 102.1 | 101.9 |
| OMe | 55.6 | 56.7 | 56.2 | — |
| NMe | 41.9 | 40.6 | 42.8 | — |
| Solvent | d | d | d | c |
| MHz | 16 | 75 | 50 | 62 |
| Ref. | *(346)* | *(355)* | *(79)* | *(76)* |

Solvent a: DMSO-$d_6$, b: $CDCl_3$-$CD_3OD$, c: $CD_3OD$, d: $CDCl_3$.

- The existence of a lactonic carbonyl group.
- The presence of a quaternary carbon signal assignable to C-10b in the chemical shift range of 42–50 ppm.

## C. MASS SPECTROMETRY

Extensive studies on the mass spectrometry of Amaryllidaceae alkaloids by electron impact were reported in the 1960s and 1970s *(94,310,328,353,378–385)*. The fragmentation patterns in the electron impact mass spectrometry (EIMS) of various skeletal types are fairly well documented and have considerable diagnostic value.

### *1. Lycorine Type*

The molecular ion appears as a quite intense peak, and generally suffers the loss of water, as well as C-1 and C-2 and their substituents, by a retro-Diels–Alder fragmentation (Fig. 12). The loss of water is not present in the spectra of acetyl derivatives.

TABLE XXII.
$^{13}C$ NMR Data of Narciclasine-Type Alkaloids.

| **Alkaloid** | **68** | **70** | **71** | **72** |
|---|---|---|---|---|
| C-1 | 124.7 | 122.0 | 120.2 | 129.9 |
| C-2 | 69.1 | 126.7 | 125.9 | 118.0 |
| C-3 | 72.3 | 128.1 | 131.0 | 129.1 |
| C-4 | 68.8 | 129.9 | 133.0 | 110.7 |
| C-4a | 52.8 | 143.8 | 136.5 | 146.7 |
| C-6 | 172.1 | 151.8 | 152.8 | 63.5 |
| C-6a | 129.2 | 123.1 | 126.6 | 134.0 |
| C-7 | 168.9 | 105.5 | 108.3 | 109.7 |
| C-8 | 152.3 | 148.1 | 159.2 | 147.5 |
| C-9 | 144.8 | 148.2 | 152.0 | 147.4 |
| C-10 | 95.7 | 99.9 | 102.2 | 110.2 |
| C-10a | 132.1 | 130.3 | 122.0 | 131.2 |
| C-10b | 133.3 | 124.3 | 135.3 | 127.2 |
| $OCH_2O$ | 102.0 | 101.9 | 105.7 | 101.3 |
| NMe | — | — | 45.9 | 30.8 |
| Solvent | a | d | c | d |
| MHz | 68 | 50 | 50 | 50 |
| Reference | (*122*) | (*99*) | (*79*) | (*79*) |

Solvent a: DMSO-$d_6$, b: $CDCl_3$-$CD_3OD$, c: $CD_3OD$, d: $CDCl_3$.

The ease of the loss of water from the molecular ion was found to be greatly dependent on the stereochemistry of the C-2 hydroxyl group. Thus, in the mass spectrum of lycorine (**1**) the relative intensity is low, while in 2-epilycorine it is the base peak (*310*).

### 2. *Homolycorine Type*

In this type of structure, the cleavage of the labile bonds in ring C by a retro-Diels–Alder reaction is dominant, generating two fragments: one, the most characteristic, represents the pyrrolidine ring (together with the substituents in position 2), and the other (a less-abundant fragment) encompasses the aromatic lactone or hemilactone moiety (Fig. 13). A further general and noteworthy feature is the low abundance of the molecular ion in all alkaloids with a double bond $\Delta^{3,4}$ (*378*).

### 3. *Hemanthamine Type*

The following observations about this type of alkaloids should be considered:

(i) In most cases, the molecular ion is the base peak.
(ii) The aromatic ring plays an important role in the stabilization of the ions, which is retained in all fragments of high mass, while the nitrogen atom is often lost.

TABLE XXIII.
$^{13}$C NMR Data of Montanine-Type Alkaloids.

| Alkaloid | 73 | 74 |
|---|---|---|
| C-1 | 115.6 | 114.7 |
| C-2 | 68.2 | 35.5 |
| C-3 | 70.4 | 72.3 |
| C-4 | 29.3 | 75.5 |
| C-4a | 58.5 | 70.0 |
| C-6 | 59.2 | 62.0 |
| C-6a | 122.2 | 125.2 |
| C-7 | 106.2 | 107.8 |
| C-8 | 146.8 | 148.3 |
| C-9 | 146.3 | 147.6 |
| C-10 | 106.8 | 108.3 |
| C-10a | 131.1 | 133.6 |
| C-11 | 44.7 | 46.4 |
| C-11a | 150.0 | 147.5 |
| C-12 | 54.5 | 56.9 |
| $OCH_2O$ | 100.7 | 102.1 |
| Solvent | c | c |
| MHz | 50 | 50 |
| Reference | (*80*) | (*80*) |

Solvent a: DMSO-$d_6$, b: $CDCl_3$-$CD_3OD$, c: $CD_3OD$, d: $CDCl_3$.

(iii) The fragmentation mechanisms are initiated by the rupture of a bond $\beta$ to the nitrogen atom, which implies the opening of the C-11/C-12 bridge (*382,383*).

a. *Alkaloids with a saturated ring C and no bridge substituent.* The configuration of the ring C substituent plays a minor role in the fragmentation process.
b. *Alkaloids with a double bond ($\Delta^{1,2}$) in ring C and no bridge substituent.* The fragmentation pattern involves ruptures of C-4a/C-10b and C-3/C-4 bonds. A characteristic feature is the loss of a nitrogen-containing moiety, $C_3H_5N$ [$M^+$-55].
c. *Alkaloids with a double bond ($\Delta^{1,2}$) in ring C and a hydroxyl substituent at C-11.* The presence of a hydroxyl group on C-11 is responsible for dramatic changes in the fragmentation pattern (Fig. 14), and it is profoundly influenced by the stereochemistry. There are three fundamental patterns of fragmentation:

- Loss of $CH_3OH$: it is more favorable when the two-carbon bridge and the substituent on C-3 are on the same side of the molecule.
- Loss of $C_2H_6N$: the relative significance of the loss of this neutral nitrogen moiety is governed by the ease with which the methanol is eliminated.
- Loss of CHO: A peak at $m/z$ [$M^+$-29] due to thc loss of an aldehyde radical is present in all alkaloids of this type.

TABLE XXIV.
$^{13}$C NMR Data of Galanthamine-Type Alkaloids.

| Alkaloid | 75 | 77 | 78 | 79 | 80 | 81 | 82 | 83 | 84 | 86 |
|---|---|---|---|---|---|---|---|---|---|---|
| C-1 | 88.1 | 86.2 | 88.2 | 88.4 | 87.9/88.7 | 88.8 | 88.7 | 88.0 | 89.8 | 90.0 |
| C-2 | 30.0 | 27.7 | 30.3 | 29.9 | 29.8 | 31.3 | 29.2 | 37.3 | 31.5 | 31.7 |
| C-3 | 62.2 | 63.2 | 61.7 | 61.9 | 61.4 | 62.6 | 61.4 | 194.4 | 65.2 | 65.5 |
| C-4 | 126.8 | 123.3 | 129.2 | 127.6 | 128.0/128.2 | 128.6 | 128.2 | 127.1 | 27.6 | 27.8 |
| C-4a | 126.0 | 121.8 | 125.9 | 127.1 | 125.8/126.2 | 128.1 | 125.1 | 144.3 | 31.7 | 37.5 |
| C-6 | 60.5 | 59.9 | 51.6 | 53.5 | 41.0/52.8 | 61.6 | 58.4 | 60.7 | 60.4 | 53.8 |
| C-6a | 129.5 | 127.2 | 123.6 | 132.1 | 127.4 | 128.0 | 127.1 | 129.4 | 129.1 | 127.2 |
| C-7 | 121.6 | 130.2 | 122.8 | 120.9 | 119.9/121.6 | 123.1 | 121.2 | 122.0 | 121.6 | 120.8 |
| C-8 | 110.5 | 111.7 | 112.2 | 111.3 | 111.4 | 116.7 | 111.3 | 111.8 | 111.3 | 110.9 |
| C-9 | 145.5 | 146.7 | 146.9 | 146.2 | 146.2/146.5 | 146.8 | 146.8 | 147.0 | 146.2 | 146.6 |
| C-10 | 144.0 | 144.4 | 145.3 | 144.1 | 144.3/144.5 | 142.6 | 144.5 | 144.0 | 144.0 | 144.2 |
| C-10a | 132.7 | 131.9 | 133.1 | 133.1 | 131.8/131.9 | 134.2 | 131.2 | 130.5 | 136.3 | 136.6 |
| C-10b | 48.2 | 47.8 | 48.2 | 48.5 | 48.0/48.1 | 47.9 | 48.1 | 49.0 | 46.7 | 47.4 |
| C-11 | 34.0 | 33.7 | 35.6 | 39.7 | 46.6/46.7 | 35.5 | 35.1 | 33.2 | 23.9 | 24.2 |
| C-12 | 54.3 | 53.4 | 45.9 | 46.8 | 35.7/39.1 | 55.2 | 35.2 | 54.1 | 54.1 | 47.4 |
| OMe | 55.5 | 56.0 | 56.2 | 55.9 | 55.8 | — | 55.6 | 56.0 | 55.9 | 56.1 |
| NMe | 42.2 | 40.9 | — | — | — | 43.1 | — | 42.4 | 41.9 | — |
| NCHO | — | — | — | — | 162.1/162.7 | — | — | — | — | — |
| $\underline{\text{OC}}$Me | — | 170.9 | — | — | — | — | 161.2 | — | — | — |
| OC$\underline{\text{Me}}$ | — | 21.4 | — | — | — | — | 21.4 | — | — | — |
| Solvent | d | d | b | d | b | c | d | d | d | d |
| MHz | — | 100 | 50 | 50 | 50 | 75 | 50 | 90 | 25 | 50 |
| Reference | (*362*) | (*167*) | (*141*) | (*107*) | (*83*) | (*377*) | (*117*) | (*181*) | (*367*) | (*107*) |

Solvent a: DMSO-$d_6$, b: $CDCl_3$-$CD_3OD$, c: $CD_3OD$, d: $CDCl_3$.

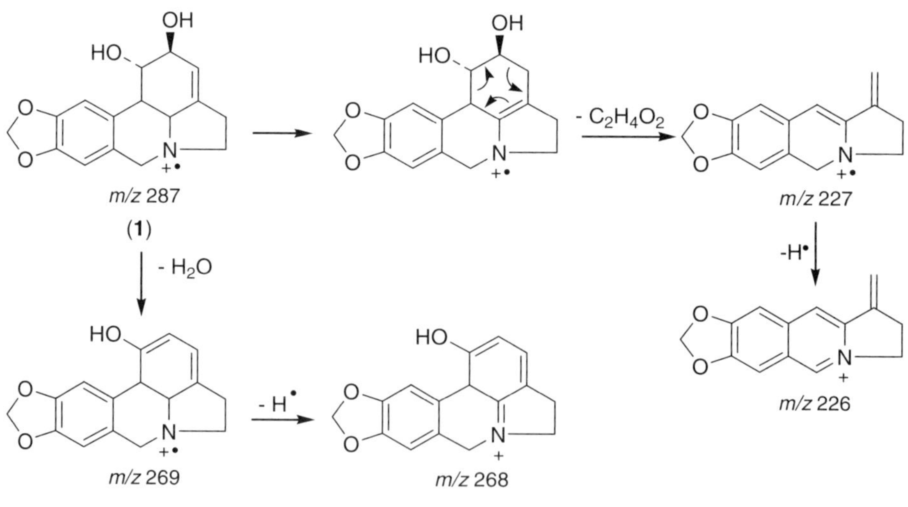


Figure 12. Mass fragmentation pattern of lycorine (**1**).

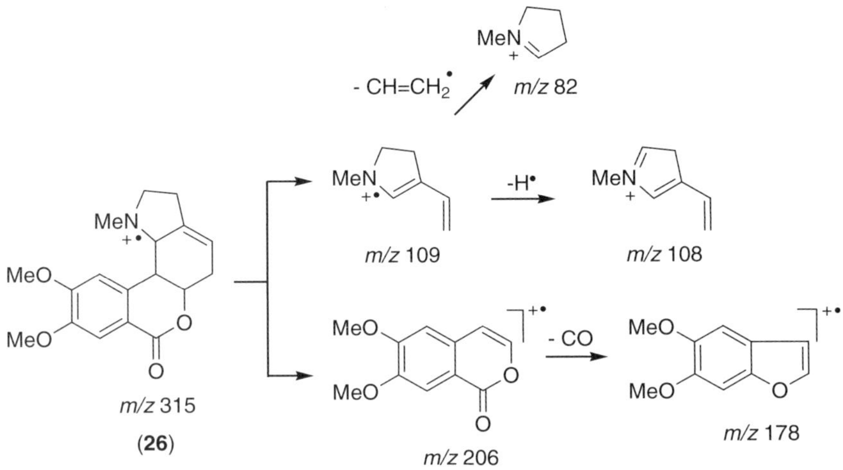


Figure 13. Mass fragmentation pattern of homolycorine (**26**).

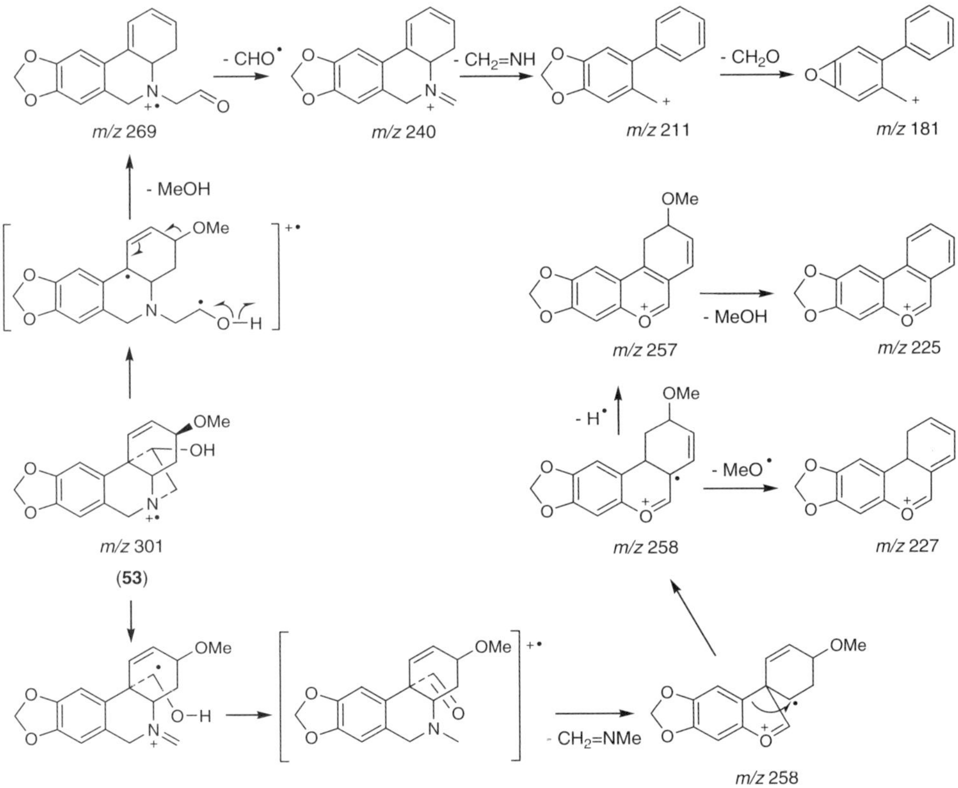


Figure 14. Mass fragmentation pattern of hemanthamine (**53**).

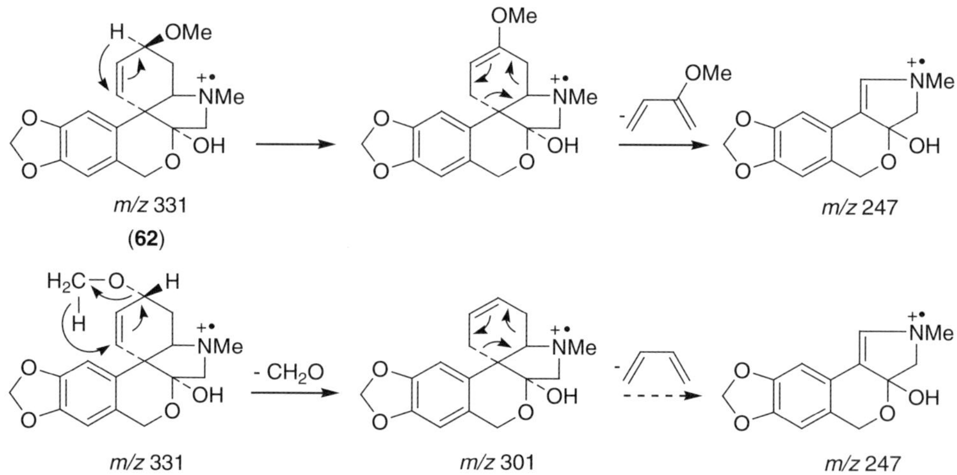


Figure 15. Mass fragmentation pattern of tazettine (**62**) and criwelline (**63**).

## 4. *Tazettine Type*

Minor changes in stereochemistry are sufficient to cause appreciable differences in the stereoisomers in this kind of structure. Thus, in the MS of tazettine (**62**), with a $\beta$-configuration of the methoxyl group at C-3, the dominant ion occurs at $m/z$ [$M^+$-84], following a C-ring fragmentation by a retro-Diels–Alder process. In contrast, the mass spectrum of its epimer criwelline (**63**) contains a peak of low abundance at $m/z$ [$M^+$-84] (Fig. 15). Ions occur in both stereoisomers owing to the successive loss of a methyl radical and water from the molecular ion (*353*).

## 5. *Montanine Type*

The mass spectral fragmentation patterns observed for alkaloids containing the 5,11-methanomorphanthridine nucleus greatly depend on the nature and particular configuration of the substituents at C-2 and C-3. Thus, all the alkaloids that possess a methoxyl group give rise to an $M^+$-31 ion.

The configuration of the C-2 substituent has a considerable effect on the extent to which the retro-Diels–Alder fragmentation ion is observed (Fig. 16). There is a definite enhancement of this fragmentation when the C-2 has an $\alpha$-configuration (*94*).

## 6. *Galanthamine Type*

In this type of structure, the intense molecular ion as well as the [$M^+$-1] peak, the breaking of ring C (losing a $C_4H_6O$ fragment), and the elimination of elements of ring B (including the nitrogen atom) are characteristic (Fig. 17). This behavior is similar for the dihydro derivatives (*380*).

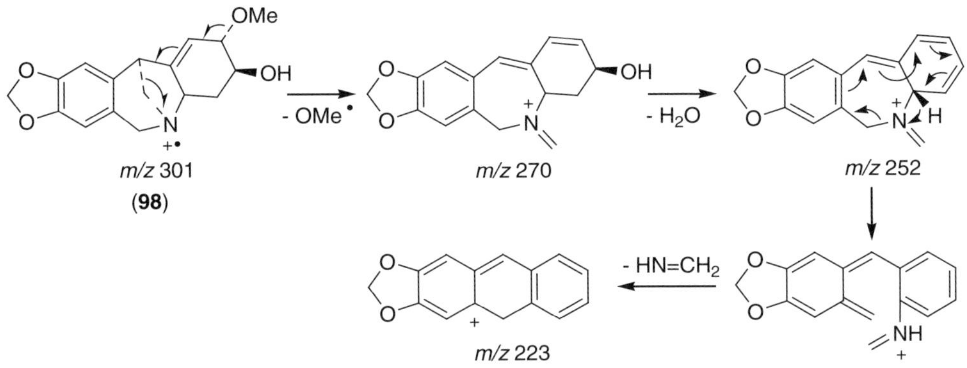


Figure 16. Mass fragmentation pattern of montanine (**98**).

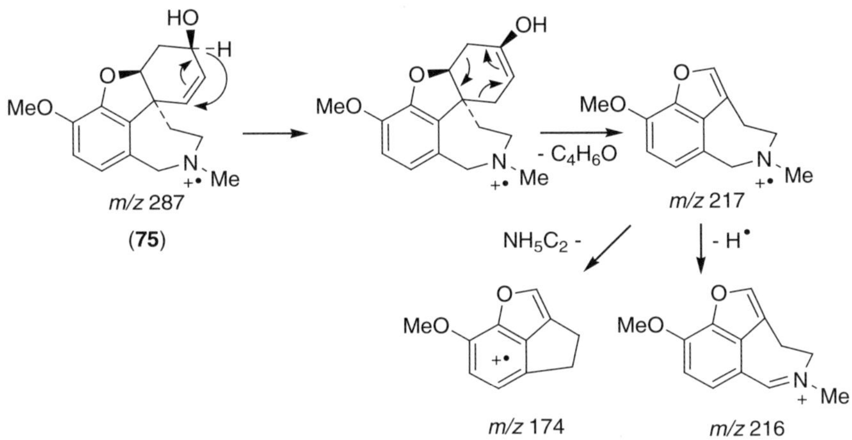


Figure 17. Mass fragmentation pattern of galanthamine (**75**).

## VI. Biological and Pharmacological Activities

### A. TRADITIONAL USES

#### *1. Traditional Medicinal Usage*

Considering that the *Narcissus* species are a rich source of alkaloids, it is not surprising that, despite their lethal potential, plants of this genus have been used throughout history in traditional medicine to treat a variety of medicinal problems (*386*). *N. poeticus*, for example, is described in the Bible as a well-established treatment for symptoms that would now be defined as cancer (*387*). In the fourth century BC, Hippocrates of Cos (the 'Father of Medicine') recommended a pessary prepared from *Narcissus* oil (probably *N. poeticus*) for the management of uterine tumors (*388*). In the first century AD, Pliny the Elder also recorded the topical use of

*N. poeticus* and *N. pseudonarcissus* for this purpose. It is now known that *N. poeticus* contains 0.012% of the antineoplastic agent narciclasine (**68**) in the fresh bulb (*14,101*). Arabian, North African, and Chinese medical practitioners of the Middle Ages continued using *Narcissus* oil in cancer treatment (*389*). For example, bulbs of *N. tazetta* L. var. *chinensis*, cultivated in China as a decorative plant, were used topically for the treatment of tumors in folk medicine. In this case, pretazettine (**64**) was established to be one of the antitumor active compounds (*133,390*). The bulbs of *N. tazetta* continue to be used in Turkey as a home remedy for the treatment of abscesses because of their antiphlogistic and analgesic properties (*391*).

*Narcissus* species have also been used as applications for wounds, hard imposthumes, strained sinews, stiff or painful joints, and other local ailments, being the basis of an ancient ointment called 'Narcissimum' (*14*). The powdered flowers have been used as an emetic, and, in the form of syrup or infusions, have been considered useful for relieving the congestive bronchial catarrh in children, and also epidemic dysentery (*392*). In France, narcissus flowers have been used as an antispasmodic (*14*). The Arabians commended the oil as a cure for baldness and as an aphrodisiac (*393*). In John K'Eogh's Irish Herbal, the roots pounded with honey were recommended to treat burns, bruised sinews, dislocations and old aches, remove freckles, heal abscesses and sores, and draw out thorns and splinters (*14,394*). The bulbs of *N. tazetta* have also been used as a contraceptive. The influence of the daffodil on the nervous system has led to the use of its flowers and bulbs for hysterical affections and even epilepsy. A homeopathic medicine is also made from the bulbs and used for respiratory disease, particularly bronchitis and whooping cough (*14*).

### 2. *Toxic and Hallucinogenic Effects*

Plants of this genus have been used throughout history as a stimulant to induce trance and hallucinations, and as an agent in suicide. Socrates called the narcissus the 'Chaplet of the infernal Gods', because of its narcotic effects. Pliny, in turn, describes the narcissus as '*narce narcissum dictum, non a fabuloso puero*', which translated means 'named narcissus from *narce*, not from the fabulous boy'. The Greek *narkao*, meaning to be numb, originates in the narcotic properties of the plant (*14*).

It has been known for a long time that daffodil ingestion is very dangerous, resulting in toxic symptoms in both man and warm-blooded animals such as cattle, goats, cats, and pigs (*393,395,396*). After ingestion of *Narcissus* species such as *N. pseudonarcissus* or *N. jonquilla* (*115*), the first visible symptoms are salivation, acute abdominal pains, nausea, vomiting, and diarrhea, followed by neurological (trembling, convulsions, paralysis, etc.) and cardiac sequels, and sometimes resulting in death, if eaten in larger quantities. There have been many cases of poisoning or death when the bulbs have been cooked by mistake in the place of leeks or onions (*393*). Recovery, however, is usually complete within a few hours without any treatment being necessary (*14,397*), but in cases of massive ingestion, activated charcoal, salts, and laxatives are administered. When symptoms are severe, atropine sulphate is given by intravenous injection and it may be necessary to induce vomiting or remove stomach contents (*398*).

The good news is that the bulb tastes awful, making it highly unlikely that anyone could even keep down one bite. In an experiment performed with several plant species that are not consumed by animals, the plant with the most repellent activity was the daffodil, specifically *N. pseudonarcissus* (*399*). As a consequence, an animal repellent containing alkaloids isolated from members of the genus *Narcissus* has been designed to repel animals from vegetation by rendering it unpalatable, being also effective against fungi, molds, and bacteria (*400,401*).

Not all *Narcissus* species are equally dangerous. The bulbs of *N. poeticus*, for example, are more dangerous than those of *N. pseudonarcissus*. Neither do all plant tissues have the same concentration or profile of alkaloids. Thus, the alkaloid content of *N. papyraceus* is five times higher in the aerial part than in the bulbs, being toxic for herbivorous mammals (*137*). The distribution of the alkaloids in the plant tissues can be related with the plant defense mechanism.

Some *Narcissus* species, such as *N. pseudonarcissus*, can produce harmful effects without being swallowed. Thus, those who pick and pack the flowers are liable to develop dermatitis, probably due to the irritant effects of the sap or an allergic reaction (*397,402–406*). The compounds responsible for the irritation are not known, but alkaloids are thought to be involved (*110*). When extracts of the bulbs are applied to open wounds they can produce staggering, numbness of the whole nervous system, and paralysis of the heart (*393*). Furthermore, the scent of flowers of species such as *N. bulbocodium* can produce headaches and even vomiting if they are placed in confined spaces. Indeed, many people refuse to have daffodils in their house, considering them to be unlucky for the way they hang their heads, which suggests tears and unhappiness (*14*).

The mucilage secreted by bulbs can also produce harmful effects in plant species such as rose, rice, and cabbage, inhibiting seed germination and seedling growth (*120,407*).

### 3. *Other Uses*

The olfactory qualities of the *Narcissus* flower have made it a valuable component of luxury perfumes since time immemorial, although the main components of the volatile part of narcissus absolute are not of alkaloidal origin. However, alkaloids are present, together with essential oils, in some *Narcissus*-derived perfumes, such as jonquil absolute (*408–410*).

## B. BIOLOGICAL ACTIVITIES OF PLANT EXTRACTS

Several *Narcissus* extracts have shown the following activities: antiviral (*390,411–417*), prophage induction (*418*), antibacterial (*418–420*), antifungal (*419, 421,422*), antimalarial (*419,423*), insecticidal (*419*), cytotoxic (*390,411, 419,424*), antitumor (*390,395,412,413,415,425*), antimitotic (*426*), antiplatelet (*419*), hypotensive (*427*), emetic (*395*), acetylcholine esterase inhibitory (*93*), antifertility (*428*), antinociceptive (*391*), chronotropic (*427*), pheromone (*429*), plant growth inhibitor, and allelopathic (*120,143,407,427*).

## C. BIOLOGICAL AND PHARMACOLOGICAL ACTIVITIES OF ALKALOIDS

The alkaloids from the genus *Narcissus* are the compounds responsible for the majority of the above-mentioned activities, although the mannosa-binding lectins have also received much interest recently (*430–435*).

In spite of the great variety of pharmacological and/or biological properties exhibited by these alkaloids, only some of the activities of a reduced number have been reported, and the most extensively studied effect is that of non-specific inhibition. The relationship of chemical structure and biological activity is largely unknown, and further studies are needed to explore the full therapeutic potential of these alkaloids. The most-studied alkaloids in this group are galanthamine (**75**), lycorine (**1**), narciclasine (**68**), and pretazettine (**64**), which possess a diversity of pharmacological activities.

### *1. Lycorine Type*

Lycorine (**1**), the most frequent and characteristic of the Amaryllidaceae alkaloids, has been reported to be a powerful inhibitor of ascorbic acid (L-Asc) biosynthesis (*436,437*), and thus has proved to be a useful tool in studying Asc-dependent metabolic reactions in L-Asc-synthesizing organisms (*438,439*). Lycorine is actually a powerful inhibitor of the activity of L-galactono-$\gamma$-lactone dehydrogenase, the terminal enzyme of L-Asc biosynthesis (*440–443*), which appears to be localized in the mitochondrial membrane (*444,445*). Galanthine (**7**) also has a high capacity to inhibit ascorbic acid biosynthesis (*437*).

It is well documented that lycorine (**1**) is a powerful inhibitor of cell growth, cell division, and organogenesis in higher plants, algae, and yeasts, inhibiting the cell cycle during interphase, which seems to be related with the L-Asc levels (*438,446–450*). In plants, it also inhibits cyanide-insensitive respiration, peroxidase activity, and protein synthesis (*451–453*). The effects of lycorine on L-Asc biosynthesis have been reported to occur at concentrations below those at which protein synthesis is affected, but it seems difficult to completely rule out non-specific effects of this alkaloid since it has been reported that, at least in yeasts, lycorine is able to interact directly with mitochondrial DNA. Thus, differing sensitivity to the alkaloid among cells devoid of mitochondrial DNA ($rho^0$) and cells with mitochondrial DNA either $rho^+$ or $rho^-$ has been found in yeasts (*440,448,454,455*), $rho^0$ cells being resistant to high concentrations of the drug (*450,456–458*). Some strains can even adapt to the presence of lycorine, because they are able to degrade the alkaloid and use its biotransformation products as growth-stimulating factors (*458*).

Lycorine-1-*O*-$\beta$-D-glucoside, on the other hand, promotes cell growth, seed germination, and the rate of development of root and root hairs in higher plants. The glucosyloxy derivatives of lycorine (**1**) and pseudolycorine (**3**) and their aglycones form stable complexes with phytosterols and with divalent metal ions, and are able to translocate them from the rhizosphere to the aerial part (*347*). Palmilycorine and some acylglucosyloxy conjugates of lycorine, in turn, are frequently encountered among the phytosterols exhibiting membrane-stabilizing action. Additionally, lycorine-1-*O*-β-D-glucoside and acylglucosyloxy conjugates of lycorine are used by plants for recognition to reject the vast majority of microorganisms and parasites (*459*).

In animals, lycorine (**1**) shows antitumor activity (*460,461*), reported to inhibit the *in vivo* and *in vitro* growth of a variety of tumor cells, such as BL6 mouse melanoma, Lewis lung carcinoma, murine ascites, or HeLa cells (*20,298,459,462–465*). It induces flat morphology in K-ras-NRK cells (transformed fibroblasts) (*466*), and reduces the cellular activity in femoral bone marrow tissue that results in granulocytic leucopenia and a decrease in the number of erythrocytes. This alkaloid's mechanism of action is thought to be through the inhibition of protein synthesis at the ribosomal level, even though the cytotoxic effects of calprotectin can also be suppressed using lycorine (*424,460,461,467,468*). Lycorine also inhibits murine macrophage production of tumor necrosis factor alpha (TNF-α) (*469*), and shows inhibitory effects on nitric oxide production and induction of inducible nitric oxide synthase (NOS) in lipopolysaccharide-activated macrophages (*470*). The molecular mechanism of lycorine against leukemia (human cell line HL-60) shows that it can suppress cell growth and reduce cell survival by arresting the cell cycle at the G2/M phase and inducing apoptosis of tumor cells (*471*). It displays pronounced cell growth inhibitory activities against both parental and multidrug resistant L5178 mouse lymphoma cell lines, but is almost inactive in inhibiting the glycoprotein responsible for the efflux-pump activity of tumor cells. Assays for interactions with tRNA revealed that the antiproliferative effects of lycorine result from its complex formation with tRNA (*73*). Interaction of lycorine (**1**), pseudolycorine (**3**), and 2-*O*-acetylpseudolycorine (**5**) with DNA has also been observed (*472,473*).

Some other alkaloids of this series, such as caranine (**10**), galanthine (**7**), pseudolycorine (**3**), and 2-*O*-acetylpseudolycorine (**5**), are also active against a variety of tumor cells (*415,463,474*). Pseudolycorine inhibits the protein synthesis in tumor cells at the step of peptide bond formation, but it has a different binding site than lycorine (*467,475*). Ungeremine (**24**), a natural metabolite of lycorine (**1**), is responsible, at least in part, for the growth-inhibitory and cytotoxic effects of lycorine, being active against leukemia (*476,477*). Lycorine-1-*O*-β-D-glucoside, in turn, has the reverse effect of lycorine, and may produce mitogenic activity in animal cells (*478*).

Lycorine (**1**) and pseudolycorine (**3**) exert antiviral effects on several RNA- and DNA-containing viruses (*479*). Antiviral activity has been observed in tests with flaviviruses, and to a slightly lesser degree, bunyaviruses. Lycorine and pseudolycorine also show inhibitory activity against the Punta Toro and Rift Valley fever viruses, but with low selectivity (*480,481*). Lycorine, in turn, acts as an anti-SARS-CoV (Severe Acute Respiratory Syndrome-associated Coronavirus) and shows pronounced activity against poliomyelitis, coxsackie, and herpes type 1 viruses (*20,482*). It possesses high antiretroviral activity accompanied by low therapeutic indices (*483*). The relationship between its structure and the mechanism of activity has been studied in the *Herpes simplex* virus, suggesting that alkaloids that may eventually prove to be antiviral agents have a hexahydroindole ring with two functional hydroxyl groups (*484*). The activity was found to be due to the inhibition of multiplication, and not to the direct inactivation of extracellular viruses, and the mechanism of the antiviral effect was partially explained as a blocking of viral DNA polymerase activity (*33,413,479,485*).

Lycorine (**1**) has appreciable inhibitory activity against acetylcholinesterase (*486*). Cholinesterase activity appears to be associated with the two free hydroxyl groups present in some of the alkaloids of this structural type (*487*). The higher cholinesterase activity of assoanine (**20**) and oxoassoanine (**21**) with respect to the other lycorine-type alkaloids could be explained by an aromatic ring C, which gives a certain planarity to those molecules (*93*). Another alkaloid, galanthine (**7**), exhibits powerful cholinergic activity and has therefore attracted much interest in the treatment of myasthenia gravis, myopathy, and diseases of the central nervous system (*488*). Caranine (**10**), pseudolycorine (**3**), ungiminorine (**17**), and in particular, ungeremine (**24**), also show an inhibitory effect on acetylcholinesterase (*93,173,320*).

Lycorine (**1**) is an analgesic, more so than aspirin, and a hypotensive (*489,490*), as are caranine (**10**) and galanthine (**7**). The analgesic activity exhibited by the Amaryllidaceae alkaloids is attributed to their resemblance to the morphine and codeine skeletons. Lycorine also has antiarrhythmic action, and lycorine hydrochloride is a strong broncholytic (*30*). In fact, lycorine shows a relaxant effect on an isolated epinephrine-precontracted pulmonary artery and increases contractility and the rate of an isolated perfused heart. These effects are mediated by stimulation of $\beta$-adrenergic receptors (*491*).

Lycorine (**1**) also has a strong inhibitory effect on parasite (*Encephalitozoon intestinalis*) development (*492*) and antifungal activity against *Candida albicans* (*326*). Additionally, lycorine has antifeedant (*493*), antimalarial (*423,494*), emetic (*495*), anti-inflammatory (*496*), antiplatelet (*419*), as well as antifertility (*490*) activities. Galanthine (**7**), in turn, shows mild *in vitro* activity against *Tripanosoma brucei rhodesiense* and *Plasmodium falciparum* (*497*).

### 2. *Homolycorine Type*

It is reported that some alkaloids of this series, such as homolycorine (**26**), 8-*O*-demethylhomolycorine (**27**), 9-*O*-demethyl-2$\alpha$-hydroxyhomolycorine (**32**), dubiusine (**33**), hippeastrine (**34**), lycorenine (**35**), or *O*-methyllycorenine (**36**), present cytotoxic effects against non-tumoral fibroblastic LMTK cells (*463*), also being moderately active in inhibiting the *in vivo* and *in vitro* growth of a variety of tumor cells, such as Molt 4 lymphoma, HepG2 human hepatoma, LNCaP human prostate cancer, or HT (*341,463,490*). Dubiusine, lycorenine, 8-*O*-demethylhomolycorine and 9-*O*-demethyl-2$\alpha$-hydroxyhomolycorine also show DNA binding activity comparable to that of vinblastine (*472*). Homolycorine possesses high antiretroviral activity, accompanied by low therapeutic indices (*483*). Hippeastrine, in turn, displays antiviral activity against *Herpes simplex* type 1 (*484*).

Dubiusine (**33**), homolycorine (**26**), 8-*O*-demethylhomolycorine (**27**), and lycorenine (**35**) have a hypotensive effect on the arterial pressure of normotensive rats (*498*). Lycorenine also shows a vasodepressor action ascribed to the maintenance of its $\alpha$-adrenergic blocking action, and produces bradycardia by modifying vagal activity (*499*). Another feature of lycorenine is its analgesic activity (*20*).

Homolycorine (**26**) and masonine (**30**) are other inductors of delayed hypersensitivity in animals (*110*). Hippeastrine (**34**), in turn, shows antifungal activity against *C. albicans* and it also possesses a weak insect antifeedant activity (*326*).

### 3. *Hemanthamine Type*

Hemanthamine (**53**), hemanthidine (**55**), crinamine (**57**), maritidine (**44**), and papyramine (**48**) display pronounced cell growth inhibitory activities against a variety of tumor cells, such as Rauscher viral leukemia, Molt 4 lymphoma, BL6 mouse melanoma, HepG2 human hepatoma, HeLa, LNCaP human prostate cancer, or HT (*298,336,341,424,462,463,500*). Some of these alkaloids, namely crinamine, hemanthamine, and papyramine, also present a cytotoxic effect against non-tumoral fibroblastic LMTK cells (*463*). The mechanism of action of hemanthamine is thought to be through the inhibition of protein synthesis, blocking the peptide bond formation step on the peptidyl transferase center of the 60S ribosomal subunit (*467,475*). Hemanthamine and hemanthidine also display the same pronounced cell growth inhibitory activities against both parental and multidrug resistant L5178 mouse lymphoma cell lines as described above for lycorine (**1**) (*73*). Crinamine, in turn, shows inhibitory effects on nitric oxide (NO) production and induction of inducible nitric oxide synthase (NOS) in lipopolysaccharide-activated macrophages (*470*).

The antimalarial activity against strains of chloroquine-sensitive *P. falciparum* observed in hemanthamine (**53**) and hemanthidine (**55**) can be attributed to the methylenedioxybenzene part of the molecule and the tertiary nitrogen without methyl (*423*). Crinamine (**57**) also exhibits moderate antimalarial activity (*494,501*). Hemanthidine also works *in vitro* against *Trypanosoma brucei rhodesiense*, and to a lesser extent against *Trypanosoma cruzi* (*497*). Vittatine (**42**) has antibacterial activity against the Gram-positive *Staphylococcus aureus* and the Gram-negative *E. coli* (*326*), and the alkaloid crinamine shows strong activity against *Bacillus subtilis* and *S. aureus* (*502*).

Like lycorine (**1**), hemanthidine (**55**) has stronger analgesic and anti-inflammatory activity than aspirin (*489,496*), and vittatine has been found to potentiate the analgesic effect of morphine (*60*). Moreover, some alkaloids of this series, such as hemanthamine (**53**) or papiramine (**48**) have a hypotensive effect (*41,498*), and hemanthamine shows strong antiretroviral activity (*483*).

### 4. *Tazettine Type*

Tazettine (**62**) is mildly active against certain tumor cell lines (*341,424,503*), with a slight cytotoxicity when tested on fibroblastic LMTK cell lines (*463*). Tazettine also displays weak hypotensive and antimalarial activities and interacts with DNA (*419,472,498*). Its chemically labile precursor, pretazettine (**64**), is far more interesting due to its antiviral and anticancer activities. In fact, when pretazettine is stereochemically rearranged to tazettine, the biological activity of the precursor is to a large extent reduced (*504,505*).

Pretazettine (**64**) shows cytotoxicity against fibroblastic LMTK cell lines and inhibits HeLa cell growth, being therapeutically effective against advanced Rauscher leukemia, Ehrlich ascites carcinoma, spontaneous AKR lymphocytic leukemia, and Lewis lung carcinoma (*412,503,506–510*). It is one of the most active of the Amaryllidaceae alkaloids against Molt4 lymphoid cells (*463*), and is used in combination with DNA-binding and alkylating agents in treating the Rauscher leukemia virus (*412,503*). In fact, pretazettine strongly inhibits the activity of reverse transcriptase from various oncogenic viruses by binding to the enzyme (*20*). It inhibits both the growth of the Rauscher virus and cellular protein synthesis in eukaryotic cells by a mechanism that does not affect DNA and RNA synthesis, even though it has a pronounced DNA-binding activity (*415,424,467,472,481,507,511*). Pretazettine has also been shown to be active against selected RNA-containing flavoviruses (Japanese encephalitis, yellow fever, and dengue) and bunyaviruses (Punta Toro and Rift Valley fever) in organ culture (*481*). It also possesses pronounced activity against *Herpes simplex* type 1 virus (*484*). This activity may reflect a general ability to inhibit protein synthesis during viral replication (*512*).

### *5. Narciclasine Type*

Narciclasine (**68**), an antimitotic and antitumoral alkaloid (*143*), affects cell division at the metaphase stage and inhibits protein synthesis in eukaryotic ribosomes by directly interacting with the 60S subunit and inhibiting peptide bond formation by preventing binding of the 3′ terminal end of the donor substrate to the peptidyl transferase center (*467,475,513–515*). It also retards DNA synthesis (*516*) and inhibits calprotectin-induced cytotoxicity at a more than 10-fold lower concentration than lycorine (**1**) (*468*). The peculiar effects of narciclasine seem to arise from the functional groups and conformational freedom of its C-ring (*122*), with the 7-hydroxyl group believed to be important in its biological activity (*253*). This alkaloid, related to pancratistatin (*516*), is one of the most important antineoplastic Amaryllidaceae alkaloids (*460*) and shows some promise as an anticancer agent. It inhibits HeLa cell growth, has antileukemic properties and is active against a variety of tumor cells, such as human and murine lymphocytic leukemia, larynx, and cervix carcinomas, and Ehrlich tumor cells (*33,77,386,388,389,516*). No effect has been observed toward solid tumors. Narciclasine-4-*O*-*β*-D-glucopyranoside shows very similar cytotoxic and antitumoral activity to narciclasine (*517*).

Narciclasine (**68**) has a prophylactic effect on the adjuvant arthritis model in rats, significantly suppressing the degree of swelling of adjuvant-treated, as well as untreated, feet (*468*). This alkaloid is also active against *Corynebacterium fascians,* inhibits the pathogenic yeast *Cryptococcus neoformans,* and modifications, like 2,3,4,7-tetra-*O*-acetylnarciclasine inhibit, the growth of the pathogenic bacterium *Neisseria gonorrhoeae* (*78*). Antiviral activity has been observed against RNA-containing flaviviruses and bunyaviruses (*481*).

At the plant level, narciclasine (**68**) is a potent inhibitor, showing a broad range of effects, including the ability to inhibit seed germination and seedling growth of some plants in a dose-dependent manner, interacting with hormones in some physiological responses (*518*). Thus, indole-3-acetic acid cannot overcome the inhibition of elongation of wheat coleoptile sections caused by narciclasine. Additionally, narciclasine suppresses the gibberellin-induced α-amylase production in barley seeds and cytokinin-induced expansion and greening of excised radish cotyledons (*120*). Like lycorine (**1**), narciclasine also inhibits ascorbic acid biosynthesis (*305*).

Narciclasine, present in daffodil mucilage, can delay tepal senescence in cut *Iris* flowers by attenuation of protease activity, which, in turn, is apparently related to the inhibition of the protein synthesis involved in senescence (*519*). At the organelle level, narciclasine inhibits both isocitrate lyase (ICL) activity in glyoxysomes and hydroxypyruvate reductase (HPR) activity in peroxysomes. It also blocks the formation of chloroplasts, markedly reducing the chlorophyll content of light-grown wheat seedlings, probably due to the inhibition of the formation of 5-aminolevulinic acid, an essential chlorophyll precursor (*520*). The formation of light-harvesting chlorophyll a/b binding protein (LHCP) is also inhibited by this alkaloid (*521*).

Some alkaloids of this series, such as trisphaeridine (**70**), possess high antiretroviral activities, accompanied by low therapeutic indices (*483*). Ismine (**72**), in turn, shows a significant hypotensive effect on the arterial pressure of normotensive rats (*498*) and is cytotoxic against Molt 4 lymphoid and LMTK fibroblastic cell lines (*463*).

### 6. *Montanine Type*

There is little information about the montanine-type alkaloids, only some data about pancracine (**73**), which shows antibacterial activity against *S. aureus* and *Pseudomonas aeroginosa* (*326*), as well as weak activity against *Tripanosoma brucei rhodesiense, T. cruzi*, and *P. falciparum* (*80*).

### 7. *Galanthamine Type*

Galanthamine (**75**), originally isolated from *Galanthus nivalis* L. in the 1940s, is a long-acting, selective, reversible and competitive inhibitor of acetylcholinesterase. This enzyme is responsible for the degradation of acetylcholine at the neuromuscular junction, in peripheral and central cholinergic synapses and in parasympathetic target organs (*522–524*). Galanthamine has the ability to cross the blood–brain barrier and act within the central nervous system (*525,526*). It binds at the base of the active site gorge of acetylcholinesterase, interacting with both the choline-binding site and the acyl-binding pocket, having a number of moderate-to-weak interactions with the protein (*527–529*). In addition, galanthamine stimulates pre- and postsynaptic nicotinic receptors which can, in turn, increase the release of neurotransmitters, thus directly stimulating neuronal function (*524,530*). It is also suggested that the stimulation of nicotinic receptors protects against apoptosis induced by $\beta$-amyloid toxicity (*524,531,532*). Its dual mode of action (*527*), coupled with the evidence that galanthamine has reduced side effects, make it a promising candidate for the treatment of nervous diseases, paralysis syndrome, schizophrenia, and other forms of dementia, as well as Alzheimer's disease (*524,527,528*).

Galanthamine (**75**) has other noteworthy pharmacological actions, including an ability to amplify the nerve-muscle transfer (*20*), affecting membrane ionic processes (*533*). It is also known to cause bradycardia or atrioventricular conduction disturbances (*41*), has long been used as a reversal agent in anesthetic practice (*181*), inhibits traumatic shock, and has been patented for use in the treatment of nicotine dependence. Besides this, galanthamine acts as a mild analeptic, shows an analgesic power as strong as morphine, compensates for the effects of opiates on respiration, relieves jet lag, fatigue syndrome, male impotence, and alcohol dependence, and when applied in eye drops, reduces the intraocular pressure (*20,81,85,534*). It also acts as a hypotensive and has a weak antimalarial activity (*419,498*).

At present, there is no preventative or curative treatment available for Alzheimer's disease, leaving the symptomatic relief offered by AChEI therapy as the only approved therapeutic option. Owing to the relative lack of alternative treatment, galanthamine (**75**) is a reasonable approximation of the ideal concept of symptomatic Alzheimer's disease therapy (*523,535*). Galanthamine hydrobromide (a third-generation cholinesterase inhibitor used against Alzheimer's disease) offers superior pharmacological profiles and increased tolerance compared to the original acetylcholinesterase inhibitors, physostigmine or tacrine (*536–540*). Galanthamine is effective and well tolerated, resulting in short-term improvements in cognition, function, and daily life activities in patients with mild to moderate symptoms (*530,541,542*). However, long-term benefits beyond 6 months are in question (*543*), because persistent elevation of acetylcholine may lead to over-stimulation of both nicotinic and muscarinic acetylcholine receptors, the former causing receptor desensitization and the latter potentially causing an increased frequency of cholinergic side effects (*524,530,544*). The safety profile of galanthamine, as well as its clinical effectiveness, will only be demonstrated after large-scale clinical trials (*544–546*).

Broadly speaking, the development of galanthamine (**75**) into a widely used Alzheimer's drug can be divided into three main periods: (1) the early development in Eastern Europe for its use in the treatment of poliomyelitis; (2) the pre-clinical development in the 1980s; and (3) the clinical development in the 1990s (*544*). Galanthamine hydrobromide was first used by Bulgarian and Russian researchers in the 1950s and exploited for a variety of clinical purposes. It has been used clinically for postsurgery reversal of tubocurarine-induced muscle relaxation and for treating post-polio paralysis, myasthenia gravis, and other neuromuscular diseases, as well as traumatic brain injuries (*547,548*). As early as 1972, Soviet researchers demonstrated that galanthamine could reverse scopolamine-induced amnesia in mice, a finding that was demonstrated in man 4 years later. However, this alkaloid was not applied to Alzheimer's disease until 1986, long after the widely accepted cholinergic hypothesis had been first postulated, when researchers in Western Europe switched their attention to galanthamine because of its ability to penetrate the blood–brain barrier and specifically to augment the central cholinergic function (*544,549*). This led to clinical trials of galanthamine in the treatment of Alzheimer's disease. In 1996, Sanochemia Pharmazeutika in Austria first launched galanthamine as 'NIVALIN®', but its strictly limited availability meant the international pharmaceutical community adopted a cautions approach (*181,550*), until Sanochemia Pharmazeutika developed a method to synthetically produce the compound in 1997 (*551*). Later, galanthamine was co-developed by Shire Pharmaceuticals (Great Britain) and the Janssen Research Foundation (Belgium), who have launched galanthamine as 'REMINIL®' in many countries (*524,544*). This renewed interest is reflected in the increasing number of scientific reviews concerned exclusively with galanthamine and its derivatives (*522,525,526,530,538,539,547,552–555*).

Sanguinine (**81**) has a more potent acetylcholinesterase inhibitory activity than galanthamine (**75**) due to an extra hydroxyl group available for potential interaction with acetylcholinesterase (*93,555*). Sanguinine, in turn, is 10-fold more selective than galanthamine for acetylcholinesterase (AChE) vs. butyrylcholinesterase (BuChE)

(*556*). The lack of AChE inhibitory activity of lycoramine (**84**) and epinorlicoramine (**86**) could be due to the occurrence of a double bond in ring C, which does not allow these alkaloids to have the same spatial configuration as the active alkaloids of this series (*93*).

Narwedine (**83**), the biogenetic precursor of galanthamine (**75**), has been studied as a respiratory stimulator. It increases the amplitude and decreases the frequency of cardiac contractions and would therefore be of value in reducing blood loss during surgery (*41*). It also inhibits the action of narcotics and hypnotics, and increases the analgesic effect of morphine (*60*) as well as the pharmacological effects of caffeine, carbazole, arecoline, and nicotine (*30*).

### 8. *Other Alkaloids*

Cherylline (**88**) is a 4-arylisoquinoline derivative, a group with several potential medicinal properties (*80*), including a weak acetylcholinesterase inhibitory activity (*486*). Mesembrenone (**90**), in turn, is mildly active against Molt 4 lymphoid and non-tumoral fibroblastic LMTK cells (*463*), has a moderate hypotensive effect on arterial pressure, and interacts slightly with DNA (*472,498*).

## References


1. A. W. Meerow and D. A. Snijman, *in* "The Families and Genera of Vascular Plants" (K. Kubitzki, ed.), vol. III, p. 83. Springer, Berlin, 1998.
2. M. Ito, A. Kawamoto, Y. Kita, T. Yukawa, and S. Kurita, *J. Plant Res.* **112**, 207 (1999).
3. G. R. Hanks, *in* "Narcissus and daffodil: the genus *Narcissus*" (G.R. Hanks, ed.), vol. 21 in the series: "Medicinal and Aromatic Plants – Industrial Profiles", p. 1. Taylor & Francis, London, 2002.
4. B. Mathew, *in* "Narcissus and daffodil: the genus *Narcissus*" (G.R. Hanks, ed.), vol. 21 in the series: "Medicinal and Aromatic Plants – Industrial Profiles", p. 30. Taylor & Francis, London, 2002.
5. S. W. Graham and S. C. H. Barrett, *Am. J. Bot.* **91**, 1007 (2004).
6. H. E. M. Dodson, J. Arroyo, G. Bergström, and I. Groth, *Biochem. Syst. Ecol.* **25**, 685 (1997).
7. J. Arroyo, S. C. H. Barrett, R. Hidalgo, and W. W. Cole, *Am. J. Bot.* **89**, 1242 (2002).
8. C. M. Herrera, *Ecology* **76**, 218 (1995).
9. A. Fernandes, *Daffodil and Tulip Year Book* (Royal Horticultural Society) **33**, 37 (1968).
10. A. Fernandes, *An. Inst. Bot. A. J. Cavanilles* **32**, 843 (1975).
11. D. A. Webb, *in* "Flora Europaea", (T. G. Tutin, V. H. Heywood, N. A. Burges, D. M. Moore, D. H. Valentine, S. M. Walters, D. A. Webb, A. O. Chater, and B. K. Richardson, eds.), vol. 5, p. 75. Cambridge University Press, Cambridge, 1980.
12. S. C. H. Barrett and L. D. Harder, *New Phytol.* **165**, 45 (2005).
13. R. Pérez, P. Vargas, and J. Arroyo, *New Phytol.* **161**, 235 (2003).
14. A. C. Dweck, *in* "Narcissus and daffodil: the genus *Narcissus*" (G.R. Hanks, ed.), vol. 21 in the series: "Medicinal and Aromatic Plants – Industrial Profiles", p. 19. Taylor & Francis, London, 2002.
15. G. R. Hanks, *in* "Narcissus and daffodil: the genus *Narcissus*" (G.R. Hanks, ed.), vol. 21 in the series: "Medicinal and Aromatic Plants – Industrial Profiles", p. 53. Taylor & Francis, London, 2002.

16. J. W. Blanchard, ,"*Narcissus*: A Guide to Wild Daffodils", Alpine Garden Society, Woking, Surrey, 1990.
17. A. R. Rees, *The Garden* (Royal Horticultural Society) **118**, 404 (1993).
18. S. Kington, ,"The International Daffodil Register and Classified List 1998", Royal Horticultural Society, London, 1998.
19. *Dictionary of Natural Products* (Net Database). Chapman & Hall/CRC Press, London, 2005.
20. S. Ghosal, K. S. Saini, and S. Razdan, *Phytochemistry* **24**, 2141 (1985).
21. I. L. Wiggins, ,"Flora of Baja California", Stanford University Press, Stanford, 1980.
22. J. Bastida, M. Sellés, C. Codina, F. Viladomat, and J. L. León de la Luz, *Planta Med.* **62**, 575 (1996).
23. P. W. Jeffs, *in* "The Alkaloids", (R. H. F. Manske and R. G. A. Rodrigo, eds.), vol. 19, p. 1. Academic Press, New York, 1981.
24. M. T. Smith, N. R. Crouch, N. Gericke, and M. Hirst, *J. Ethnopharmacol.* **50**, 119 (1996).
25. W. Döpke, E. Sewerin, and Z. Trimiño, *Z. Chem.* **20**, 298 (1980).
26. J. Bastida, F. Viladomat, J. M. Llabrés, G. Ramirez, C. Codina, and M. Rubiralta, *J. Nat. Prod.* **52**, 478 (1989).
27. J. A. Seijas, M. P. Vázquez-Tato, M. T. Linares, P. Ramil-Rego, and M. I. Buján, *8th International Electronic Conference on Synthetic Organic Chemistry (ECSOC-8)*, p. 625 (2004).
28. G. I. Kaya, N. Unver, B. Gözler, and J. Bastida, *Biochem. Syst. Ecol.* **32**, 1059 (2004).
29. D. A. Mulholland, N. Crouch, B. Decker, and M. T. Smith, *Biochem. Syst. Ecol.* **30**, 183 (2002).
30. O. A. Cherkasov and O. N. Tolkachev, *in* "Narcissus and daffodil: the genus *Narcissus*" (G.R. Hanks, ed.), vol. 21 in the series: "Medicinal and Aromatic Plants – Industrial Profiles", p. 242. Taylor & Francis, London, 2002.
31. S. Ghosal, S. K. Singh, and S. Unnikrishnan, *Phytochemistry* **29**, 805 (1990).
32. S. Ghosal, K. Datta, S. K. Singh, and Y. Kumar, *J. Chem. Res.-S* 334 (1990).
33. O. Hoshino, *in* "The Alkaloids: Chemistry and Biology", (G. A. Cordell, ed.), vol. 51, p. 323. Academic Press, New York, 1998.
34. Z. Jin, *Nat. Prod. Rep.* **20**, 606 (2003).
35. T. Hudlicky, U. Rinner, D. Gonzalez, H. Akgun, S. Schilling, P. Siengalewicz, T. A. Martinot, and G. R. Pettit, *J. Org. Chem.* **67**, 8726 (2002).
36. M. F. Grundon, *Alkaloids* (London) **9**, 137 (1979).
37. M. F. Grundon, *Alkaloids* (London) **10**, 135 (1981).
38. M. F. Grundon, *Alkaloids* (London) **11**, 131 (1981).
39. M. F. Grundon, *Alkaloids* (London) **12**, 151 (1982).
40. M. F. Grundon, *Alkaloids* (London) **13**, 187 (1983).
41. S. F. Martin, *in* "The Alkaloids", (A. Brossi, ed.), vol. 30, p. 251. Academic Press, New York, 1987.
42. R. Polt, *in* "Organic Synthesis: Theory and Applications", (H. Hudlicky, ed.), vol. 3, p. 109. JAI Press, Greenwich, CT, 1996.
43. J. R. Lewis, *in* "Rodd's Chemistry of Carbon Compounds (2nd Edition)", (M. Sainsbury, ed.), vol. 4, p. 165. Elsevier, Amsterdam, 1997.
44. F. Viladomat, J. Bastida, C. Codina, J. J. Nair, and W. E. Campbell, *in* "Recent Research Developments in Phytochemistry", (S. G. Pandalai, ed.), vol. 1, p. 131. Research Signpost, Trivandrum, 1997.
45. J. Bastida, F. Viladomat, and C. Codina, *in* "Studies in Natural Products Chemistry", (Atta-ur-Rahman, ed.), vol. 20, p. 20, Elsevier, Amsterdam, 1998.
46. P. Quayle, *Annu. Rep. Prog. Chem., Sect. B: Org. Chem.* **96**, 259 (2000).
47. S. Prabhakar and M. R. Tavares, *in* "Alkaloids: Chemical and Biological Perspectives", (S. W. Pelletier, ed.), vol. 20, p. 433. Pergamon Press, Oxford, 2001.

48. J. Bastida and F. Viladomat, ksd lakdj, *in* "Narcissus and daffodil: the genus *Narcissus*" (G.R. Hanks, ed.), vol. 21 in the series: "Medicinal and Aromatic Plants – Industrial Profiles", p. 141. Taylor & Francis, London, 2002.
49. M. F. Grundon, *Nat. Prod. Rep.* **1**, 247 (1984).
50. M. F. Grundon, *Nat. Prod. Rep.* **2**, 249 (1985).
51. M. F. Grundon, *Nat. Prod. Rep.* **4**, 89 (1987).
52. M. F. Grundon, *Nat. Prod. Rep.* **6**, 79 (1989).
53. J. R. Lewis, *Nat. Prod. Rep.* **7**, 549 (1990).
54. J. R. Lewis, *Nat. Prod. Rep.* **9**, 183 (1992).
55. J. R. Lewis, *Nat. Prod. Rep.* **10**, 291 (1993).
56. J. R. Lewis, *Nat. Prod. Rep.* **11**, 329 (1994).
57. J. R. Lewis, *Nat. Prod. Rep.* **12**, 339 (1995).
58. J. R. Lewis, *Nat. Prod. Rep.* **13**, 171 (1996).
59. J. R. Lewis, *Nat. Prod. Rep.* **14**, 303 (1997).
60. J. R. Lewis, *Nat. Prod. Rep.* **15**, 107 (1998).
61. J. R. Lewis, *Nat. Prod. Rep.* **16**, 389 (1999).
62. J. R. Lewis, *Nat. Prod. Rep.* **17**, 57 (2000).
63. J. R. Lewis, *Nat. Prod. Rep.* **18**, 95 (2001).
64. J. R. Lewis, *Nat. Prod. Rep.* **19**, 223 (2002).
65. Z. Jin, Z. Li, and R. Huang, *Nat. Prod. Rep.* **19**, 454 (2002).
66. Z. Jin, *Nat. Prod. Rep.* **22**, 111 (2005).
67. J. Bastida, C. Codina, F. Viladomat, M. Rubiralta, J. C. Quirion, and B. Weniger, *J. Nat. Prod.* **55**, 134 (1992).
68. J. Bastida, J. M. Fernández, F. Viladomat, C. Codina, and G. de la Fuente, *Phytochemistry* **38**, 549 (1995).
69. L. M. Stark, X. F. Lin, and L. A. Flippin, *J. Org. Chem.* **65**, 3227 (2000).
70. J. Bastida, C. Codina, F. Viladomat, M. Rubiralta, J. C. Quirion, and B. Weniger, *J. Nat. Prod.* **55**, 122 (1992).
71. G. R. Almanza, J. M. Fernández, E. W. T. Wakori, F. Viladomat, C. Codina, and J. Bastida, *Phytochemistry* **43**, 1375 (1996).
72. J. Bastida, J. M. Llabrés, F. Viladomat, C. Codina, M. Rubiralta, and M. Feliz, *Phytochemistry* **27**, 3657 (1988).
73. J. Hohmann, P. Forgo, J. Molnár, K. Wolfard, A. Molnár, T. Thalhammer, I. Máthé, and D. Sharples, *Planta Med.* **68**, 454 (2002).
74. J. Labraña, G. Choy, X. Solans, M. Font-Bardia, G. de la Fuente, F. Viladomat, C. Codina, and J. Bastida, *Phytochemistry* **50**, 183 (1999).
75. W. C. Wildman and D. T. Bailey, *J. Am. Chem. Soc.* **89**, 5514 (1967).
76. F. Viladomat, J. Bastida, C. Codina, M. Rubiralta, and J. C. Quirion, *J. Nat. Prod.* **55**, 804 (1992).
77. G. R. Pettit, N. Melody, and D. L. Herald, *J. Org. Chem.* **66**, 2583 (2001).
78. G. R. Pettit, N. Melody, D. L. Herald, J. M. Schmidt, R. K. Pettit, and J. C. Chapuis, *Heterocycles* **56**, 139 (2002).
79. F. Viladomat, J. Bastida, G. Tribó, C. Codina, and M. Rubiralta, *Phytochemistry* **29**, 1307 (1990).
80. J. Labraña, A. K. Machocho, V. Kricsfalusy, R. Brun, C. Codina, F. Viladomat, and J. Bastida, *Phytochemistry* **60**, 847 (2002).
81. M. Kreh, *in* "Narcissus and daffodil: the genus *Narcissus*" (G.R. Hanks ed.), vol. 21 in the series: "Medicinal and Aromatic Plants – Industrial Profiles", p. 256. Taylor & Francis, London, 2002.
82. J. Bastida, J. M. Llabrés, F. Viladomat, C. Codina, M. Rubiralta, and M. Feliz, *J. Nat. Prod.* **50**, 199 (1987).
83. J. Bastida, F. Viladomat, J. M. Llabrés, C. Codina, M. Feliz, and M. Rubiralta, *Phytochemistry* **26**, 1519 (1987).
84. S. López, J. Bastida, F. Viladomat, and C. Codina, *Planta Med.* **69**, 1166 (2003).

85. R. M. Moraes-Cerdeira, *in* "Narcissus and daffodil: the genus *Narcissus*" (G.R. Hanks ed.), vol. 21 in the series: "Medicinal and Aromatic Plants – Industrial Profiles", p. 273. Taylor & Francis, London, 2002.
86. C. Codina, F. Viladomat, J. Bastida, M. Rubiralta, and J. C. Quirion, *Phytochemistry* **29**, 2685 (1990).
87. F. Viladomat, J. M. Llabrés, J. Bastida, R. M. Cusidó, and C. Codina, *Physiol. Plant.* **68**, 657 (1986).
88. O. A. Cherkasov, G. M. Tokhtabaeva, N. N. Margvelashvili, N. I. Maisuradze, and O. N. Tolkachev, *Rastit. Resur.* **24**, 414 (1988).
89. H. G. Boit and W. Döpke, *Chem. Ber.* **89**, 2462 (1956).
90. H. G. Boit and W. Stender, *Chem. Ber.* **87**, 624 (1954).
91. W. Döpke and E. Sewerin, *Z. Chem.* **21**, 71 (1981).
92. R. D. Harken, C. P. Christensen, and W. C. Wildman, *J. Org. Chem.* **41**, 2450 (1976).
93. S. López, J. Bastida, F. Viladomat, and C. Codina, *Life Sci.* **71**, 2521 (2002).
94. W. C. Wildman and C. L. Brown, *J. Am. Chem. Soc.* **90**, 6439 (1968).
95. J. J. Willaman and H. L. Li, *Lloydia* **33S**, 1 (1970).
96. H. G. Boit, *Chem. Ber.* **87**, 681 (1954).
97. W. Döpke, *Naturwissenschaften* **50**, 354 (1963).
98. W. Döpke and T. D. Nguyen, *Z. Chem.* **14**, 57 (1974).
99. F. Viladomat, M. Sellés, C. Codina, and J. Bastida, *Planta Med.* **63**, 583 (1997).
100. E. D. Bazhenova, K. U. Aliev, and U. B. Zakirov, *Farmakol. Alkaloidov Serdech. Glikozidov.* 100 (1971).
101. F. Piozzi, M. L. Marino, C. Fuganti, and A. Di Martino, *Phytochemistry* **8**, 1745 (1969).
102. J. Bastida, F. Viladomat, J. M. Llabrés C. Codina, C. Codina, and M. Rubiralta, *Planta Med.* **54**, 362 (1988).
103. H. G. Boit, W. Döpke, and A. Beitner, *Chem. Ber.* **90**, 2197 (1957).
104. C. Codina, J. Bastida, F. Viladomat, J. M. Fernández, S. Bergoñon, M. Rubiralta, and J. C. Quirion, *Phytochemistry* **32**, 1354 (1993).
105. C. Codina, F. Viladomat, J. Bastida, and M. Rubiralta, *Abstracts of the 36th Annual Congress on Medicinal Plant Research*, p. 49. Freiburg (1988).
106. C. Codina, F. Viladomat, J. Bastida, M. Rubiralta, and J. C. Quirion, *Nat. Prod. Lett.* **1**, 85 (1992).
107. J. Bastida, F. Viladomat, S. Bergoñon, J. M. Fernández, C. Codina, M. Rubiralta, and J. C. Quirion, *Phytochemistry* **34**, 1656 (1993).
108. J. Bastida, S. Bergoñon, F. Viladomat, and C. Codina, *Planta Med.* **60**, 95 (1994).
109. J. W. Cook, J. D. Loudon, and P. McCloskey, *J. Chem. Soc.* **4176** (1954).
110. M. Gude, B. M. Hausen, H. Heitsch, and W. A. König, *Contact Dermatitis* **19**, 1 (1988).
111. J. Bastida, J. M. Llabrés, F. Viladomat, C. Codina, M. Rubiralta, and M. Feliz, *Planta Med.* **54**, 524 (1988).
112. J. A. Seijas, M. P. Vázquez-Tato, J. Seijo-Muras, P. Ramil-Rego, and M. I. Buján, *8th International Electronic Conference on Synthetic Organic Chemistry (ECSOC-8)*, p. 623 (2004).
113. J. M. Llabrés, F. Viladomat, J. Bastida, C. Codina, M. Serrano, M. Rubiralta, and M. Feliz, *Phytochemistry* **25**, 1453 (1986).
114. J. M. Llabrés, F. Viladomat, J. Bastida, C. Codina, and M. Rubiralta, *Phytochemistry* **25**, 2637 (1986).
115. C. Vigneau, J. Tsao, R. Ducluzeae, and J. Galzot, *J. Toxicol. Med.* **4**, 21 (1984).
116. S. H. Hung, Z. X. Chen, and C. F. Li, *Yao Hsueh Hsueh Pao (Acta Pharm. Sin.)* **9**, 719 (1962).
117. O. M. Abdallah, *Phytochemistry* **34**, 1447 (1993).
118. K. A. Abduazimov and S. Y. Yunusov, *Khim. Prir. Soedin.* **3**, 64 (1967).

119. A. H. Abou-Donia, F. A. Darwish, and N. M. Ghazy, *Alex. J. Pharm. Sci.* **3**, 122 (1989).
120. Y. R. Bi, K. H. Yung, and Y. S. Wong, *Plant Sci.* **135**, 103 (1998).
121. H. G. Boit and W. Döpke, *Naturwissenschaften* **47**, 109 (1960).
122. A. Evidente, *Planta Med.* **57**, 293 (1991).
123. A. Evidente, R. Lanzetta, A. H. Abou-Donia, M. E. Amer, F. F. Kassem, and F. M. Harraz, *Arch. Pharm.* **327**, 595 (1994).
124. E. Furusawa, S. Furusawa, S. Tani, H. Irie, K. Kitamura, and W. C. Wildman, *Chem. Pharm. Bull.* **24**, 336 (1976).
125. F. Piozzi, C. Fuganti, R. Mondelli, and G. Ceriotti, *Tetrahedron* **24**, 1119 (1968).
126. E. Späth and L. Kahovec, *Chem. Ber.* **67**, 1501 (1934).
127. E. Späth, H. Kondo, and F. Kuffner, *Chem. Ber.* **69**, 1086 (1936).
128. S. Tani, N. Kobayashi, H. Fujiwara, T. Shingu, and A. Kato, *Chem. Pharm. Bull.* **29**, 3381 (1981).
129. D. T. Youssef and A. A. Khalifa, *Pharmazie* **56**, 818 (2001).
130. I. Orhan and B. Sener, *Chem. Nat. Compd.* **39**, 383 (2003).
131. S. H. Hung, G. E. Ma, and M. Chu, *Yao Hsueh Hsueh Pao (Acta Pharm. Sin.)* **13**, 499 (1966).
132. M. Laing and R. C. Clark, *Tetrahedron Lett.* **583** (1974).
133. G. E. Ma, H. Y. Li, C. E. Lu, X. M. Yang, and S. H. Hong, *Heterocycles* **24**, 2089 (1986).
134. J. Bastida, C. Codina, F. Viladomat, and M. Rubiralta, *Proceedings of the 18th IUPAC Symposium Chem. Nat. Prod.*, p. 189. Strasbourg (1992).
135. J. Bastida, C. Codina, F. Viladomat, M. Rubiralta, J. C. Quirion, H. P. Husson, and G. E. Ma, *J. Nat. Prod.* **53**, 1456 (1990).
136. S. H. Hung, G. E. Ma, and S. Q. Sung, *Acta Chim. Sin.* **39**, 529 (1981).
137. R. Suau, R. Rico, A. I. García, and A. I. Gómez, *Heterocycles* **31**, 517 (1990).
138. J. Bastida, C. Codina, F. Viladomat, M. Rubiralta, J. C. Quirion, and H. P. Husson, *Phytochemistry* **29**, 2683 (1990).
139. A. Vrondeli, P. Kefalas, and E. Kokkalou, *Pharmazie* **60**, 559 (2005).
140. J. Bastida, J. L. Contreras, C. Codina, C. W. Wright, and J. D. Phillipson, *Phytochemistry* **40**, 1549 (1995).
141. J. Bastida, F. Viladomat, J. M. Llabrés, S. Quiroga, C. Codina, and M. Rubiralta, *Planta Med.* **56**, 123 (1990).
142. H. G. Boit, W. Stender, and A. Beitner, *Chem. Ber.* **90**, 725 (1957).
143. G. Ceriotti, *Nature* **213**, 595 (1967).
144. O. A. Cherkasov, N. I. Maisuradze, G. S. Glyzina, and A. V. Gayevskii, *Khim.-Farm. Zh.* **23**, 621 (1989).
145. A. Barra and G. López-González, *An. Jard. Bot. Madr.* **40**, 345 (1984).
146. A. Barra and G. López-González, *An. Jard. Bot. Madr.* **40**, 369 (1984).
147. E. Dorda and J. Fernández-Casas, *Fontqueria* **5**, 15 (1984).
148. E. Dorda and J. Fernández-Casas, *Fontqueria* **6**, 7 (1984).
149. E. Dorda and J. Fernández-Casas, *Fontqueria* **27**, 103 (1989).
150. E. Dorda and J. Fernández-Casas, *Fontqueria* **30**, 235 (1990).
151. E. Dorda and J. Fernández-Casas, *Fontqueria* **39**, 69 (1994).
152. E. Dorda, M. A. Rivas-Ponce, and J. Fernández-Casas, *Fontqueria* **31**, 235 (1991).
153. A. Fernandes, *Fontqueria* **31**, 141 (1991).
154. J. Fernández-Casas, *Fontqueria* **3**, 23 (1983).
155. J. Fernández-Casas, *Fontqueria* **5**, 35 (1984).
156. J. Fernández-Casas, *Fontqueria* **6**, 35 (1984).
157. J. Fernández-Casas, *Fontqueria* **11**, 15 (1986).
158. H. G. Boit and H. Ehmke, *Chem. Ber.* **89**, 163 (1956).
159. H. M. Fales, L. D. Giuffrida, and W. C. Wildman, *J. Am. Chem. Soc.* **78**, 4145 (1956).

160. C. Fuganti, D. Ghiringhelli, and P. Grasselli, *Tetrahedron Lett.* 2261 (1974).
161. E. Tojo, *J. Nat. Prod.* **54**, 1387 (1991).
162. J. K. Bastos, L. Xu, N. P. D. Nanayakkara, C. L. Burandt, R. M. Moraes-Cerdeira, and J. D. McChesney, *J. Nat. Prod.* **59**, 638 (1996).
163. G. Ceriotti, L. Spandrio, and G. A. Gazzani, *Tumori* **53**, 359 (1967).
164. R. M. Moraes-Cerdeira, C. L. Burandt, J. K. Bastos, N. P. D. Nanayakkara, J. Mikell, J. Thurn, and J. D. McChesney, *Planta Med.* **63**, 472 (1997).
165. M. Kreh and R. Matusch, *Phytochemistry* **38**, 1533 (1995).
166. M. Kreh, R. Matusch, and L. Witte, *Phytochemistry* **38**, 773 (1995).
167. M. Kreh, R. Matusch, and L. Witte, *Phytochemistry* **40**, 1303 (1995).
168. N. R. Mustafa, I. K. Rhee, and R. Verpoorte, *J. Liq. Chromatogr. Relat. Technol.* **26**, 3217 (2003).
169. G. M. Gorbunova, V. I. Sheichenko, and O. N. Tolkachev, *Khim. Prir. Soedin.* 800 (1984).
170. G. M. Tokhtabaeva, V. I. Sheichenko, I. V. Yartseva, and O. N. Tolkachev, *Khim. Prir. Soedin.* 872 (1987).
171. R. M. Moraes-Cerdeira, J. K. Bastos, C. L. Burandt, N. P. D. Nanayakkara, J. Mikell, and J. D. McChesney, *Planta Med.* **63**, 92 (1997).
172. A. H. Abou-Donia, F. A. Darwish, M. E. Amer, F. F. Kassem, H. M. Hammoda, and Y. G. Shin, *Alex. J. Pharm. Sci.* **16**, 96 (2002).
173. K. Ingkaninan, A. Hazekamp, C. M. de Best, H. Irth, U. R. Tjaden, R. van der Heijden, J. van der Greef, and R. Verpoorte, *J. Nat. Prod.* **63**, 803 (2000).
174. D. H. R. Barton and G. W. Kirby, *J. Chem. Soc.* 806 (1962).
175. G. W. Kirby and J. Michael, *J. Chem. Soc.- Perkin Trans. I* 115 (1973).
176. G. W. Kirby and H. P. Tiwari, *J. Chem. Soc.-C* 676 (1966).
177. W. Döpke, *Naturwissenschaften* **50**, 645 (1963).
178. W. Döpke, *Planta Med.* **11**, 154 (1963).
179. W. Döpke, *Arch. Pharm.* **297**, 39 (1964).
180. W. Döpke and H. Dalmer, *Naturwissenschaften* **52**, 60 (1965).
181. J. Eichhorn, T. Takada, Y. Kita, and M. H. Zenk, *Phytochemistry* **49**, 1037 (1998).
182 R. J. Suhadolnik, A. G. Fischer, and J. Zulalian, *Biochem. Biophys. Res. Commun.* **11**, 208 (1963).
183. R. H. Wightman, J. Staunton, A. R. Battersby, and K. R. Hanson, *J. Chem. Soc., Perkin Trans. I* 2355 (1972).
184. S. Ghosal, Y. Kumar, S. K. Singh, and A. Shanthy, *J. Chem. Res.-S* 28 (1986).
185. S. Ghosal, A. Shanthy, and S. K. Singh, *Phytochemistry* **27**, 1849 (1988).
186. D.H.R. Barton and T. Cohen, *Festschrift Arthur Stoll*, p. 117. Birkhäuser Verlag, Basel, 1957.
187. W. R. Bowman, I. T. Bruce, and G.W. Kirby, *J. Chem. Soc., Chem. Commun.* 1075 (1969).
188. I. T. Bruce and G. W. Kirby, *J. Chem. Soc., Chem. Commun.* 207 (1968).
189. W. C. Wildman and N. E. Heimer, *J. Am. Chem. Soc.* **89**, 5265 (1967).
190. C. Fuganti and M. Mazza, *J. Chem. Soc., Chem. Commun.* 936 (1972).
191. A. R. Battersby, R. Binks, S. W. Breuer, H. M. Fales, W. C. Wildman, and R. J. Highet, *J. Chem. Soc.* 1595 (1964).
192. C. Fuganti, D. Ghiringhelli, and Grasselli, P. *J. Chem. Soc., Chem. Commun.* 350 (1974).
193. M. Kihara, L. Xu, K. Konishi, Y. Nagao, S. Kobayashi, and T. Shingu, *Heterocycles* **34**, 1299 (1992).
194. C. Fuganti and M. Mazza, *J. Chem. Soc., Perkin Trans. I* 954 (1973).
195. K. R. Hanson, *J. Am. Chem. Soc.* **88**, 2731 (1966).
196. C. Fuganti and M. Mazza, *J. Chem. Soc., Chem. Commun.* 1196 (1971).
197. A. I. Feinstein and W. C. Wildman, *J. Org. Chem.* **41**, 2447 (1976).
198. C. Fuganti, *Chim. Ind. (Milan)* **51**, 1254 (1969).

199. A. R. Battersby, J. E. Kelsey, and J. Staunton, *J. Chem. Soc., Chem. Commun.* 183 (1971).
200. H. M. Fales and W. C. Wildman, *J. Am. Chem. Soc.* **86**, 294 (1964).
201. W. C. Wildman and D. T. Bailey, *J. Am. Chem. Soc.* **91**, 150 (1969).
202. C. Fuganti, J. Staunton, and A. R. Battersby, *J. Chem. Soc., Chem. Commun.* 1154 (1971).
203. C. Fuganti and M. Mazza, *J. Chem. Soc., Chem. Commun.* 1388 (1971).
204. C. Fuganti and M. Mazza, *J. Chem. Soc., Chem. Commun.* 239 (1972).
205. C. Fuganti, *Tetrahedron Lett.* 1785 (1973).
206. C. Fuganti and M. Mazza, *J. Chem. Soc., Chem. Commun.* 1466 (1970).
207. D. H. R. Barton, G. W. Kirby, J. B. Taylor, and G. M. Thomas, *J. Chem. Soc.* 4545 (1963).
208. M. P. Vázquez-Tato, L. Castedo, and R. Riguera, *Heterocycles* **27**, 2833 (1988).
209. J. G. Bhandarkar and G. W. Kirby, *J. Chem. Soc.-C* 1224 (1970).
210. T. Hudlicky, *J. Heterocycl. Chem.* **37**, 535 (2000).
211. U. Rinner and T. Hudlicky, *Synlett* 365 (2005).
212. L. Czollner, M. Treu, J. Froehlich, B. Kueenburg, and U. Jordis, *ARKIVOC* **2**, 191 (2001).
213. M. Hermetsberger, M. Treu, U. Jordis, K. Mereiter, C. Hametner, and J. Froehlich, *Monatsh. Chem.* **135**, 1275 (2004).
214. A. H. Lewin, J. Szewczyk, J. W. Wilson, and F. I. Carroll, *Tetrahedron* **61**, 7144 (2005).
215. M. Node, S. Kodama, Y. Hamashima, T. Baba, N. Hamamichi, and K. Nishide, *Angew Chem., Int. Ed.* **40**, 3060 (2001).
216. S. Kodama, Y. Hamashima, K. Nishide, and M. Node, *Angew. Chem., Int. Ed.* **43**, 2659 (2004).
217. S. Kodama, H. Takita, T. Kajimoto, K. Nishide, and M. Node, *Tetrahedron* **60**, 4901 (2004).
218. H. E. Pelish, N. J. Westwood, Y. Feng, T. Kirchhausen, and M. Shair, *J. Am. Chem. Soc.* **123**, 6740 (2001).
219. C.-A. Fan, Y.-Q. Tu, Z.-L. Song, E. Zhang, L. Shi, M. Wang, B. Wang, and S.-Y. Zhang, *Org. Lett.* **6**, 4691 (2004).
220. Z. L. Song, B. M. Wang, Y. Q. Tu, C. A. Fan, and S. Y. Zhang, *Org. Lett.* **5**, 2319 (2003).
221. S. V. Ley, O. Schucht, A. W. Thomas, and P. J. Murray, *J. Chem. Soc., Perkin Trans. I* 1251 (1999).
222. I. R. Baxendale, S. V. Ley, and C. Piutti, *Angew. Chem., Int. Ed.* **41**, 2194 (2002).
223. I. R. Baxendale, S. V. Ley, M. Nessi, and C. Piutti, *Tetrahedron* **58**, 6285 (2002).
224. I. R. Baxendale and S. V. Ley, *Ind. Eng. Chem. Res.* **44**, 8588 (2005).
225. M. M. Abelman, L. E. Overman, and V. D. Tran, *J. Am. Chem. Soc.* **112**, 6959 (1990).
226. R. Grigg, V. Santhakumar, V. Sridharan, M. Thornton-pett, and A. W. Bridge, *Tetrahedron* **49**, 5177 (1993).
227. C. Guillou, J.-L. Beunard, E. Gras, and C. Thal, *Angew. Chem., Int. Ed.* **40**, 4745 (2001).
228. C. Bru, C. Thal, and C. Guillou, *Org. Lett.* **5**, 1845 (2003).
229. E. Gras, C. Guillou, and C. Thal, *Tetrahedron Lett.* **40**, 9243 (1999).
230. P.-H. Liang, J.-P. Liu, L.-W. Hsin, and C.-Y. Cheng, *Tetrahedron* **60**, 11655 (2004).
231. J. Jin and S. M. Weinreb, *J. Am. Chem. Soc.* **119**, 5773 (1997).
232. D. Enders, A. Lenzen, and G. Raabe, *Angew. Chem., Int. Ed.* **44**, 3766 (2005).
233. P. Sahakitpichan and S. Ruchirawat, *Tetrahedron Lett.* **44**, 5239 (2003).
234. B. M. Trost and F. D. Totse, *J. Am. Chem. Soc.* **122**, 11262 (2000).
235. B. M. Trost and W. Tang, *Angew. Chem., Int. Ed.* **41**, 2795 (2002).
236. B. M. Trost and W. Tang, *J. Am. Chem. Soc.* **124**, 14542 (2002).

237. B. M. Trost and S. R. Pulley, *J. Am. Chem. Soc.* **117**, 10143 (1995).
238. T. Nishimata, Y. Stao, and M. Mori, *J. Org. Chem.* **69**, 1837 (2004).
239. T. Harayama, A. Hori, H. Abe, and Y. Takeuchi, *Tetrahedron* **60**, 1611 (2004).
240. J. C. Torres, A. C. Pinto, and S. J. Garden, *Tetrahedron* **60**, 9889 (2004).
241. H.-J. Knölker and S. Filali, *Synlett* 1752 (2003).
242. Z. Shao, J. Chen, R. Huang, C. Wang, L. Li, and H. Zhang, *Synlett* 2228 (2003).
243. K. Hiroya, S. Itoh, and T. Sakamoto, *J. Org. Chem.* **69**, 1126 (2004).
244. M. G. Banwell, D. W. Lupton, X. Ma, J. Renner, and M. O. Sydnes, *Org. Lett.* **6**, 2741 (2004).
245. C.-K. Sha, A.-W. Hong, and C.-M. Huang, *Org. Lett.* **3**, 2177 (2001).
246. C. Hoarau, A. Couture, E. Deniau, and P. Grandclaudon, *J. Org. Chem.* **67**, 5846 (2004).
247. P. A. Patil and V. Snieckus, *Tetrahedron Lett.* **39**, 1325 (1998).
248. C. G. Hartung, A. Fecher, B. Chapell, and V. Snieckus, *Org. Lett.* **5**, 1899 (2003).
249. D. F. Taber and Y. He, *J. Org. Chem.* **70**, 7711 (2005).
250. T. Hudlicky, D. Gonzalez, and D. T. Gibson, *Aldrichim. Acta* **32**, 35 (1999).
251. T. Hudlicky, H. F. Olivo, and B. McKibben, *J. Am. Chem. Soc.* **116**, 5108 (1994).
252. T. Hudlicky, X. Tian, K. Königsberger, R. Maurya, J. Rouden, and B. Fan, *J. Am. Chem. Soc.* **118**, 10752 (1996).
253. D. Gonzalez, T. Martinot, and T. Hudlicky, *Tetrahedron Lett.* **40**, 3077 (1999).
254. U. Rinner, P. Siengalewicz, and T. Hudlicky, *Org. Lett.* **4**, 115 (2002).
255. T. Hudlicky, U. Rinner, D. Gonzalez, H. Akgun, S. Schilling, P. Siengalewicz, T. A. Martinot, and G. Pettit, *J. Org. Chem.* **67**, 8726 (2002).
256. U. Rinner, T. Hudlicky, H. Gordon, and G. R. Pettit, *Angew. Chem., Int. Ed.* **43**, 5342 (2004).
257. U. Rinner, H. L. Hillebrenner, D. R. Adams, T. Hudlicky, and G. R. Pettit, *Bioorg. Med. Chem. Lett.* **14**, 2911 (2004).
258. T. Hudlicky, U. Rinner, K. J. Finn, and I. Ghiviriga, *J. Org. Chem.* **70**, 3490 (2005).
259. A. N. Phung, M. T. Zannetti, G. Whited, and W.-D. Fessner, *Angew. Chem., Int. Ed.* **42**, 4821 (2003).
260. P. Magnus and I. K. Sebhat, *Tetrahedron* **54**, 15509 (1998).
261. P. Magnus, J. M. Bailey, and M. J. Porter, *Tetrahedron* **55**, 13927 (1999).
262. A. F. Parsons and D. A. J. Williams, *Tetrahedron* **56**, 7217 (2000).
263. F. González-López de Turiso and D. P. Curran, *Org. Lett.* **7**, 151 (2005).
264. G. E. Keck, T. T. Wager, and S. F. McHardy, *J. Org. Chem.* **63**, 9164 (1998).
265. G. E. Keck, T. T. Wager, and J. F. Duarte-Rodríquez, *J. Am. Chem. Soc.* **121**, 5176 (1999).
266. M. Ikeda, M. Hamada, T. Yamashita, F. Ikegami, T. Sato, and H. Ishibashi, *Synlett* 1246 (1998).
267. M. Ishikazi, K. Kurihara, E. Tanazawa, and O. Hoshino, *J. Chem. Soc., Perkin Trans. I* 101 (1993).
268. G. L. Barclay, B. Quiclet-Sire, G. Sanchez-Jimenez, and S. Z. Zard, *Org. Biomol. Chem.* **3**, 823 (2005).
269. A. M. Rosa, A. M. Lobo, P. S. Branco, S. Prabhakar, and M. Sá-da-Costa, *Tetrahedron* **53**, 299 (1997).
270. L. E. Overman and J. Shim, *J. Am. Chem. Soc.* **58**, 4662 (1993).
271. H. Ko, E. Kim, J. E. Park, D. Kim, and S. Kim, *J. Org. Chem.* **69**, 112 (2004).
272. J. H. Rigby, U. S. M. Maharoof, and M. E. Mateo, *J. Am. Chem. Soc.* **122**, 6624 (2000).
273. J. H. Rigby and M. E. Mateo, *J. Am. Chem. Soc.* **119**, 12655 (1997).
274. J. H. Rigby and W. Dong, *Org. Lett.* **2**, 1673 (2000).
275. S. E. Wolkenberg and D. L. Boger, *J. Org. Chem.* **67**, 7361 (2002).
276. A. Padwa, M. A. Brodney, and S. M. Lynch, *J. Org. Chem.* **66**, 1716 (2001).
277. Q. Wang and A. Padwa, *Org. Lett.* **6**, 2189 (2004).

278. T. J. Doyle, M. Hendrix, D. VanDerveer, S. Javanmard, and J. Haseltine, *Tetrahedron* **53**, 11153 (1997).
279. J. L. Aceña, O. Arjona, M. L. León, and J. Plumet, *Org. Lett.* **2**, 3683 (2000).
280. S. Gao, Y. Q. Tu, Z. Song, A. Wang, X. Fan, and Y. Jiang, *J. Org. Chem.* **70**, 6523 (2005).
281. T. Yasuhara, K. Nishimura, M. Yamashita, N. Fukuyama, K. Yamada, O. Muraoka, and K. Tomioka, *Org. Lett.* **5**, 1123 (2003).
282. M. Manpadi and A. Kornienko, *Tetrahedron Lett.* **46**, 4433 (2005).
283. O. N. Nadein and A. Kornienko, *Org. Lett.* **6**, 831 (2004).
284. S. P. Chavan, D. A. Khobragade, A. B. Pathak, and U. R. Kalkote, *Tetrahedron Lett.* **45**, 5263 (2004).
285. D. F. Taber and T. D. Neubert, *J. Org. Chem.* **66**, 143 (2001).
286. A. Padwa, M. D. Danca, K. I. Hardcastle, and M. S. McClure, *J. Org. Chem.* **68**, 929 (2003).
287. S. Lebrun, A. Couture, E. Deniau, and P. Grandclaudon, *Org. Biomol. Chem.* **1**, 1701 (2003).
288. S. Elango and T.-H. Yan, *J. Org. Chem.* **67**, 6954 (2002).
289. S. Elango and T.-H. Yan, *Tetrahedron* **58**, 7335 (2002).
290. S. Elango, Y.-C. Wang, C.-L. Cheng, and T.-H. Yan, *Tetrahedron Lett.* **43**, 3757 (2002).
291. G. R. Pettit, B. Orr, and S. Ducki, *Anticancer Drug Des.* **15**, 389 (2000).
292. G. R. Pettit and N. Melody, *J. Nat. Prod.* **68**, 207 (2005).
293. W.-C. Shieh and J. A. Carlson, *J. Org. Chem.* **59**, 5463 (1994).
294. S. Ibn-Ahmed, M. Khaldi, F. Chrétien, and Y. Chapleur, *J. Org. Chem.* **69**, 6722 (2004).
295. S. W. Baldwin and J. S. Debenham, *Org. Lett.* **2**, 99 (2000).
296. M. Bohno, H. Imase, and N. Chida, *Chem. Commun.* 1086 (2004).
297. G. Pandey, A. Murugan, and M. Balakrishnan, *Chem. Commun.* 624 (2002).
298. K. Likhitwitayawuid, C. K. Angerhofer, H. Chai, J. M. Pezzuto, and G. A. Cordell, *J. Nat. Prod.* **56**, 1331 (1993).
299. M. Spohn, V. Brecht, and A. W. Frahm, *Arch. Pharm.* **327**, 123 (1994).
300. J. Wagner, H. L. Pham, and W. Döpke, *Tetrahedron* **52**, 6591 (1996).
301. E. M. Gopalakrishna, W. H. Watson, P. Pacheco, and M. Silva, *Cryst. Struct. Commun.* **5**, 795 (1976).
302. W. Döpke and H. Dalmer, *Naturwissenschaften* **52**, 61 (1965).
303. A. Evidente, I. Iasiello, and G. Randazzo, *J. Nat. Prod.* **47**, 1003 (1984).
304. S. Kobayashi, H. Ishikawa, M. Kihara, T. Shingu, and T. Hashimoto, *Chem. Pharm. Bull.* **25**, 2244 (1977).
305. A. Evidente, O. Arrigoni, R. Liso, G. Calabrese, and G. Randazzo, *Phytochemistry* **25**, 2739 (1986).
306. K. Kuriyama, T. Iwata, M. Moriyama, K. Kotera, Y. Hamada, R. Mitsui, and K. Takeda, *J. Chem. Soc., Sec. B* 46 (1967).
307. H. G. Boit, H. Ehmke, S. Uyeo, and H. Yajima, *Chem. Ber.* **90**, 363 (1957).
308. F. Sandberg and K. H. Michel, *Acta Pharm. Suec.* **5**, 61 (1968).
309. H. M. Fales and W. C. Wildman, *J. Am. Chem. Soc.* **80**, 4395 (1958).
310. T. H. Kinstle, W. C. Wildman, and C. L. Brown, *Tetrahedron Lett.* 4659 (1966).
311. J. C. Clardy, W. C. Wildman, and F. M. Hauser, *J. Am. Chem. Soc.* **92**, 1781 (1970).
312. A. Immirzi and C. Fuganti, *J. Chem. Soc.-B* 1218 (1971).
313. M. Kihara, T. Ozaki, S. Kobayashi, and T. Shingu, *Chem. Pharm. Bull.* **43**, 318 (1995).
314. P. Richomme, V. Pabuccuoglu, T. Gözler, A. J. Freyer, and M. Shamma, *J. Nat. Prod.* **52**, 1150 (1989).
315. R. Suau, A. I. Gómez, R. Rico, M. P. Vazquez-Tato, L. Castedo, and R. Riguera, *Phytochemistry* **27**, 3285 (1988).

316. J. Bastida, C. Codina, P. Peeters, M. Rubiralta, M. Orozco, F. J. Luque, and S. C. Chhabra, *Phytochemistry* **40**, 1291 (1995).
317. A. Evidente, G. Randazzo, G. Surico, P. Lavermicocca, and O. Arrigoni, *J. Nat. Prod.* **48**, 564 (1985).
318. A. H. Abou-Donia, A. A. Abib, A. S. El Din, A. Evidente, M. Gaber, and A. Scopa, *Phytochemistry* **31**, 2139 (1992).
319. J. Bastida, C. Codina, C. L. Porras, and L. Paiz, *Planta Med.* **62**, 74 (1996).
320. I. K. Rhee, N. Appels, B. Hofte, B. Karabatak, C. Erkelens, L. M. Stark, L. A. Flippin, and R. Verpoorte, *Biol. Pharm. Bull.* **27**, 1804 (2004).
321. P. W. Jeffs, A. Abou-Donia, D. Campau, and D. Staiger, *J. Org. Chem.* **50**, 1732 (1985).
322. F. Viladomat, J. Bastida, C. Codina, X. Solans, and M. Font-Bardia, *Acta Crystallogr.-C* **55**, 385 (1999).
323. A. Latvala, M. A. Önür, T. Gözler, A. Linden, B. Kivçak, and M. Hesse, *Phytochemistry* **39**, 1229 (1995).
324. A. Latvala, M. A. Önür, T. Gözler, A. Linden, B. Kivçak, and M. Hesse, *Tetrahedron Asym.* **6**, 361 (1995).
325. H. Irie, Y. Tsuda, and S. Uyeo, *J. Chem. Soc.* 1446 (1959).
326. A. Evidente, A. Andolfi, A. H. Abou-Donia, S. M. Touema, H. M. Hammoda, E. Shawky, and A. Motta, *Phytochemistry* **65**, 2113 (2004).
327. T. Kitagawa, W. I. Taylor, S. Uyeo, and H. Yajima, *J. Chem. Soc.* 1066 (1955).
328. T. Ibuka, H. Irie, A. Kato, S. Uyeo, K. Kotera, and Y. Nakagawa, *Tetrahedron Lett.* 4745 (1966).
329. J. C. Clardy, J. A. Chan, and W. C. Wildman, *J. Org. Chem.* **37**, 49 (1972).
330. A. W. Frahm, A. A. Ali, and M. A. Ramadan, *Magn. Reson. Chem.* **23**, 804 (1985).
331. V. Pabuççuoglu, P. Richomme, T. Gözler, B. Kivçak, A. J. Freyer, and M. Shamma, *J. Nat. Prod.* **52**, 785 (1989).
332. G. G. De Angelis and W. C. Wildman, *Tetrahedron Lett.* 729 (1969).
333. K. Tomioka, K. Koga, and S. I. Yamada, *Chem. Pharm. Bull.* **25**, 2681 (1977).
334. V. Zabel, W. H. Watson, P. Pacheco, and M. Silva, *Cryst. Struct. Commun.* **8**, 371 (1979).
335. S. Ghosal, A. Razdan, and S. Razdan, *Phytochemistry* **24**, 635 (1985).
336. D. T. A. Youssef and A. W. Frahm, *Planta Med.* **64**, 669 (1998).
337. M. Kihara, T. Koike, Y. Imakura, K. Kida, T. Shingu, and S. Kobayashi, *Chem. Pharm. Bull.* **35**, 1070 (1987).
338. G. Baudouin, F. Tillequin, and M. Koch, *Heterocycles* **38**, 965 (1994).
339. W. H. Watson, J. Galloy, and M. Silva, *Acta Crystallogr.-C* **40**, 156 (1984).
340. Z. Trimiño, M. Castillo, and I. Spenglers, *Rev. Cubana Farm.* **23**, 147 (1989).
341. M. D. Antoun, N. T. Mendoza, and Y. R. Rios, *J. Nat. Prod.* **56**, 1423 (1993).
342. F. Viladomat, J. Bastida, C. Codina, W. E. Campbell, and S. Mathee, *Phytochemistry* **35**, 809 (1994).
343. F. Viladomat, G. R. Almanza, C. Codina, J. Bastida, W. E. Campbell, and S. Mathee, *Phytochemistry* **43**, 1379 (1996).
344. R. Roques, J. P. Declercq, and G. Germain, *Acta Crystallogr.-B* **33**, 3696 (1977).
345. W. C. Wildman and C. J. Kaufman, *J. Am. Chem. Soc.* **76**, 5815 (1954).
346. W. O. Crain, W. C. Wildman, and J. D. Roberts, *J. Am. Chem. Soc.* **93**, 990 (1971).
347. S. Ghosal, Y. Kumar, and S. Sinhg, *Phytochemistry* **23**, 1167 (1984).
348. Z. Trimiño, C. Iglesias, M. Sánchez, and I. Spengler, *Rev. Cubana Quím.* **3**, 57 (1987).
349. W. Döpke, L. H. Pham, E. Gründemann, M. Bartoszek, and S. Flatau, *Planta Med.* **61**, 564 (1995).
350. L. H. Pham, E. Gründemann, J. Wagner, M. Bartoszek, and W. Döpke, *Phytochemistry* **51**, 327 (1999).
351. S. Ide, B. Sener, H. Temizer, and S. Könükol, *Cryst. Res. Technol.* **31**, 617 (1996).
352. H. Hauth and D. Stauffacher, *Helv. Chim. Acta* **47**, 185 (1964).

353. A. M. Duffield, R. T. Aplin, H. Budzikiewicz, C. Djerassi, C. F. Murphy, and W. C. Wildman, *J. Am. Chem. Soc.* **87**, 4902 (1965).
354. G. G. De Angelis and W. C. Wildman, *Tetrahedron* **25**, 5099 (1969).
355. J. Razafimbelo, M. Andriantsiferana, G. Baudouin, and F. Tillequin, *Phytochemistry* **41**, 323 (1996).
356. A. Linden, G. Akineri, S. Noyan, T. Gozler, and M. Hesse, *Acta Crystallogr.-C* **54**, 1653 (1998).
357. I. Spenglers and Z. Trimiño, *Rev. Cubana Farm.* **23**, 151 (1989).
358. R. Suau, A. I. Gómez, and R. Rico, *Phytochemistry* **29**, 1710 (1990).
359. D. T. A. Youssef, *J. Nat. Prod.* **64**, 839 (2001).
360. F. Viladomat, C. Codina, J. Bastida, X. Solans, and M. Font-Bardia, *Acta Crystallogr.-C* **54**, 81 (1998).
361. A. A. Ali, M. K. Mesbah, and A. W. Frahm, *Planta Med.* **50**, 188 (1984).
362. O. M. Abdallah, A. A. Ali, and H. Itokawa, *Phytochemistry* **28**, 3248 (1989).
363. O. M. Peeters, N. M. Blaton, and C. J. De Ranter, *Acta Crystallogr.-C* **53**, 1284 (1997).
364. T. Kametani, K. Yamaki, H. Yagi, and K. Fukumoto, *J. Chem. Soc.-C* 2602 (1969).
365. R. Vlahov, D. Krikorian, G. Spassov, M. Chinova, I. Vlahov, S. Parushev, G. Snatzke, L. Ernst, K. Kieslich, W. R. Abraham, and W. S. Sheldrick, *Tetrahedron* **45**, 3329 (1989).
366. S. M. Laiho and H. M. Fales, *J. Am. Chem. Soc.* **86**, 4434 (1964).
367. H. Y. Li, G. E. Ma, Y. Xu, and S. H. Hong, *Planta Med.* **53**, 259 (1987).
368. R. Roques and J. Lapasset, *Acta Crystallogr.-B* **32**, 579 (1976).
369. S. Kobayashi, K. Satoh, A. Numata, T. Shingu, and M. Kihara, *Phytochemistry* **30**, 675 (1991).
370. S. Kobayashi, T. Tokumoto, M. Kihara, K. Imakura, T. Shingu, and Z. Taira, *Chem. Pharm. Bull.* **32**, 3015 (1984).
371. P. W. Jeffs, T. Capps, D. B. Johnson, J. M. Karle, N. H. Martin, and B. Rauckman, *J. Org. Chem.* **39**, 2703 (1974).
372. P. W. Jeffs, G. Ahmann, H. F. Campbell, D. S. Farrier, G. Ganguli, and R. L. Hawks, *J. Org. Chem.* **35**, 3512 (1970).
373. S. E. Denmark and L. R. Marcin, *J. Org. Chem.* **62**, 1675 (1997).
374. P. W. Jeffs, L. Mueller, A. H. Abou-Donia, A. A. Seif el-Din, and D. Campau, *J. Nat. Prod.* **51**, 549 (1988).
375. L. Zetta, G. Gatti, and C. Fuganti, *J. Chem. Soc., Perkin Trans. II* 1180 (1973).
376. A. Evidente, *J. Nat. Prod.* **49**, 168 (1986).
377. A. Mary, D. Z. Renko, C. Guillou, and C. Thal, *Bioorg. Med. Chem.* **6**, 1835 (1998).
378. H. K. Schnoes, D. H. Smith, A. L. Burlingame, P. W. Jeffs, and W. Döpke, *Tetrahedron* **24**, 2825 (1968).
379. H. M. Fales, G. W. A. Milne, and M. L. Vestal, *J. Am. Chem. Soc.* **91**, 3682 (1969).
380. R. Razakov, V. N. Bochkarev, Kh. A. Abduazimov, N. S. Vul'fson, and S. Y. Yunusov, *Khim. Prir. Soedin.* **5**, 280 (1969).
381. H. M. Fales, H. A. Lloyd, and G. W. A. Milne, *J. Am. Chem. Soc.* **92**, 1590 (1970).
382. P. Longevialle, D. H. Smith, A. L. Burlingame, H. M. Fales, and R. J. Highet, *Org. Mass Spectrom.* **7**, 401 (1973).
383. P. Longevialle, H. M. Fales, R. J. Highet, and A. L. Burlingame, *Org. Mass Spectrom.* **7**, 417 (1973).
384. E. H. C. Samuel, *Org. Mass Spectrom.* **10**, 427 (1975).
385. O. S. Onyiriuka and A. H. Jackson, *Isr. J. Chem.* **17**, 185 (1978).
386. G. R. Pettit, G. R. Pettit, III, G. Groszek, R. A. Backhaus, D. L. Doubek, R. J. Barr, and A. W. Meerow, *J. Nat. Prod.* **58**, 756 (1995).
387. G. R. Pettit, G. M. Cragg, S. B. Singh, J. A. Duke, and D. L. Doubek, *J. Nat. Prod.* **53**, 176 (1990).

388. G. R. Pettit, V. Gaddamidi, D. L. Herald, S. B. Singh, G. M. Cragg, J. M. Schmidt, F. E. Boettner, M. Williams, and Y. Sagawa, *J. Nat. Prod.* **49**, 995 (1986).
389. G. R. Pettit, G. R. Pettit, III, R. A. Backhaus, M. R. Boyd, and A. W. Meerow, *J. Nat. Prod.* **56**, 1682 (1993).
390. E. Furusawa, N. Suzuki, S. Tani, S. Furusawa, G. Y. Ishioka, and J. Motobu, *Proc. Soc. Exp. Biol. Med.* **143**, 33 (1973).
391. I. Çakici, H. Y. Ulug, S. Inci, B. Tunçtan, N. Abacioglu, I. Kanzik, and B. Sener, *J. Pharm. Pharmacol.* **49**, 828 (1997).
392. G. Grigson, ,"The Englishman's Flora", Helicon Publishing Ltd., Oxford, 1996.
393. M. Grieve, ,"A Modern Herbal – the Medicinal, Culinary, Cosmetic and Economic Properties, Cultivation and Folklore of Herbs, Grasses, Fungi, Shrubs and Trees with All Their Modern Scientific Uses", Tiger Books International, London, 1998.
394. M. Scott, ,"An Irish Herbal. The Botananalogia Universalis Hibernica by JohnK'Eogh (1681–1754)", The Aquarium Press, Wellingborough, 1986.
395. T. C. Wu, S. T. Chen, T. Y. Wang, and P. Hsu, *Yao Hsueh Hsueh Pao (Acta Pharm. Sin.)* **21**, 31 (1965).
396. R. Jaspersen-Schib, L. Theus, M. Guirguis-Oeschger, B. Gossweiler, and P. J. Meier-Abt, *Schweiz. Med. Wochenschr.* **126**, 1085 (1996).
397. C. G. Julian and P. W. Bowers, *in* "Narcissus and daffodil: the genus *Narcissus*" (G.R. Hanks ed.), vol. 21 in the series: "Medicinal and Aromatic Plants – Industrial Profiles", p. 399. Taylor & Francis, London, 2002.
398. I. Junko, T. Akito, K. Yumiko, and O. Noriyoshi, *Pharm. Mon.* **36**, 855 (1994).
399. S. Ries, R. Baughan, M. G. Nair, and R. Schutzki, *HortTechnology* **11**, 302 (2001).
400. R. K. O'Leary, U.S. pat. Appl. 6565867 B1 20030520 (CA 138: 364205) (2003).
401. R. K. O'Leary, U.S. pat. Appl. 2003202998 A1 20031030 (CA 139: 303299) (2003).
402. D. P. Bruynzeel, E. M. De Boer, E. J. Brouwer, F. A. De Wolff, and P. De Haan, *Contact Dermatitis* **29**, 11 (1993).
403. D. P. Bruynzeel, *Contact Dermatitis* **37**, 70 (1997).
404. N. W. De Jong, A. M. Vermeulen, R. Gerth van Wijk, and H. De Groot, *Allergy* **53**, 204 (1998).
405. S. Gonçalo, J. D. Freitas, and I. Sousa, *Contact Dermatitis* **16**, 115 (1987).
406. S. Güneser, A. Atici, I. Cengizler, and N. Alparslan, *Allergol. Immunopathol.* **24**, 116 (1996).
407. W. G. Van Doorn, *J. Am. Soc. Hort. Sci.* **123**, 146 (1998).
408. B. Maurer, *Perfum. Flavor.* **19**(19–22), 24 (1994).
409. M. Reuveni, Z. Sgi, D. Evnor, and A. Hetzroni, *Plant Sci.* **147**, 19 (1999).
410. C. Remy, *in* "Narcissus and daffodil: the genus *Narcissus*" (G.R. Hanks ed.), vol. 21 in the series: "Medicinal and Aromatic Plants – Industrial Profiles", p. 392. Taylor & Francis, London, 2002.
411. M. Abou-Karam and W. T. Shier, *J. Nat. Prod.* **53**, 340 (1990).
412. E. Furusawa, N. Suzuki, S. Furusawa, and J. Y. B. Lee, *Proc. Soc. Exp. Biol. Med.* **149**, 771 (1975).
413. T. S. Papas, L. Sandhaus, M. A. Chirigos, and E. Furusawa, *Biochem. Biophys. Res. Commun.* **52**, 88 (1973).
414. S. Ramanathan, E. Furusawa, M. Kroposki, S. Furusawa, and W. Cutting, *Chemotherapy* **13**, 121 (1968).
415. N. Suzuki, S. Tani, S. Furusawa, and E. Furusawa, *Proc. Soc. Exp. Biol. Med.* **145**, 771 (1974).
416. J. P. Vacik, W. B. Davis, C. S. Kelling, L. J. Schermeister, and I. A. Schipper, *Adv. Ophtalmol.* **38**, 281 (1979).
417. D. A. Van den Bergue, M. Ieven, F. Mertens, J. Vlietinck, and E. Lammens, *Lloydia* **41**, 463 (1978).
418. K. Dornberger and H. Lich, *Pharmazie* **37**, 215 (1982).

419. B. Sener, M. Koyuncu, F. Bingöl, and F. Muhtar, ,"IUPAC International Conference on Biodiversity and Bioresources: Conservation and Utilization", Phuket, Thailand, 1997.
420. M. Ieven, D. A. Van den Bergue, F. Mertens, A. Vlietinck, and E. Lammens, *Planta Med.* **36**, 311 (1979).
421. J. P. Chaumont and J. M. Senet, *Plant. Med. Phytother.* **12**, 186 (1978).
422. H. Chaumont, H. Schemaker, and J. Rouseau, *Plant. Med. Phytother.* **12**, 157 (1978).
423. B. Sener, I. Orham, and J. Satayavivad, *Phytother. Res.* **17**, 1220 (2003).
424. E. Furusawa, H. Irie, D. Combs, and W. C. Wildman, *Chemotherapy* **26**, 36 (1980).
425. E. Furusawa, N. Suzuki, S. Ramanathan, S. Furusawa, and W. Cutting, *Proc. Soc. Exp. Biol. Med.* **140**, 1034 (1972).
426. M. Ikram, *Fitoterapia* **54**, 123 (1983).
427. K. W. Chiu, Y. C. Lee, and K. H. Yung, *Phytother. Res.* **6**, 231 (1992).
428. A. D. S. Matsui, J. Rogers, Y. K. Woo, and W. C. Cutting, *Med. Pharmacol. Exp.* **16**, 414 (1967).
429. I. Keiser, E. J. Harris, D. H. Miyashita, M. Jacobson, and R. E. Perdue, *Lloydia* **38**, 141 (1975).
430. A. Barre, E. J. M. Van Damme, W. J. Peumans, and P. Rougé, *Plant Physiol.* **112**, 1531 (1996).
431. E. J. M. Van Damme, W. J. Peumans, A. Barre, and P. Rougé, *Crit. Rev. Plant Sci.* **17**, 575 (1998).
432. E. J. M. Van Damme and W. J. Peumans, *in* "Narcissus and daffodil: the genus *Narcissus*" (G.R. Hanks ed.), vol. 21 in the series: "Medicinal and Aromatic Plants – Industrial Profiles", p. 380. Taylor & Francis, London, 2002.
433. S. López, C. Codina, J. Bastida, F. Viladomat, E. Davidson, and D. Stewart, *J. Agric. Food Chem.* **50**, 2507 (2002).
434. S. López, M. Armand-Ugon, J. Bastida, F. Viladomat, J. A. Este, D. Stewart, and C. Codina, *Planta Med.* **69**, 109 (2003).
435. L. S. M. Ooi, F. Liu, V. E. C. Ooi, T. B. Ng, and M. C. Fung, *Biochem. Cell Biol.* **80**, 271 (2002).
436. O. Arrigoni, R. Arrigoni-Liso, and G. Calabrese, *Nature* **256**, 513 (1975).
437. A. Evidente, M. R. Cicala, G. Randazzo, R. Riccio, G. Calabrese, R. Liso, and O. Arrigoni, *Phytochemistry* **22**, 2193 (1983).
438. O. Arrigoni, G. Calabrese, L. De Gara, M. B. Bitonti, and R. Liso, *J. Plant Physiol.* **150**, 302 (1997).
439. M. C. De Tullio, L. De Gara, C. Paciolla, and O. Arrigoni, *Plant Physiol. Biochem.* **36**, 433 (1998).
440. M. W. Davey, G. Persiau, A. De Bruyn, J. Van Damme, G. Bauw, and M. Van Montagu, *Anal. Biochem.* **257**, 80 (1998).
441. L. De Gara, C. Paciolla, F. Tommasi, R. Liso, and O. Arrigoni, *J. Plant Physiol.* **144**, 649 (1994).
442. T. Imai, S. Karita, G. Shiratori, M. Hatturi, T. Nunome, and K. Ôba, *Plant Cell Physiol.* **39**, 1350 (1998).
443. E. Kuzniak, *Acta Physiol. Plant.* **26**, 327 (2004).
444. O. Arrigoni, C. Paciolla, and L. De Gara, *B. Soc. Ital. Biol. Sper.* **75**, 37 (1996).
445. O. Arrigoni, L. De Gara, C. Paciolla, A. Evidente, M. C. De Pinto, and R. Liso, *J. Plant Physiol.* **150**, 362 (1997).
446. O. Arrigoni, *J. Bioenerg. Biomembr.* **26**, 407 (1994).
447. M. C. Córdoba-Pedregosa, J. A. González-Reyes, M. S. Cañadillas, P. Navas, and F. Córdoba, *Plant Physiol.* **112**, 1119 (1996).
448. A. Del Giudice, D. R. Massardo, F. Manna, N. Koltovaya, H. Hartings, L. Del Giudice, and K. Wolf, *Curr. Microbiol.* **34**, 382 (1997).

449. R. Liso, G. Calabrese, M. B. Bitonti, and O. Arrigoni, *Exp. Cell Res.* **150**, 314 (1984).
450. S. Onofri, E. Poerio, P. Serangeli, S. Tosi, I. Garuccio, and O. Arrigoni, *Antonie Leeuwenhoek* **71**, 277 (1997).
451. O. Arrigoni, R. Arrigoni-Liso, and G. Calabrese, *Science* **194**, 332 (1976).
452. R. Liso, L. De Gara, F. Tommasi, and O. Arrigoni, *FEBS Lett.* **187**, 141 (1985).
453. M. Kukhanova, L. Victorova, and A. Krayevsky, *FEBS Lett.* **160**, 129 (1983).
454. A. Del Giudice, D. R. Massardo, F. Manna, A. Evidente, G. Randazzo, and K. Wolf, *Curr. Genet.* **8**, 493 (1984).
455. D. R. Massardo, F. Manna, B. Schafer, K. Wolf, and L. Del Giudice, *Curr. Genet.* **25**, 80 (1994).
456. D. R. Massardo, F. Manna, L. Del Giudice, and K. Wolf, *Curr. Genet.* **17**, 455 (1990).
457. D. R. Massardo, S. G. Zweifel, N. Gunge, I. Miyakawa, N. Sando, A. Del Giudice, K. Wolf, and L. Del Giudice, *Can. J. Microbiol.* **46**, 1058 (2000).
458. S. Onofri, D. Barreca, and I. Garuccio, *Antonie Leeuwenhoek* **83**, 57 (2003).
459. S. Ghosal, A. Shanthy, A. Kumar, and Y. Kumar, *Phytochemistry* **24**, 2703 (1985).
460. D. H. Hua, S. Saha, and D. J. Takemoto, *Anticancer Res.* **17**, 2435 (1997).
461. S. Yui, M. Mikami, M. Kitahara, and M. Yamazaki, *Immunopharmacology* **40**, 151 (1998).
462. L. Z. Lin, S. F. Hu, H. B. Chai, T. Pengsuparp, J. M. Pezzuto, G. A. Codell, and N. Ruangrungsi, *Phytochemistry* **40**, 1295 (1995).
463. B. Weniger, L. Italiano, J. P. Beck, J. Bastida, S. Bergoñon, C. Codina, A. Lobstein, and R. Anton, *Planta Med.* **61**, 77 (1995).
464. J. J. Nair, W. E. Campbell, D. W. Gammon, C. F. Albrecht, F. Viladomat, C. Codina, and J. Bastida, *Phytochemistry* **49**, 2539 (1998).
465. B. S. Min, J. J. Gao, N. Nakamura, Y. H. Kim, and M. Hattori, *Chem. Pharm. Bull.* **49**, 1217 (2001).
466. N. Kushida, S. Atsumi, T. Koyano, and K. Umezawa, *Drug Exp. Clin. Res.* **23**, 151 (1997).
467. A. Jimenez, A. Santos, G. Alonso, and D. Vázquez, *Biochim. Biophys. Acta* **425**, 342 (1976).
468. M. Mikami, M. Kitahara, M. Kitano, Y. Ariki, Y. Mimaki, Y. Sashida, M. Yamazaki, and S. Yui, *Biol. Pharm. Bull.* **22**, 674 (1999).
469. S. Yui, M. Mikami, Y. Mikami, Y. Sashida, and M. Yamazaki, *Yakugaku Zasshi* **121**, 167 (2001).
470. O. B. Abdel-Halim, T. Morikawa, S. Ando, H. Matsuda, and M. Yoshikawa, *J. Nat. Prod.* **67**, 1119 (2004).
471. J. Liu, W.-X. Hu, L.-F. He, M. Ye, and Y. Li, *FEBS Lett.* **578**, 245 (2004).
472. G. Schmeda-Hirschmann, L. Astudillo, J. Bastida, F. Viladomat, and C. Codina, *Bol. Soc. Chil. Quím.* **45**, 515 (2000).
473. H. Karadeniz, B. Gulmez, F. Sahinci, A. Erdem, G. I. Kaya, N. Unver, B. Kivcak, and M. Ozsoz, *J. Pharm. Biomed. Anal.* **32**, 295 (2003).
474. G. R. Pettit, V. Gaddamidi, A. Goswami, and G. M. Cragg, *J. Nat. Prod.* **47**, 796 (1984).
475. A. Jimenez, L. Sánchez, and D. Vázquez, *FEBS Lett.* **60**, 66 (1975).
476. S. Ghosal, S. K. Singh, Y. Kumar, S. Unnikrishnan, and S. Chattopadhyay, *Planta Med.* **54**, 114 (1988).
477. R. K. Y. Zee-Cheng, S. J. Yan, and C. C. Cheng, *J. Med. Chem.* **21**, 199 (1978).
478. U. Chattopadhyay, L. Chaudhuri, S. Das, Y. Kumar, and S. Ghosal, *Pharmazie* **39**, 855 (1984).
479. M. Ieven, D. A. Van den Bergue, and A. Vlietinck, *Planta Med.* **49**, 109 (1983).
480. M. Ieven, A. Vlietinck, D. A. Van den Bergue, J. Totte, R. Dommisse, E. Esmans, and F. Alderweireldt, *J. Nat. Prod.* **45**, 564 (1982).

481. B. Gabrielsen, T. P. Monath, J. W. Huggins, D. F. Kefauver, G. R. Pettit, G. Groszek, M. Hollingshead, J. J. Kirsi, W. M. Shannon, E. M. Schubert, J. Dare, B. Ugarkar, M. A. Ussery, and M. J. Phelan, *J. Nat. Prod.* **55**, 1569 (1992).
482. S. Y. Li, C. Chen, H. Q. Zhang, H. Y. Guo, H. Wang, L. Wang, X. Zhang, S. N. Hua, J. Yu, P. G. Xiao, R. S. Li, and X. Tan, *Antiviral Res.* **67**, 18 (2005).
483. L. Szlávik, A. Gyuris, J. Minárovits, P. Forgo, J. Molnár, and J. Hohmann, *Planta Med.* **70**, 871 (2004).
484. J. Renard-Nozaki, T. Kim, Y. Imakura, M. Kihara, and S. Kobayashi, *Res. Virol.* **140**, 115 (1989).
485. R. Vrijsen, D. A. Van der Berghe, A. J. Vlietinck, and A. Boeyé, *J. Biol. Chem.* **261**, 505 (1986).
486. E. E. Elgorashi, G. I. Stafford, and J. van Staden, *Planta Med.* **70**, 260 (2004).
487. P. J. Houghton, J. M. Agbedahunsi, and A. Adegbulugbe, *Phytochemistry* **65**, 2893 (2004).
488. G. A. Cordell, ,"Introduction to Alkaloids: A Biogenetic Approach", Wiley Interscience, New York, 1981.
489. M. Tanker, G. Çitoglu, B. Gumusel, and B. Sener, *Int. J. Pharmacogn.* **34**, 194 (1996).
490. C. W. Fennell and J. van Staden, *J. Ethnopharmacol.* **78**, 15 (2001).
491. S. Abdalla, M. Abu Zarga, and S. Sabri, *Fitoterapia* **64**, 518 (1993).
492. M. Q. Amara, J. F. Franetich, N. Bouladoux, D. Mazier, G. Eisenbrand, D. Marko, L. Meijer, C. Doerig, and I. Desportes-Livage, *J. Eukaryot. Microbiol.* **99S**(Suppl.) (2001).
493. R. P. Singh and N. C. Pant, *Experientia* **36**, 552 (1980).
494. W. E. Campbell, J. J. Nair, D. W. Gammon, C. Codina, J. Bastida, F. Viladomat, P. J. Smith, and C. F. Albrecht, *Phytochemistry* **53**, 587 (2000).
495. A. G. Schultz, M. A. Holoboski, and M. S. Smyth, *J. Am. Chem. Soc.* **118**, 6210 (1996).
496. G. Çitoglu, M. Tanker, and B. Gümüsel, *Phytother. Res.* **12**, 205 (1998).
497. M. R. Herrera, A. K. Machocho, R. Brun, F. Viladomat, C. Codina, and J. Bastida, *Planta Med.* **67**, 191 (2001).
498. G. Schmeda-Hirschmann, J. A. Rodriguez, J. I. Loyola, L. Astudillo, J. Bastida, F. Viladomat, and C. Codina, *Pharm. Pharmacol. Commun.* **6**, 309 (2000).
499. K. Miyasaka and Y. Hiramatsu, *Jpn. J. Pharmacol.* **30**, 655 (1980).
500. M. Alarcón, G. Cea, and G. Weigert, *Bull. Environ. Contam. Toxicol.* **37**, 508 (1986).
501. F. Viladomat, J. Bastida, C. Codina, W. E. Campbell, and S. Mathee, *Phytochemistry* **40**, 307 (1995).
502. S. A. Adesanya, T. A. Olugbade, O. O. Odebiyi, and J. A. Aladesanmi, *Int. J. Pharmacogn.* **30**, 303 (1992).
503. E. Furusawa, S. Furusawa, J. Y. B. Lee, and S. Patanavanich, *Proc. Soc. Exp. Biol. Med.* **152**, 186 (1976).
504. S. Kobayashi, M. Kihara, T. Shingu, and K. Shingu, *Chem. Pharm. Bull.* **28**, 2924 (1980).
505. J. H. Rigby, A. Cavezza, and M. J. Heeg, *J. Am. Chem. Soc.* **120**, 3664 (1998).
506. E. Furusawa and S. Furusawa, *Chemotherapy* **32**, 521 (1986).
507. E. Furusawa and S. Furusawa, *Oncology* **45**, 180 (1988).
508. E. Furusawa, R. H. Lockwood, S. Furusawa, M. K. M. Lum, and J. Y. B. Lee, *Chemotherapy* **25**, 308 (1979).
509. E. Furusawa, M. K. M. Lum, and S. Furusawa, *Chemotherapy* **27**, 277 (1981).
510. E. Furusawa, E. Furusawa, and L. Sokugawa, *Chemotherapy* **29**, 294 (1983).
511. E. Furusawa, S. Furusawa, J. Y. B. Lee, and S. Patanavanich, *Chemotherapy* **24**, 259 (1978).

512. D. Brown, *in* "Narcissus and daffodil: the genus *Narcissus*" (G.R. Hanks, ed.), vol. 21 in the series: "Medicinal and Aromatic Plants – Industrial Profiles", p. 332. Taylor & Francis, London, 2002.
513. A. Baez and D. Vazquez, *Biochim. Biophys. Acta* **518**, 95 (1978).
514. A. Jimenez, L. Sánchez, and D. Vázquez, *FEBS Lett.* **55**, 53 (1975).
515. C. Rodríguez-Fonseca, R. Amils, and R. A. Garret, *J. Mol. Biol.* **247**, 224 (1995).
516. G. R. Pettit, G. R. Pettit, III, R. A. Backhaus, and F. E. Boettner, *J. Nat. Prod.* **58**, 37 (1995).
517. A. A. Abou-Donia, A. De Giulio, A. Evidente, M. Gaber, A. A. Habib, R. Lanzetta, and A. A. Seif El Din, *Phytochemistry* **30**, 3445 (1991).
518. S. Ghosal, S. Singh, Y. Kumar, and S. Srivastava, *Phytochemistry* **28**, 611 (1989).
519. W. G. Van Doorn, A. Sinz, and M. M. Tomassen, *Phytochemistry* **65**, 571 (2004).
520. Y. Bi, J. Guo, L. Zhang, and Y. Wong, *J. Plant Physiol.* **160**, 1041 (2003).
521. Y. Bi, L. Zhang, J. K. Guo, K. Yung, and Y. Wong, *New Zeal. J. Crop Hort. Sci.* **31**, 335 (2003).
522. B. Fulton and P. Benfield, *Drug Aging* **9**, 60 (1996).
523. G. Wilcock and D. Wilkinson, *in* "Alzheimer's Disease: Biology, Diagnosis and Therapeutics", (K. Iqbal, B. Winblad, T. Nishimura, M. Takeda, and H. M. Wisniewski, eds.), p. 661. Wiley, West Sussex, 1997.
524. J. J. Sramek, E. J. Frackiewicz, and N. R. Cutler, *Expert Opin. Inv. Drug* **9**, 2393 (2000).
525. H. A. M. Mucke, *Drug Today* **33**, 251 (1997).
526. H. A. M. Mucke, *Drug Today* **33**, 259 (1997).
527. H. M. Greenblatt, G. Kriger, T. Lewis, I. Silman, and J. L. Sussman, *FEBS Lett.* **463**, 321 (1999).
528. H. M. Greenblatt, C. Guillou, D. Guenard, A. Argaman, S. Botti, B. Badet, C. Thal, I. Silman, and J. L. Sussman, *J. Am. Chem. Soc.* **126**, 15405 (2004).
529. C. Guillou, A. Mary, D. Z. Renko, E. Gras, and C. Thal, *Bioorg. Med. Chem. Lett.* **10**, 637 (2000).
530. L. J. Scott and K. L. Goa, *Drugs* **60**, 1095 (2000).
531. E. Arias, E. Ales, N. H. Gabilan, M. F. Cano-Abad, M. Villarroya, A. G. Garcia, and M. G. López, *Neuropharmacology* **46**, 103 (2004).
532. T. Kihara, H. Sawada, T. Nakamizo, R. Kanki, H. Yamashita, A. Maelicke, and S. Shimohama, *Biochem. Biophys. Res. Commun.* **325**, 976 (2004).
533. N. Radicheva, K. Mileva, N. Stoyanova, and B. Georgieva, *Method. Find. Exp. Clin. Pharmacol.* **21**, 5 (1999).
534. C. V. R. Blacker, D. T. Greenwood, K. A. Wesnes, R. Wilson, C. Woosward, I. Howe, and T. Ali, *J. Am. Med. Assoc.* **292**, 1195 (2004).
535. M. Rainer, *Drug Today* **33**, 273 (1997).
536. G. Ezio, *Neurochem. Int.* **32**, 413 (1998).
537. A. M. Hermann and P. D. Mucke, *Drug Today* **33**, 251 (1997).
538. A. Nordberg and A. L. Svensson, *Drug Saf.* **19**, 465 (1998).
539. H. Kewitz, *Drug Today* **33**, 265 (1997).
540. M. Rainer, *CNS Drugs* **7**, 89 (1997).
541. M. A. Raskind, E. R. Peskind, T. Wessel, W. Yuan, and the Galanthamine USA-1 Study Group, *Neurology* **54**, 2261 (2000).
542. Q. Zhao, M. Brett, N. van Osselaer, F. Huang, A. Raoult, A. van Peer, T. Verhaeghe, and R. Hust, *J. Clin. Pharmacol.* **42**, 1002 (2002).
543. C. E. Patterson, A. P. Passmore, and V. L. S. Crawford, *Int. J. Clin. Pract.* **58**, 144 (2004).
544. M. Heinrich and H. L. Teoh, *J. Ethnopharmacol.* **92**, 147 (2004).
545. R. W. Jones, H. Soininen, K. Hager, D. Aarsland, P. Passmore, A. Murthy, R. Zhang, and R. Bahra, *Int. J. Geriatr. Psych.* **19**, 58 (2004).

546. G. S. J. Mannens, C. A W. Snel, J. Hendrickx, T. Verhaeghe, L. Le Jeune, W. Bode, L. van Beijsterveldt, K. Lavrijsen, J. Leempoels, N. van Osselaer, A. van Peer, and W. Meuldermans, *Drug Metab. Dispos.* **30**, 553 (2002).
547. G. M. Bores and R. W. Kosley, *Drug Future* **21**, 621 (1996).
548. N. Radicheva, M. Vydevska, and K. Mileva, *Method. Find. Exp. Clin. Pharmacol.* **18**, 301 (1996).
549. H. Allain, D. Bentué-Ferrer, S. Belliard, and C. Derouesné, *in* "Progress in Medicinal Chemistry", (G. P. Ellis and D. K. Luscombe, eds.), vol. 34, p. 1. Elsevier, Amsterdam, 1997.
550. A. M. Hermann and P. D. Mucke, *Drug Today* **33**, 259 (1997).
551. L. Czollner, W. Frantsits, B. Kuenburg, U. Hedenig, J. Frohlich, and U. Jordis, *Tetrahedron Lett.* **39**, 2087 (1998).
552. A. L. Svensson and A. Nordberg, *in* "Alzheimer's Disease: Biology, Diagnosis and Therapeutics", (K. Iqbal, B. Winblad, T. Nishimura, M. Takeda, and H. M. Wisniewski, eds.), p. 751. Wiley, West Sussex, 1997.
553. R. Bullock, *Expert Rev. Neurotherapeutics* **4**, 153 (2004).
554. A. N. Dengiz and P. Kershaw, *CNS Spectr.* **9**, 377 (2004).
555. G. M. Bores, F. P. Huger, W. Petko, A. E. Mutlib, F. Camacho, D. K. Rush, D. E. Selk, V. Wolf, R. W. Kosley, L. Davis, and H. M. Vargas, *J. Pharmacol. Exp. Ther.* **227**, 728 (1996).
556. R. Bachus, U. Bickel, T. Thomsen, I. Roots, and H. Kewitz, *Pharmacogenetics* **9**, 661 (1999).

—CHAPTER 4—

# BISINDOLE ALKALOIDS

TOH-SEOK KAM AND YEUN-MUN CHOO

*Department of Chemistry, University of Malaya, 50603 Kuala Lumpur, Malaysia*



THE ALKALOIDS, Vol. 63
ISSN: 1099-4831
DOI: 10.1016/S1099-4831(06)63004-6

## I. Introduction

The last review in this series dealing exclusively with the bisindole alkaloids was that by Cordell and Saxton in Volume 20 (1981) (*1*). In the intervening period of more than 20 years, the literature dealing with the bisindoles has multiplied profusely, in tandem with vastly improved analytical tools, in particular NMR spectroscopy, and a second review appears timely. This review will be confined to the bisindoles from plants and will cover the period from the previous review (1981) to the present. In the intervening period however, there have also been several reviews covering the same topic and over varying periods (*2–4*). The organization of the present chapter will follow that adopted in the earlier volume (*1*), i.e., according to their constituent monomers and approximately along the lines of a progressing biosynthetic pathway.

## II. Indole–Indole and Tryptamine–Tryptamine Type, and Related Alkaloids

### A. The Annonidines and Caulindoles

The light petroleum extract of the stem-bark of the Annonaceous plant, *Annonidium mannii* Engl. & Diels, provided five prenylated bisindoles, annonidines A–E (**1–5**), which are dimers of prenylated indoles (*5,6*). Annonidines A and B (**1,2**) are considered as 1,1-annonidines (linked *via* one carbon atom), while annonidines C, D, and E (**3–5**) are considered as 1,3-annonidines (linked *via* a three-carbon bridge). The common monomeric unit in the annonidines is a 7-prenylated indole, bridged from the nucleophilic C(3) to a second prenylated indole. In annonidines A (**1**), C (**3**), and E (**5**), both the constituent monomers are 7-prenylindoles, hence, from a biogenetic viewpoint, these bisindoles can be regarded as the dimerization products of the 7-prenylindole, **6**, which also occurs as a major constituent of the stem extract. In annonidines B (**2**) and D (**4**), the prenyl side chain in the other indole unit is attached at C(3″) and C(6″), respectively, which were determined largely based on the $^{1}$H and $^{13}$C NMR spectral data. Annonidine E (**5**) is notable for the presence of an epoxide function on the prenyl substituent in place of the usual double bond, as was evident on comparison of its NMR spectral data with those of annonidine C (**3**). Reaction of **6** with the dienyl derivative **7** in the presence of TFA gave a mixture of annonidines A (**1**) and C (**3**) (*7*). Reaction of **6** with **8** (both prepared from 2,3-dihydro-7-iodoindole), in the presence of HCl/THF, followed by dehydrogenation, gave only annonidine A (**1**) (*8*). Annonidines A (**1**) and C (**3**) were also found in the roots of *Esenbeckia leiocarpa* (*9*), while recently, annonidine F (**9**) has been obtained from the Tanzanian plant, *Monodora angolensis* (Annonaceae) (*10*). Monodoroindole (**10**) and isomonodoroindole (**11**) are bisindoles constituted from identical 6-prenylated indole monomers, which were isolated from the seeds of *M. myristica* (Annonaceae) (*11*).

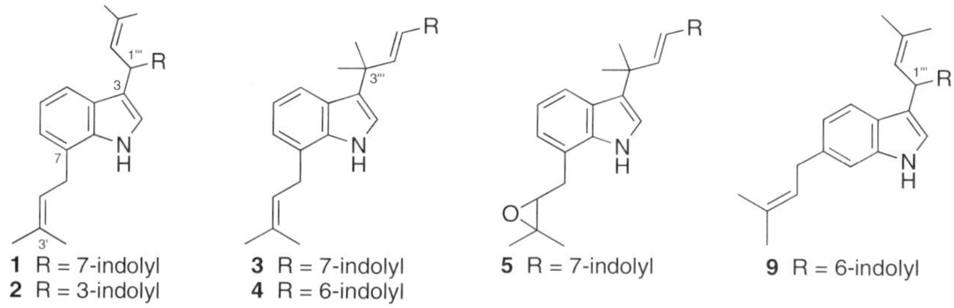


**1** R = 7-indolyl
**2** R = 3-indolyl

**3** R = 7-indolyl
**4** R = 6-indolyl

**5** R = 7-indolyl

**9** R = 6-indolyl

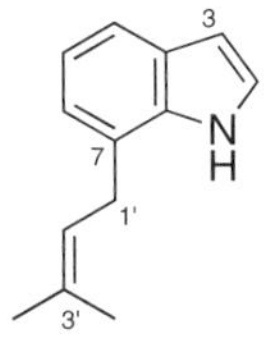

6

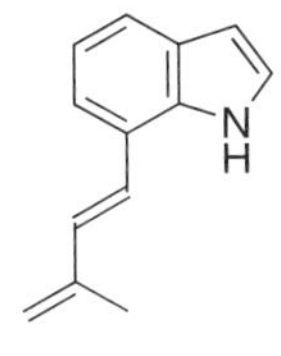

7

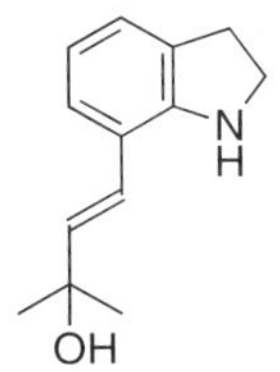

8

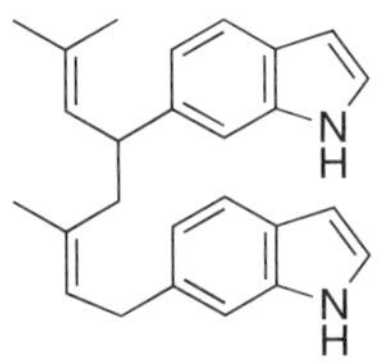

10

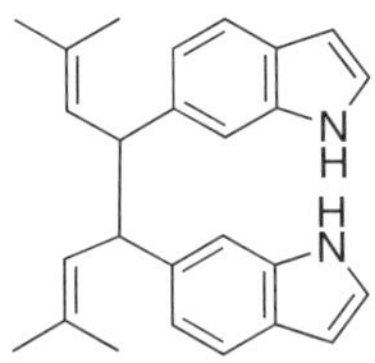

11

The caulindoles A–D (**12**–**15**) are dimeric prenylindoles, occurring in diastereomeric pairs, isolated from the stem and root-bark of *Isolona cauliflora*, an ecologically endangered Annonaceae species in Tanzania (*12*). The mass spectra of caulindoles A and B (**12**,**13**), displayed peaks due to the mono-prenylindole fragment ion, corresponding to **16**, which originates from *retro*-Diels–Alder fragmentation of the parent ions of **12** and **13**. Similar *retro*-Diels–Alder fragments corresponding to the bisprenylindole unit, **17**, were seen in the mass spectra of caulindoles C (**14**) and D (**15**). These observations raise the possibility that the caulindoles could have been formed through Diels–Alder cycloaddition processes, involving the mono-prenylindole **16**, which also occurs in the plant. However, when **16** was left to stand for a long period under the isolation conditions, it decomposed with no trace of any cycloaddition product (such as **12**) detected, indicating that the caulindoles were true natural products, and not artifacts of the isolation procedure.

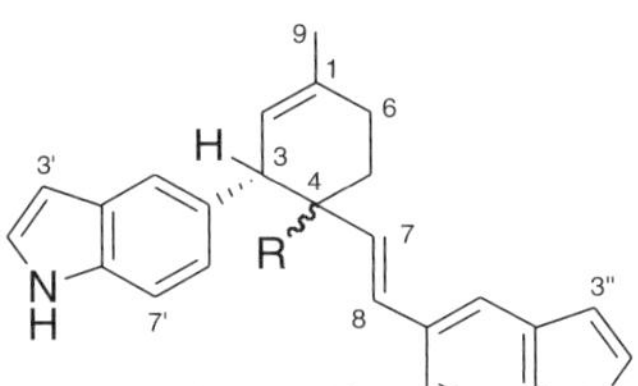

**12** R = α-Me
**13** R = β-Me

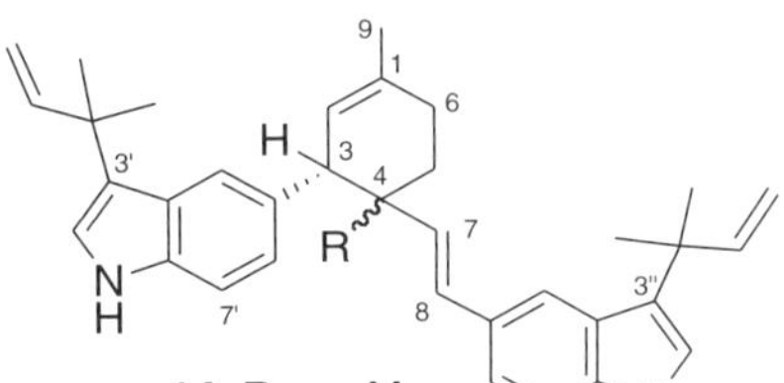

**14** R = α-Me
**15** R = β-Me

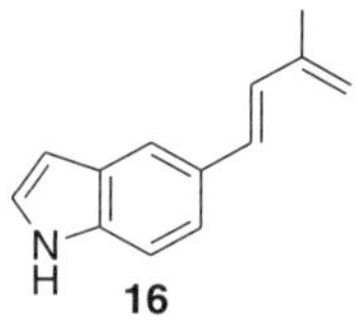

**16**

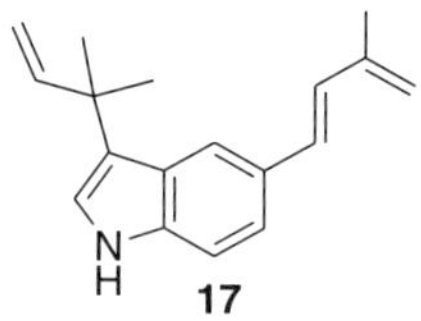

**17**

## B. THE *MURRAYA* ALKALOIDS

The main alkaloids from the genus *Murraya* (Rutaceae), in addition to the monomeric indoles, are the biscarbazoles and the prenylated biscarbazoles. The first biscarbazole to be reported was murrafoline A (**18**), from *Murraya euchrestifolia* Hayata, collected in Taiwan. The alkaloid was obtained in racemic form and the structure was elucidated by X-ray diffraction analysis, which showed it to be constituted from girinimbine (**19**) and cyclomahanimbine (**20**), which were also present in the same plant (*13,14*). Subsequently, a whole series of bisindole alkaloids of the same basic structure, have been isolated, mainly from *M. euchrestifolia* and *M. koenigii*. The bismurrayafolines A–E (**21**–**25**) are dimeric biscarbazoles constituted from the respective monomeric carbazoles (*15–17*). This was readily established in the case of **21**, where hydrogenolysis (5% Pd–C, MeOH/HCOOH) gave murrayafoline A (**26**) as the sole product (*15*).

The murrafolines B–E, G, H (**27**–**30**, **32**, **33**) are characterized by the presence of a common girinimbine unit (*14,18,19*). In murrafolines C (**28**) and H (**33**), the other unit is another girinimbine, while in murrafolines B (**27**), D (**29**), E (**30**), and G (**32**), the other unit is murrayafoline A (**26**). Murrafoline F (**31**) on the other hand, is constituted from the union of murrayafoline A and a *N*-methoxycarbazole derivative (*19*). Attempted dimerization of girinimbine (**19**) using the perfluorosulfonic acid cation exchange membrane, Nafion 117, did not lead to murrafoline C (**28**), but instead gave the regioisomeric 12-6′ dimerization product corresponding to murrafoline H (**33**). Similar treatment of **26** and **19** with Nafion 117 gave a mixture comprising murrafolines D (**29**), G (**32**), and H (**33**) (*14,18*).

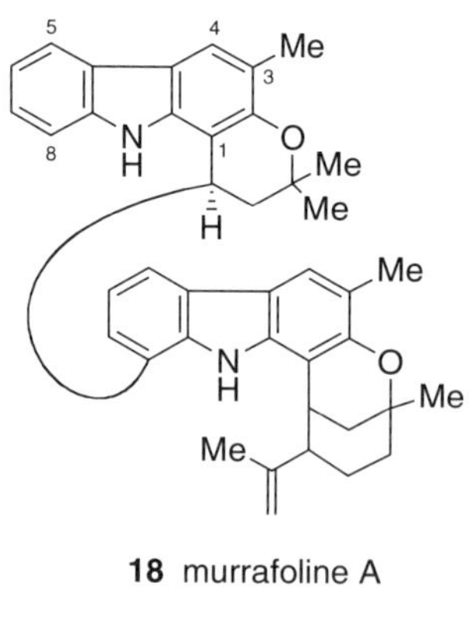

**18** murrafoline A

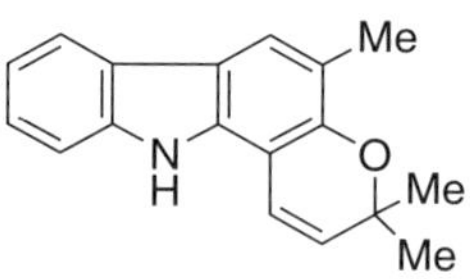

**19** girinimbine

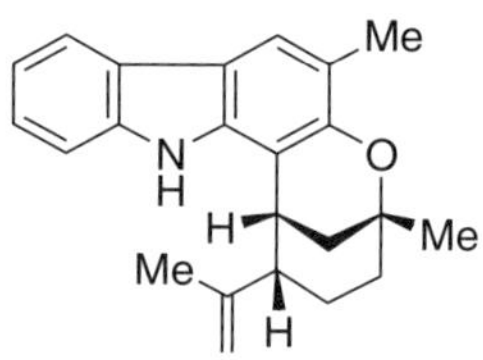

**20** cyclomahanimbine

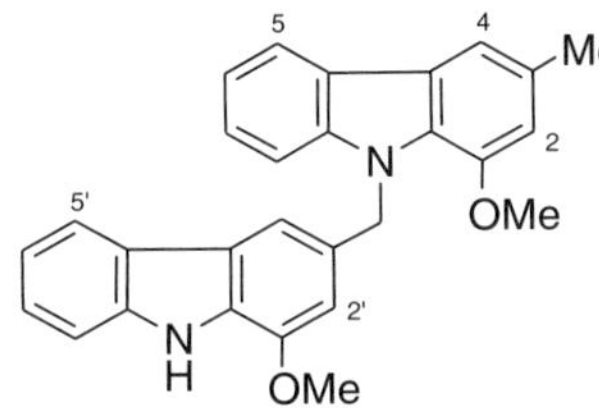

**21** bismurrayafoline A

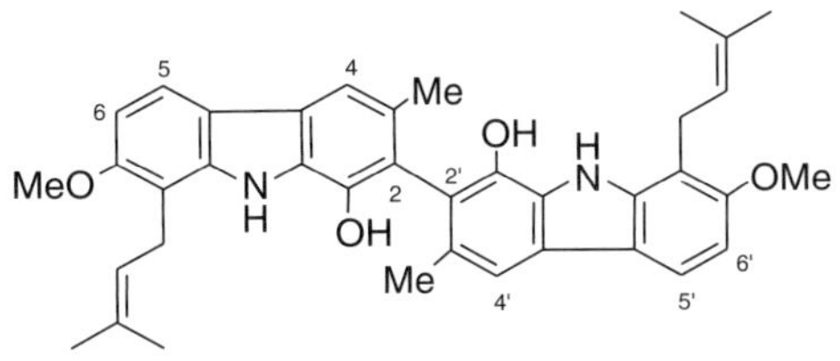

**22** bismurrayafoline B

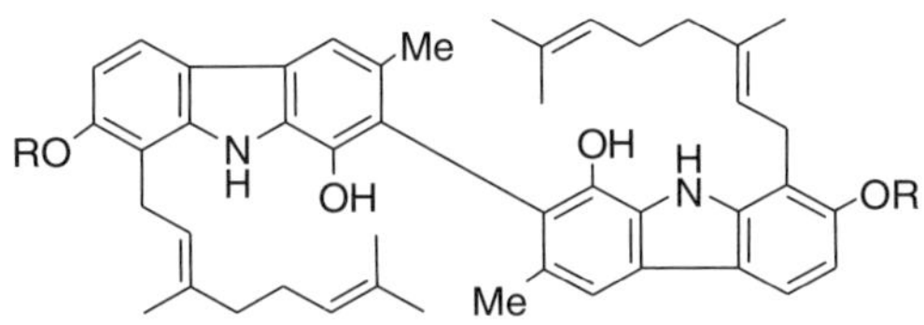

**23** R = H, bismurrayafoline C
**24** R = Me, bismurrayafoline D

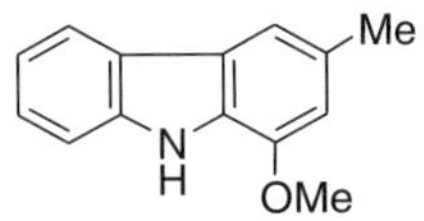

**26** murrayafoline A

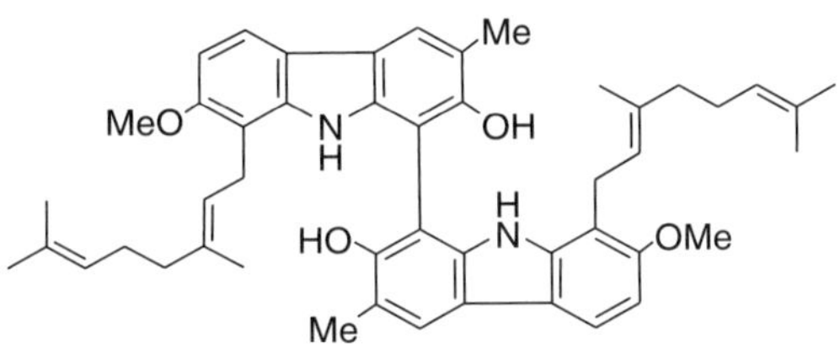

**25** bismurrayafoline E

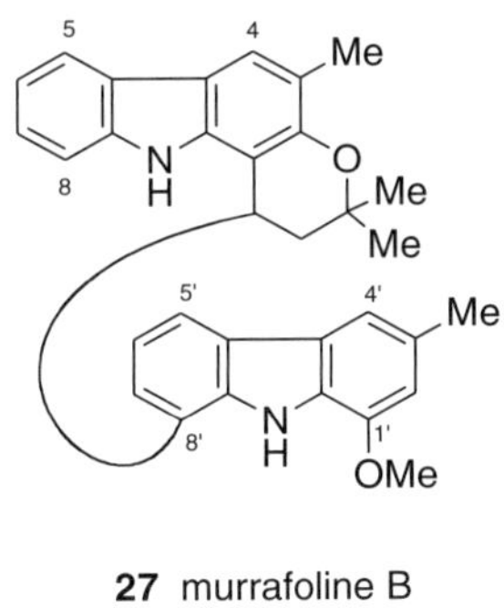

**27** murrafoline B

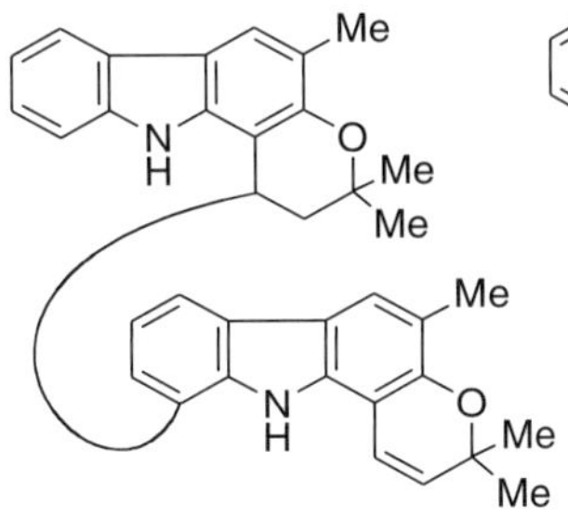

**28** murrafoline C

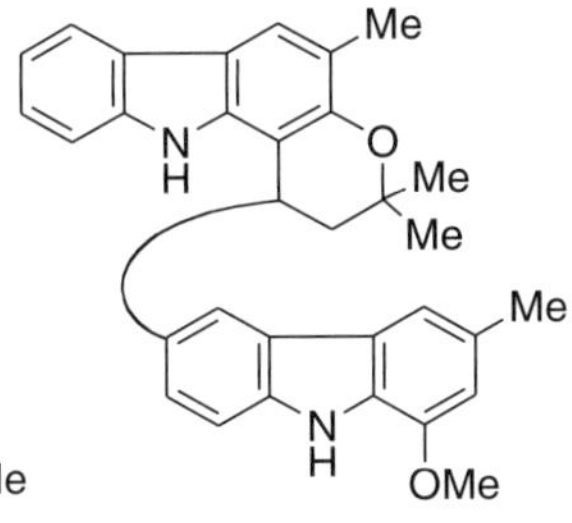

**29** murrafoline D

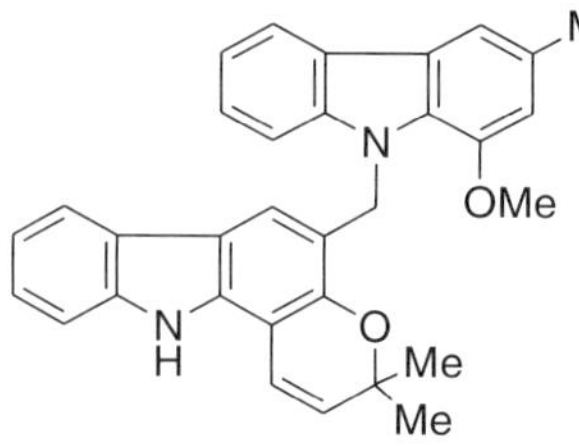

**30** murrafoline E

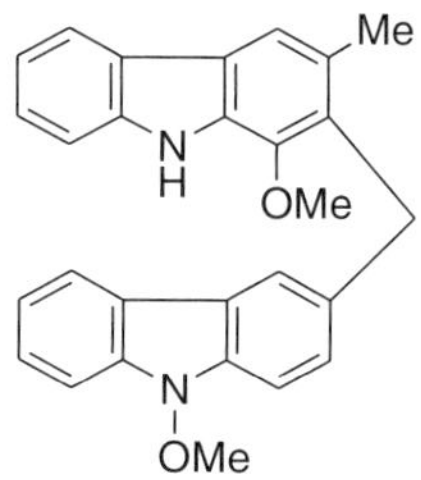

**31** murrafoline F

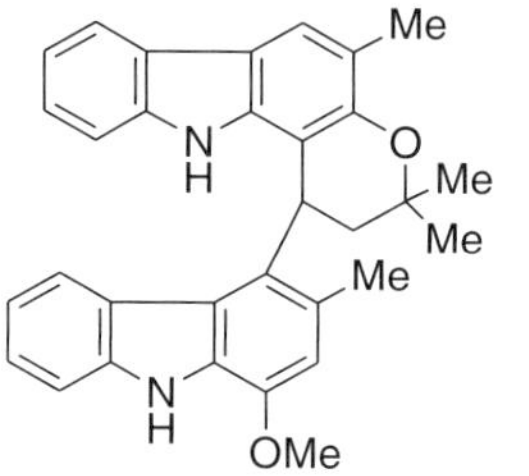

**32** murrafoline G

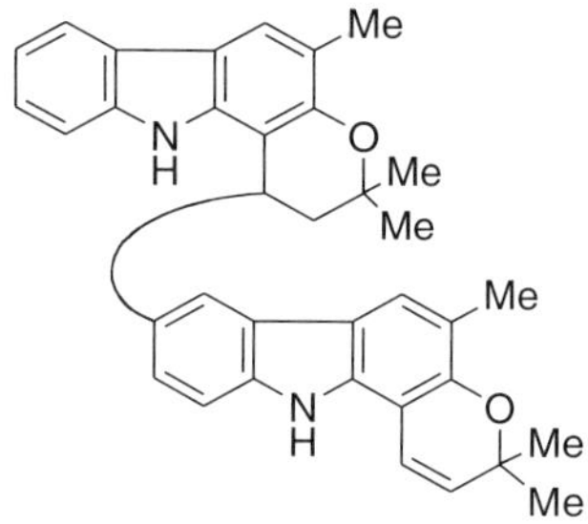

**33** murrafoline H

In the murrastifolines A–F (**34**–**39**), and the chrestifolines A–D (**40**–**43**), the common unit present is murrayafoline A (**26**). In the murrastifolines, the point of branching from the murrayafoline A unit is from the indolic nitrogen, while in the case of the chrestifolines, the branching from the murrayafoline A unit is from C(3) *via* a methylene bridge to the second unit (*20–22*). Chrestifoline D (**43**), being the C(3) formyl derivative of **21**, was readily obtained from the latter *via* DDQ oxidation (*23*). Bismurrayafolinol (**44**) is similar in all respects to bismurrayafoline A (**21**) except for replacement of the methyl at C(3) with a hydroxymethyl group, while oxydimurrayafoline (**45**) constitutes the first example of a dimeric carbazole alkaloid possessing an ether linkage. Hydride reduction ($NaBH_4$/MeOH) of the monomeric murrayanine (**46**), which also occurs in the plant, followed by acid-induced coupling (HCl/$CHCl_3$), and subsequent acetylation, gave the acetylated derivative of bismurrayafolinol (**47**) (*24*).

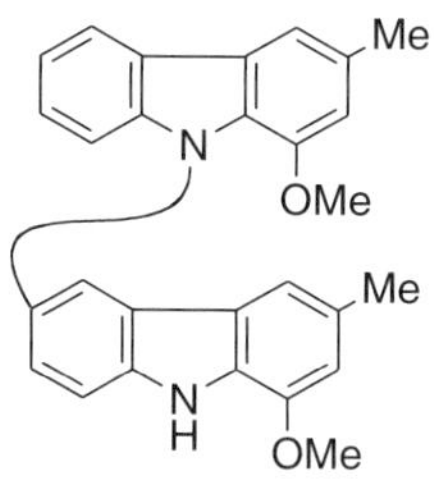

**34** murrastifoline A

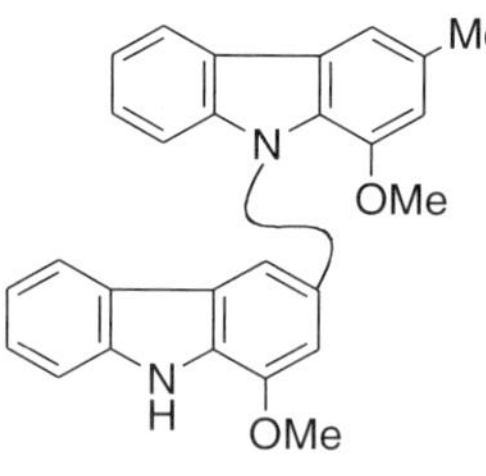

**35** murrastifoline B

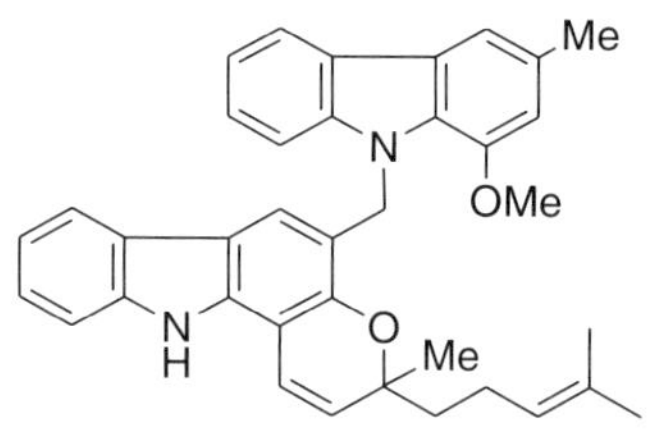

**36** murrastifoline C

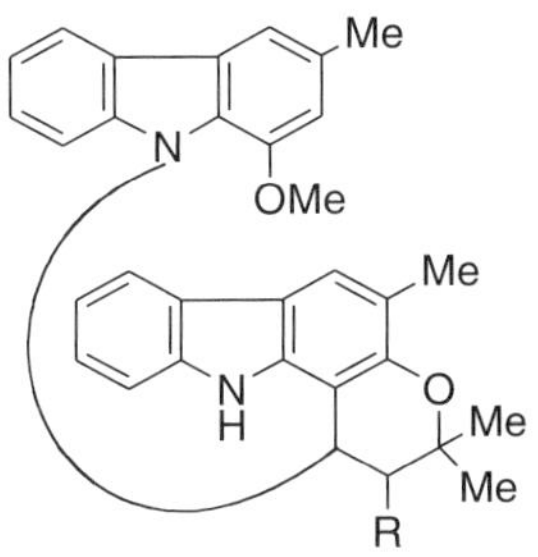

**37** R = OH murrastifoline D

**38** R = H murrastifoline E

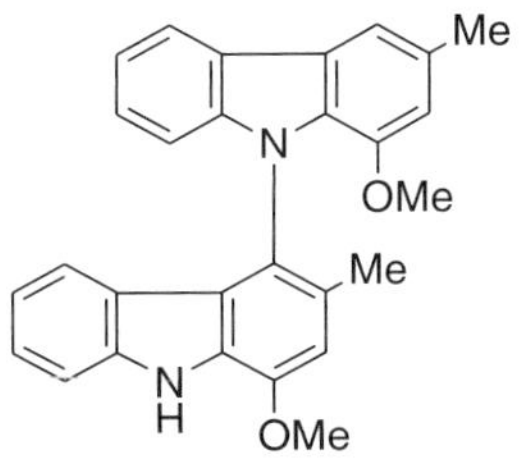

**39** murrastifoline F

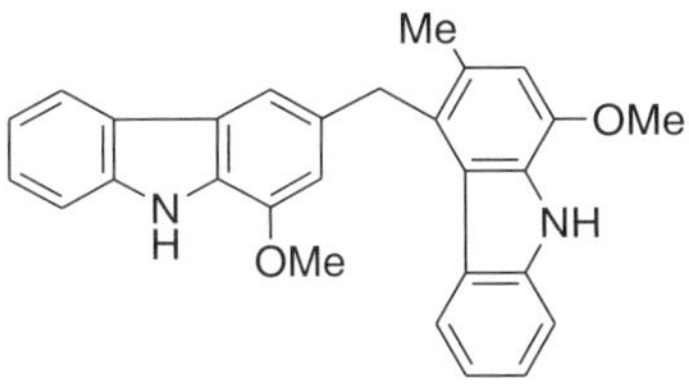

**40** chrestifoline A

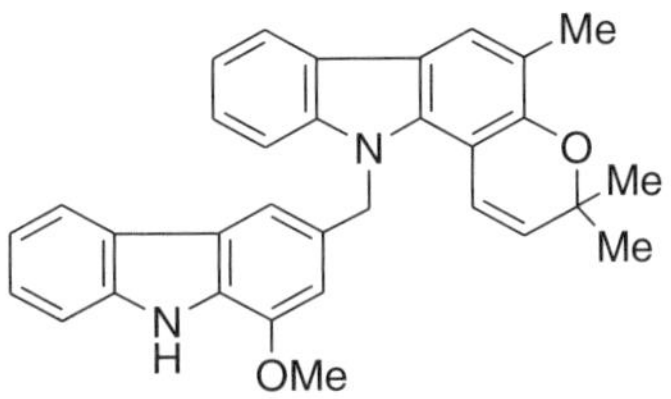

**41** chrestifoline B

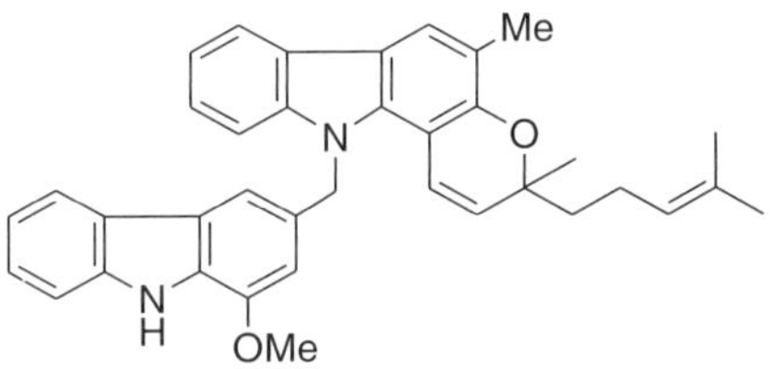

**42** chrestifoline C

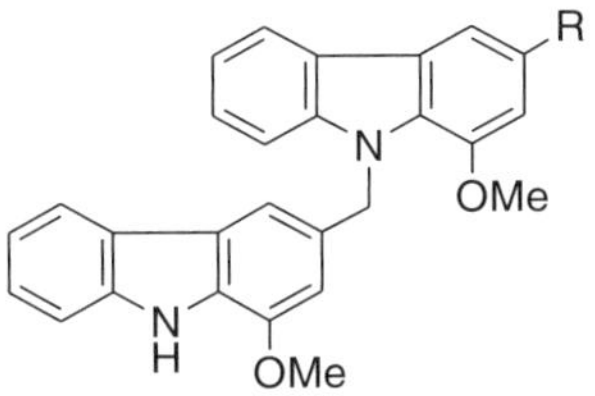

**43** R = CHO chrestifoline D

**44** R = $CH_2OH$ bismurrayafolinol

**47** R = $CH_2OAc$

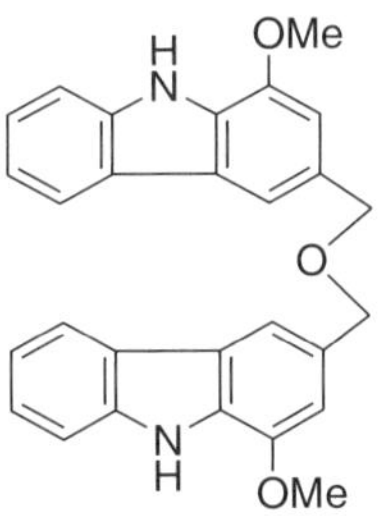

**45** oxydimurrayafoline

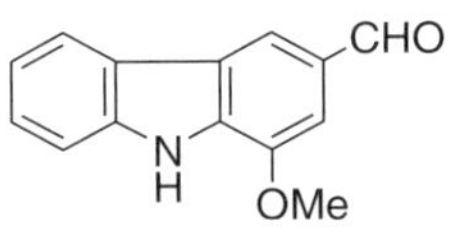

**46** murrayanine

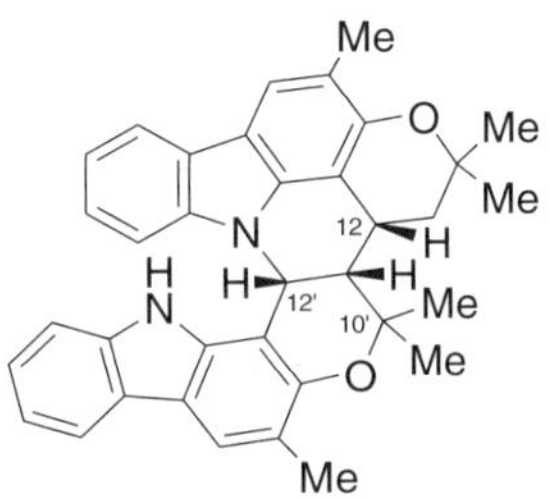

**48** murranimbine

Murranimbine (**48**) is constituted from the union of two girinimbine units, which are branched from N(1) and C(12) of one unit, to C(12′) and C(11′), respectively, of the other (*25*). The mode of linking of the two units was determined from the HMBC spectrum, while the relative stereochemistry was established from NOE experiments. The phenolic bis-7-hydroxygirinimbines A (**49**) and B (**50**), differ in the point of branching of the hydroxygirinimbine units, from C(8) to C(8′) in the case of **49** giving rise to a symmetric dimer, and from C(8) to C(6′) in the case of **50** (*26*). The bismethoxylated derivative of the former, **51** was obtained from *M. exotica*, cultivated in Egypt (*27*).

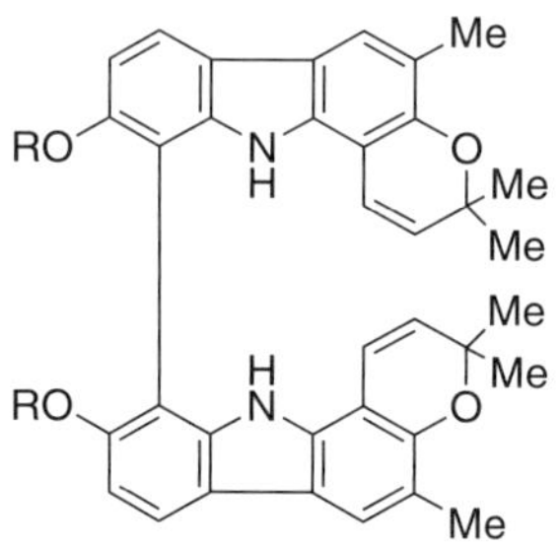

**49** R = H bis-7-hydroxygirinimbine A

**51** R = Me bis-7-methoxygirinimbine

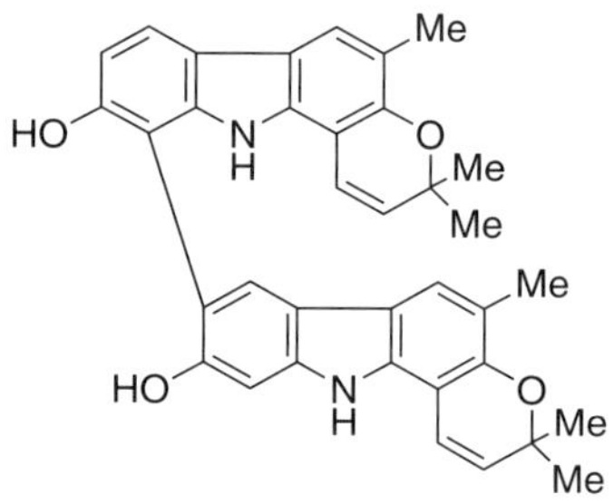

**50** bis-7-hydroxygirinimbine B

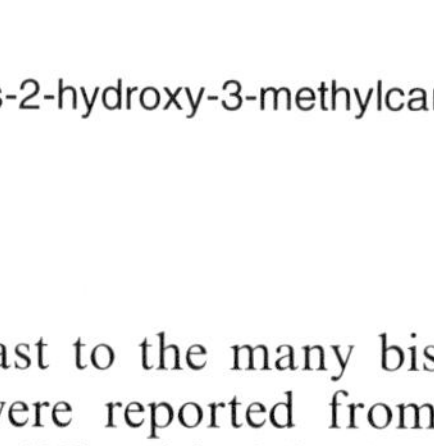

**52** bis-2-hydroxy-3-methylcarbazole

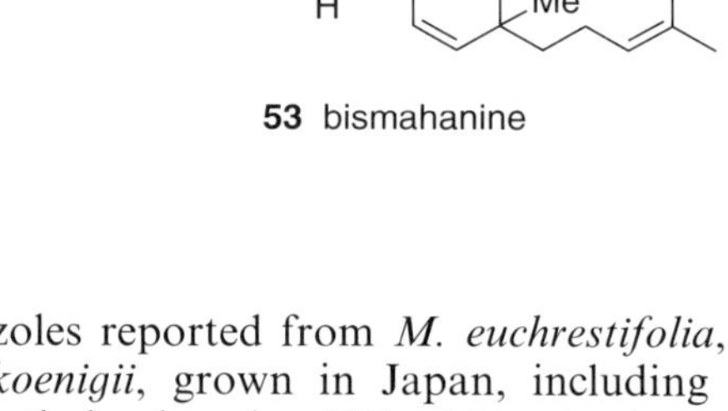

**53** bismahanine

In contrast to the many biscarbazoles reported from *M. euchrestifolia*, five biscarbazoles were reported from *M. koenigii*, grown in Japan, including murrastifoline F (**39**), bis-2-hydroxy-3-methylcarbazole (**52**), bismahanine (**53**), bikoeniquinone A (**54**), and bismurrayaquinone A (**55**) (*22*). A new biscarbazole, 8,8″-biskoenigine (**56**) was isolated from the same plant collected from Yunnan, China (*28*), while the biscarbazole, **57** and bispyrayafoline (**58**) were obtained from *M. koenigii* occurring in Malaysia (*29*). Biskoenigine (**56**) was readily prepared *via* $FeCl_3$-mediated oxidative coupling of koenigine (**59**) ($FeCl_3.6H_2O$, in solid state, rt, 30 days), and showed antiosteoporotic activity in the CAT-B model ($IC_{50}$ 1.3 μg $mL^{-1}$) (*28*).

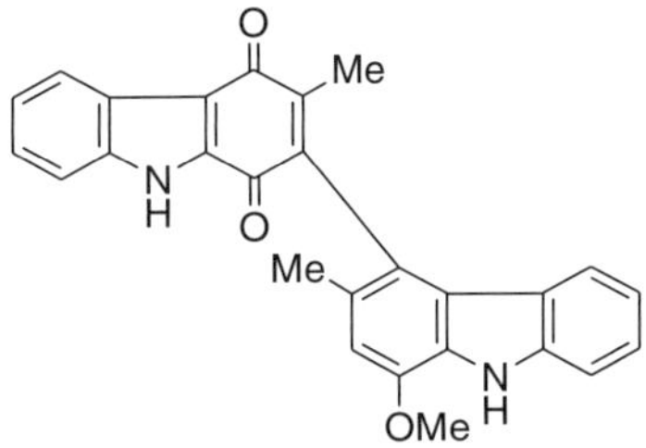

**54** bikoeniquinone A

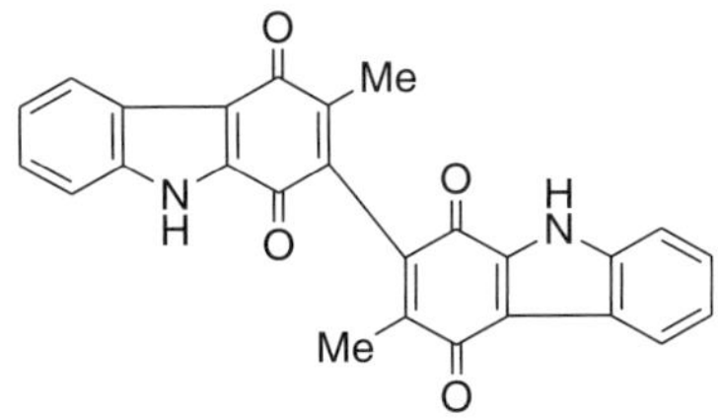

**55** bismurrayaquinone A

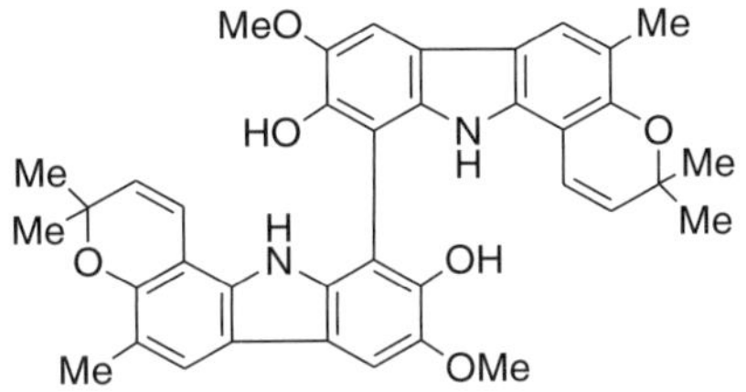

**56** 8,8''-biskoenigine

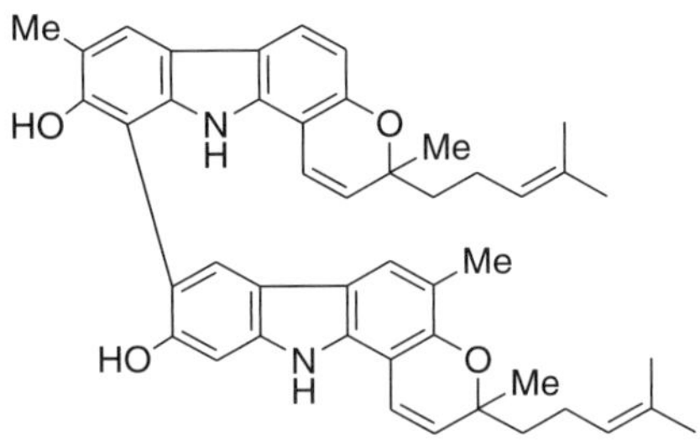

**57**

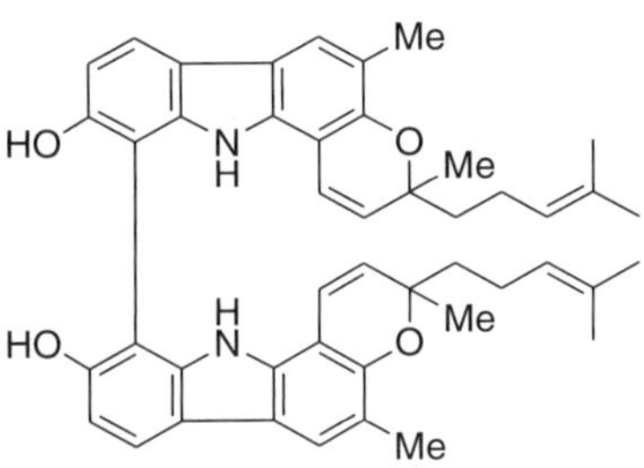

**58**

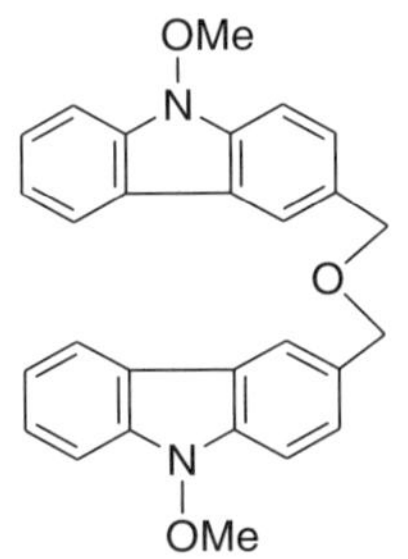

**60**

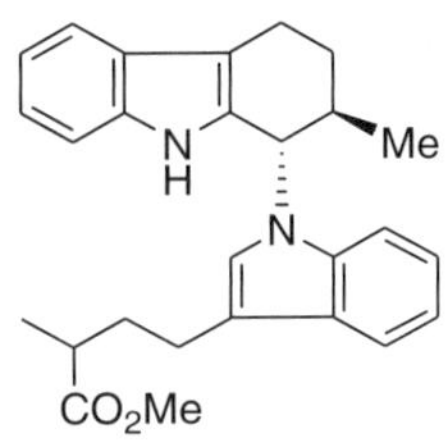

**61**

A new ether-linked biscarbazole, 3,3′-[oxybis(methylene)]bis(9-methoxy-9*H*-carbazole) (**60**), incorporating unusual *N*-methoxy substituents, was isolated from a sample of *M. koenigii* collected from Bangladesh. The biscarbazole **60** showed strong antimicrobial activity against the Gram-negative bacteria, *Escherichia coli* and *Proteus vulgaris*, and antifungal activity against *Aspergilus niger* and *Candida albicans* (*30*). The roots of *M. gleniei* provided a new bisindole, branched from the indolic nitrogen of a 3-prenylindole unit to C(1) of a tetrahydrocarbazole unit as shown in **61** (*31*), while the root extract of *Glycosmis stenocarpa*, a Rutaceous plant occurring in Vietnam provided in addition to the monomeric carbazole alkaloids, murrayafoline A and murrayanine, a new bispyranocarbazole alkaloid, bisisomahanine (**62**), constituted from the union of two isomahanine units (*32*). Another new biscarbazole, clausenamine A (**63**), in addition to a novel carbazole-pyranocoumarin, carbazomarin A (**64**), were obtained from *Clausena excavata* (*33*).

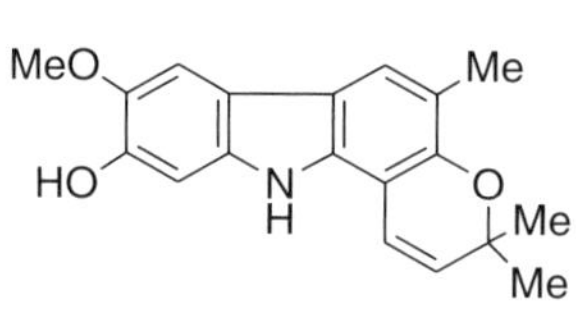

**59** koenigine

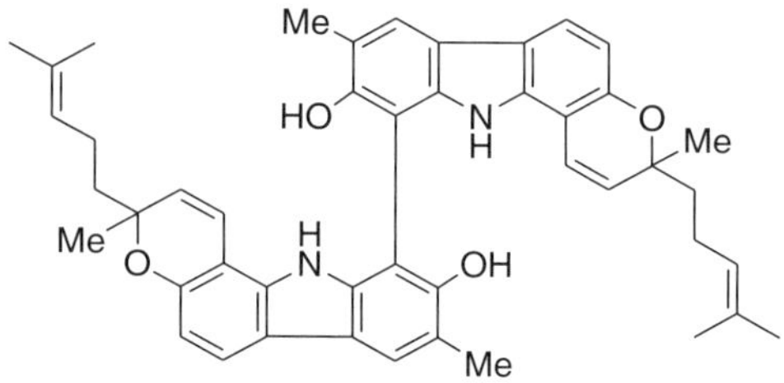

**62** bisisomahanine

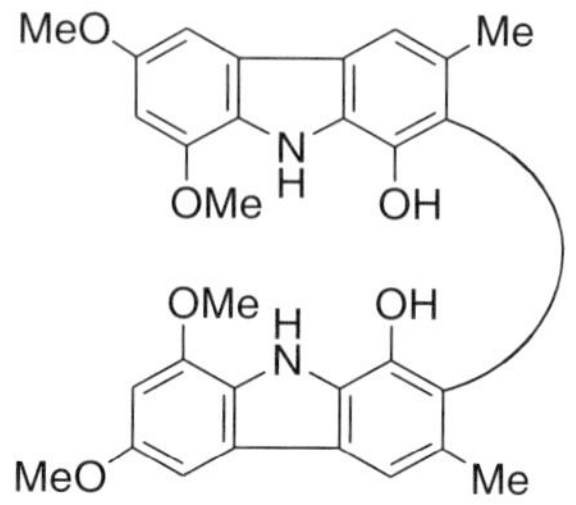

**63** clausenamine A

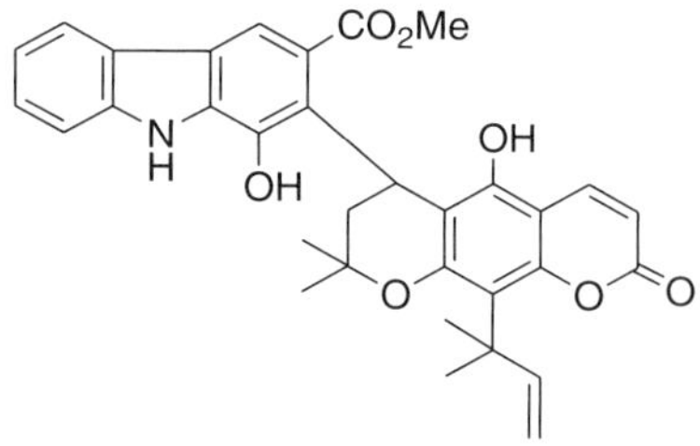

**64** carbazomarin A

The *N*,*C*-coupled biscarbazole, murrastifoline F (**39**) has been prepared by lead tetraacetate-mediated oxidative coupling of the monomer, murrayafoline A (**26**), which was in turn synthesized from indole-3-carbaldehyde (Scheme 1) (*34*,*35*). The existence of **39** as a pair of stable atropo-enantiomers was demonstrated by LC-CD studies. Racemate resolution was achieved by *O*-demethylation, derivatization with Mosher's acid, followed by chromatographic separation of the resulting diastereomers, following which, the absolute configurations of the atropisomers of **39** were assigned by CD spectroscopy in combination with quantum-chemical calculations. The absolute configurations of the diastereomeric Mosher derivatives were assigned by ROESY experiments. Murrastifoline F occurring in the root extract of *M. koenigii* was found by LC-CD analysis to exist as a 56:44 mixture in favor of the *M*-enantiomer (*34*).

Bringmann (*36*) has also reported the synthesis of the methylene-bridged biscarbazole alkaloids, bismurrayafoline A (**21**) and chrestifoline A (**40**). The former

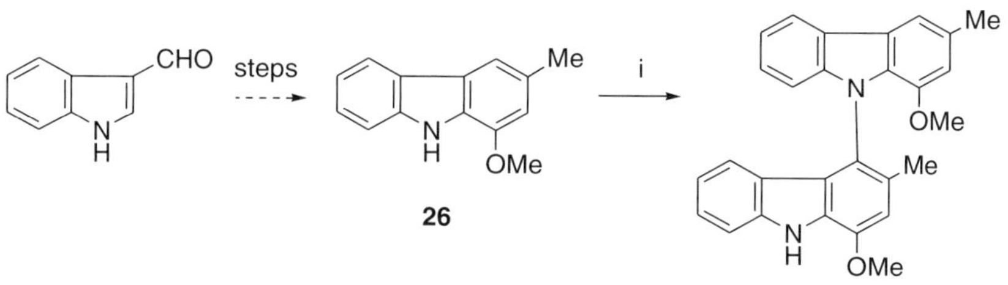

**26**

**39**

Scheme 1. Reagents: (i) $Pb(OAc)_4$, $BF_3$ etherate, $CH_3CN$, rt, 3.5 h.

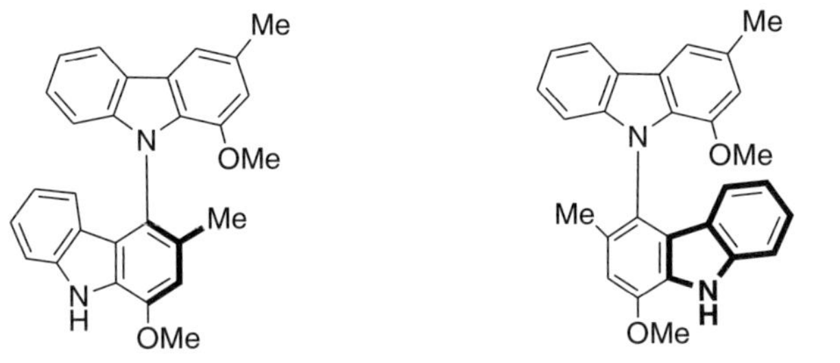

**39** *M*-enantiomer **39** *P*-enantiomer

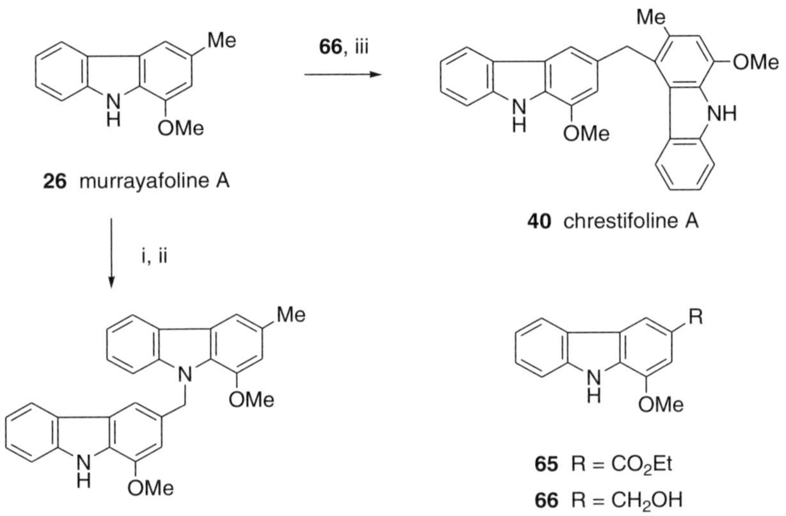


**26** murrayafoline A

**40** chrestifoline A

**21** bismurrayafoline A

**65** R = $CO_2Et$

**66** R = $CH_2OH$

Scheme 2. Reagents: (i) $LiAlH_4$, $Et_2O/CH_2Cl_2$, rt; (ii) **65**; (iii) conc. HCl in $Et_2O/CH_2Cl_2$.

was prepared *via* $LiAlH_4$-mediated coupling of murrayafoline A (**26**) and the ester **65**, while the latter was obtained *via* coupling of **26** with koenoline (**66**) under acidic conditions (Scheme 2). Bismurrayaquinone A (**55**) was obtained by PCC oxidation of the biscarbazole **67**, in turn prepared by oxidative dimerization ($t$-$Bu_2O_2/C_6H_5Cl$) of the monomeric carbazole **68** (*37,38*). The racemic product **55** was resolved by chiral phase HPLC, and the absolute configurations of the atropo-enantiomers were assigned by comparison of their theoretical and experimental CD spectra.

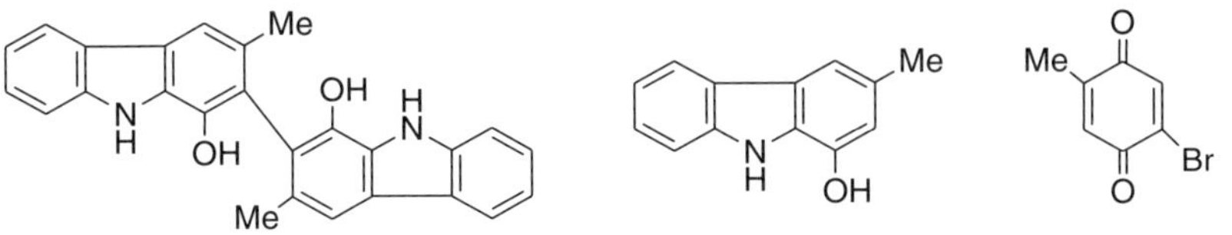

**67** **68** **69**

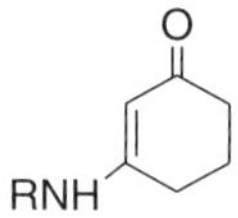


70

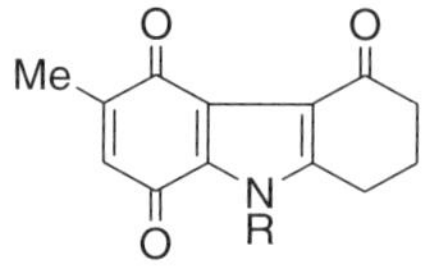


71

R = $p$-MeOC$_6$H$_4$CH$_2$

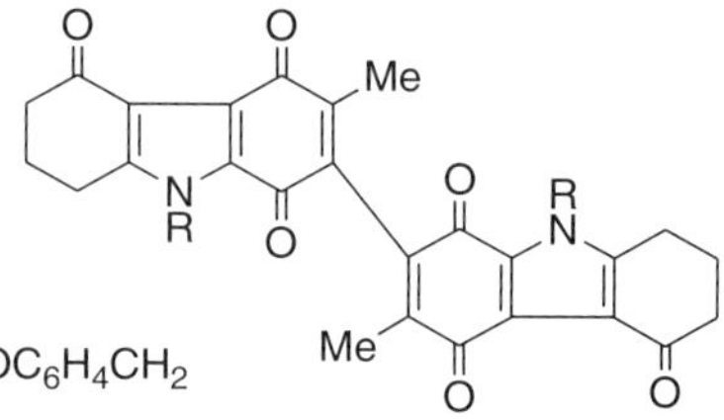


72

In a synthesis of murrayaquinone A, the initial annulation reaction between bromo-1,4-benzoquinone **69**, and the N-protected enamino ketone **70**, gave in addition to the desired product **71**, minor amounts (7%) of a dimerization side product, **72**, which was processed in several steps to bismurrayaquinone A (**55**) (*39*).

In a study of the cytotoxic effects of the alkaloids of *M. euchrestifolia* on human cancer cell lines, it was found that while the monomeric alkaloids murrayaquinone and murrayafoline showed significant cytotoxicity against SK-MEL-5 and Colo-205 cells, the biscarbazole alkaloids, bismurrayafoline A (**21**) and chrestifoline A (**40**) by comparison, showed only marginal cytotoxicity against HOP-92, a nonsmall cell lung cancer cell line, and LOX IMVI, a melanoma cell line, whereas chrestifoline C (**42**), bismurrayafoline B (**22**), and murranimbine (**48**), were inactive toward all the cell lines tested (*40*).

## C. YUEHCHUKENE

Another *Murraya* alkaloid of note is yuehchukene (YCK) (**73**), a bisprenylindole with strong anti-implantation activity in rats and hamsters, first isolated as a racemate from *M. paniculata* (L.) Jack (*41–43*), and subsequently from other *Murraya* species (*44,45*). The structure was elucidated based primarily on the NMR spectral data (*41*), and confirmed by X-ray analysis of its *N*-acetyl derivative (*46*). The potential application of YCK as an antifertility agent, in the light of the observed biological activity, stimulated interest in its total synthesis as well as the synthesis of its analogues. It was first obtained in low yield (10%) *via* acid-induced dimerization of $\beta$-(dehydroprenyl)indole, **74** (*47*). A twofold improvement in the yield (25–28%) was subsequently achieved by the use of the tertiary allylic alcohol **75**, or its isomer **76**, as a synthetic equivalent of the diene **74** (*48–51*). In another variation by the same group, the $\beta$-prenylindoles, **77** and **78** (prepared from the reaction of indole-3-carbaldehyde (**79**) and dimethylvinylidene carbene), on heating in ethylene glycol at 165–170°C, underwent smooth elimination of **79** (as its anion) to produce the diene **74**, *en route* to YCK in 42% yield (*52*). In the course of studies directed at the preparation of various prenylated indoles, Wenkert developed an efficient synthesis of diene **74** from $\beta$-bromoindole **80**. During drying of the alcohol **81** with a large excess of magnesium sulfate, it unexpectedly rearranged to tertiary alcohol **82**, which yielded the desired diene **74** on bisulfate-promoted dehydration, followed by hydrolysis with potassium hydroxide and 18-crown-6 in acetonitrile. Heating the diene in ethylene glycol solution at 155°C in air gave YCK in 24% yield (*53,54*).

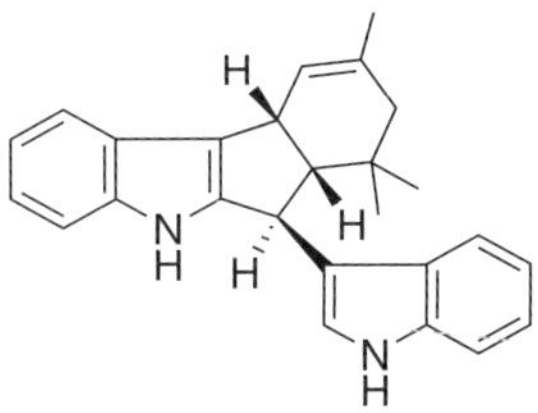

**73**

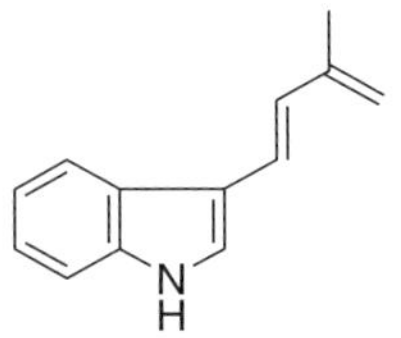

**74**

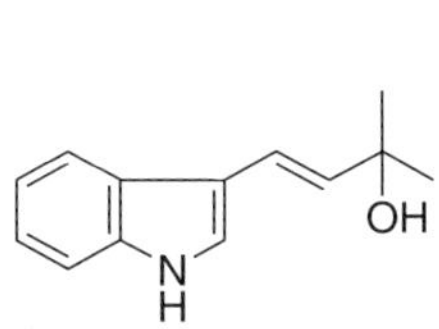

**75**

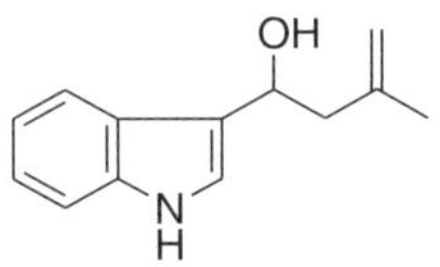

**76**

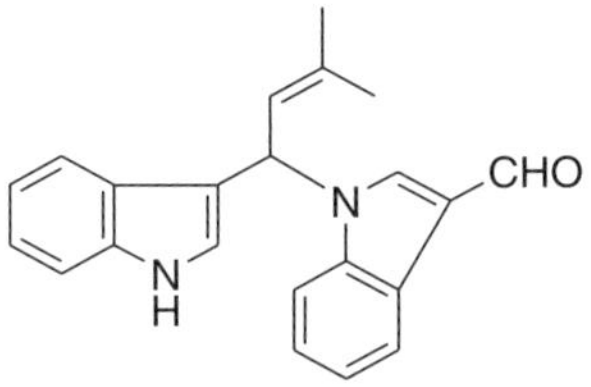

**77**

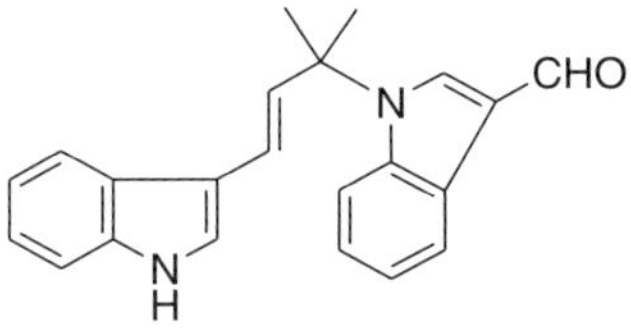

**78**

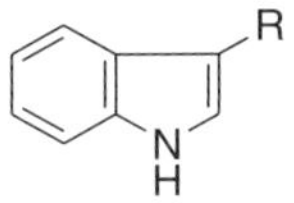

**79** R = CHO

**80** R = Br

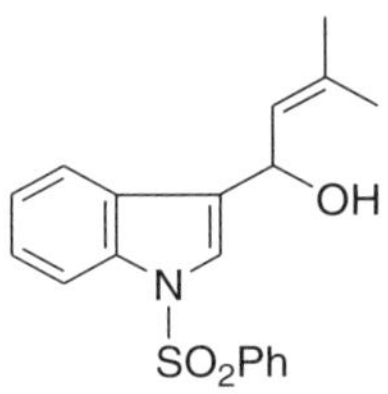

**81**

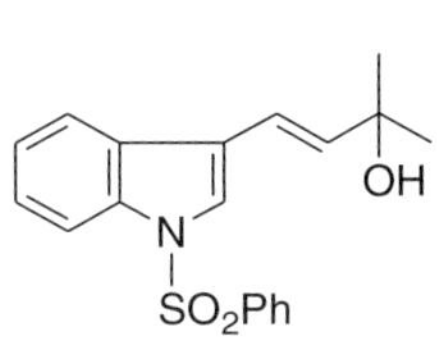

**82**

Subsequent syntheses of YCK focussed on the elaboration of the tetracyclic tetrahydroindeno[2,1-*b*]indole unit. In Bergman's concise synthesis (Scheme 3), the monoterpenoid aldehyde **83** was reacted with indole employing Katritsky's method for entry into 2-substituted indoles, using $CO_2$ for N-protection and carbanion stabilization. Oxidation to the ketone followed by TFA-mediated cyclization gave the tetracyclic acyl indole, which on reduction with $LiEt_3BH$, followed by acid-catalyzed

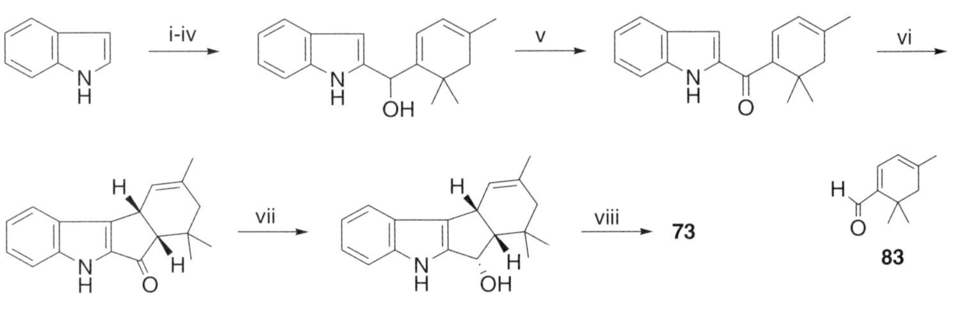


Scheme 3. Reagents: (i) BuLi; (ii) $CO_2$; (iii) *t*-BuLi; (iv) **83**; (v) $MnO_2$; (vi) TFA; (vii) $LiEt_3BH$; (viii) indole, HCl.

condensation of the alcohol with indole, furnished YCK as the sole product in 34% overall yield (*55,56*).

Kutney's approach started with isophorone **84**, which was converted into the dibenzoate **85**, and thence to the acid **86** on treatment with indolylmagnesium iodide. Conversion to the acid chloride followed by treatment with indolylmagnesium iodide, induced cyclization to the *trans*-fused tetracycle **87**. Base-catalyzed epimerization to the *cis*-fused tetracyclic ketone **88**, followed successively by hydride reduction, conversion into the benzoate, treatment with the indole Grignard reagent, and deprotection, gave YCK (**73**) (Scheme 4). The overall yield was however low (4%), but the method allows access to 6α-*epi*-YCK *via* the ketone **87** (*57,58*).

Grieco has applied the use of the unusual solvent system $LiClO_4$-$Et_2O$ to effect a Brønsted acid-catalyzed electrophilic substitution of indole by *β,β*-disubstituted allylic alcohols *en route* to YCK as shown in Scheme 5 (*59*). The synthesis commenced with the diol **89** (available in two steps from isophorone **84**), which on treatment with indole in $LiClO_4$-$Et_2O$ containing 0.01 equivalent of acetic acid furnished the substitution product **90** in 86% yield. In the face of the failure of standard methods, oxidation to the aldehyde required conversion to the alkoxymagnesium bromide, followed by treatment

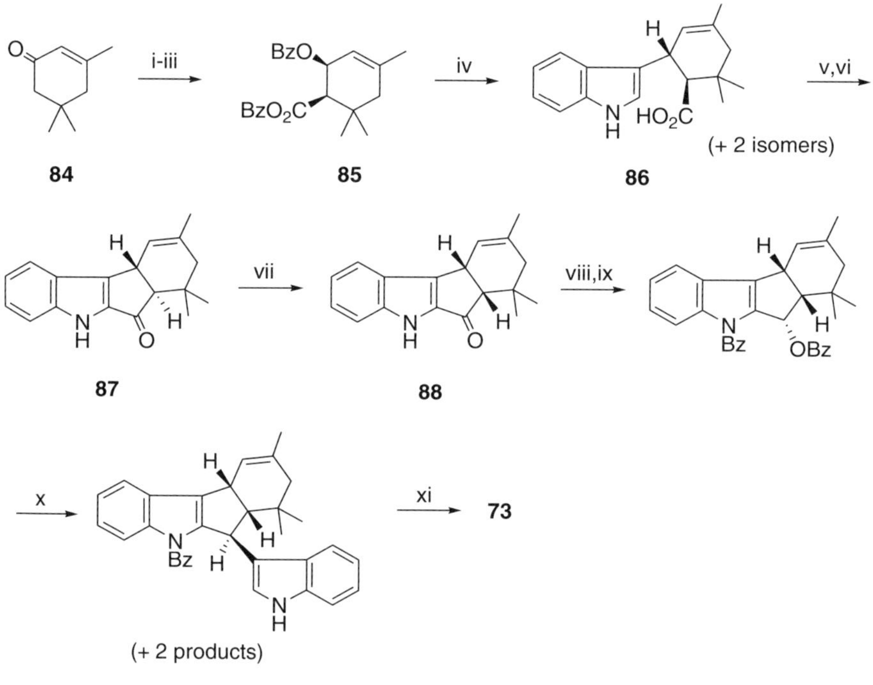


Scheme 4. Reagents: (i) Lithium 2,6-di-*t*-butyl-4-methylphenoxide, $CO_2$, $Et_2O$; (ii) $NaBH_4$; (iii) PhCOCl, DMAP; (iv) indolylmagnesium iodide; (v) $(COCl)_2$; (vi) indolylmagnesium iodide; (vii) NaOMe/MeOH; (viii) $LiAlH_4$; (ix) PhCOCl, DMAP; (x) indolylmagnesium iodide; (xi) NaOMe/MeOH.

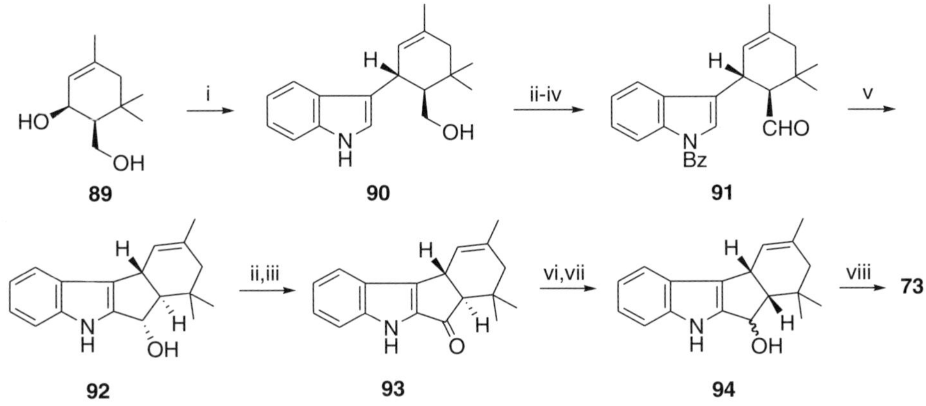

Scheme 5. Reagents: (i) $LiClO_4$–$Et_2O$, AcOH, indole; (ii) *t*-BuOMgBr; (iii) 1,1′-(azodicarbonyl)dipiperidine; (iv) BzCl, $Et_3N$, DMAP; (v) LDA/THF; (vi) NaOMe/MeOH; (vii) $LiAlH_4$; (viii) $LiClO_4$-$Et_2O$, indole.

with 1,1′-(azodicarbonyl)dipiperidine, to provide 81% yield of the aldehyde **91**, which was then processed *via* **92** and **93** to the alcohol **94**, obtained as a 2:1 mixture of diastereomers. In the formation of **92**, deprotonation occurred preferentially at the C(2) position of the indole, as the rotamer required for proton abstraction adjacent to the aldehyde was not accessible. Both epimers of **94** were readily converted into YCK in high yields (>90%), using the same protocol as in the first step (indole in $LiClO_4$-$Et_2O$), but without requiring the presence of a Brønsted acid. Another synthesis of YCK (and its analogues) was based on a palladium-catalyzed carbonylative cross coupling between the indolylborate **95** and cyclohexadienyl triflate **96** as the key step, as shown in Scheme 6 (*60–62*).

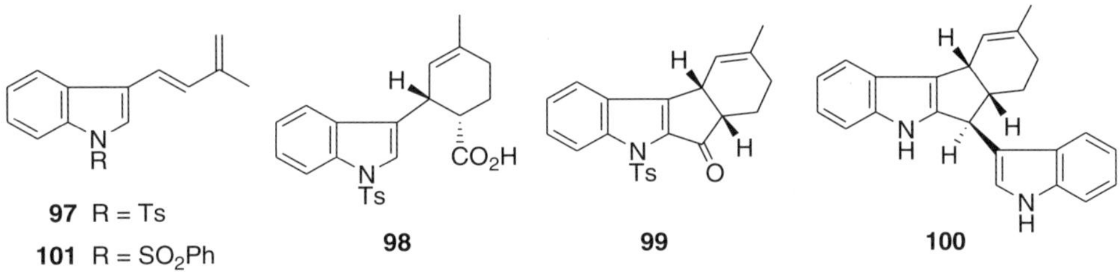

Several YCK analogues were prepared by Cheng in the context of structure-reactivity studies. Reaction of diene **97** with acrylic acid gave a 4:1 mixture of adducts. Treatment of the major adduct **98** with polyphosphate ester (PPE) furnished the tetracyclic ketone (**99**) which led eventually to the YCK analogue, 7,7-bis-nor-YCK (**100**) (*63*).

PPE-induced cyclization was also employed in the synthesis of a spirofused YCK analogue (**103**) (Scheme 7) (*64*). The precursor spiro acid **102** was obtained from the Diels–Alder reaction between the diene (**101**) and ethyl cyclopropylideneacetate. Diels–Alder reaction of diene **97** with maleic anhydride afforded adduct **104**, which on

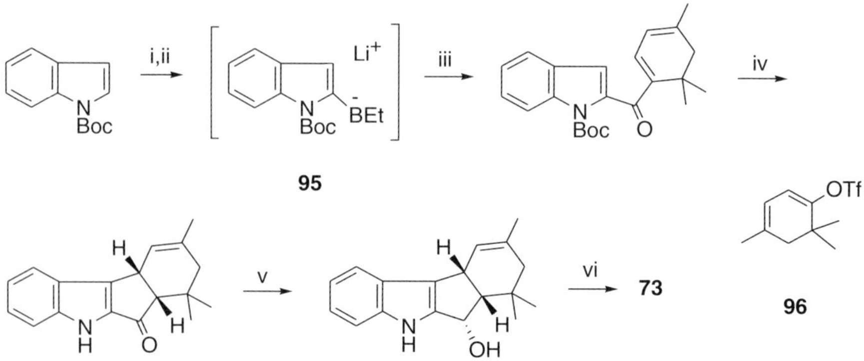


Scheme 6. Reagents: (i) *t*-BuLi/THF; (ii) $BEt_3$; (iii) **96**, CO (10 atm), $PdCl_2(PPh_3)_2$/THF; (iv) TFA/$CH_2Cl_2$; (v) DIBAL-H/THF; (vi) indole, $Eu(OTf)_3$.

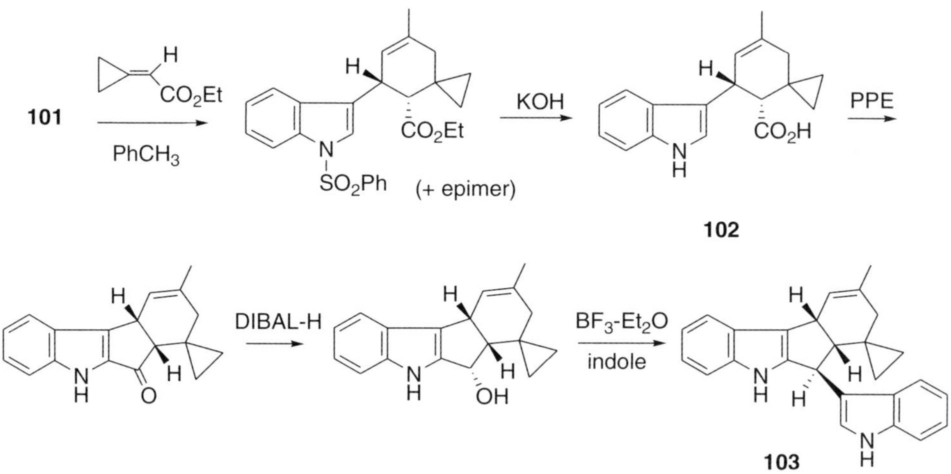


Scheme 7

exposure to aluminium trichloride resulted in intramolecular acylation, followed by rearrangement to give a mixture of the bridged keto acids, **105** and **106**, which were then converted into the bridged YCK analogues **107** and **108** (*65*).

Another synthesis of YCK analogues **112** and **113** by Cheng (Scheme 8) involved coupling of a cyclohexene vinyl anion **109** (generated by the Shapiro reaction), with the N-protected indole-2-carbaldehyde, followed by subsequent Nazarov cyclization of the tricyclic ketone **110** to the tetracyclic *cis*-fused ketone **111** as the key steps (*66*). Using essentially the same approach, but starting from (−)- and (+)-camphor, Cheng synthesized both enantiomers of the bridged YCK analogues (**114**,**115**) and showed that the biologically active analogue (**115**), possessed

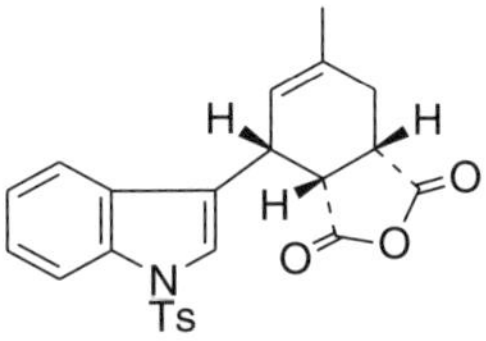

**104**

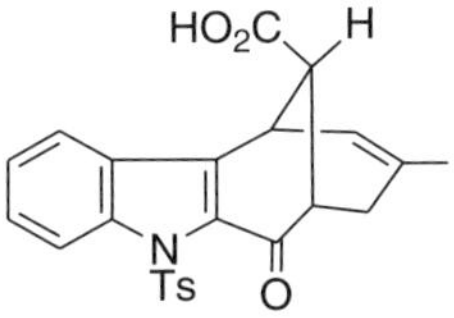

**105**

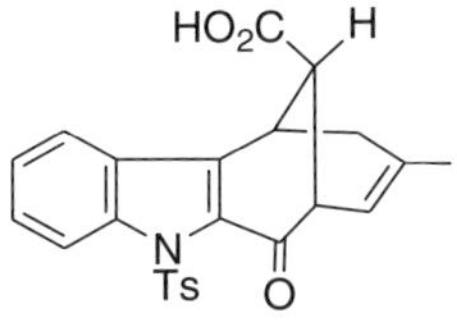

**106**

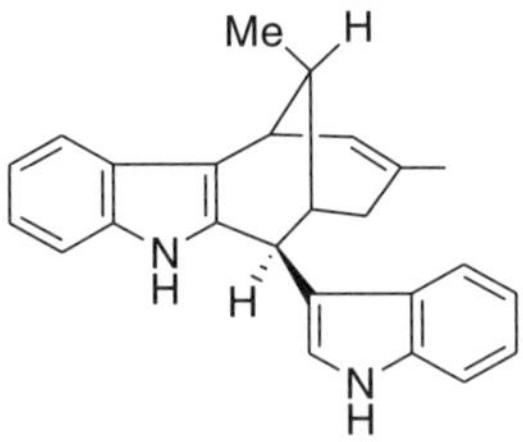

**107**

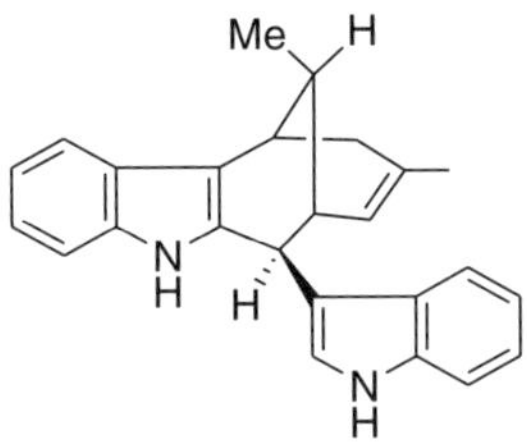

**108**

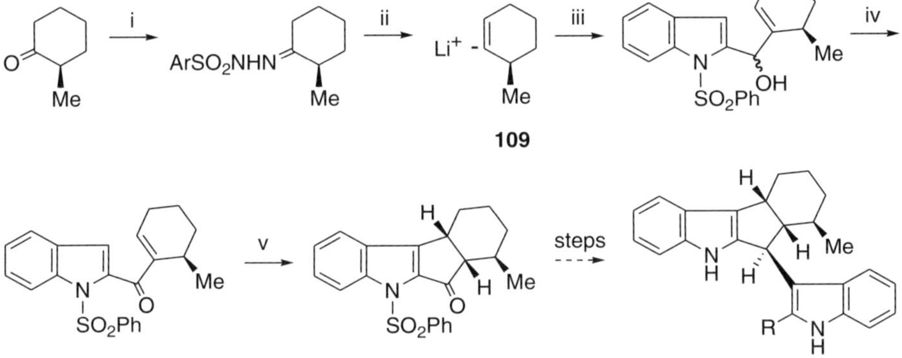

**109** **110** **111** (major product) **112** R = H; **113** R = Me

Scheme 8. Reagents: (i) Trisylhydrazine, HCl, MeOH; (ii) *t*-BuLi, THF, −70 to 0°C; (iii) *N*-phenylsulfonylindole-2-carbaldehyde, THF; (iv) $MnO_2$, PhH, rt; (v) $AlCl_3$, PhH, rt, 20 h.

the 6(*R*) configuration, whereas the 6(*S*) compound was inactive, from which it seemed reasonable to conclude that the biologically active enantiomer of YCK is (6*R*)-**73** (*67,68*).

The same approach was also used to prepare the truncated YCK analogues, **116**–**119** (*69,70*). Cheng also prepared inverto-YCK (**124**) from the acid chloride **120** (Scheme 9). The key steps involved the coupling of 3-indolezinc reagent with the acid chloride **120**, followed by Nazarov cyclization of the divinyl ketone **121**, to the tetracyclic ketone intermediate, **122**. The indolyl moiety was introduced by Pd(0)-mediated cross coupling between the indolezinc reagent and the acetate derivative, **123** (*71*).

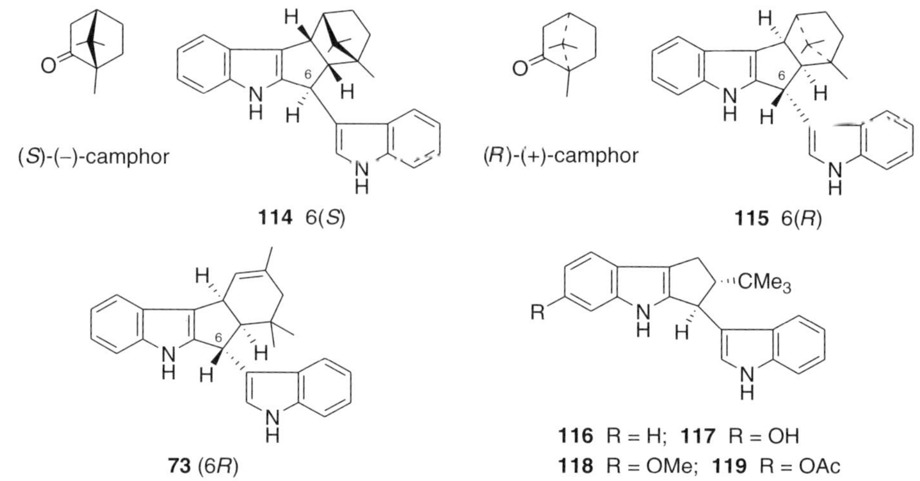


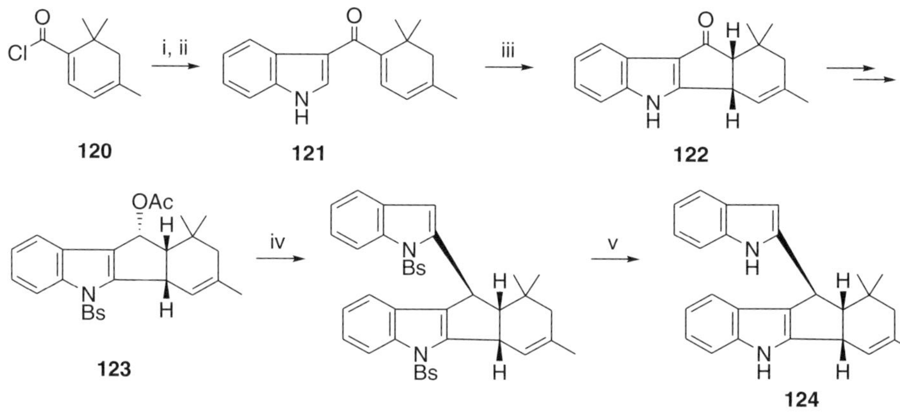


Scheme 9. Reagents: (i) Indole, EtMgBr; (ii) $ZnCl_2$; (iii) conc. HCl, dioxane; (iv) Zinc (*N*-Bs-indol-2-yl) chloride, $Pd(PPh_3)_2Cl_2$, DIBAL-H, THF; (v) Na–Hg, aq. $Na_2HPO_4$, $Et_2O$-MeOH.

Structure-activity studies on YCK derivatives indicated that substitutions on the indolic nitrogens, or on the 2- and 5′-positions of the indole units abolished the anti-implantation activity, while methylation of N(1′) resulted in appreciably diminished activity. Saturation of the cyclohexenyl double bond had no adverse effect, whereas hydroxylation at the aromatic C(2) or C(5′), or at the 9,10-double bond, resulted in complete loss of activity (*72*). The relative coplanarity or otherwise of the two indole moieties also appeared to be important. The presence of a bulky group at C(7), which serves to restrict the rotation of the C(6) indole substituent, appeared to be a prerequisite for the biological activity. The absence of both the methyl groups at C(7), as in bis-nor-YCK (**100**), rendered the analogue virtually inactive, while reintroduction of bulky substituents at C(7), or at C(2′), increased the potency in weak analogues (*73*). Yuehchukene showed competitive binding to the estrogen receptor, in addition to induction of estradiol-2-hydroxylase activity, which, taken together, may account for its anti-estrogenic effect (*74,75*). In a recent study, the response of MCF-7, an estrogen-receptor-positive breast cancer cell line, was tested against different combinations of estradiol, cyclophosphamide, and YCK (*76*). It was found that cyclophosphamide inhibited MCF-7 cell growth at a concentration of $10^{-2}$ M, but lost its cytotoxic effect on reducing the concentration to about $1 \times 10^{-3}$ M. Low concentrations of YCK ($10^{-8}$–$10^{-9}$ M) however, were found to potentiate the cytotoxic effect of cyclophosphamide on the MCF-7 cell line. Such an effect was absent in the estrogen-receptor-negative cell line, MDA-MB-231. In view of the known mixed estrogen and anti-estrogen effect of YCK (*74,75*), these findings suggested that YCK and cyclophosphamide can be considered as a potential combination in chemohormonal therapy for breast cancer (*76*).

## D. ARUNDO ALKALOIDS

Several bisindole alkaloids were isolated from *Arundo donax*, a perennial giant grass found in Uzbekistan, in addition to the simple bisindoles, arundine (3,3′-diindolylmethane, **125**) (*77*) and ardine (**126**) (*78*).

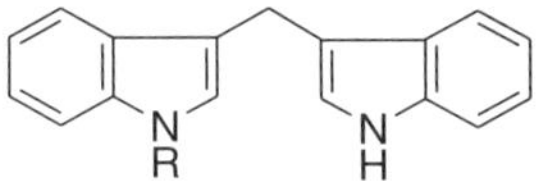

**125** R = H
**126** R = Me

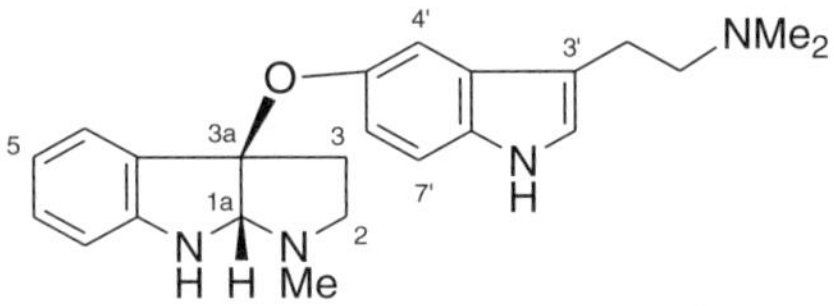

**127**

Arundinine (**127**), was obtained from the epigeal parts, and the structure was established by spectroscopic methods as well as by X-ray analysis. The molecule is constituted from the union of two tryptamine moieties, a physostigmine-type unit and another tryptamine unit. The two units are linked by an ether bridge from C(3a) of the former to the aromatic C(5′) of the other unit (*79*). A number of related bisindoles have been subsequently isolated from the root extract of the same plant, viz., arundamine (**128**) (*80,81*), arundacine (**129**) (*82*), arundanine (**130**) (*83*), arundavine (**131**) (*84*), arundaphine (**132**) (*85*), and arundarine (**133**) (*86*). The structures of **128**, **131**, and **132** have also been confirmed by X-ray analysis, while the NMR spectrum of arundacine (**129**) showed the existence of two equilibrating conformers due to restricted rotation about the amide C–N bond.

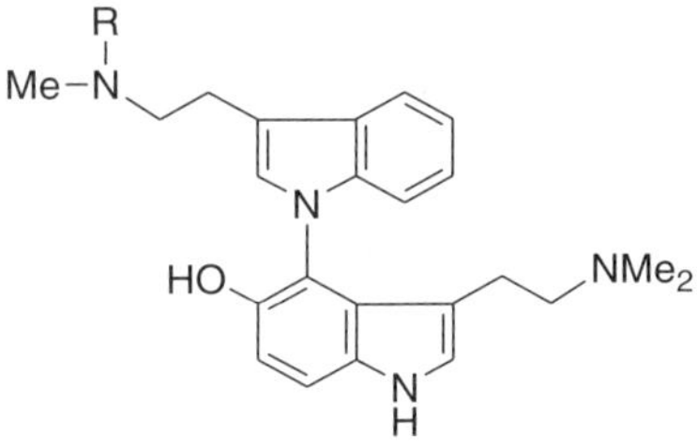

**128** R = H
**129** R = COMe
**130** R = Me

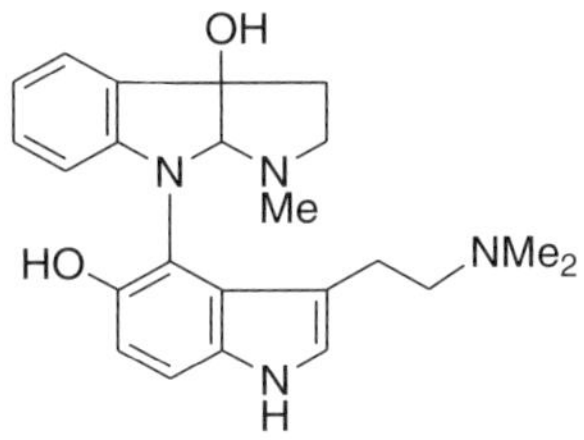

**131**

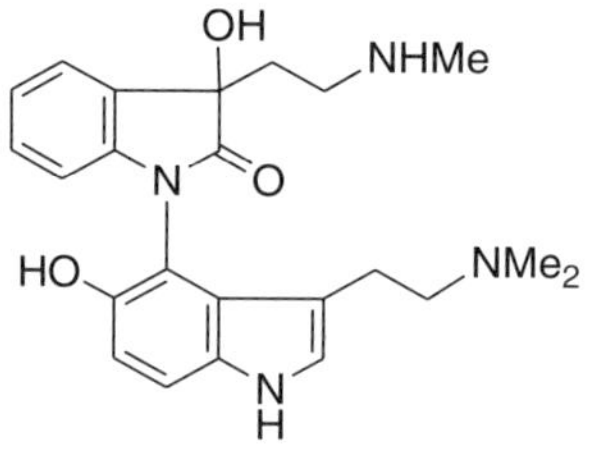

**132**

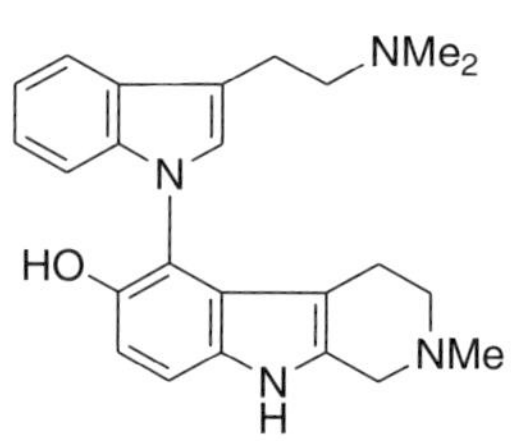

**133**

## E. The Picrasidines and Related Alkaloids

The bark and root-bark of *Picrasma quassioides* (Simaroubaceae), a plant cultivated in Japan, furnished a series of new bis-β-carbolines, viz., picrasidines A (**134**) (*87*), C (**135**) (*88*), F (**136**) (*89*), G (**137**) (*90*), H (**138**), R (**139**) (*91*), S (**140**) (*90*), and T (**141**) (*92*), and canthinone-β-carboline alkaloids, including, picrasidines M (**142**) (*93*), N (**143**) (*94*), and U (**144**) (*95*). Picrasidines F, G, S, and T were isolated in the form of their hydrochloride salts, and as racemic mixtures. Picrasidine G was also found in the bark extract of *P. javanica* occurring in Flores Island, Indonesia (*96*). The structure and absolute configuration of picrasidine F were confirmed by X-ray analysis (*89*), while a detailed NMR study of picrasidine G has allowed unambiguous assignment of the $^{1}$H and $^{13}$C NMR spectral data (*97*). The two bis-β-carboline alkaloids, kumujansine and kumujantine, subsequently reported from the same plant found in China (*98*), have the same structures as picrasidines G and S, respectively.

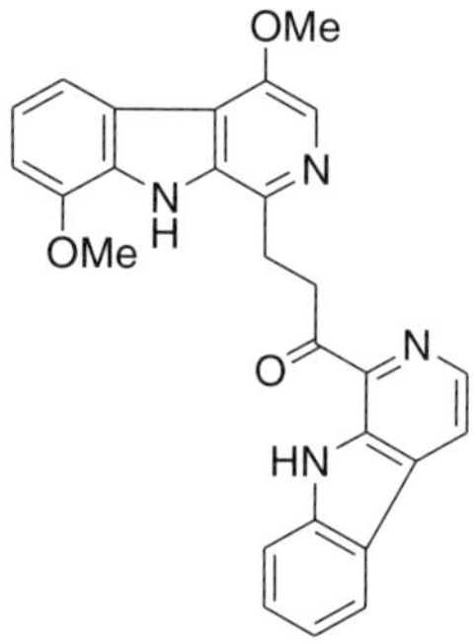

**134**

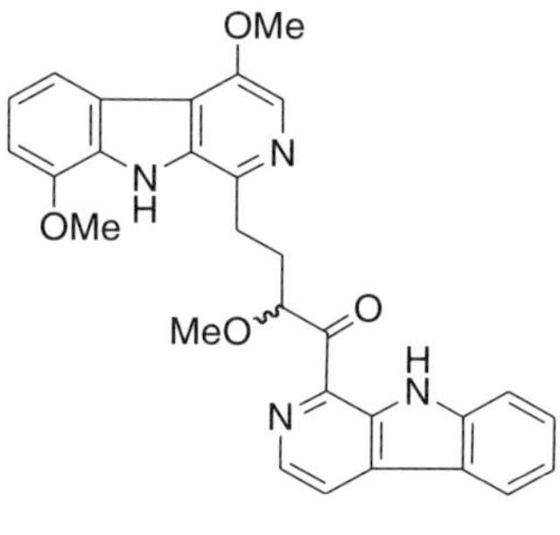

**135**

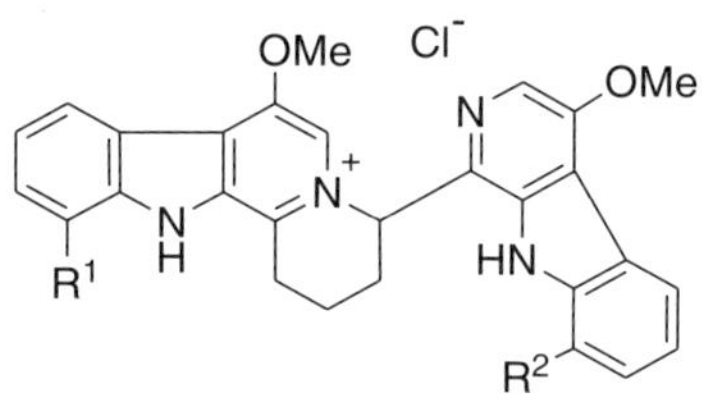

**136** $R^1$ = H, $R^2$ = OMe
**137** $R^1$ = $R^2$ = H
**140** $R^1$ = $R^2$ = OMe
**141** $R^1$ = $R^2$ = OH

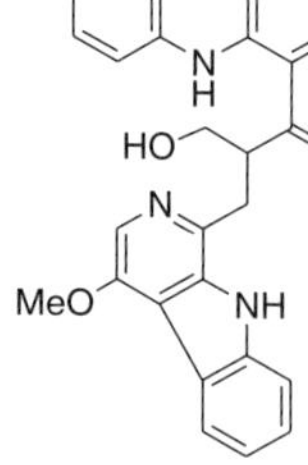

**138**

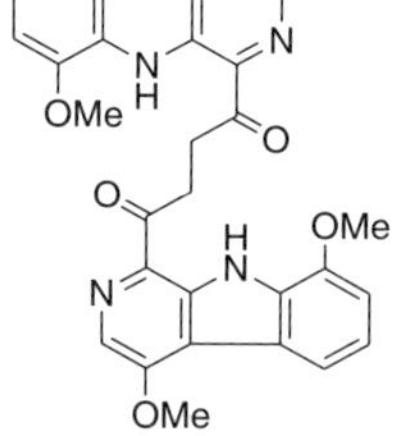

**139**

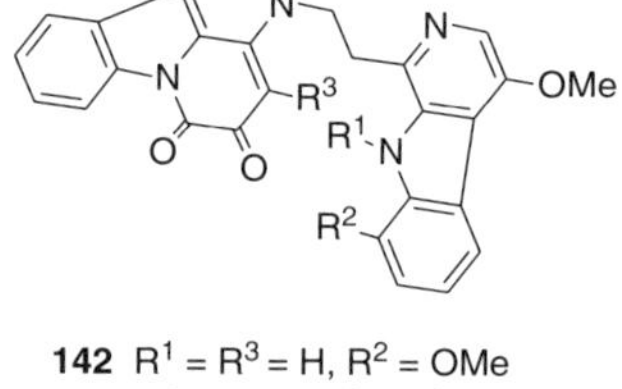

**142** $R^1$ = $R^3$ = H, $R^2$ = OMe
**143** $R^1$ = OMe, $R^2$ = $R^3$ = H
**144** $R^1$ = H, $R^2$ = $R^3$ = OMe

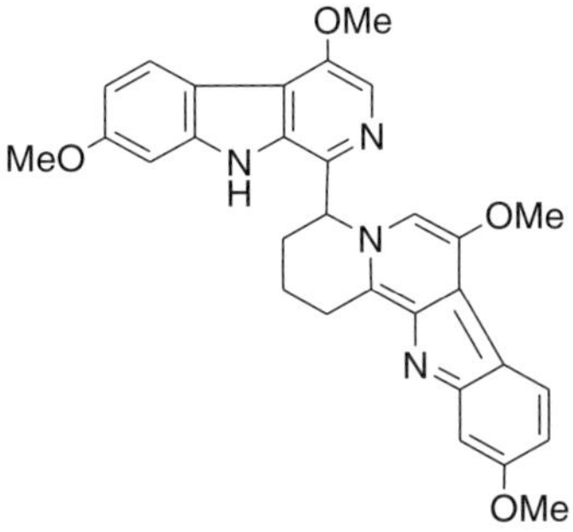

**145**

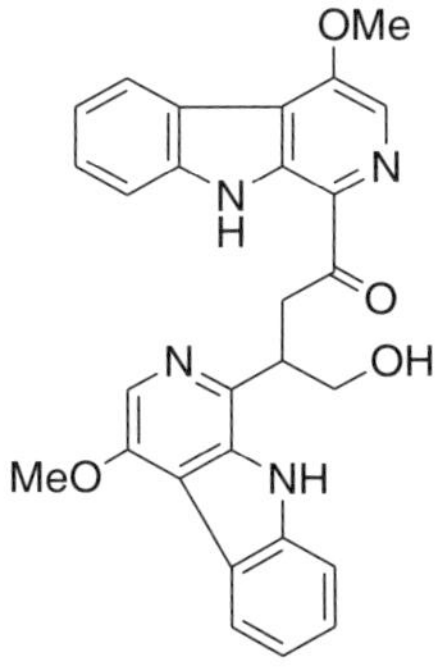

**148**

Another picrasidine congener, kirondrine (**145**), was isolated from the Madagascan plant, *Perriera madagascariensis* (Simaroubaceae) (*99*). An attempted synthesis (Scheme 10) of picrasidine H (**138**) *via* $Mg(OEt)_2$-mediated enolate alkylation involving the two fragments, **146** and **147**, led on subsequent $BH_3$ reduction, followed by benzylic oxidation ($MnO_2$), to the product **138**, which was however found to be different from the natural compound (differences in mp, TLC $R_f$, EIMS, $^1H$ and $^{13}C$ NMR), suggesting the possibility that an incorrect structure had been previously assigned for picrasidine H, athough the authors proposed an alternative structure, **148**, which is a regioisomer of **138**, a corresponding synthesis has not been reported (*100*).

## F. CALYCANTHINE–CHIMONANTHINE AND RELATED ALKALOIDS

Three calycanthine-type alkaloids, (−)-calycanthine (**149**) (the optical antipode of (+)-calycanthine (**150**)), (−)-isocalycanthine (**151**), and *meso*-chimonanthine (**152**) were isolated from *Psychotria forsteriana*, from New Caledonia, and complete $^1H$ and $^{13}C$ NMR assignments were reported (*101*). This represents the second isolation of the levorotatory isomer of calycanthine (the first was from *Pausinystalia macroceras*, Rubiaceae, (*102*)) as well as the first isolation of *meso*-chimonanthine. (−)-Calycanthine and (+)-chimonanthine (**153**) were also found to occur in the skin of the Colombian poison dart frog, *Phyllobates terribilis* (*103*).

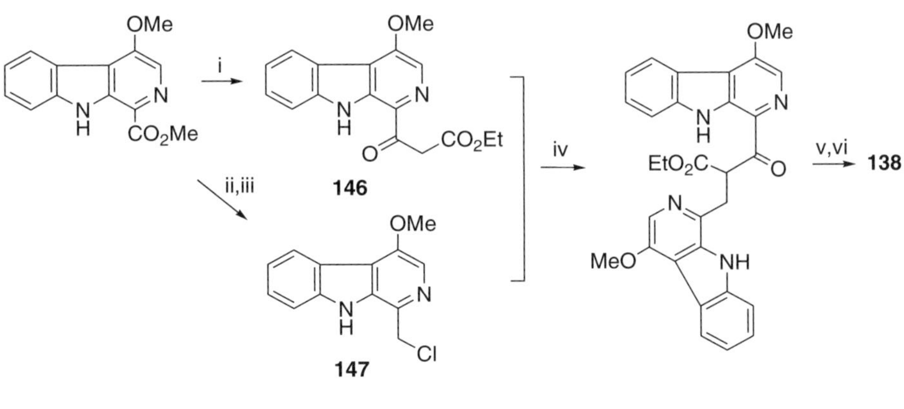

Scheme 10. Reagents: (i) $CH_3CO_2Et$/LiHMDS; (ii) $LiAlH_4$, THF; (iii) $SOCl_2$; (iv) $Mg(OEt)_2$, KI, DMF; (v) $BH_3$–THF; (vi) $MnO_2$, $CH_2Cl_2$–MeOH.

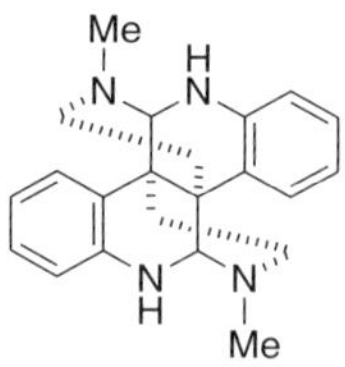

**149** (–)-calycanthine

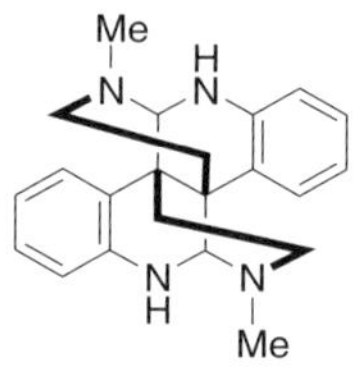

**150** (+)-calycanthine

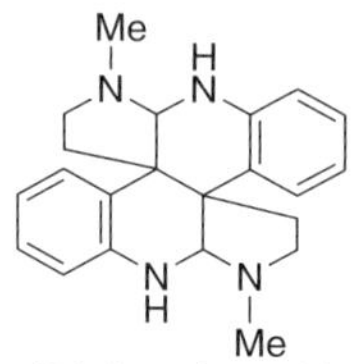

**151** isocalycanthine

Three new dehydrocalycanthine derivatives, glomerulatines A, B, and C (**155**–**157**), were isolated from the aerial parts of *Psychotria glomerulata* from Panama (*104*). The major alkaloid, glomerulatine A (**155**) was identified as (8-8a),(8′-8′a)-tetradehydro-(−)-calycanthine, while the two minor alkaloids, glomerulatines B (**156**) and C (**157**), were identified as (8-8a),(8′-8′a)-tetradehydro-*N*′-demethyl-(−)-calycanthine and (8-8a)-didehydro-(−)-calycanthine, respectively, on the basis of the spectral data.

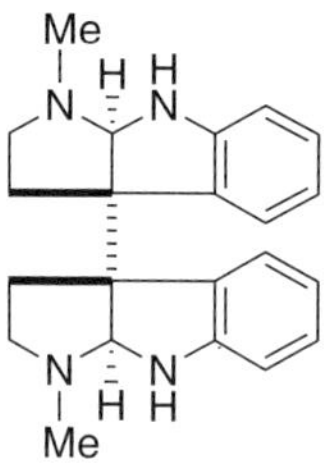

**152** *meso*-chimonanthine

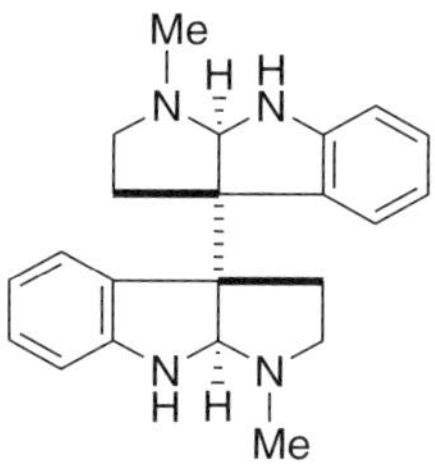

**153** (+)-chimonanthine

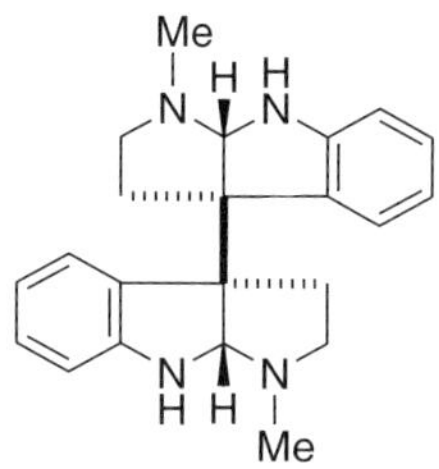

**154** (–)-chimonanthine

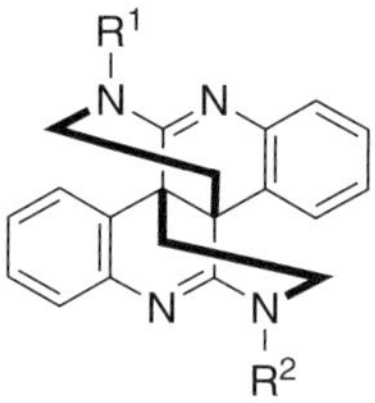

**155** $R^1 = R^2 = Me$

**156** $R^1 = H$, $R^2 = Me$

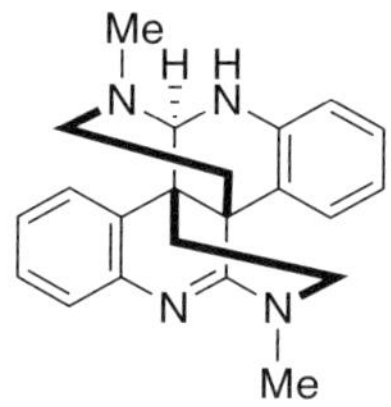

**157**

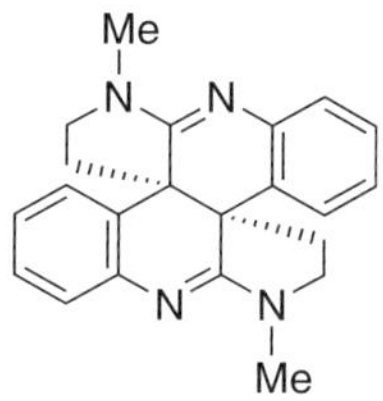

**158**

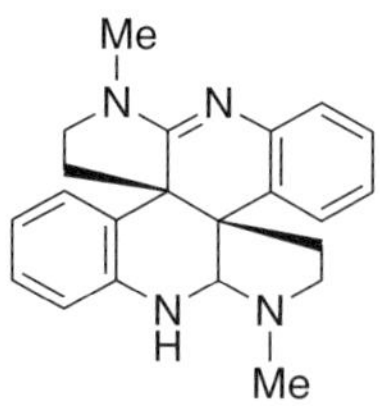

**160** bhesine

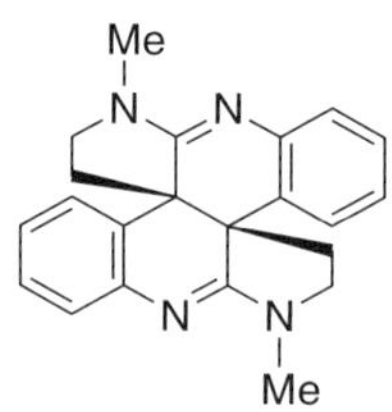

**161** dehydrobhesine

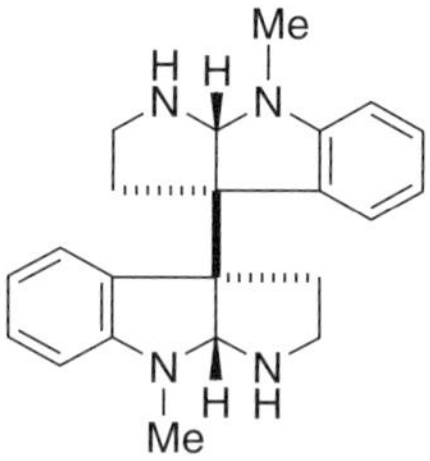

**162** (+)-isochimonanthine

A new dehydroisocalycanthine derivative, (8-8a),(8′-8′a)-tetradehydroisocalycanthine 3a(*R*), 3′a(*R*) (**158**), in addition to (−)-calycanthine, isocalycanthine, (+)-chimonanthine, hodgkinsine, and quadrigemine C, were isolated from another *Psychotria* species, *P. colorata*. The molecule is characterized by the presence of a $C_2$ axis, as indicated from the spectral data, and confirmed by X-ray analysis of the dihydrobromide derivative (*105*), which also established its absolute configuration. This report also provided for the first time full NMR assignments for hodgkinsine (**159**), (+)-chimonanthine, and (−)-calycanthine as well as their CD values. Compound **158** is identical in all respects, except for the sign of the specific rotation, to dehydrobhesine (**161**), isolated earlier, together with bhesine ((8-8′a)-dehydroisocalycanthine) (**160**), from the bark extract of the Malayan plant, *Bhesa paniculata*. The structure and relative configuration of these compounds were established in the earlier study based on an X-ray diffraction analysis of bhesine (*106*). Dehydrobhesine (**161**) is therefore the optical antipode of **158**. The leaves of *Argostemma yappii* (Rubiaceae) from Indonesia, gave a new chimonanthine isomer, (+)-isochimonanthine (**162**). The $^{1}$H NMR spectrum at ambient temperature consisted of a series of broadened peaks, suggestive of conformational mobility, and unlike the $^{13}$C NMR spectrum, bore no resemblance to that of chimonanthine. The $^{1}$H NMR spectrum at lowered temperature (248 K), however, showed sharper signals, and indicated the existence of two conformers with a ratio of 53:47, with the major conformer having the transoid structure as shown in **162**, and the minor conformer assigned the cisoid structure **163** (*107*). A new chimonanthine derivative, chimonanthidine (**164**), was isolated from the seeds of *Chimonanthus praecox*, occurring in Japan. The NMR spectral data were very similar to those of (−)-chimonanthine (**154**), (−)-folicanthine (**165**), and (−)-calycanthidine (**166**) which were also obtained, the major difference from folicanthine being replacement of an NMe with NH (*108*).

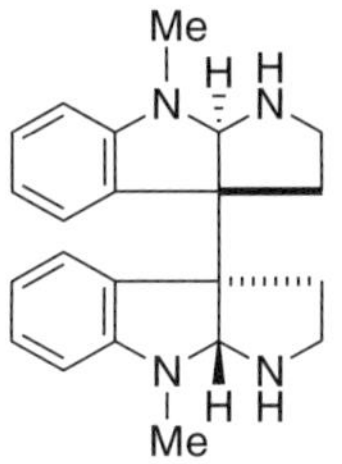

**163**

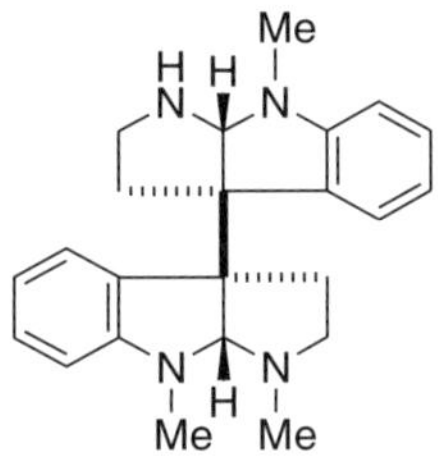

**164** (–)-chimonanthidine

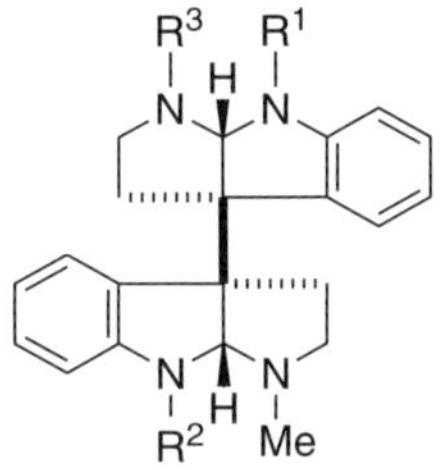

**165** $R^1 = R^2 = R^3 = Me$

**166** $R^1 = R^3 = Me$, $R^2 = H$

Three new tryptamine oligomers, quadrigemine C (**167**), isopsychotridines A (**168**) and B (**169**), in addition to the known tetrameric and pentameric bases, hodgkinsine (**159**) and psychotridine (**170**), respectively, were isolated from *Psychotria oleoides* from New Caledonia (*109*). The primary structures of these alkaloids were determined mainly from the spectral data, and without stereochemical assignments. Hodgkinsine (**159**) and calycosidine (**171**), the latter at the time known only as a hodgkinsine isomer, were obtained from *Calycodendron milnei* from Vanuatu in the New Hebrides (*109*). The stereochemistry of quadrigemine C (**167**) was later established based on analysis of the NMR data ($^{1}$H and $^{13}$C NMR and ROESY) as well as comparison of the CD spectrum with that of hodgkinsine (**159**) (*110*), while the subsequent transformation of hodgkinsine under mild acidic conditions (reflux in dilute AcOH) to a product, constituted from *meso*-calycanthine and an *N*-methylpyrrolidinoindoline subunit, and identical (spectral data and optical rotation) to the previously isolated calycosidine, permitted definitive structure assignment for calycosidine (**171**) (*111*). Tentative stereochemical assignments were later made for psychotridine (**170**) and isopsychotridine B (**169**) based on NMR spectral data, CD analysis, and chemical correlations (*112*).

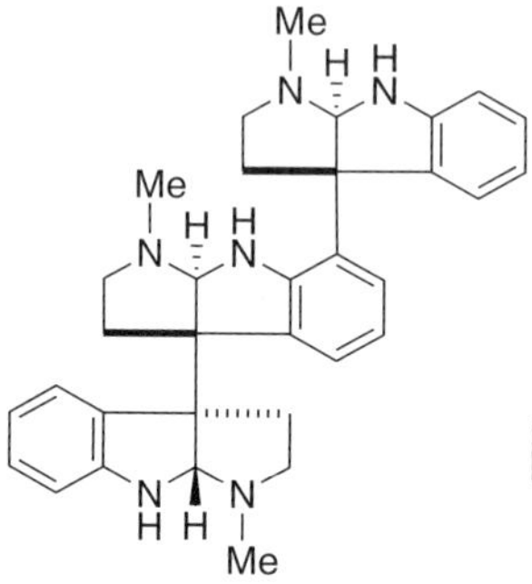

**159** hodgkinsine

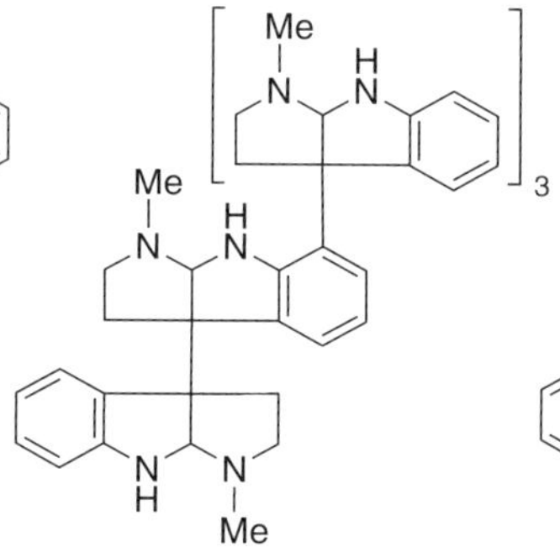

**168** isopsychotridines A, C

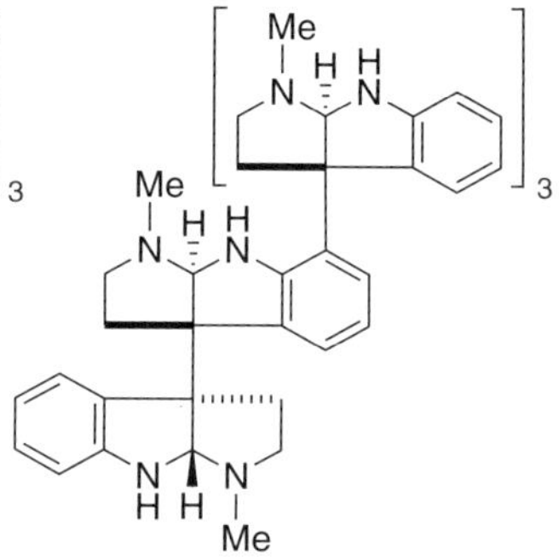

**169** isopsychotridine B

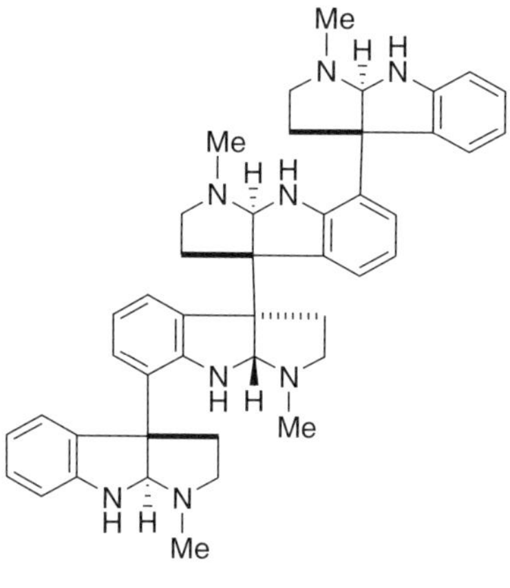

**167** quadrigemine C

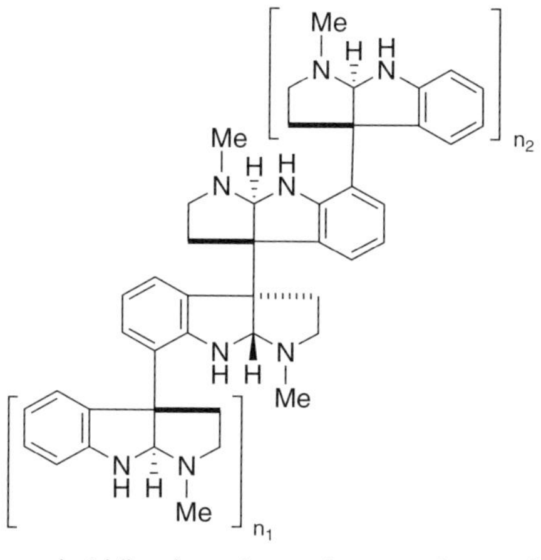

**170** psychotridine ($n_1 = 1$, $n_2 = 2$ or $n_1 = 2$, $n_2 = 1$)

Another new tryptamine tetramer, psycholeine (**172**), was also isolated from *P. oleoides* using bioassay-guided fractionation (*110*). The structure was assigned by comparison of the spectral data (MS, NMR) with those of chimonanthine (**153**), hodgkinsine (**159**), quadrigemine C (**167**), and calycosidine (**171**) as well as by chemical correlation with quadrigemine C (acid-induced isomerization of the latter to psycholeine). Comparison of the mass-spectral data revealed analogy to calycosidine, while the NMR spectral data revealed constitution from a central *meso*-calycanthine subunit, linked at C(7′) and C(7″) to two *N*-methylpyrrolidinoindoline units.

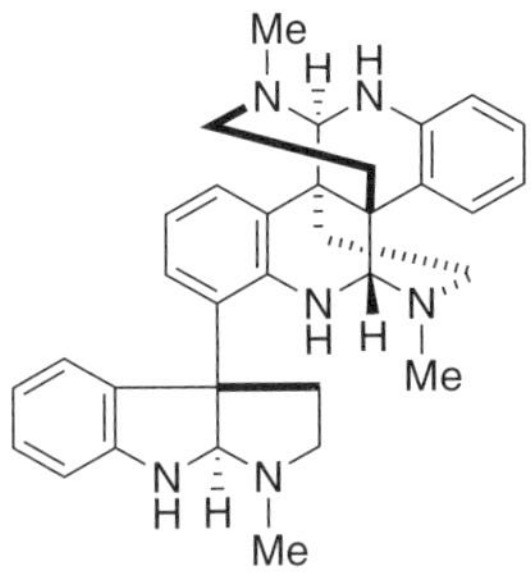

**171** calycosidine

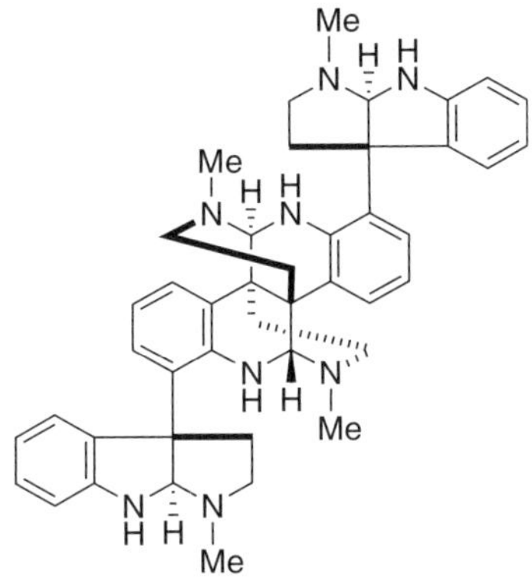

**172** psycholeine

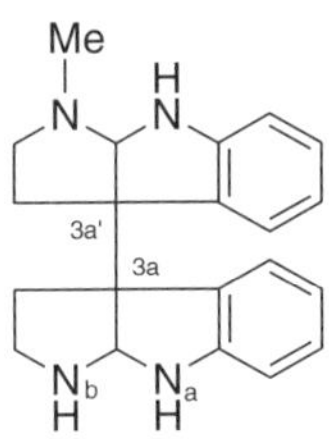

**173**

$N_b$-demethyl-*meso*-chimonanthine

(3a*S*, 3a'*R* or 3a*R*, 3a'*S*)

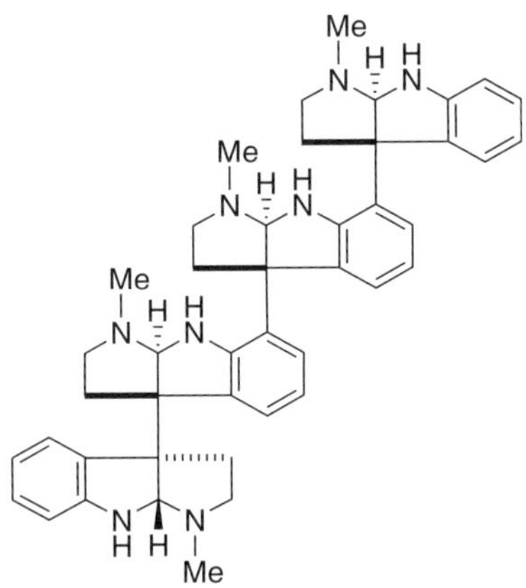

**174** quadrigemine I

A reinvestigation of *P. oleides*, and *P. lyciiflora*, prompted by the earlier observation of the effects of **167** and **172** on the neuroendocrine system, led to the isolation of additional new pyrrolidinoindoline alkaloids, the dimeric demethyl-*meso*-chimonanthine derivative, **173**, from *P. lyciiflora*, and three new oligomers from *P. oleoides*, quadrigemine I (**174**), oleoidine (**175**), and caledonine (**176**). The structure elucidation of the dimeric **173** was based on spectral data and *via* chemical correlation with *meso*-chimonanthine (**152**). The structures and tentative stereochemistries of quadrigemine I (**174**), oleoidine (**175**), and caledonine (**176**) were provisionally assigned by a combination of two-dimensional (2-D) NMR analysis, CD spectra comparisons, and chemical correlations (*112*).

As part of a search for naturally occurring substances acting on neurochemical transmission, a new trimeric tryptamine alkaloid, idiospermuline (**177**), was isolated from the seeds of *Idiospermum australiense*, occurring in North Queensland, Australia (*113*). The structure elucidation was based on spectral data and an X-ray crystallographic analysis of its trimethiodide derivative, which also provided the absolute configuration.

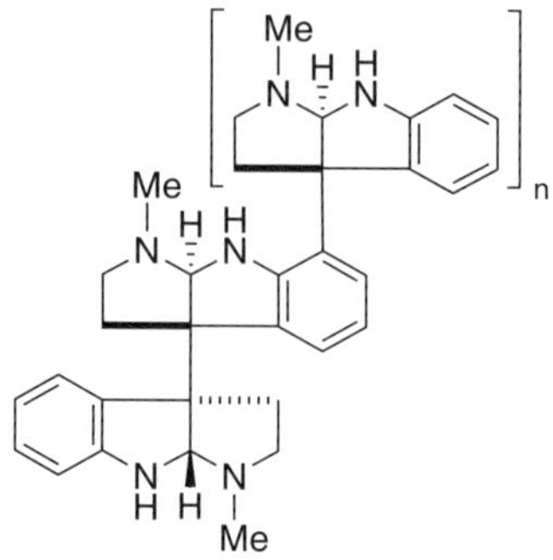

**175** n = 4 oleoidine

**176** n = 5 caledonine

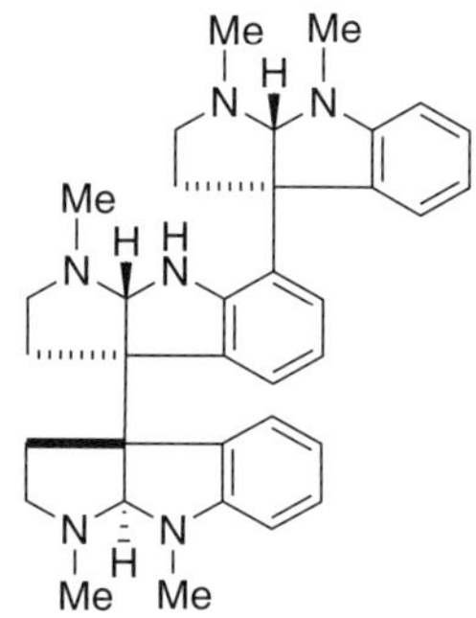

**177** idiospermuline

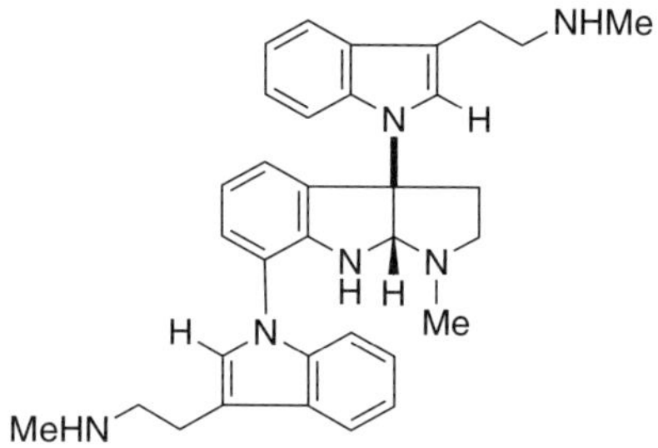

**178** psychotrimine

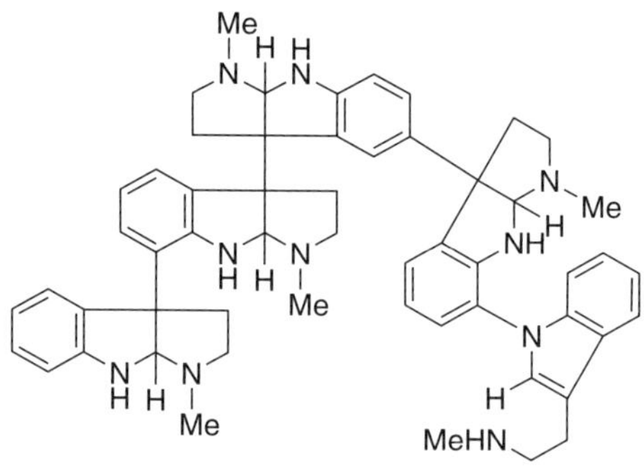

**179** psychopentamine

An earlier study of the Malaysian species, *Psychotria rostrata* gave quadrigemine B as the major alkaloid, accompanied by hodgkinsine, (−)-calycanthine, (+)-chimonanthine, and calycosidine. This study represents the first isolation of (+)-chimonanthine from a plant (*114*). A recent reinvestigation of the same species gave two new tryptamine oligomers, psychotrimine (**178**), and psychopentamine (**179**) (*115*). The spectral data showed the former to be constituted from a central *N*-methylpyrrolidinoindoline unit, linked to two *N*-methyltryptamines, from C(7′) and C(3a′), to N(1) and N(1″), respectively. Psychopentamine, on the other hand, was shown to possess a novel bent-type structure, in which a trimeric pyrrolidinoindoline unit is linked to a dimeric pyrrolidinoindoline–tryptamine unit, as shown in **179**. These alkaloids represent the first examples of tryptamine oligomers incorporating tryptamine units (as opposed to previously known tryptamine-derived oligomers, which are constituted from pyrrolidinoindoline units), and in the case of psychopentamine (**179**), the first to possess the bent-type structure, due to its novel mode of union of the constituent moieties. Stereochemical data for these compounds are however still pending, as are those for several of the higher order pyrrolidinoindoline alkaloids, including, hodgkinsine A (**180**) (*116,118*), quadrigemines A (**181**), B (**182**) (*119*), G (**181**) (*116,117*), and H (**182**) (*116,118*), psychotridine C (**183**) (*116–118*), isopsychotridines C (**168**) (*119*), D (**168**) (*116,117*), E (**168**) (*118*), vatine (**184**), vatine A (stereoisomer of vatine), vatamine (**185**), and vatamidine (**186**) (*116–118*).

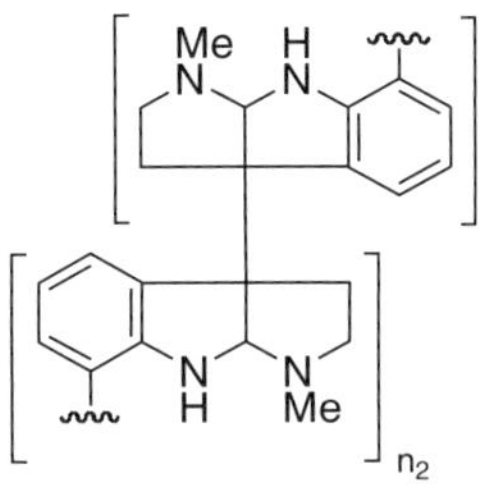


**180** $n_1 = 1, n_2 = 2$ hodgkinsine A
**181** $n_1 = 2, n_2 = 2$ quadrigemines A, G
**182** $n_1 = 1, n_2 = 3$ quadrigemines B, H
**183** $n_1 = 2, n_2 = 3$ psychotridine C
**168** $n_1 = 1, n_2 = 4$ isopsychotridines D, E
**184** $n_1 = 2, n_2 = 4$ vatine, vatine A
**185** $n_1 = 2, n_2 = 5$ vatamine
**186** $n_1 = 2, n_2 = 6$ vatamidine

Earlier synthetic efforts directed toward alkaloids of this group were based on the 3,3-dimerization of indole and oxindole precursors, and were invariably characterized by poor yields. A marked improvement in the yield was achieved by Rodrigo, whose route to (±)-folicanthine featured a tetraiodomethane-mediated dimerization of the enolate of the 3-substituted oxindole **187** as the key step (Scheme 11). A radical anion chain mechanism was proposed for the crucial coupling step (*120*).

A significant improvement in the yield was also achieved in Takayama's concise synthesis of *meso*-chimonanthine (*121*), *via* hypervalent iodine-mediated dimerization of a tryptamine precursor, giving *meso*-chimonanthine in 30% yield over three steps (Scheme 12). This approach has also been applied by Takayama to the synthesis of chimonanthidine (*108*). A similar three-step procedure to *meso*-chimonanthine, albeit in lower overall yield, involving thallium trifluoroacetate-mediated oxidative coupling of the same tryptamine precursor has also been reported (*122*).

Subsequent efforts directed toward the stereocontrolled construction of these molecules have been virtually dominated by Overman, who released a steady stream of papers, beginning in 1996 (*123*) with the synthesis of *meso*-chimonanthine and

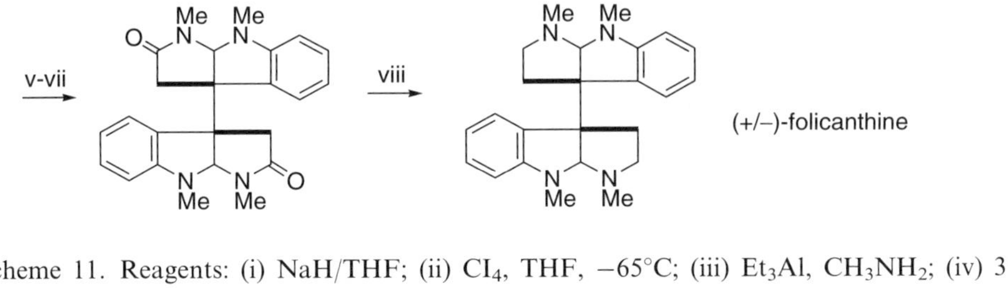


Scheme 11. Reagents: (i) NaH/THF; (ii) $CI_4$, THF, −65°C; (iii) $Et_3Al$, $CH_3NH_2$; (iv) 3% HCl; (v) LDA/THF; (vi) DIBAL-H; (vii) CSA, $CH_2Cl_2$; (viii) Red-Al.

*meso*-calycanthine from N-protected isoindigo **188**, featuring a novel samarium-mediated reductive dialkylation *via* the chelated intermediate **189** (Scheme 13). Reduction of the cyclohexene derivative **190** with Red-Al gave the unusual hexacycle **191**, whose structure was confirmed by X-ray analysis, as was that of the final *meso*-chimonanthine product. Cleavage of the cyclohexene ring and subsequent

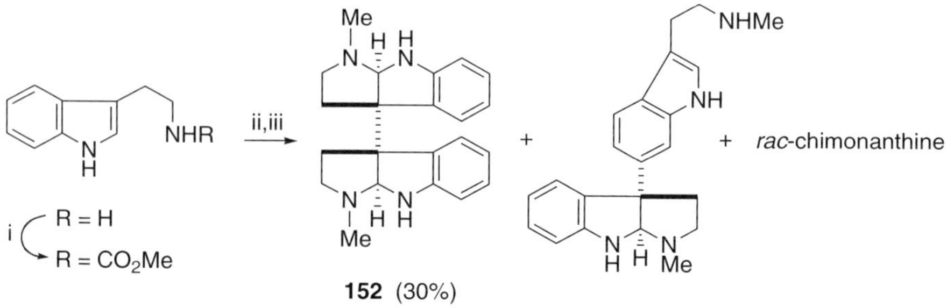


Scheme 12. Reagents: (i) $CH_3CO_2Cl$, NaOH/$H_2O$, $CH_2Cl_2$; (ii) $PhI(CF_3COO)_2$, $CF_3CH_2OH$, $-30$°C; (iii) Red-Al, toluene, reflux.

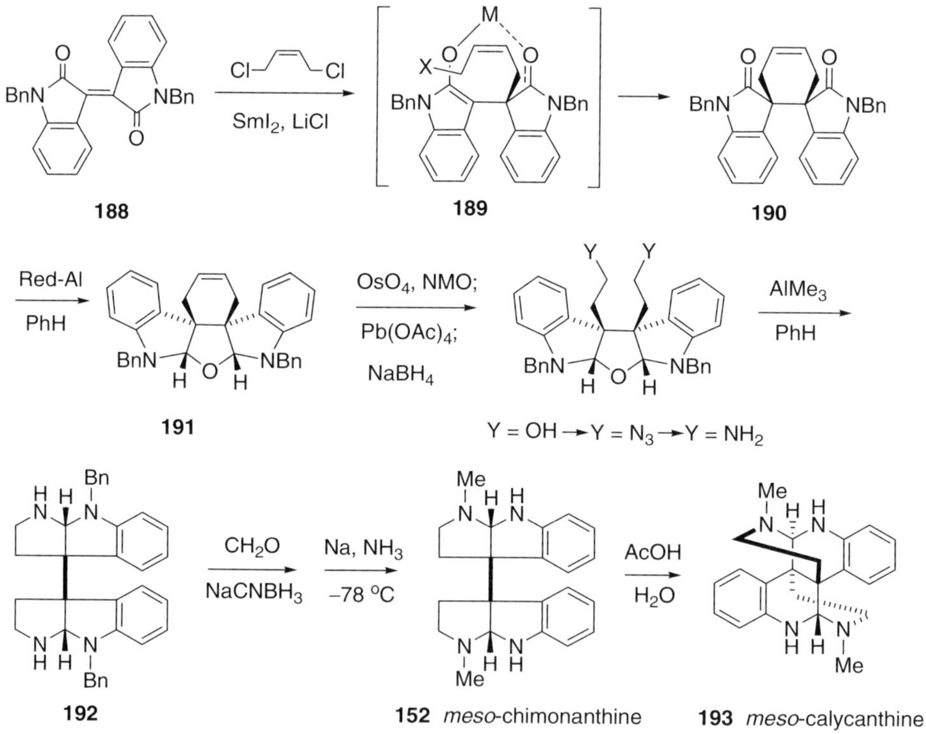


Scheme 13

Mitsunobu reaction led to the diazide, which upon reduction to the corresponding diamine, followed by exposure to excess $Me_3Al$, provided the bispyrroloindoline **192**, *en route* to *meso*-chimonanthine (**152**) and *meso*-calycanthine (**193**).

This was followed shortly by a stereo- and enantiocontrolled synthesis of (−)-chimonanthine (**154**) and calycanthine (**150**) as well as a second route to *meso*-chimonanthine (**152**). The central step in this synthesis features the use of a double Heck cyclization to create vicinal quaternary carbon centers in high yields and with complete stereocontrol (*124*). The synthesis commenced with a double alkylation of the lithium dienolate of dimethyl succinate **194** and tartrate-derived diiodide **195** to give a diastereomeric mixture of the saturated diesters. Subsequent oxidation of the diesters, followed in succession by aminolysis, *N*-benzylation, removal of the benzyl ethers, and silylation, provided the cyclization substrate **197**, which on Heck cyclization yielded a single product, **198**, a pentacyclic bisoxindole, subsequently shown to have the *meso* relationship of the two oxindole groups. Further manipulations of **198** led eventually to the diazide derivative **199**, which can be processed to *meso*-chimonanthine (**152**), following the procedure established in the preceding synthesis (Scheme 14).

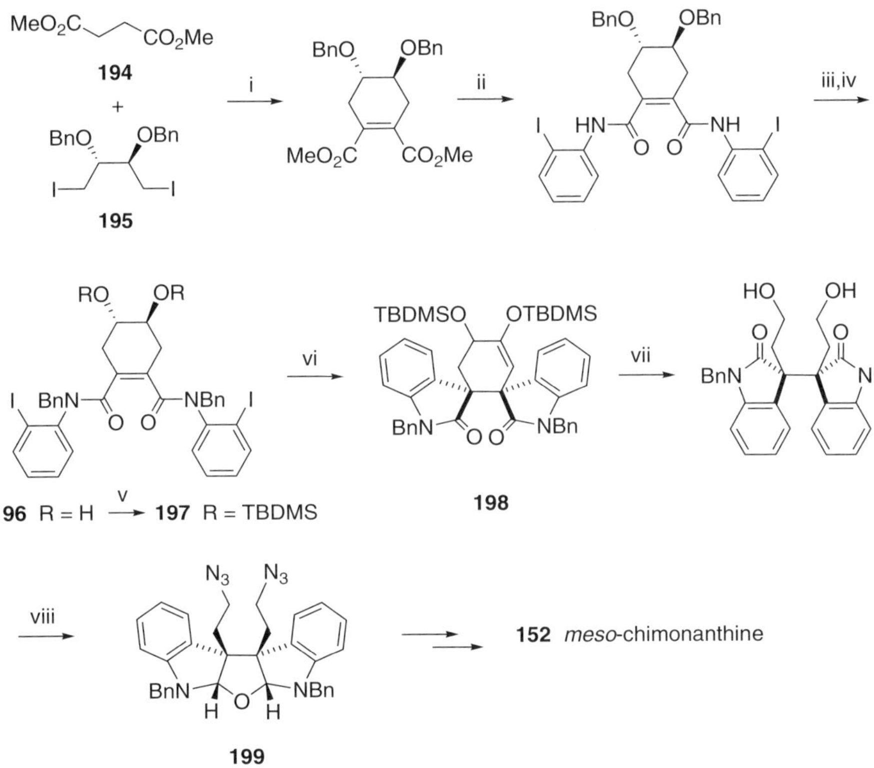


Scheme 14. Reagents: (i) LDA, THF/HMPA, −78°C; LDA, THF, $I_2$, −78°C; (ii) 2-iodoaniline, $Me_3Al$, PhMe; (iii) NaH, BnBr; (iv) $BCl_3$, −78°C; (v) TBDMSCl, imidazole, $CH_2Cl_2$; (vi) 10% $(Ph_3P)_2PdCl_2$, $Et_3N$, DMA, 100°C; (vii) HF, MeCN; $NaBH_4$, MeOH; $Pb(OAc)_4$, PhH; $NaBH_4$, MeOH; (viii) Red-Al, THF; $HN_3$, $Ph_3P$, $EtO_2CN{=}NCO_2Et$, THF.

It was found that protection of the cyclohexanediol **196** as an acetonide **200** dramatically altered the course of the double Heck cyclization, providing instead the bisoxindole **201** as the sole product, whose structure was established by X-ray analysis. Compound **201** was then processed as in the previous cases to the diazide, which was then transformed *via* the diamine **202** into the bispyrroloindoline **203**, and thence to (−)-chimonanthine (**154**) and (+)-calycanthine (**150**) (Scheme 15). This paper also includes a correction of the absolute configuration of (−)-chimonanthine, previously incorrectly represented in the literature.

In another disclosure (*125*), isoindigo **188**, used earlier as starting compound in the samarium-mediated reductive alkylation, was this time reduced to the dihydroisoindigo derivative **204**, and then alkylated *via* its disodium dienolate with the tartrate-derived ditriflate **205** to give the pentacycle **206**, possessing a *cis* relationship of the two spiroindoxyl groups (Scheme 16). Further manipulations led to the diol **207**, which could be processed into *meso*-chimonanthine in six steps, following the earlier

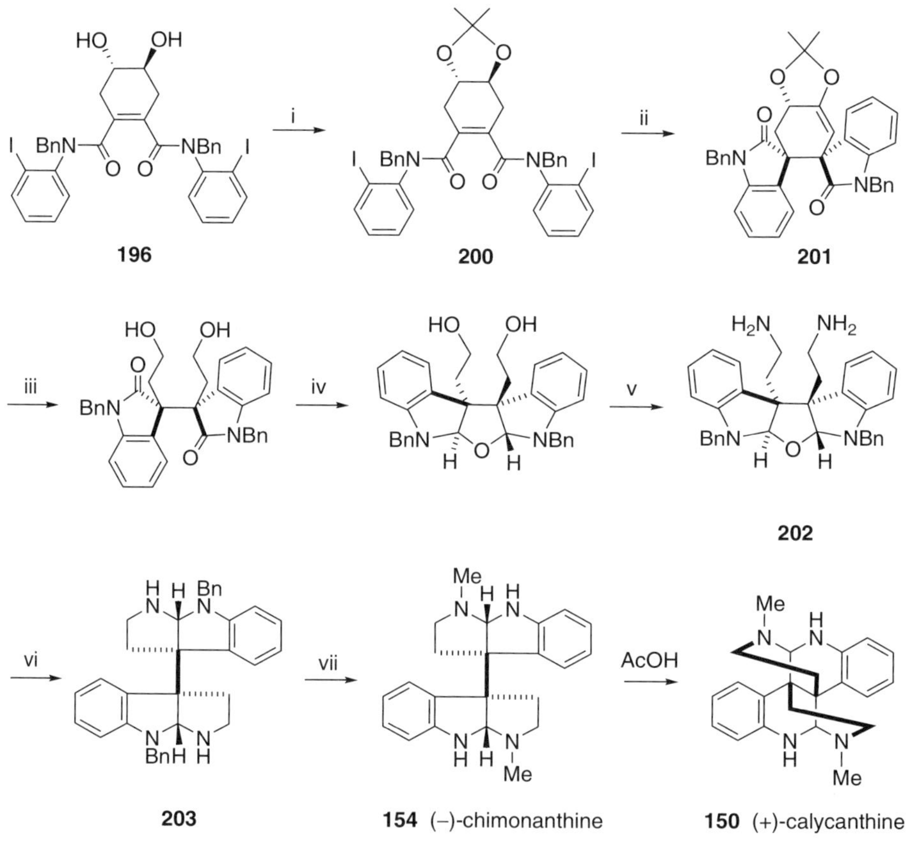


Scheme 15. Reagents: (i) 2,2-dimethoxypropane, CSA; (ii) 10% $(Ph_3P)_2PdCl_2$, $Et_3N$, DMA, 100°C; (iii) CSA; $NaBH_4$; $Pb(OAc)_4$; $NaBH_4$; (iv) Red-Al, THF; (v) $(PhO)_2P(O)N_3$, $Ph_3P$, $EtO_2CN{=}NCO_2Et$, THF; $H_2$, 10% Pd/C, EtOH; (vi) MeOH, 110°C; (vii) $CH_2O$, $NaBH(OAc)_3$, MeOH; Na, $NH_3$/THF.

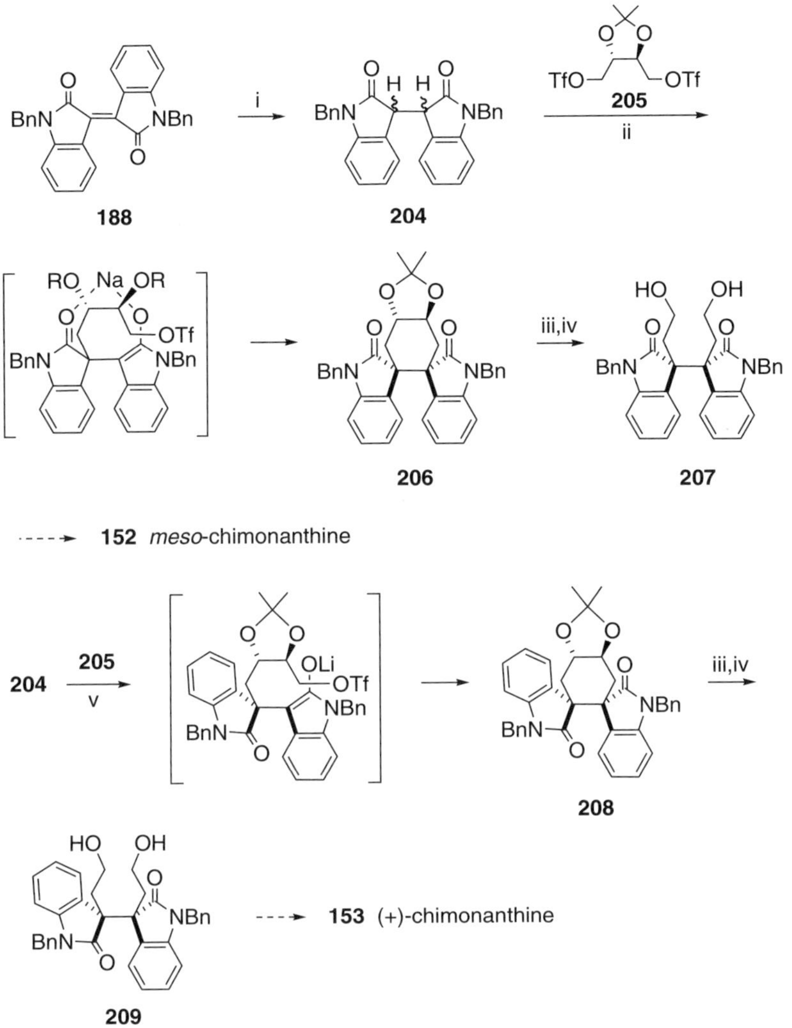


Scheme 16. Reagents: (i) Zn, AcOH; (ii) NaHMDS (2.1 equiv.), THF, −78°C; (iii) CSA, $MeOH/CH_2Cl_2$; (iv) $Pb(OAc)_4$, PhH; $NaBH_4$, MeOH; (v) LiHMDS (2.1 equiv.), THF/DMPU, −78°C.

protocol established in the samarium diiodide-mediated route (*123*). An advantage of this route over the earlier samarium-mediated alkylation is that the key alkylation step is more amenable to scale-up. In addition, changing the dienolate counterion from sodium to lithium resulted in a dramatic change in the product, leading instead to the $C_2$-symmetric bisoxindole **208**, which was converted in two steps into the diol **209**, and thence to (+)-chimonanthine (**153**) in six additional steps by the earlier protocol as before.

With the crucial methodologies in place as showcased in the above syntheses, the higher-order polypyrrolidinoindolines would be expected to be the next logical targets, and Overman duly disclosed elegant syntheses for quadrigemine C (**167**) and psycholeine (**172**) (Scheme 17), idiospermuline (**171**) (Scheme 18), and hodgkinsine (**159**) and hodgkinsine B (**223**) (Scheme 19). The quadrigemine C synthesis (*126*)

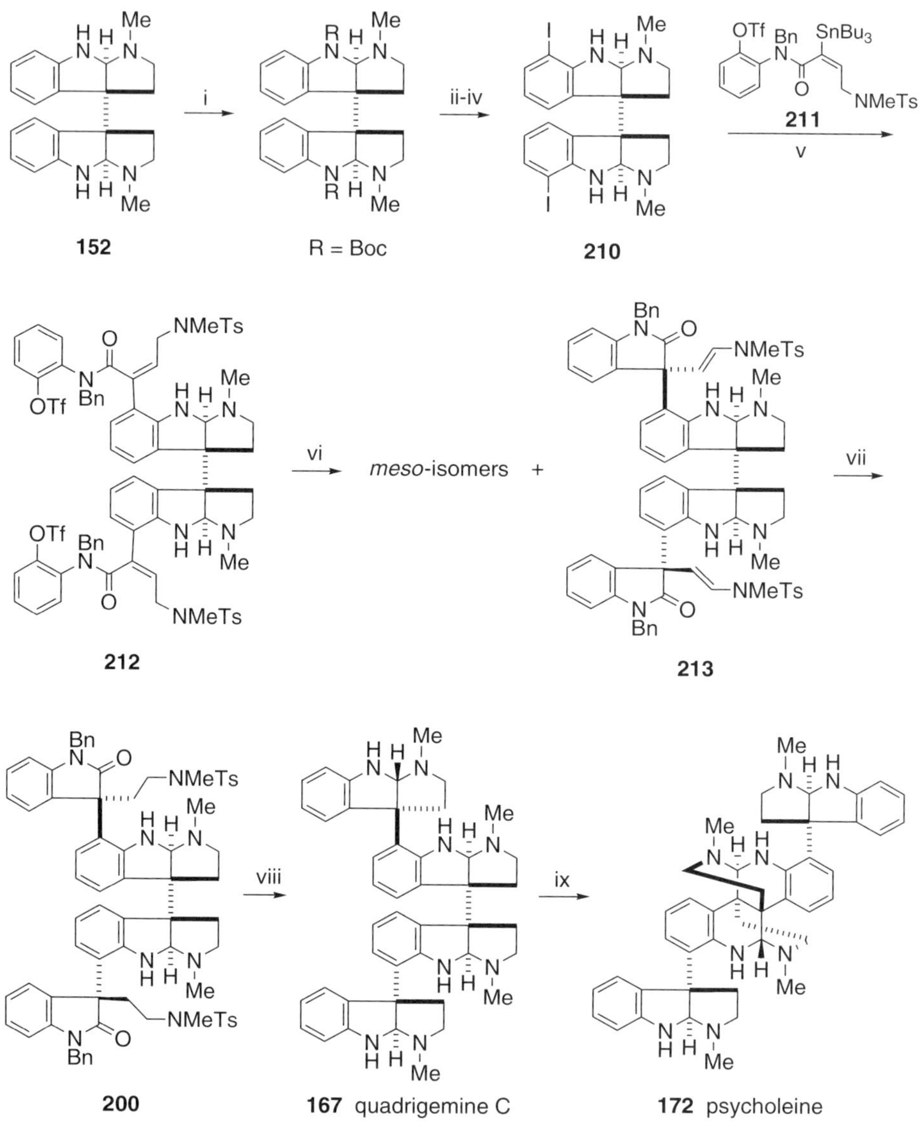


Scheme 17. Reagents: (i) $Boc_2O$, THF; NaHMDS; (ii) *s*-BuLi, TMEDA, $Et_2O$, $-78°C$; (iii) $ICH_2CH_2I$, $Et_2O$; (iv) TMSOTf, $CH_2Cl_2$; (v) $Pd_2(dba)_3 \cdot CHCl_3$, P(2-furyl)$_3$, CuI, NMP; (vi) $Pd(OAc)_2$, (*R*)-Tol-BINAP, PMP, MeCN, 80°C; (vii) $Pd(OH)_2$, EtOH, MeOH, $H_2$(100 psi), 80°C; (viii) Na, $NH_3$,THF, $-78°C$; $NH_4Cl$; (ix) 0.1 N AcOH, 100°C.

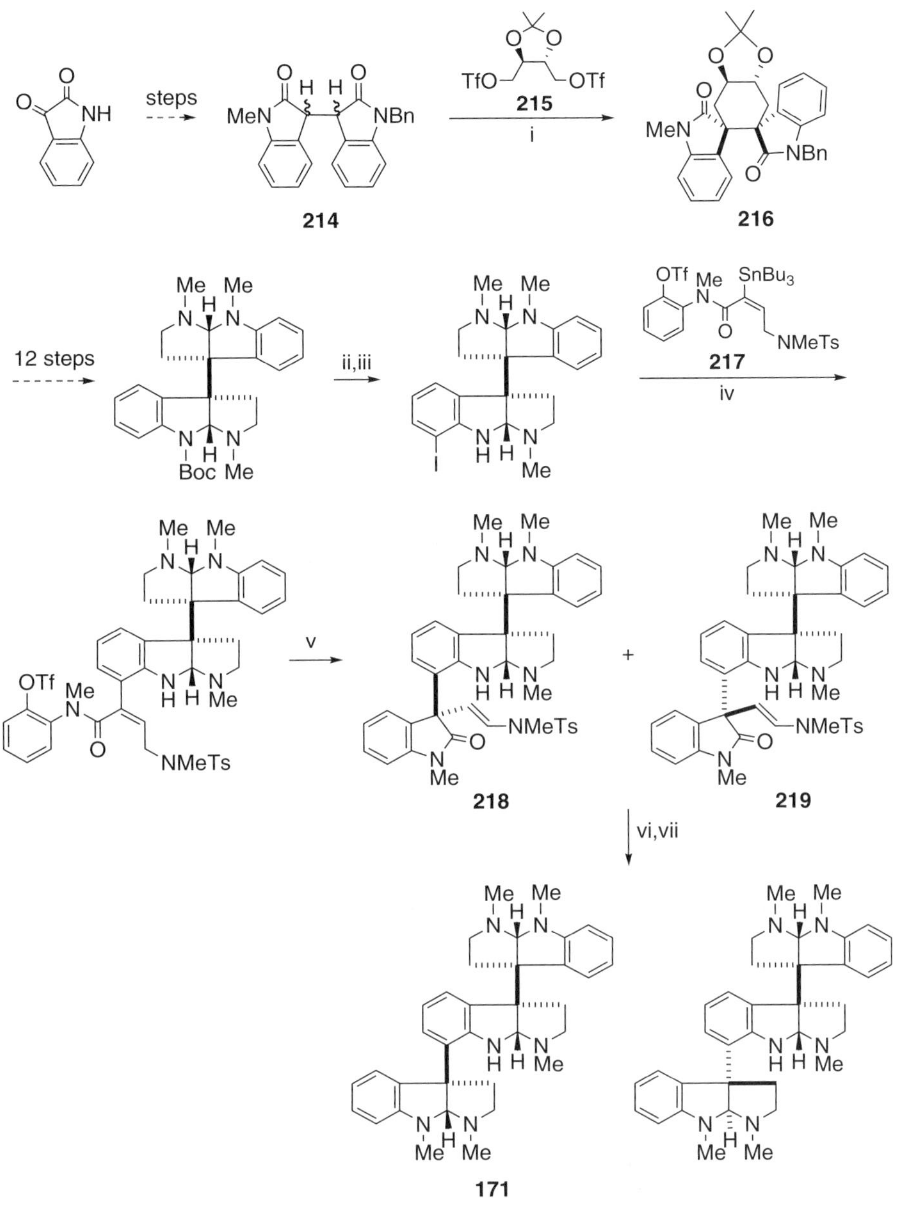

Scheme 18. Reagents: (i) LiHMDS (2 equiv.), THF/HMPA, **215**, −40°C; (ii) *s*-BuLi, TMEDA, $Et_2O$, −78°C; diiodoethane; (iii) TMSOTf, $CH_2Cl_2$; (iv) **217**, $Pd_2(dba)_3 \cdot CHCl_3$, P(2-furyl)$_3$, CuI, NMP; (v) 10 mol% $Pd(OAc)_2$, 20 mol% (*S*)-Tol-BINAP, PMP, MeCN, 80°C; (vi) $Pd(OH)_2$, 1500 psi $H_2$; (vii) Red-Al, PhMe; Na, $NH_3$, THF, −78°C.

started with *meso*-chimonanthine (**152**) (previously synthesized from commercially available oxindole and isatin in 13 steps and in 35% overall yield, or alternatively, *via* Takayama's three-step, stereorandom, hypervalent iodine-mediated dimerization

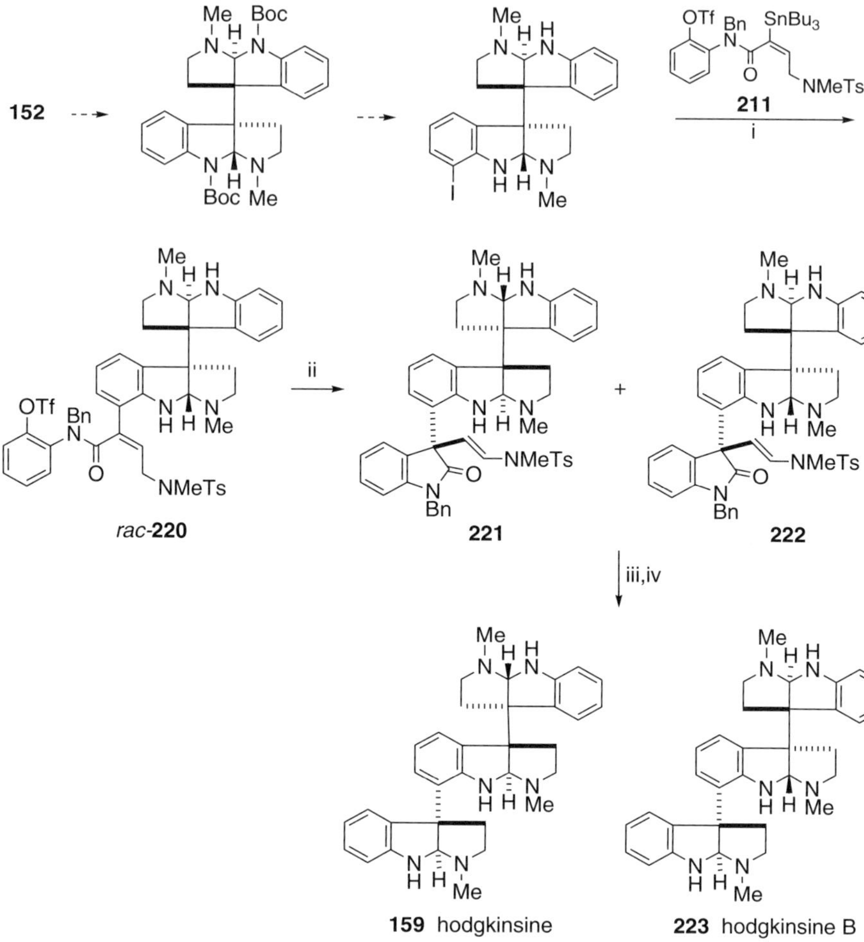


Scheme 19. Reagents: (i) **211**, $Pd_2(dba)_3 \cdot CHCl_3$, P(2-furyl)$_3$, CuI, DMA; (ii) 10 mol% $Pd(OAc)_2$, 20 mol% (*R*)-Tol-BINAP, PMP, MeCN, 80°C; (iii) $Pd(OH)_2$, 80 psi $H_2$, EtOH, 80°C; (iv) Na, $NH_3$, −78°C, $NH_4Cl$ quench.

of a tryptamine precursor giving *meso*-chimonanthine in 30% yield) and features a chemoselective double Stille cross-coupling of the diiodide **210** with stannane **211** to give the *meso* dibutenanilide **212**, in addition to the crucial, catalytic, asymmetric double Heck cyclization to deliver the $C_1$-symmetric dioxindole **213**, *en route* to quadrigemine C and psycholeine (Scheme 17).

The idiospermuline synthesis (*127,128*) (Scheme 18) required preparation of the differentially functionalized dihydroisoindigo derivative **214**, followed by the crucial alkylation of the lithium dienolate with enantiopure ditriflate **215**, to selectively produce the desired pentacycle **216** (four distinct $C_1$-symmetric dialkylation products are possible). It was found that the stereoselection in this step is crucially dependent on the choice of solvent, temperature, and concentration, to deliver the dielectrophile from the *re*-face of the dienolate, followed by cyclization of the monoalkylated product, without chelate formation. The other key steps involved attachment of the stannyl butenanilide **217** *via* a Stille reaction, followed by subsequent Heck cyclization to give the epimers **218** and **219** (6:1 mixture), which, as they were not readily separable, were advanced to the end of the synthesis, whereupon, HPLC separation gave idiospermuline **171** in 47% overall yield from the mixture of the Heck products.

*meso*-Chimonanthine (**152**), was also the starting point in the hodgkinsine and hodgkinsine B synthesis (*129*) (Scheme 19). It was converted in several steps into the racemic intramolecular Heck precursor, *rac*-**220**, which on Heck cyclization under optimized conditions (solvent, temperature, and catalyst system), provided a separable 1:1 mixture of diastereomers, **221** (48% yield, 79% *ee*) and **222** (45% yield, 83% *ee*), which could then be readily transformed into hodgkinsine (**159**) and hodgkinsine B (**223**), respectively, each in 29% overall yield. An unprecedented feature of this synthesis was the use of the asymmetric intramolecular Heck reaction for resolution (*rac*-**220** to enantioenriched **221** and **222**), resulting in the alkaloid products being obtained in enantioenriched form (ca. 80% *ee*) and in 1.5% overall yield. This first total synthesis also establishes the relative and absolute configuration of hodgkinsine B (**223**), a previously undisclosed alkaloid isolated from an Amazon *Psychotria* species (*130*).

The alkaloids of this group show a wide spectrum of biological activity. Quadrigemine C (**167**) and psycholeine (**172**) were found to be weak antagonists of the somatostatin (SRIF) receptor, constituting the first examples of non-peptide, plant-derived antagonists of the somatostatin family of receptors (*110,131*). The isolation of psycholeine was in fact guided by its inhibition of GH release as well as its binding to the somatostatin receptors (*110*). *P. colorata* has been used by the indigenous peoples of the Amazon for treating pain (*105*), and examination of the alkaloid extracts confirmed the analgesic activity (*132*). Further investigations revealed that hodgkinsine (**159**), one of the alkaloids present, produced a dose-dependent, naloxone reversible, analgesic effect in thermal modes of nociception (implicating opioid receptors), and in capsaicin-induced pain (implicating NMDA receptors) (*133*). Later SAR studies showed strong binding affinities for $\mu$ opioid receptors by (−)- and (+)-chimonanthine monourethanes, while (−)-, (+)-, and *meso*-chimonanthine, and hodgkinsine (**169**), displayed low affinity (*122*). Cytotoxic activity has been observed against rat hepatoma (HTC) and human leukemia $L_{1210}$ cells by the polyindolenine alkaloids of *P. forsteriana* and *C. milnei*, including quadrigemines A (**181**) and B (**182**), psychotridine (**170**), and isopsychotridine C (**168**) (*134,135*), hodgkinsine (**159**) and hodgkinsine A (**180**), quadrigemines G (**181**) and H (**182**), psychotridine C (**183**), isopsychotridines D and E (**168**), vatine, vatine A (**184**), vatamine (**185**), vatamidine (**186**), and calycanthine (*116*). Hodgkinsine A, quadrigemines C and H, psychotridine C, isopsychotridine E, and the four vatine alkaloids (from *C. milnei*) showed cytotoxic activity against proliferating and non-proliferating Vero African green monkey kidney cells, in addition to antibacterial, antifungal, and anticandidal activities, while only hodgkinsine A displayed substantial antiviral activity against herpes simplex type 1 and vesicular stomatitis viruses (*118*). Quadrigemines A, B, psychotridine, and isopsychotridine C from *P. forsteriana* were also found to be potent inhibitors of the aggregation of human platelets (*136*), while idiospermuline (**177**) was shown to be a cholinergic antagonist at the $\mu$M level (*113*).

## III. Tryptamine–Tryptamine Type with an Additional Monoterpene Unit and Related Alkaloids

Previous studies of *Borreria verticillata* (Rubiaceae) yielded the indole alkaloid borrerine (**224**), and the bisindole alkaloids, borreverine (**225**), and isoborreverine (**226**). A further investigation of this plant gave, in addition to these alkaloids, another new bisindole, spermacoceine (**227**), which was readily shown from the spectral data to be a derivative of borreverine, possessing a hydroxyethyl substituent at C-7, instead of the $N_b$-methyl tryptamine bridge in **225**. The clearest departure was shown by the resonances of the oxymethylene C-5 in both the $^1$H and $^{13}$C NMR spectral data of **227** when compared to those of **225** (*137*).

The bisaristone group of alkaloids, though in a strict sense not part of this group, are considered here for convenience as they are dimers of aristone (**228**), which itself originates from condensation of tryptamine with a monoterpene. The first two members of this group, bisaristones A (**229**) and B (**230**), were isolated from the Australian plant, *Aristotelia australasica* (Elaeocarpaceae) (*138*). The structures were determined from NMR and MS data. The NMR spectral data pointed to a dimeric structure from union of two aristone monomers. The point of attachment is from the indoline N(1) to C(7′) in the case of bisaristone A, and from N(1) to C(5′) in the case of bisaristone B. A possible biogenetic route to the bisaristones was proposed based on a radical-type coupling between an amino radical derived from the monomer alkaloid and the carbon-centered radicals which are mesomeric forms of the former (Scheme 20). Several partial syntheses of bisaristone B (**230**) based on the proposed radical-mediated pathway were attempted, of which only one, involving potassium ferricyanide-mediated oxidative coupling in a two-phase system, gave a satisfactory yield (71%) of **230** (*138*).

Another new bisindole belonging to this group was subsequently reported from the same plant, viz., aristoaristone (**231**), which is an isomer of the other two dimers (*139*). The NMR spectral data showed that it is constituted from the union of an aristone unit and aristolasicone (**232**) which had been previously isolated from the same plant, and which was shown (with 17-*epi*-aristoteline) to be the first examples of inverted indole alkaloids (*140*). The NMR data showed branching from the indolic N(1) of the aristone unit to C(8′) of the aristolasicone unit. Attempted semisynthesis using potassium ferricyanide, as before in the case of bisaristone B, with **228** and **232**, did not furnish the anticipated bisindole product, but instead led to a mixture of bisaristone B (**230**) and a new monomeric compound, **233**, which was shown to be the oxidation product of **232**. Similar oxidation using **232** alone gave **233** as the sole product. The $^{13}$C NMR spectral data of borrerine, borreverine, isoborreverine, and 15′-hydroxy-14′,15′-dihydroborreverine, isolated from *Flindersia fournieri*, have been reported (*141*).

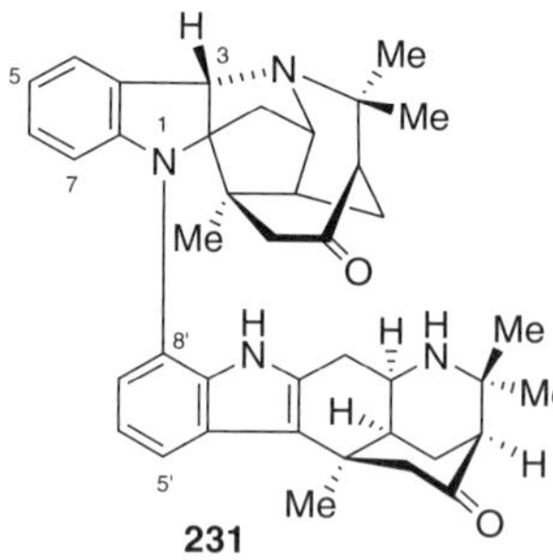

**231**

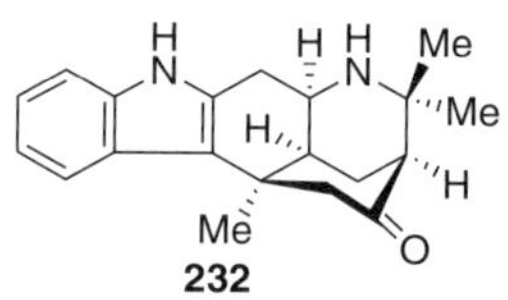

**232**

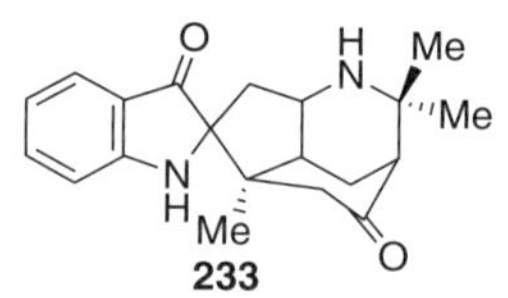

**233**

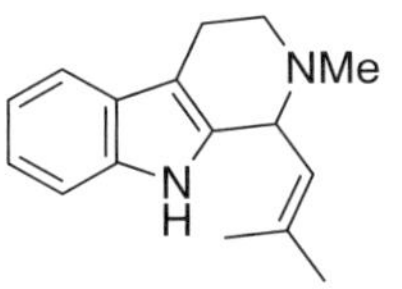

**224**

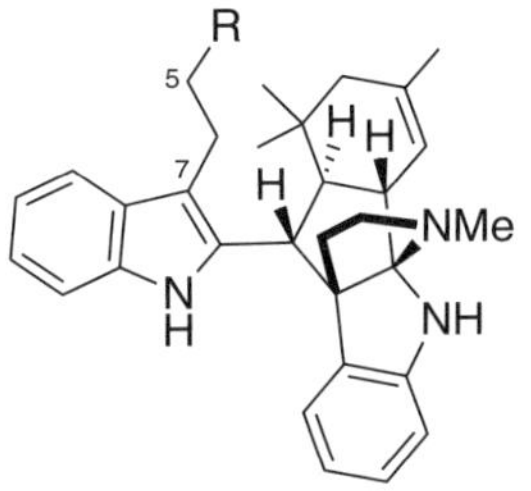

**225** R = NHMe

**227** R = OH

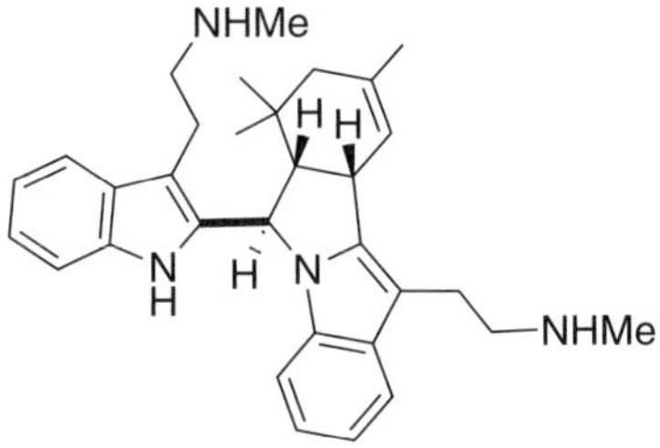

**226**

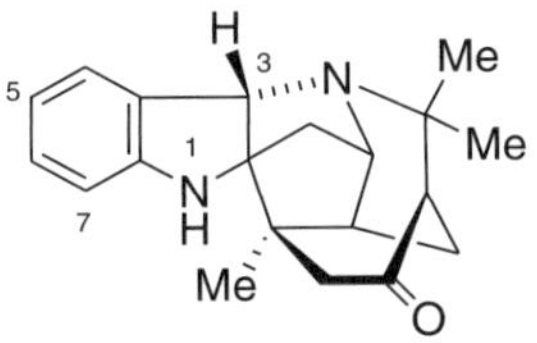

**228**

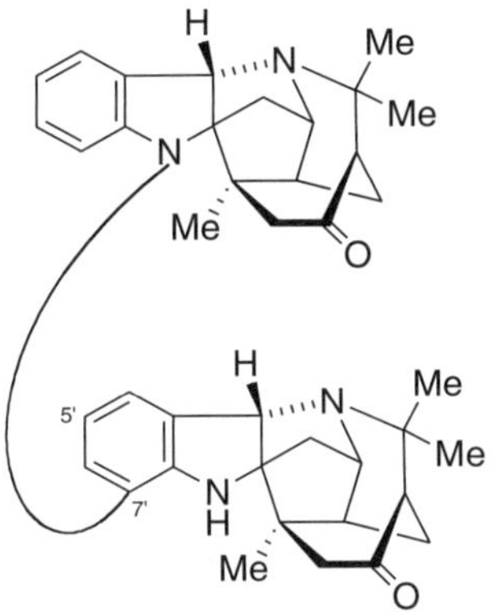

**229**

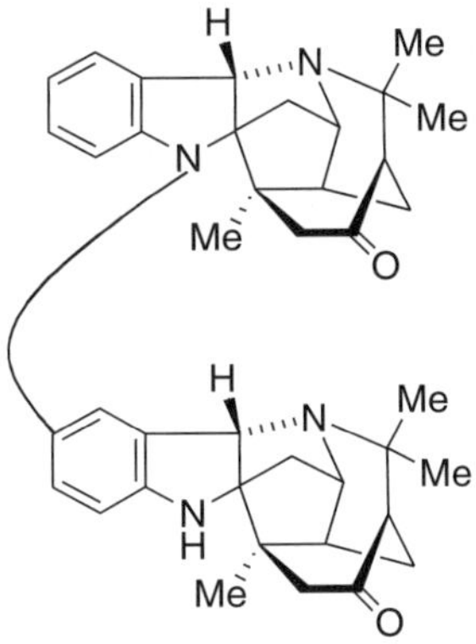

**230**

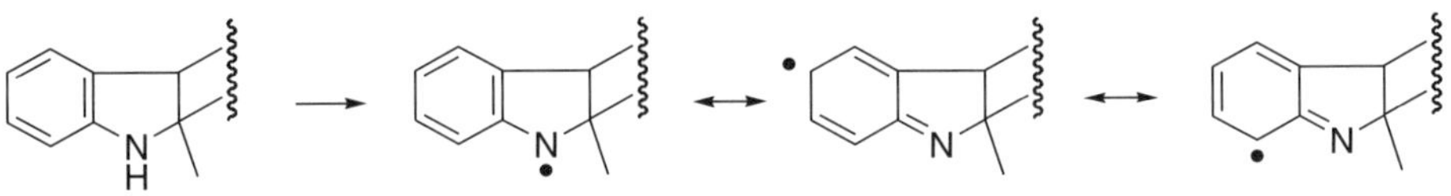

Scheme 20

## IV. Corynanthe–Tryptamine Type

Alkaloids of this group are constituted from the union of a tryptamine unit and a corynanthe moiety. They are found mainly in plants of the genus *Strychnos*, but have also been isolated from other genera of the Apocynaceae, Rubiaceae, and Loganiaceae families. Investigations of the leaves and roots of the African *Strychnos* species, *S. usambarensis* (used by the Banyambo tribesmen of Rwanda as the main ingredients of an arrow poison) have provided alkaloids of the usambarane group, exemplified by usambarine (**234**) and usambarensine (**235**), and their derivatives (*1*).

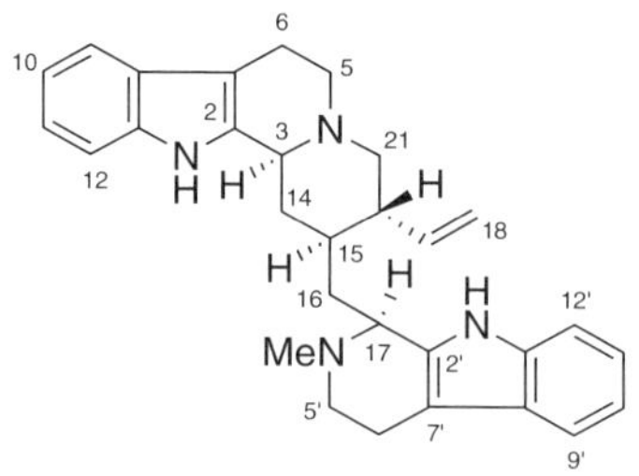

**234** usambarine

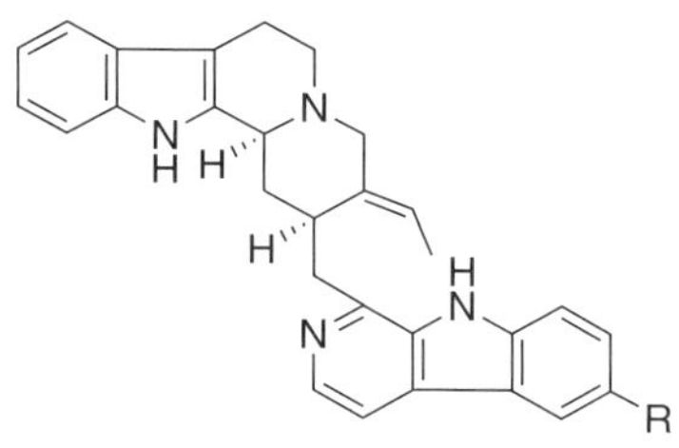

**235** R = H usambarensine
**238** R = OH

Later studies yielded the new derivatives, $N_b$-methyl-10-hydroxyusambarine (**236**), $N_b$-methyl-11-hydroxyusambarine (**237**), (*142*), and 10′-hydroxyusambarensine (**238**) from *S. usambarensis* (*143*), 4′,17-dihydro-17α-tchibangensine (17*S*-tetrahydrousambarensine) (**239**), 4′,17-dihydro-17*β*-tchibangensine (17*R*-tetrahydrousambarensine) (**240**), 10′-hydroxy-4′,17-dihydro-17α-tchibangensine (**241**), and 10′-hydroxy-4′,17-dihydro-17*β*-tchibangensine (**242**) from *S. ngouniensis* (*144*). The former two alkaloids (**241**,**242**) also occur in *Aspidosperma marcgravianum*, which also provided three other new usambarensine derivatives, $N_{b'}$-carbomethoxy-17*R*-tetrahydrousambarensine (**243**), $N_{b'}$-carboethoxy-17*R*-tetrahydrousambarensine (**244**), and the N-oxide of 17(*R*)-tetrahydrousambarensine (**245**) (*145*). These alkaloids share several common features, viz., absence of Bohlmann bands in the IR spectrum, a deshielded H(3) downfield of ca. δ 4.0, and characteristic carbon shifts, which allowed assignment of the C(3) configuration as 3α-*cis* (*S*). The $^{13}$C NMR spectral data of usambarensine, usambarine, and their derivatives have been reported (*146*,*147*), from which it was shown that nigritanine isolated from *S. nigritana* and *S. barteri* (*148*) was identical to 18,19-dihydrousambarine.

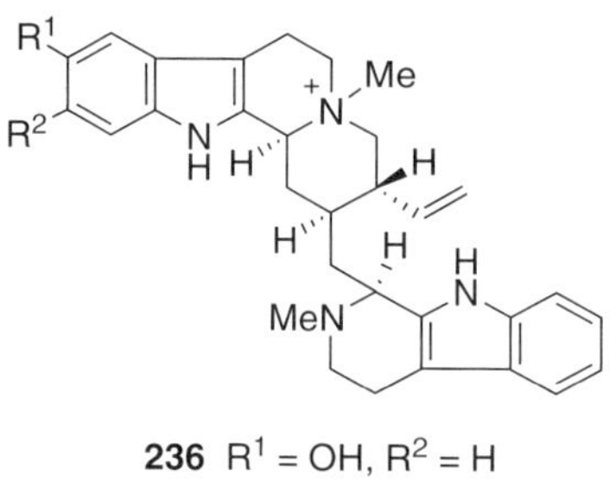

**236** $R^1$ = OH, $R^2$ = H
**237** $R^1$ = H, $R^2$ = OH

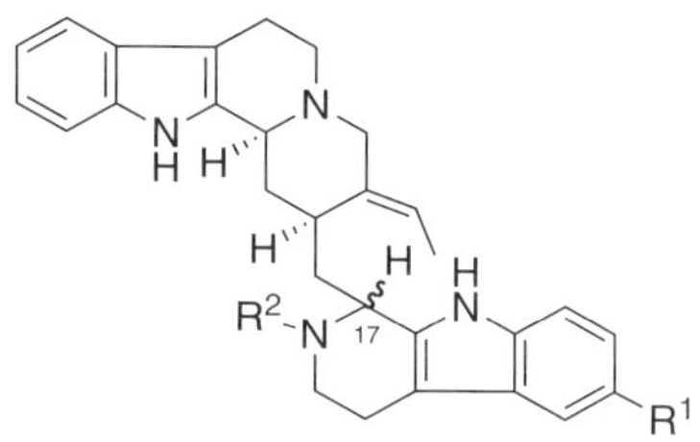

**239** $R^1 = R^2$ = H, 17-αH
**240** $R^1 = R^2$ = H, 17-βH
**241** $R^1$ = OH, $R^2$ = H, 17-αH
**242** $R^1$ = OH, $R^2$ = H, 17-βH
**243** $R^1$ = H, $R^2$ = $CO_2Me$, 17-βH
**244** $R^1$ = H, $R^2$ = $CO_2Et$, 17-βH
**245** $R^1 = R^2$ = H, 17-βH, *N*(4*S*)→O

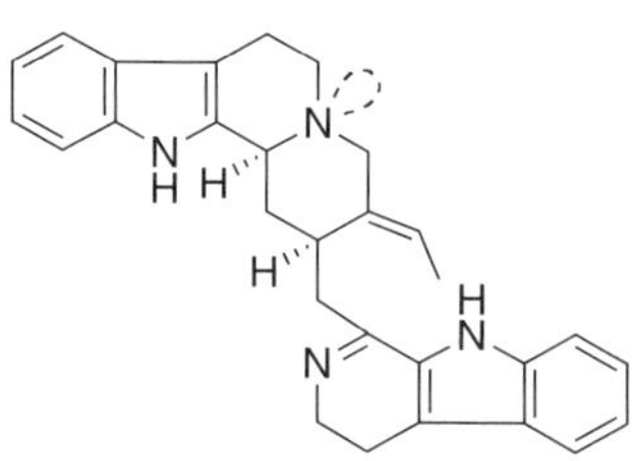

**246** tchibangensine

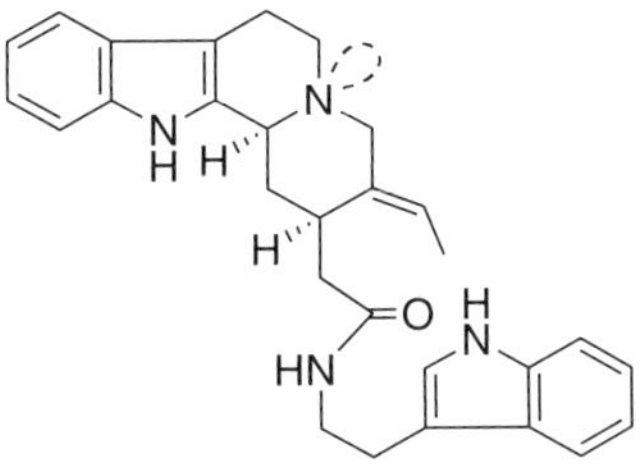

**247**

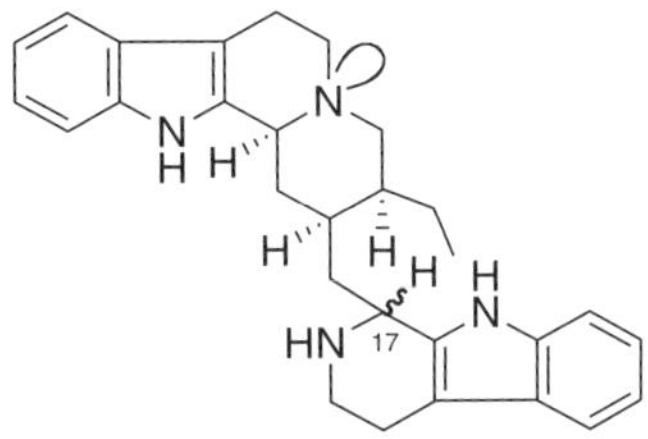

**248** H17α ochrolifuanine C
**249** H17β ochrolifuanine D

A partial synthesis of tchibangensine (**246**) has been reported (*149*) along similar lines as that described by Yamada for the synthesis of 3,4-dihydrousambarensine (*150*), i.e., by exposure of the condensation product of tryptamine and geissoschizoic acid, **247**, with $POCl_3/CHCl_3$ under reflux. Reduction of tchibangensine (**246**) with $KBH_4$ gave two derivatives, epimeric at C(17), which were separated by chromatography. On catalytic hydrogenation ($PtO_2$/MeOH), the less polar derivative gave ochrolifuanine C (**248**), while the other furnished ochrolifuanine D (**249**). Both transformations were accompanied by inversion of the nitrogen lone pair, as was previously observed in the hydrogenation of tchibangensine.

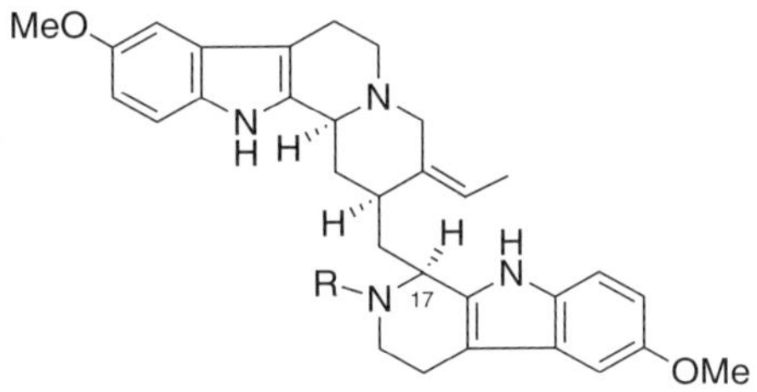

**250** R = H
**251** R = Me

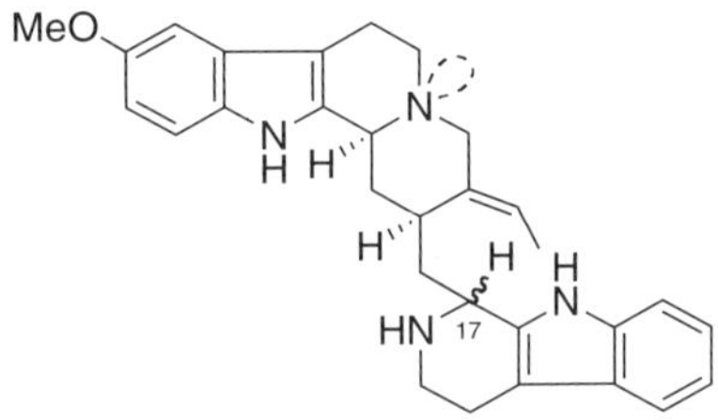

**252** H17α
**253** H17β

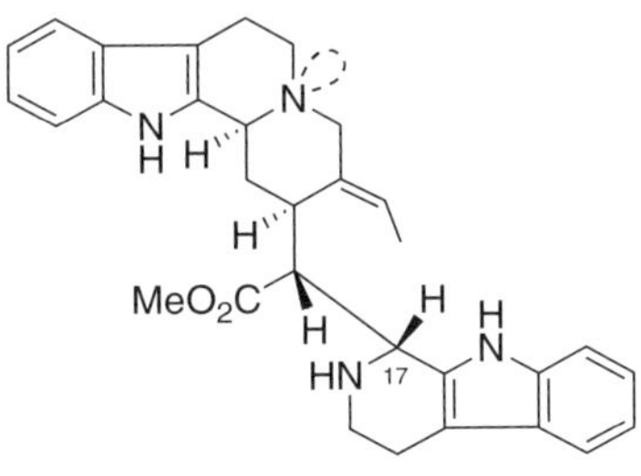

**254** buchtienine

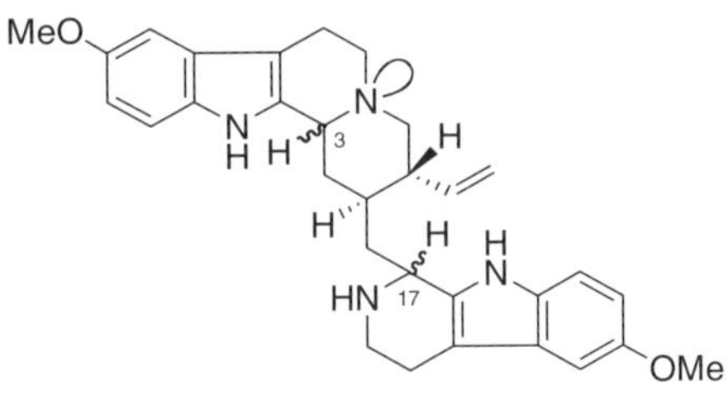

**255** H3α, H17α
**256** H3α, H17β
**257** H3β, H17α
**258** H3β, H17β

Another African *Strychnos* species, *S. dale*, furnished two new derivatives, 10,10′-dimethoxy-3*S*,17*S*-*Z*-tetrahydrousambarensine (**250**), and 10,10′-dimethoxy-$N_b$′-methyl-3*S*,17*S*-*Z*-tetrahydrousambarensine (**251**) (*151*). These two alkaloids showed spectral data (in particular the presence of 19,20-unsaturation), which placed them in the tetrahydrousambarensine group. However, the observed shift of H(3) upfield from $\delta$ 4.0 was not in agreement with previous observations of the deshielded H(3) at ca. $\delta$ 4.3 for these alkaloids possessing the 3$\alpha$-*cis* conformation of the C/D quinolizidine ring system, from which it was concluded that either a 3$\alpha$- or a 3$\beta$-*trans* quinolizidine system was present. The *trans* conformation was supported by the $^{13}$C NMR spectral data as well as the presence of Bohlmann bands in the IR spectrum, while the similarity of the CD and $^{13}$C NMR spectra to those of 3*S* and 17*S*-cinchophylline indicated that C(3) and C(17) both have the *S* configuration (H-3$\alpha$). The $^{13}$C NMR of these alkaloids are, however, different from that of 3*S*,17*S*-tetrahydrousambarensine (*145*), and moreover the 3$\alpha$-*trans* conformation is only found in alkaloids with either an ethyl or vinyl side chain. These difficulties were resolved when NOE experiments showed that the configuration of the 19,20-double bond in these compounds is no longer *E* as in the usambarensine compounds, but *Z* (NOE between H(18)/H(21); H(19)/H(16), H(17)).

Two new alkaloids ramiflorines A (**252**) and B (**253**), which are 10-methoxy derivatives of tetrahydrousambarensine, were isolated from *Aspidosperma ramiflorum* (*152*). The structure and absolute configuration of these alkaloids were established by NMR analysis and CD measurements.

Buchtienine (**254**), another new quasidimer, was isolated from the Bolivian plant, *Peschiera buchtieni* (*Tabernaemontana buchtieni*) (*153*). It has also been subsequently found in *Kopsia griffithii* from Malaysia (*154,155*). The spectral data indicated affinity with 4′,17-dihydro-17$\beta$-tchibangensine, except for the presence of a methyl ester substituent at C(16). The configuration at C(16) was assigned as *R* based on comparison of the hydrogen chemical shifts with those of 16(*R*)- and 16(*S*)-isositsirikine.

Four cinchophylline-type alkaloids, including three previously known ones, were isolated from *Cinchona ledgeriana*, viz., isocinchophylline (3$\alpha$,17$\alpha$-cinchophylline) (**255**), cinchophylline (3$\alpha$,17$\beta$-cinchophylline) (**256**), 3$\beta$,17$\alpha$-cinchophylline (**257**), and cinchophyllamine (3$\beta$,17$\beta$-cinchophylline) (**258**). The stereochemistry of these compounds was assigned based on IR, NMR, and CD spectra (*156*). The structure of isocinchophylline was previously established by X-ray diffraction analysis (*157*). Three additional new cinchophylline alkaloids, 17,4′,5′,6′-tetradehydro-3$\alpha$-cinchophylline (**259**), 17,4′-dehydro-3$\alpha$-cinchophylline (**260**) and 18,19-dihydro-3$\beta$,17$\beta$-cinchophylline (**261**) were also obtained from the same plant (*158*).

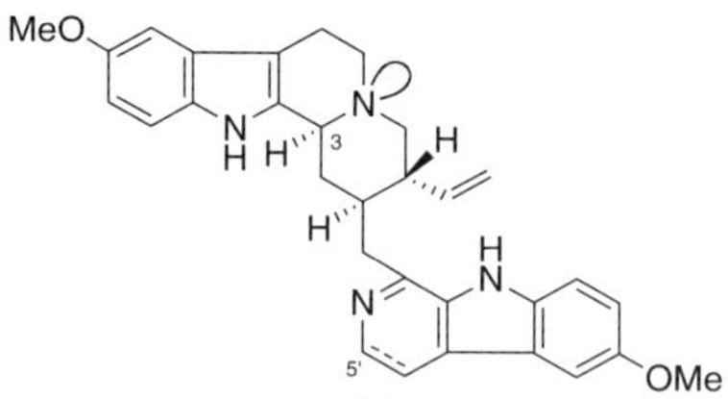


**259** $\Delta^{5',6'}$
**260**

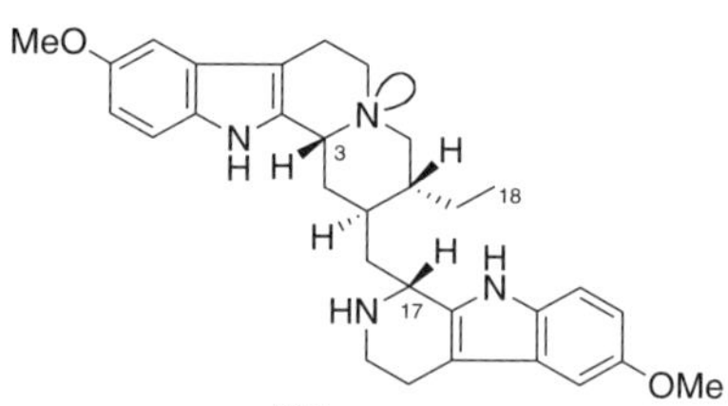


**261**

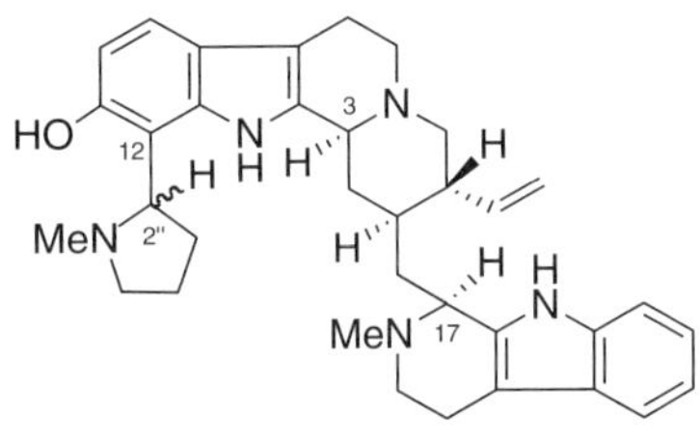

**262** H2″α strychnopentamine
**263** H2″β isostrychnopentamine

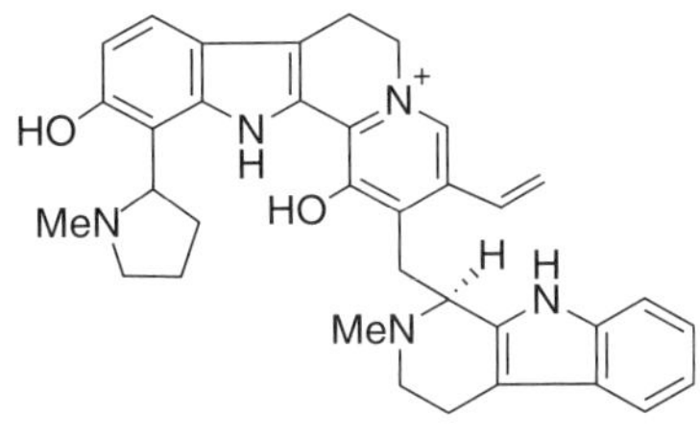

**264** chrysopentamine

*S. usambarensis* also provided two unusual quasidimeric alkaloids of the cinchophylline type, strychnopentamine (**262**) and isostrychnopentamine (**263**), which are distinguished by the presence of a pyrrolidine ring at C(12). The structure of the former was previously established by X-ray diffraction analysis (*159*), while that of isostrychnopentamine was deduced to be the C(2″) epimer of strychnopentamine based on analysis of the NMR spectral data (*160*). A third derivative, chrysopentamine (**264**), incorporating an additional hydroxy substituent at C(14), as well as aromatization of ring D, leading to anhydronium base properties, was later obtained from the same plant. Chrysopentamine showed strong antiplasmodial activity against *Plasmodium falciparum*, with $IC_{50}$ ca. 500 nM (*161*).

Six alkaloids of the quasidimeric type were isolated from *Dyera costulata*, a plant introduced into Zaire from Malaysia, the previously known ochrolifuanine A (**265**), the new ochrolifuanines E (**266**) and F (**267**), and the 18-dehydroochrolifuanines A (**268**), E (**269**), and F (**270**) (*162*). Compound **268** was readily shown to be the dehydro derivative of **265** from its spectral data as well as from its facile hydrogenation to **265**. The 3α configuration and *trans* C/D ring junction were supported by the presence of Bohlmann bands in the IR and by a positive Cotton effect in the 280–300 nm region of the CD spectrum. Alkaloids **266** and **267** are isomers of **265**, while **269** and **270** are isomers of **268**, which can be readily established through catalytic hydrogenation. Whereas C(15) is considered as a biogenetically invariant center, and the configuration at C(3) can be assigned as *β* based on the CD curves, the available data did not, however, permit definitive assignment of the stereochemistry at C(17) and C(20).

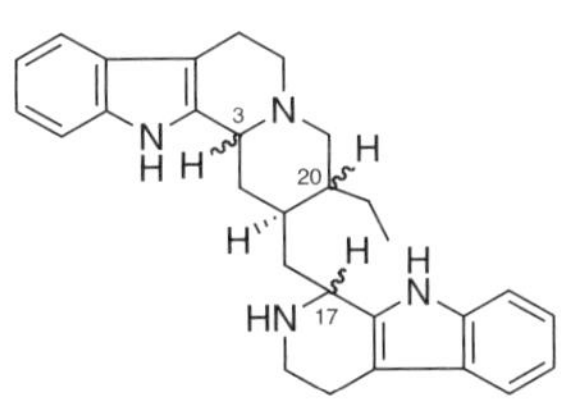

**265** H3α, H17β, H20β ochrolifuanine A
**266** H3β ochrolifuanine E
**267** H3β ochrolifuanine F
**268** $\Delta^{18,19}$, H3α, H17β, H20β
**269** $\Delta^{18,19}$, H3β
**270** $\Delta^{18,19}$, H3β

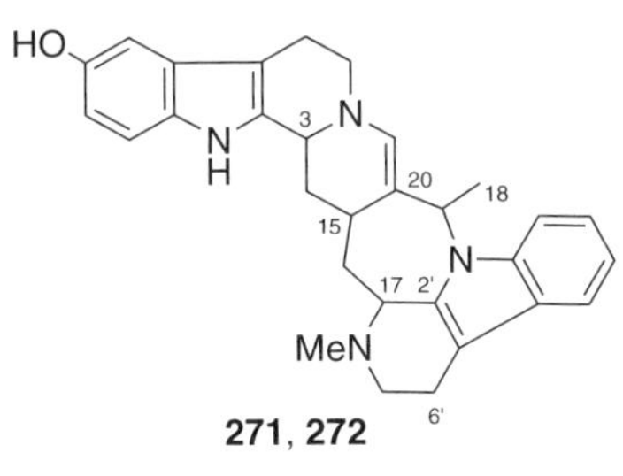

**271**, **272**

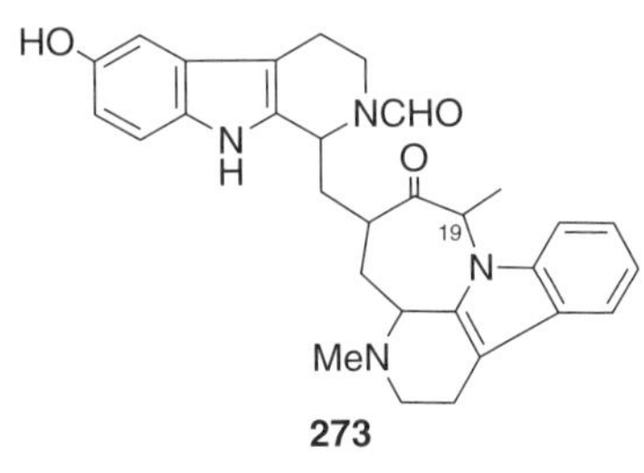

**273**

*Strychnos johnsonii*, a liana from Zaire, Central Africa, provided the novel janussines A (**271**) and B (**272**) and their oxo-derivative, oxojanussine (**273**). From a biogenetic viewpoint, these alkaloids may be considered as arising from the union of decussine and 10-hydroxytryptamine, or alternatively, from 10-hydroxycathenamine and *N*(b)-methyltryptamine. The structures were elucidated based on the spectral data, which revealed a novel arrangement comprising eight contiguous rings. However, despite intensive investigations, the stereochemistry of these compounds could not be assigned with certainty (*163,164*).

A total synthesis of strychnofoline (**277**), a bisindole constituted from the union of a tetracyclic oxindole and a tryptamine unit, and previously isolated from *S. usambarensis* (*165*), has been recently reported (*166*). A key feature of the synthesis is the convergent assembly of the tetracyclic core containing the oxindole moiety **276** through the ring-expansion reaction of a spiro[cyclopropan-1,3-oxindole] **274** and a cyclic imine **275**, as shown in Scheme 21.

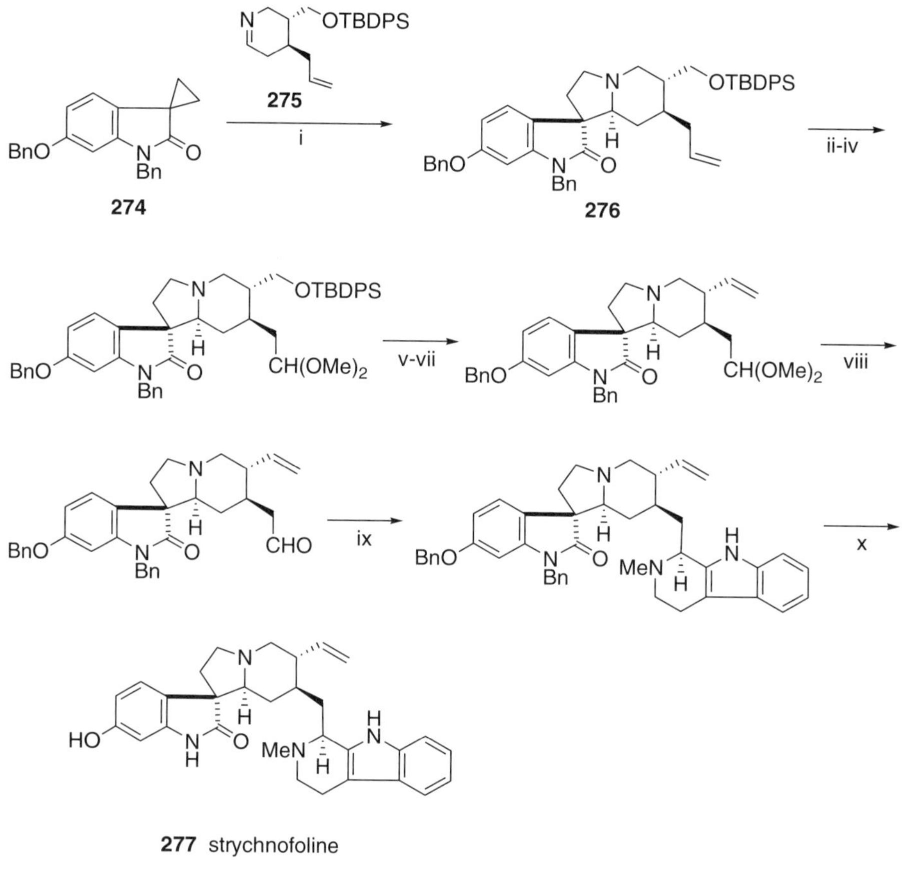


Scheme 21. Reagents: (i) **275**, $MgI_2$, THF; (ii) NMO, $OsO_4$, $H_2O$, dioxane, *t*-BuOH; (iii) $NaIO_4$, $H_2O$, dioxane, *t*-BuOH; (iv) *p*-TsOH, MeOH, $CH(OMe)_3$; (v) TBAF, THF; (vi) IBX, DMSO; (vii) *t*-BuOK, $Ph_3PMeBr$, THF; (viii) 10% aq. HCl, acetone; (ix) *N*-methyltryptamine, AcOH, PhMe; (x) Na, $NH_3$, THF, *t*-BuOH, −78°C to −45°C.

The alkaloids of this group, not surprisingly in view of their structural affinity to the well-known amoebicidal agent emetine (from *Cephaelis ipecuanha*), have been subjected to intensive screening for antiamoebic and antiplasmodial activity. Eleven ochrolifuanine- and cinchophylline-type compounds were tested against *Entamoeba histolytica*, and were found to be about 10–30 times less active compared to emetine (*167,168*). Strychnopentamine (**262**) and 5′,6′-dihydrousambarensine were highly active against *P. falciparum in vitro*, but were inactive and non-toxic against *P. berghei in vivo*. Usambarensine (**235**), usambarine (**234**), and 18,19-dihydrousambarine were highly active against *E. histolytica in vitro*, but were less active against *P. falciparum in vitro* (*169,170*). Strychnopentamine (**262**), isostrychnopentamine (**263**), and ochrolifuanine A (**265**), showed high *in vitro* potency against both chloroquine-sensitive, as well as chloroquine-resistant, strains of *P. falciparum*, while 5′,6′-dihydrousambarensine showed a 30-fold increase in potency against the resistant strain compared to the drug-sensitive strain (*171–173*). Buchtienine (**254**) from *K. griffithii* was found to show significant activity against *Leishmania donovani* (*155*). The observed leishmanicidal activity of the alkaloid is consistent with the use of the Bolivian plant for the treatment of leishmaniasis (*153*).

A total of 34 alkaloids of the quasidimeric type from *S. dale*, *S. tchibangensis*, and *S. ngouniensis*, were tested for antimicrobial activity. Almost all of the alkaloids showed activity against the Gram-positive bacteria (*Staphylococcus aureus*, *Bacillus subtilis*), while only three were active against *E. coli* (*174*). The alkaloids of the usambarane group also showed cytotoxic activity against B16 melanoma, L1210, and HeLa cells, with strychnopentamine (262) being the most potent (*175,176*).

The cytotoxicity of strychnopentamine (**262**) has been further assessed against various cell lines (Hepatoma HW165, ascites Ehrlich, B16 melanoma) *in vitro* (*177–179*) as well as against Ehrlich ascites tumor (ELT) *in vivo* (*180*). The latter study showed a significant decrease in the number of tumor cells as well as a significant increase in survival of the treated mice. Isostrychnopentamine (**263**), on the other hand, was found to induce cell cycle arrest as well as apoptosis in human colon cancer cells (*181*), while usambarensine (**235**) was shown to intercalate into DNA, inducing apoptosis in human leukemia HL60 cells (*182*).

## V. Corynanthe–Corynanthe Type

*Uncaria callophylla* provided three bisindole alkaloids, callophylline (**278**), callophylline A (**279**), and callophylline B (**280**), constituted from union of heteroyohimbine and yohimbine moieties (*183*). Gambirine (**281**), a tetracyclic heteroyohimbine unique to *U. callophylla* (as is callophylline) (*183,184*), was the common unit present in all three bisindoles. The structures of these alkaloids were established based on their spectral data. Callophylline has been previously reported by Goh (*185,186*), and is identical to uncaramine, for which a detailed NMR study has also firmly established its structure (*187*). A novel bisindole strychnofuranine (**282**) was isolated from *S. matopensis*. The structure was determined based on the spectral data, which revealed its constitution from the union of two geissoschizal units, with C(16) and C(17) condensed into a furan ring (*188*).

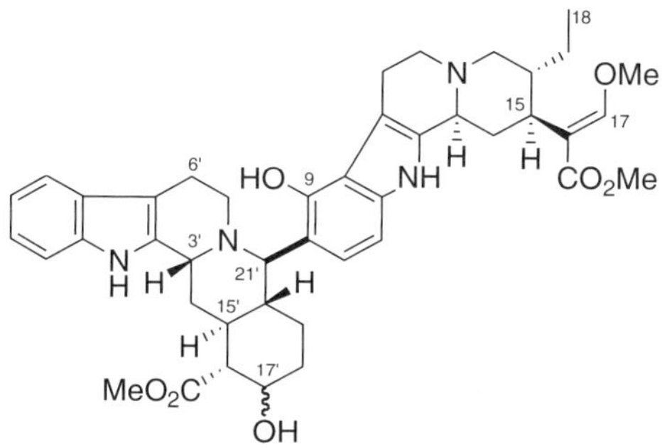

**278** 17-αOH callophylline (uncaramine)
**279** 17-βOH callophylline A

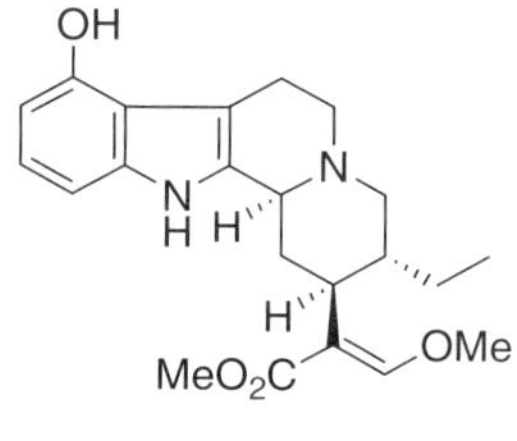

**281** gambirine

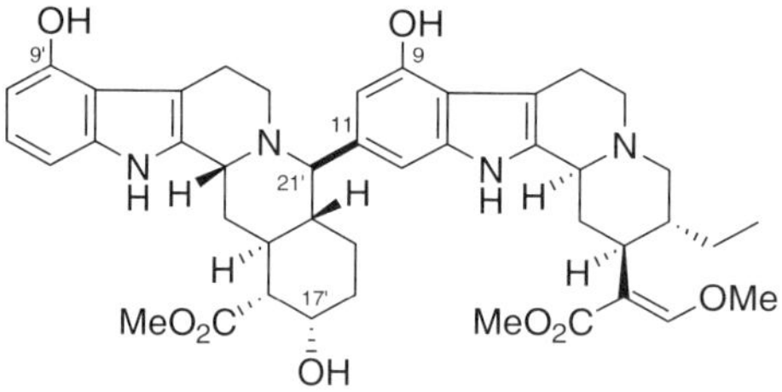

**280** callophylline B

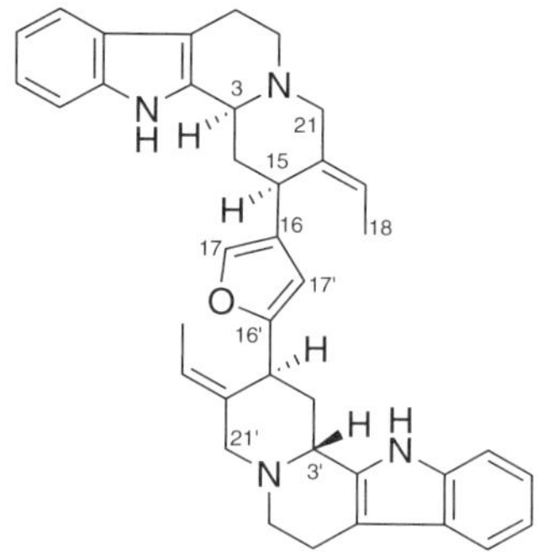

**282** strychnofuranine

## VI. Corynanthe–*Strychnos* Type

The bisindole alkaloid longicaudatine (**283**), first detected in various *Strychnos* species by its characteristic reaction with ferric chloride or Ce(IV) reagents (blue coloration in TLC), was obtained in a number of instances with its isomer, bisnor-C-alkaloid H (**284**), which also showed the same chromatographic and chromogenic properties. This, coupled with the small amounts obtained in the earlier studies, has been the cause of some initial confusion, but the structures of both alkaloids were eventually published in a joint disclosure (*189*). The structure of **283** was elucidated mainly on the basis of the NMR spectral data, which revealed constitution from strychnine (**285**) and geissoschizine (**286**) halves. In addition, the carbon shifts of the strychnan half displayed resemblance to the spectrum of bisnordihydrotoxiferine (**287**). The NMR data also indicated that the relative configuration of the bisindole follows that present in the constituent monomers from which it is derived, consistent with the suggested origin of longicaudatine as arising *via* the condensation of geissoschizal (**288**) and Wieland–Gumlich aldehyde (**289**). The same report also included the confirmation of the structure of bisnor-C-alkaloid H (**284**) as the 18-hydroxy derivative of bisnordihydrotoxiferine (**287**). Longicaudatine (**283**) has been subsequently found in various other *Strychnos* species, and was the most abundant alkaloid of the root-bark of *S. longicaudata* (*144*). It showed potent *in vitro* antiplasmodial activity against both chloroquine-sensitive, as well as chloroquine-resistant strains of *P. falciparum* (*172*).

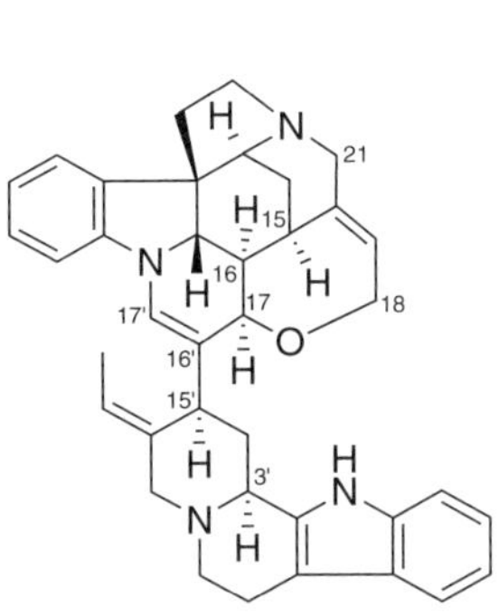

**283** longicaudatine

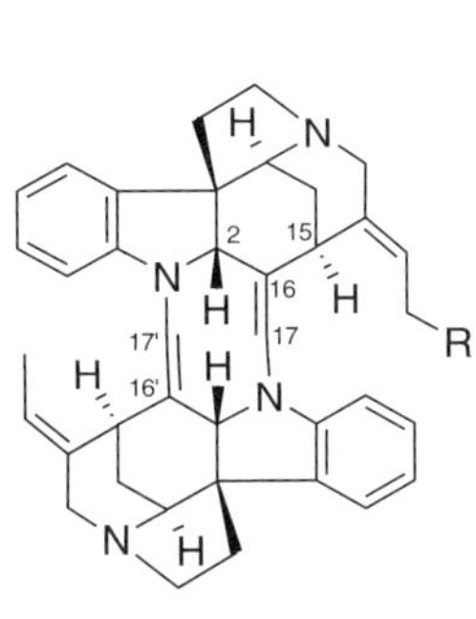

**284** R = OH bisnor-C-alkaloid H
**287** R = H bisnordihydrotoxiferine

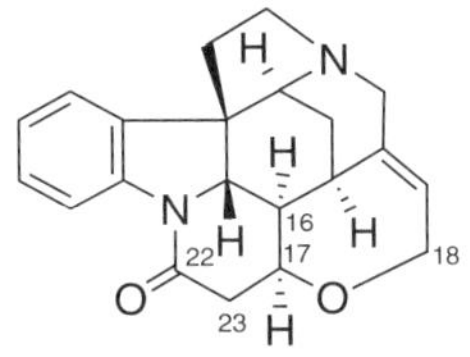

**285** strychnine

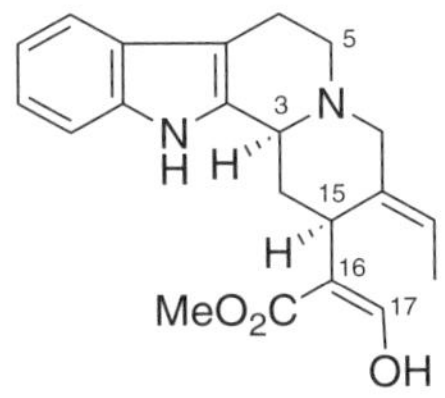

**286** geissoschizine

Two additional congeners of longicaudatine have been isolated from *S. longicaudata*, longicaudatines Y (**290**) and F (**291**) (*144*). Longicaudatine Y (**290**) differs from longicaudatine by the presence of two ethylidene side chains, in addition to a change in the stereochemistry of C(17), from the observed $J_{16-17}$ coupling of 10 Hz (compared to 2.5 Hz in longicaudatine).

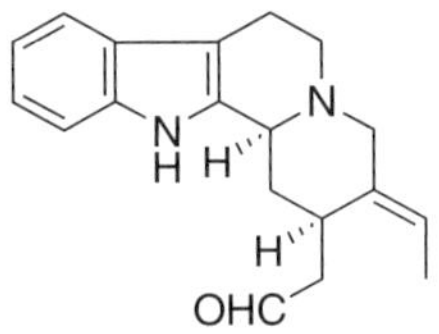

**288** geissoschizal

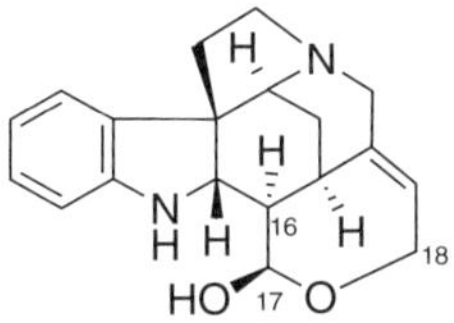

**289** Wieland-Gumlich aldehyde

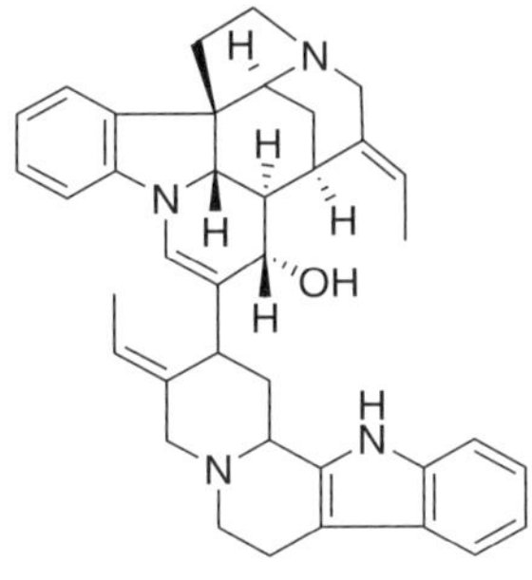

**290** longicaudatine Y

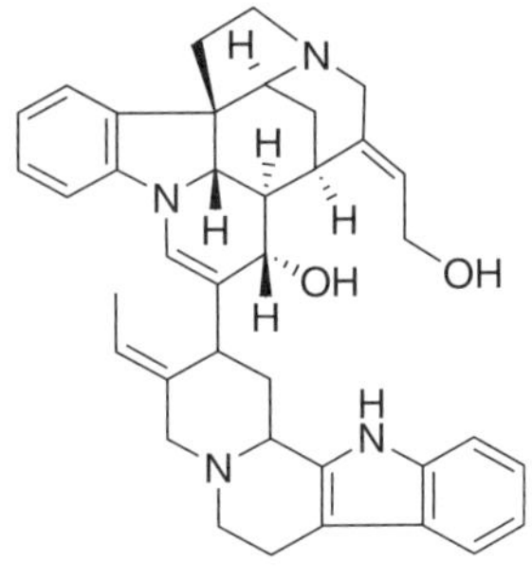

**291** longicaudatine F

Longicaudatine F (**291**) showed the same change in the stereochemistry at C(17), but has one ethylidene group replaced by a C=CHCH$_2$OH unit. The dihydroderivatives of both longicaudatine and longicaudatine Y, (**292**) and (**293**), respectively, have been subsequently found in *S. potatorium* (*190*). The $^1$H NMR spectra of both alkaloids differed from that of longicaudatine by the presence of signals due to an ethyl side chain in place of an ethylidene, while the main difference in the spectra of the two alkaloids concerns the signals for H(18) and H(19) of the strychnan half, which allowed assignment of structures **292** and **293** to dihydrolongicaudatine and dihydrolongicaudatine Y, respectively.

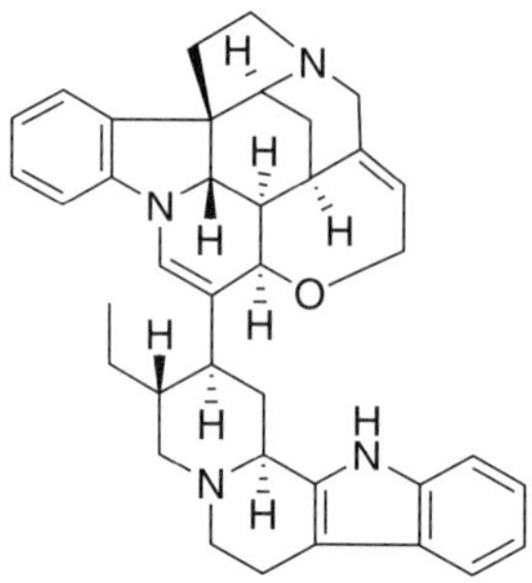

**292** dihydrolongicaudatine

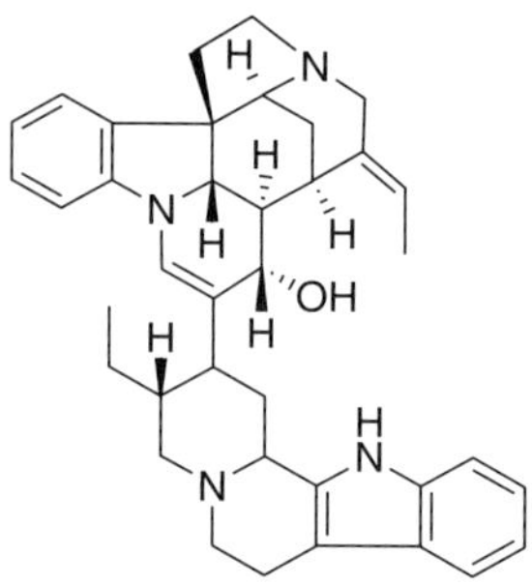

**293** dihydrolongicaudatine Y

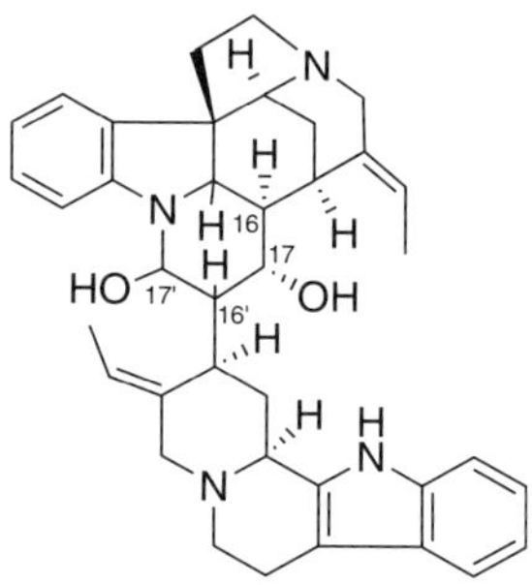

**294** longicaudatine Z

Longicaudatine Z (**294**) from *S. matopensis* differs from the other longicaudatines by the presence of a hydroxy function on C(17′) as shown by the presence of oxymethine signals in the NMR spectrum (*188*). Assignment of the stereochemistry (17′*S*, 16′*S*, 17*S*) was from the observed $J_{16'-17'}$ value of 3 Hz, and the long range interaction between H(17′) and H(17) in the COSY spectrum. The N(4)-oxide of longicaudatine (*N*-oxylongicaudatine) previously found in *S. chrysophylla* (*191*), was also obtained from this plant.

*Strychnos* species were also sources of a number of quaternary alkaloids of the longicaudatine type, for example, 3′,4′,5′,6′-tetradehydrolongicaudatine Y (**295**), from *S. usambarensis* (*192*). The UV spectrum indicated a *β*-carboline chromophore in neutral or acidic solutions, while a bathochromic shift suggestive of a zwitterionic species was observed in alkaline solutions (*193*). The NMR spectral data showed the presence of vicinal, pyridine-type hydrogens, in place of the usual H(5′) and H(6′) signals of the longicaudatines, and two deshielded geminal hydrogens, attributed to the two H(21′) which are adjacent to the quaternary nitrogen. The $J_{16-17}$ coupling of 10 Hz was similar to that in longicaudatines Y, F, and Z, indicating a similar stereochemistry at C(17). Finally, treatment of **295** with $NaBH_4$ gave longicaudatine Y (**290**). The structure and stereochemistry of afrocurarine (**296**), a bisquaternary alkaloid, previously isolated from *S. usambarensis* (*194*), have been established based primarily on analysis of the $^1H$ NMR data. Its difficult isolation was greatly facilitated by the application of ion-pair reverse-phase column chromatography (*195*). A reinvestigation of the famed *S. nux-vomica* provided a new quaternary alkaloid, the orange-colored strychnochrysine (**297**) from the roots (*196*). The UV spectrum is similar to that of afrocurarine (**296**) and is consistent with a highly conjugated chromophore. The elucidation of the structure was based mainly on the 2-D NMR spectral data. A number of new quaternary bisindoles were isolated from *S. guianensis* from the Brazilian Amazonia, including the zwitterionic guianensine (**298**) (*197*), and the bisquaternary alkaloids, guiaflavine (**299**) (*198*), guiachrysine (**300**) (*199*), 5′,6′-dehydroguiachrysine (**301**), and 5′,6′-dehydroguiaflavine (**302**) (*200*).

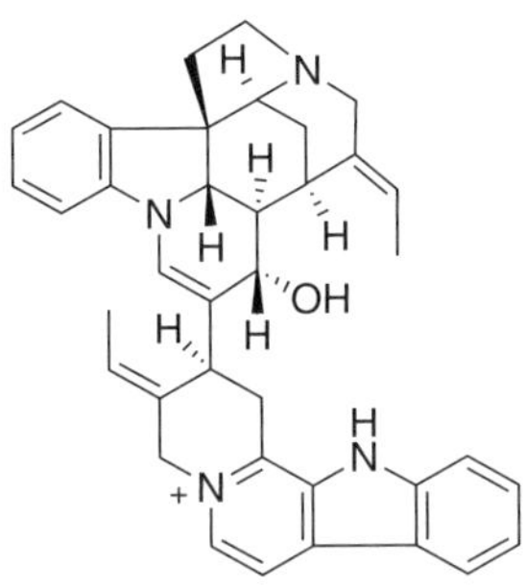

**295**

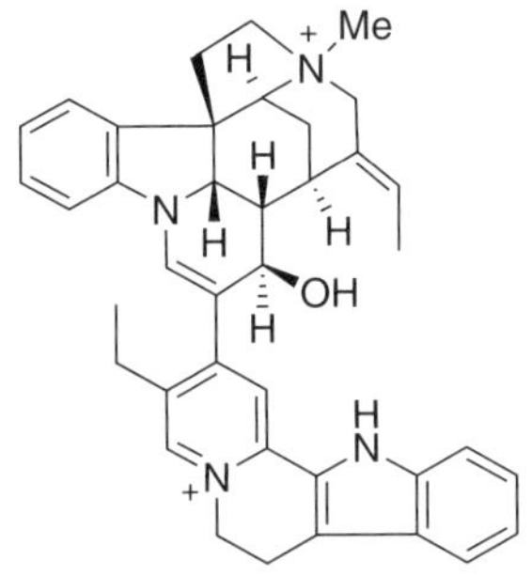

**296** afrocurarine

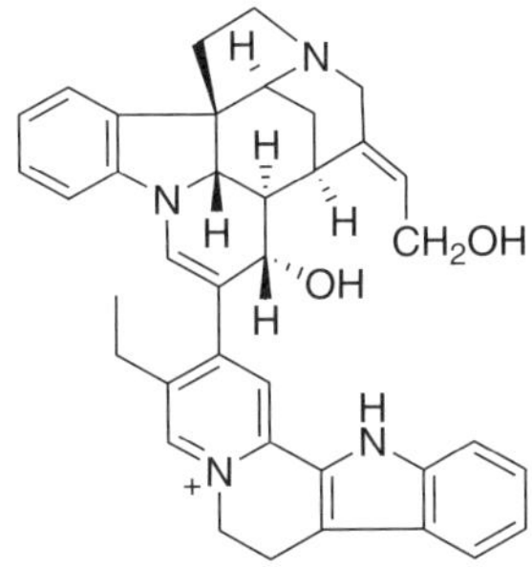

**297** strychnochrysine

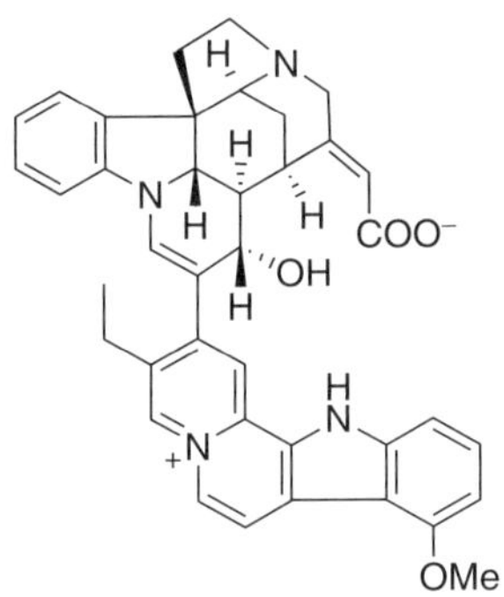

**298** guianensine

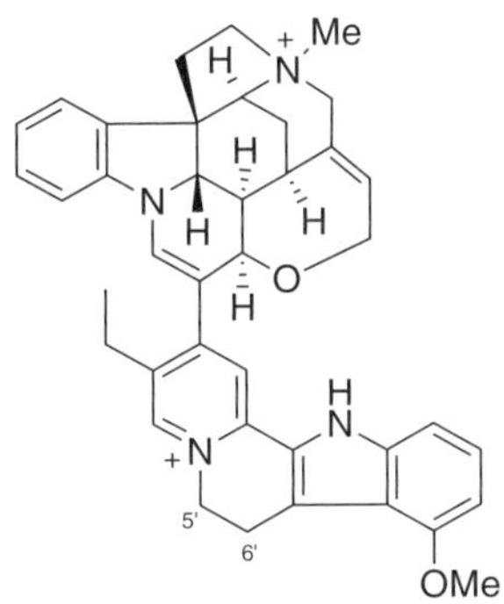

**299** guiaflavine
**302** $\Delta^{5',6'}$

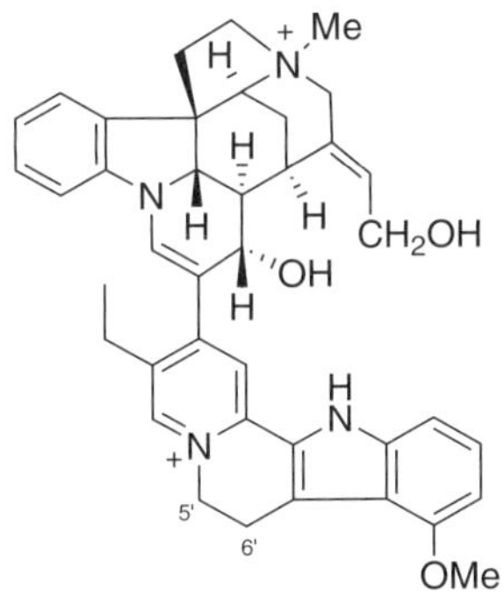

**300** guiachrysine
**301** $\Delta^{5',6'}$

Several of the quaternary bisindoles from *S. guianensis*, guiachrysine (**300**), guiaflavine (**299**), and their respective 5′,6′-dehydro analogues **301** and **302**, were found to show inhibitory effects on neuromuscular transmission in isolated mouse hemidiaphragm preparations, with guiachrysine (**300**) showing the greatest potency, although the potency was only about half of that shown by a Venezuelan calabash curare (based on *S. toxifera*) (*200*). These bisindoles were also found to be effective antagonists of nicotinic acetylcholine receptors in cultured human cells (*201*). Using a human clonal cell line (TE671) expressing muscle-type nicotinic acetylcholine receptors, it was shown that guiachrysine (**300**), guiaflavine (**299**), and 5′,6′-dehydroguiaflavine (**302**), were effective antagonists of the carbamylcholine-dependent, receptor-mediated, $^{86}Rb^+$ efflux, with the former showing the highest potency ($IC_{50}$ ca. 0.43 μM in the presence of 0.5 mM carbamylcholine). Longicaudatine (**283**) was found to antagonize carbachol- and histamine-induced contractions of the guinea pig ileum and bradykinin responses in the rat uterus, but had no effect on voltage-dependent calcium ion channels. In the aorta, the alkaloid antagonized the intracellular calcium-dependent transient contractions of noradrenaline, and was found to be about 133 times more potent than procaine, a known inhibitor of the release of calcium ions from intercellular stores (*202*).

## VII. Vobasine–Tryptamine Type

Ceridimine (**303**), from *Pagiantha cerifera* was the first example of a quasidimeric-type alkaloid in which the additional tryptamine unit has a free side chain and is linked by its aromatic moiety to the monoterpenoid indole unit (*203*). The primary amine function was readily acetylated and methylated, and its partial synthesis *via* coupling of vobasinol with tryptamine confirmed the structure assignment. Demethylceridimine (**304**) was later obtained from *P. buchtieni* (*T. buchtieni*), which also provided another quasidimer, buchtienine (**254**) (*vide supra*) (*153*). The third member of this group, hunteriatryptamine (**305**), was obtained from *Hunteria zeylanica* from Thailand. It differs from the previous two alkaloids by the presence of a hydroxymethyl function on the vobasinyl C(16) in place of H (*204*).

## VIII. Vobasine–Vobasine Type

The structure of the alkaloid gardmultine (**306**), previously isolated from the Japanese plant *Gardneria multiflora* (Loganiaceae) (*205*), has been confirmed in a subsequent full report which also included the isolation of the demethoxy derivative (**307**) (*206*). Concurrently, an X-ray crystallographic study was also carried out, providing additional confirmation of the structure and absolute configuration (*207*). The alkaloid is constituted from the union of chitosenine (**308**) and gardneramine (**309**) moieties, linked by a spirocyclic five-membered ring. A striking feature of the $^1$H NMR spectrum is the signal due to the highly shielded 18′-methyl ($\delta$ 0.88) which arises as a consequence of the structure and stereochemistry of the alkaloid, resulting in placement of this methyl within the shielding zone of the aromatic ring of the gardneramine unit (*205–207*).

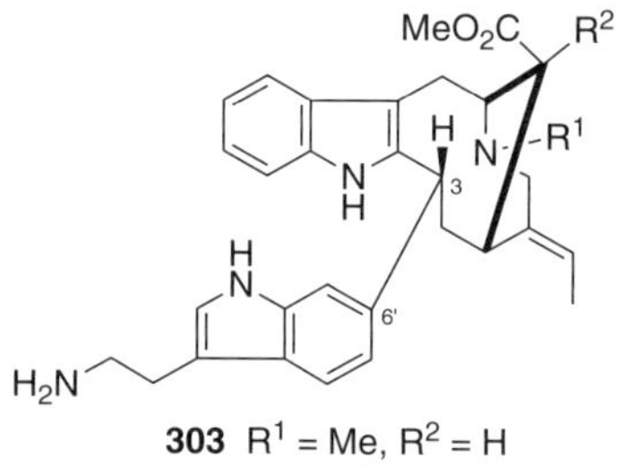


**303** $R^1$ = Me, $R^2$ = H
**304** $R^1$ = $R^2$ = H
**305** $R^1$ = Me, $R^2$ = $CH_2OH$

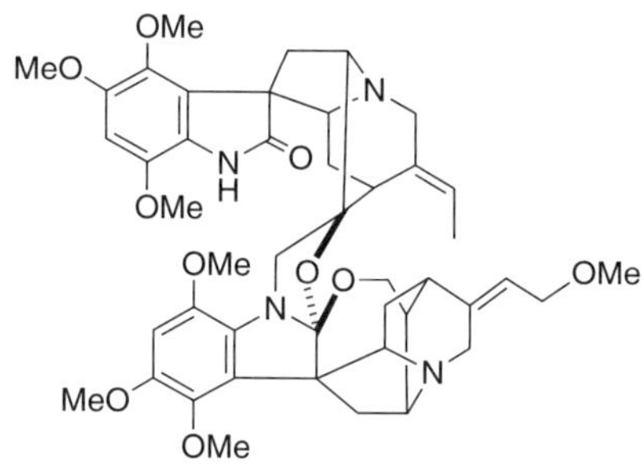


**306** gardmultine

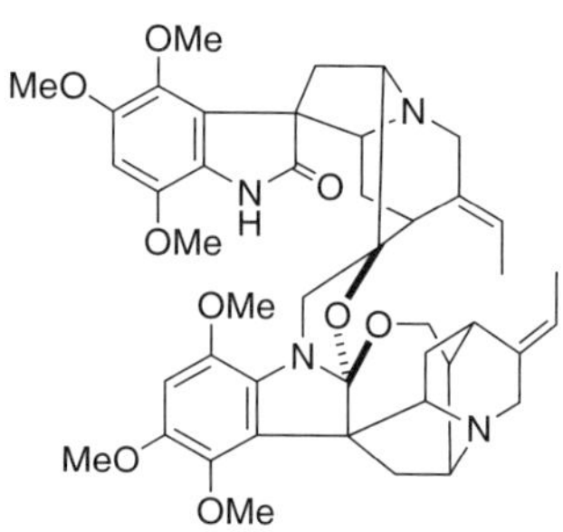


**307** demethoxygardmultine

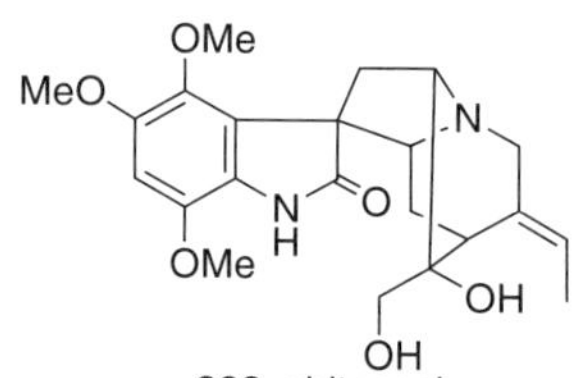


**308** chitosenine

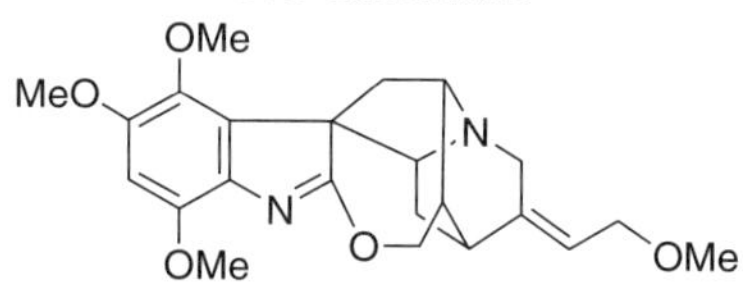


**309** gardneramine

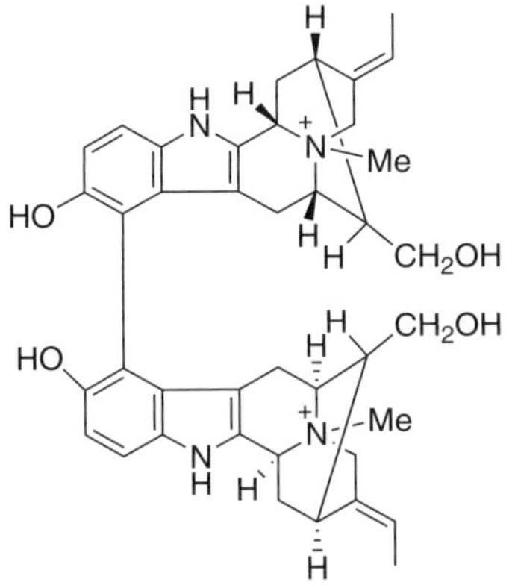

**310** dispegatrine

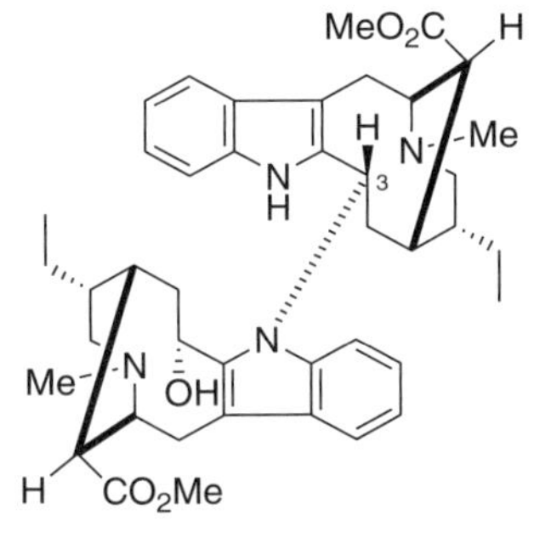

**311** hazuntamine

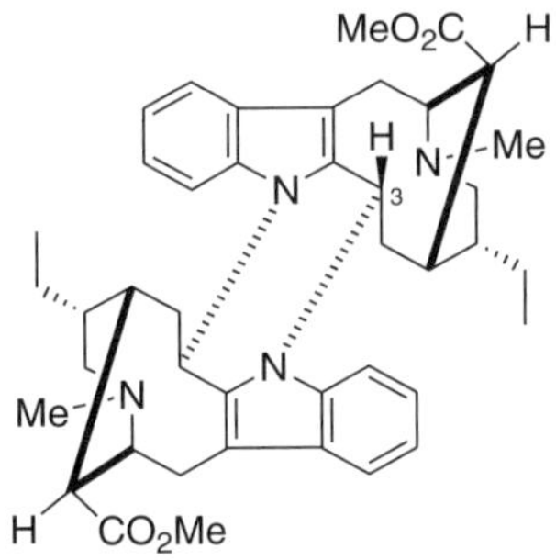

**312** anhydrohazuntamine

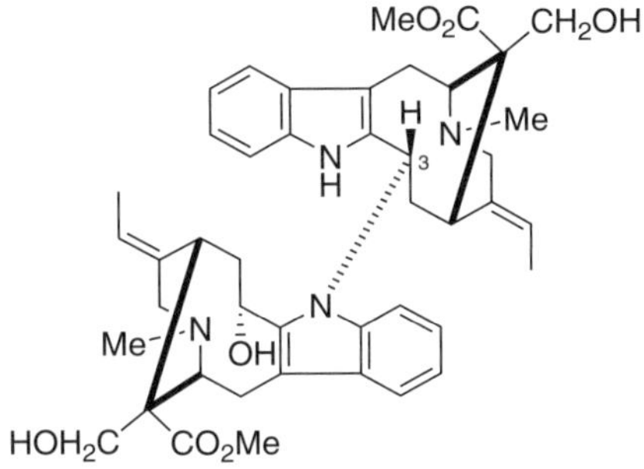

**313** vobasonidine

A bisquaternary alkaloid, dispegatrine (**310**), was isolated from the water-soluble fraction of the roots of *Rauvolfia verticillata hainannensis* (*208*).

The root-bark extract of *Hazunta modesta* from Madagascar provided a new bisindole, hazuntamine (**311**), constituted from the union of two dregamine-like units (*209*). The NMR data indicated that the monomeric units are linked from N(1′) of dregamin-3-α-ol, a new indole alkaloid which also occurred in the plant, to C(3) of a 3-deoxydregamine (or 3-deoxy-19,20-dihydrovobasinol) unit. The C(3) substitution was deduced to be α from a consideration of the observed coupling for H(3). In the event, treatment of hazuntamine with trifluoroacetic acid gave the symmetric bisindole, named anhydrohazuntamine (**312**), a dehydration product of hazuntamine, its formation providing further support for the proposed structure of hazuntamine. A second N(1′)–C(3) linked vobasine–vobasine bisindole, vobasonidine (**313**), was obtained from *Tabernaemontana corymbosa* from Malaysia (*210*). The NMR and MS data indicated constitution from two vobasine-type units. One unit of the bisindole corresponds to that of a 3-deoxy-16-hydroxymethylvobasinol moiety, similar to that found in conodiparine A, another bisindole previously reported from the same plant (*211*), while the other, except for the presence of a hydroxymethyl substituent at C(16′), corresponds to vobasinol. The presence of only one H(3) in the former, and the presence of only one indolic NH, indicated a C(3)–N(1′) linkage, which was also confirmed by the observed H(3) to C(2′) long-range heteronuclear correlation in the HMBC spectrum. The substitution at C(3) was deduced to be α from the observed $J_{3-14}$ coupling of 13.5 Hz and the observed NOE between H(3) and NH.

*P. buchtieni* (*T. buchtieni*) also provided the 3-10′ linked, perivinol-affinisine bisindole, demethylaccedinisine (**314**), in addition to buchtienine (**254**) and demethylceridimine (**304**) (*153*). This alkaloid was also isolated from *P. van heurkii* and the structure confirmed by methylation, which yielded accedinisine (*212*).

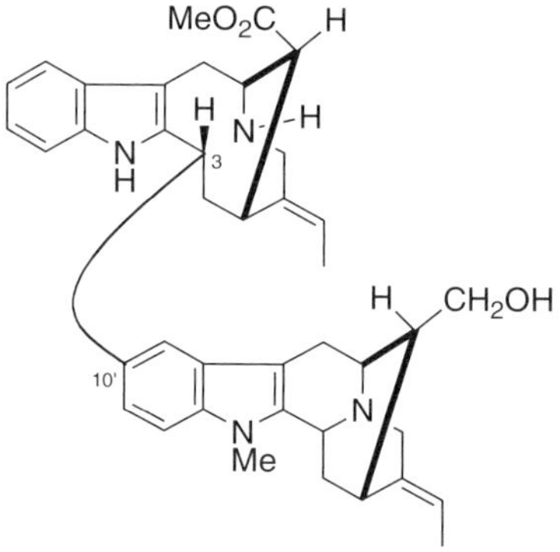


**314** demethylaccedinisine

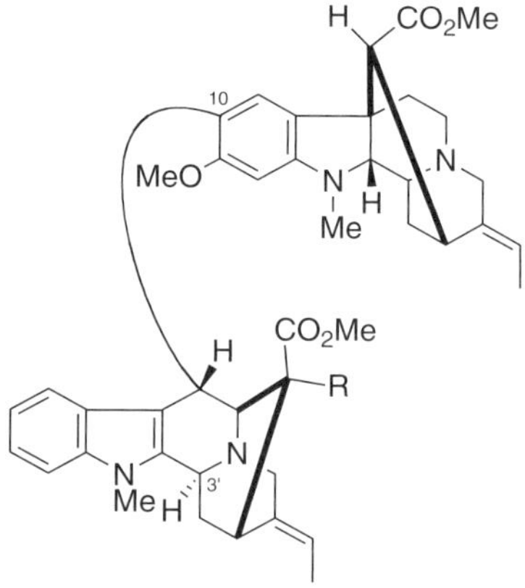


**315** R = $CH_2OH$ undulatine
**316** R = H deformylundulatine

## IX. Vobasine–Akuammiline Type

*Alstonia sphaerocapitata* and *A. undulata* provided undulatine (**315**) and deformylundulatine (**316**), respectively (*213,214*). The alkaloids are constituted from the union of a cabucraline unit and a *N*-methylpericyclivine unit, *via* C(10) of the former to C(6′) of the latter. The structures were elucidated on the basis of mass spectral and NMR data (*213*). An undulatine-type bisindole **317** has been prepared *via* acid-mediated coupling of a pericyclivine derivative **318** with the electron-rich cabucraline (**319**) (*215*).

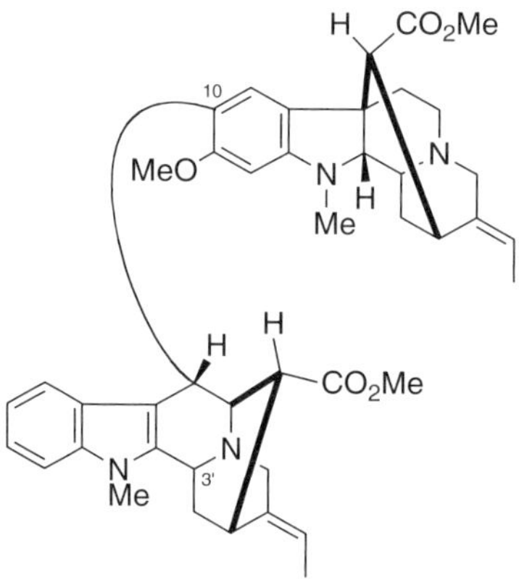


**317**

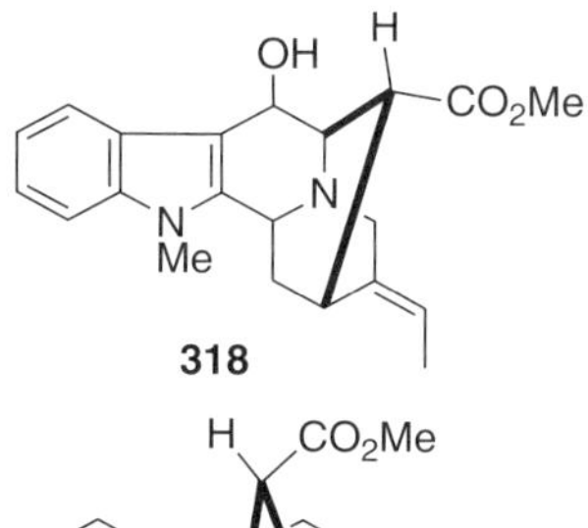


**318**

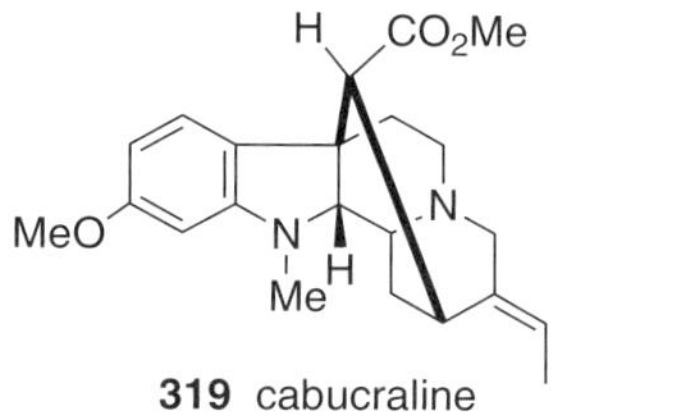


**319** cabucraline

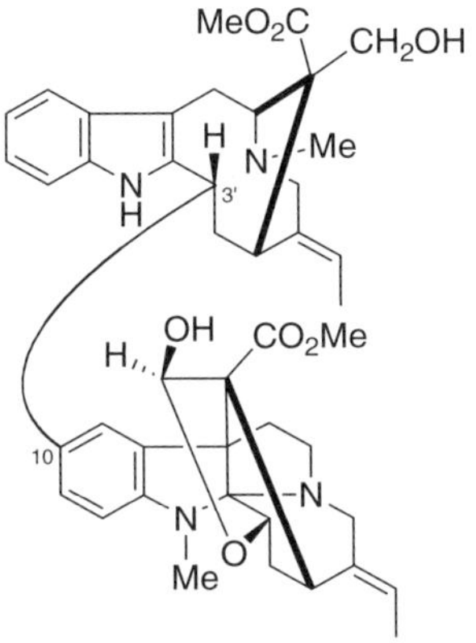


**320** coryzeylamine

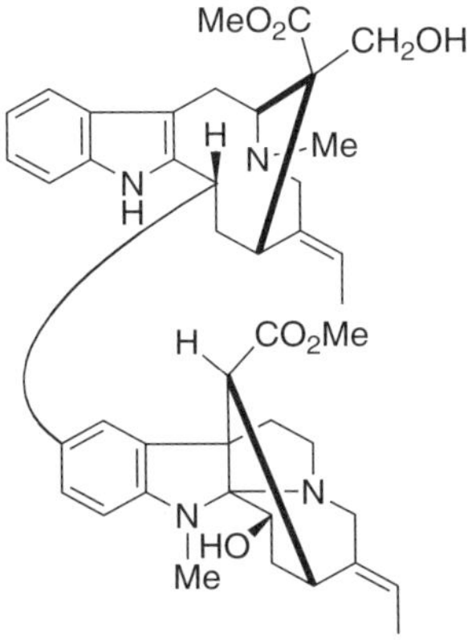


**321** deformylcoryzeylamine

## X. Vobasine–Vincorine Type

*H. zeylanica* from Thailand provided coryzeylamine (**320**) and deformylcoryzeylamine (**321**), to date the only known examples of vobasine–vincorine bisindoles (*216*). They were obtained, together with corymine, which was the main alkaloid present. The NMR spectra were obtained in pyridine-$d_5$ and revealed the constituent units to be vincorine and a 3-deoxy-16-hydroxymethylvobasinol moiety similar to that found in vobasonidine (*vide supra*). The substitution at C(3′) of the vobasinyl moiety was deduced to be α, for steric reasons, as well as from the observed $J_{3-14}$ coupling of 12 Hz. The observed NOE between H(9) and H(17) allowed the C(17) configuration to be assigned as *R*. The structure of **321** was readily assigned from comparison of its spectral data with those of **320**. A distinct departure in the NMR spectra was the absence of signals due to the hemiacetal function (H(17) and C(17)), while the observed NOE between H(18) and the ester OMe allowed assignment of the stereochemistry at C(16). In the event, following previous precedent, **320** was readily converted into **321** by treatment with 1% KOH in dioxane.

## XI. Bisindole Alkaloids from *Alstonia*

Plants of the genus *Alstonia* are rich sources of alkaloids and are characterized by a preponderance of the macroline skeleton. *A. angustifolia* from Malaysia provided eight new bisindoles, of which five were simple derivatives of the previously known bisindoles, villalstonine (**322**), and macrocarpamine (**323**), e.g., villalstonine-*N*(4′)-oxide (**324**), 10-methoxyvillalstonine (**325**) and its *N*(4′)-oxide (**326**), and 10-methoxymacrocarpamine (**327**) and its *N*(4′)-oxide (**328**). The other three, angusticraline (**329**), alstocraline (**330**), and foliacraline (**331**) are new macroline–cabucraline-type bisindoles (*217*).

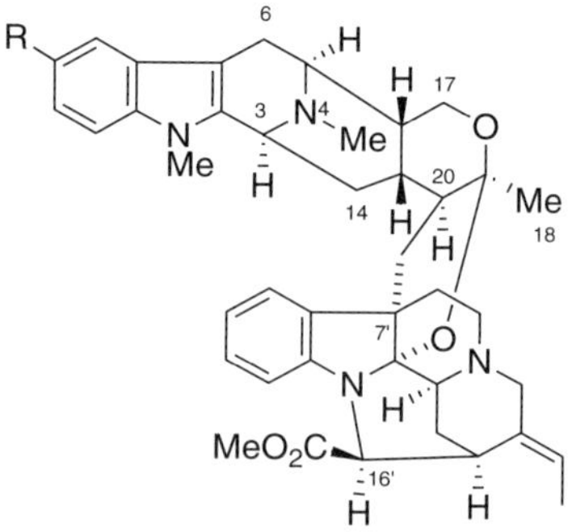

**322** R = H villalstonine
**324** R = H, *N*(4')→O
**325** R = OMe
**326** R = OMe, *N*(4')→O

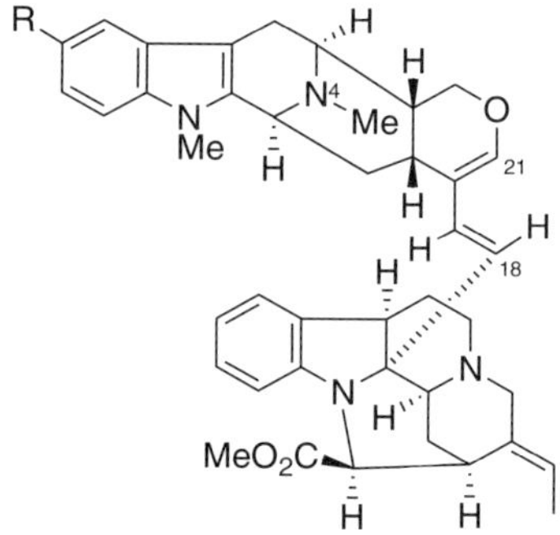

**323** R = H macrocarpamine
**327** R = OMe
**328** R = OMe, *N*(4')→O

The mass spectra of angusticraline (**329**) and alstocraline (**330**) displayed prominent ions due to methoxymacroline (*m/z* 368, 200) and cabucraline (*m/z* 194) moieties. The NMR spectral data confirmed the presence of these units and further unravelled the point of attachment of the monomeric units. The substitution on the cabucraline moiety is at the aromatic C(10′), which is linked to C(19) of the macroline part, a conclusion supported by the absence of signals normally due to these hydrogens, and the presence of a four-carbon fragment corresponding to the C(18)–C(21) unit in the $^{13}$C NMR spectrum. It is also consistent with the presence of a hemiacetal function, which accounts for the observation of an OH in the IR spectrum as well as a peak due to loss of water in the mass spectrum of **329**.

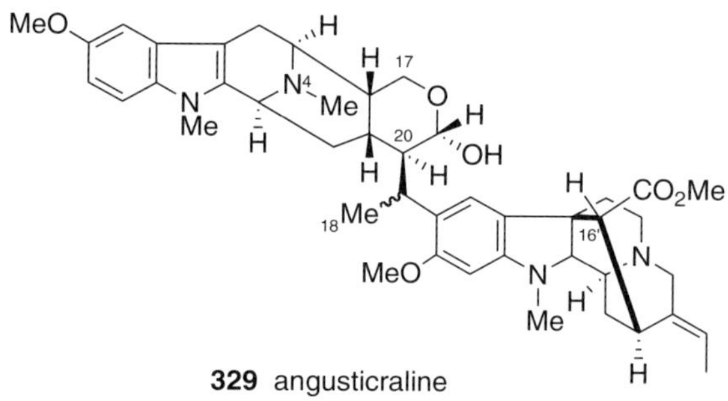

**329** angusticraline

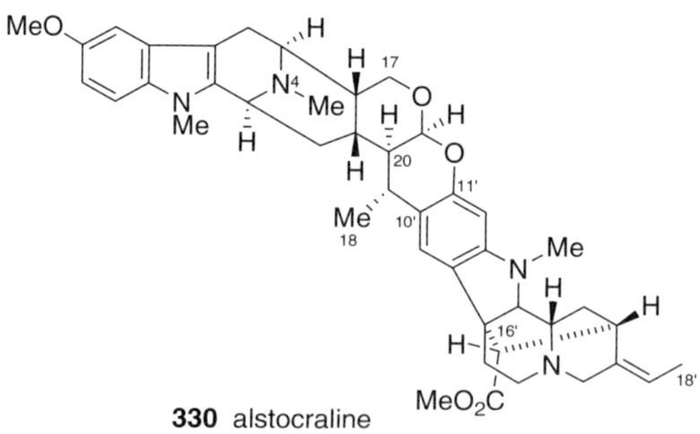

**330** alstocraline

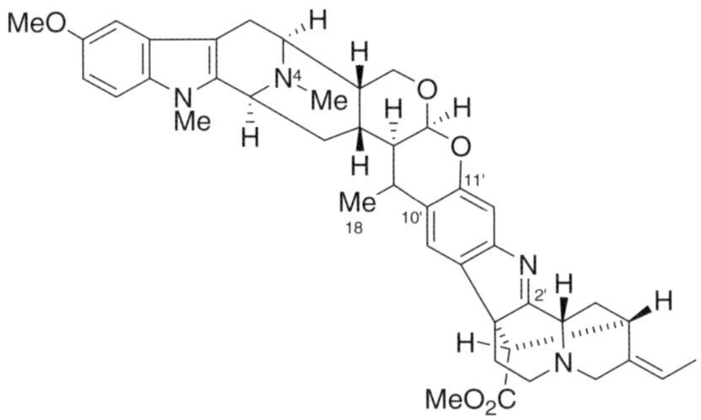

**331** foliacraline

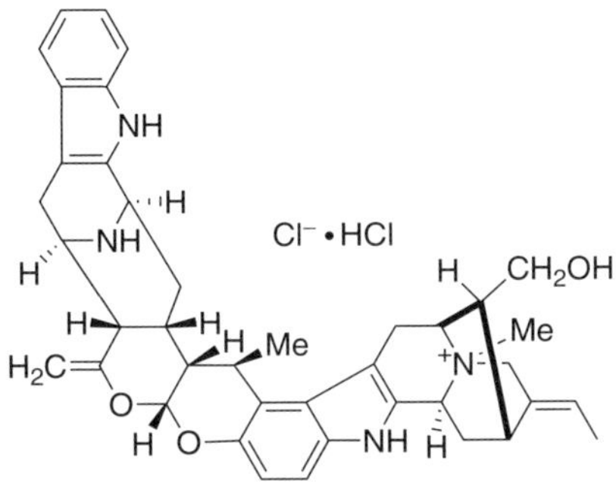

**332** macrospegatrine

The observed molecular ion for alstocraline (**330**) indicated a difference in the molecular formula from angusticraline (**329**) by $CH_4O$, corresponding to demethylation of a methoxy group and loss of water, an inference supported by the NMR data, which also indicated the loss of an aromatic methoxy function, in addition to formation of an additional dihydropyranyl ring contiguous to the macroline ring E. The E/F ring junction stereochemistry was deduced to be *cis* from the observed coupling constants of the ring junction hydrogens. The mass spectrum of the third bisindole, foliacraline (**331**), showed that it differs from alstocraline by a loss of 16 mass units, a difference which the NMR data showed to be due to loss of the indolic methyl and additional unsaturation of the cabucraline unit. The observation of an imine quaternary carbon signal at $\delta$ 185 confirmed the presence of an indolenine, and thus the replacement of cabucraline with strictamine for the second part of the molecule.

Macrospegatrine (**332**), a quaternary bisindole alkaloid, constituted from macroline and vobasinyl halves, was obtained from the water-soluble fraction of the roots of *R. verticillata*. Its structure was established by spectroscopic and X-ray crystallographic analysis (*218,219*).

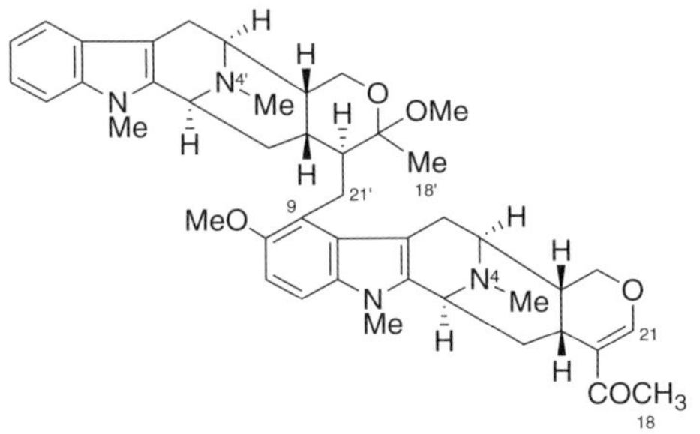

**333** alstomacrophylline

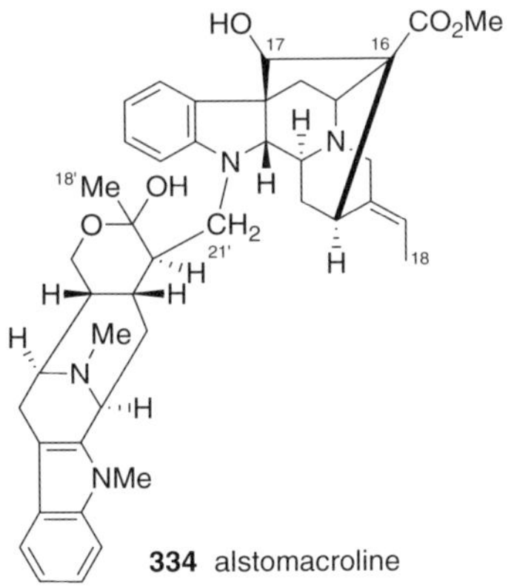

**334** alstomacroline

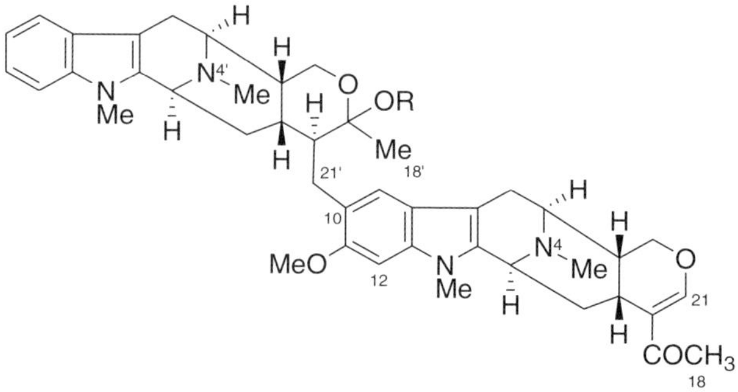

**335** R = Me *O*-methylmacralstonine
**336** R = H macralstonine

The root-bark of *Alstonia macrophylla* from Thailand provided several new bisindoles, viz., alstomacrophylline (**333**), alstomacroline (**334**), and *O*-methylmacralstonine (**335**), in addition to the known bisindoles, macralstonine (**336**), macrocarpamine (**323**), villalstonine (**322**), and its *N*(4)-oxide (**324**) (*220,221*). The HRFAB mass spectrum of alstomacrophylline (**333**) yielded a molecular formula which was identical to that of *O*-methylmacralstonine. The NMR data indicated that the major difference between the two compounds concerns the point of attachment of the two macroline units. In *O*-methylmacralstonine, the lower macroline unit is branched from C(10), giving the characteristic aromatic singlets for the aromatic H(9) and H(12), while in alstomacrophylline, a characteristic AB doublet due to the *ortho*-coupled H(11) and H(12) was seen instead, indicating branching from C(9). The stereochemistry at C(20′) was deduced to be *S* (H(20′α)), from the observed NOEs of H(20′)/H(14′) and H(21′)/H(16′).

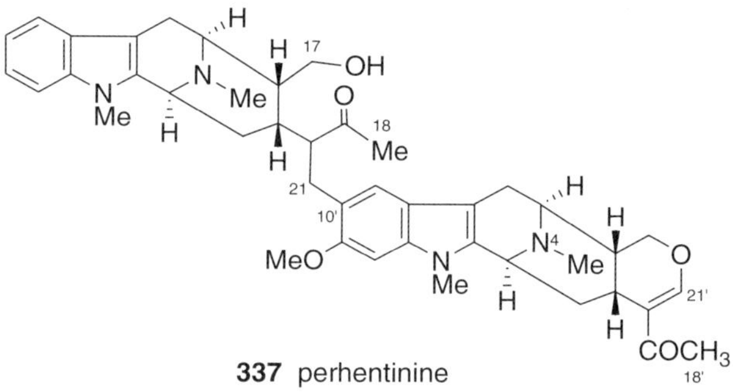

**337** perhentinine

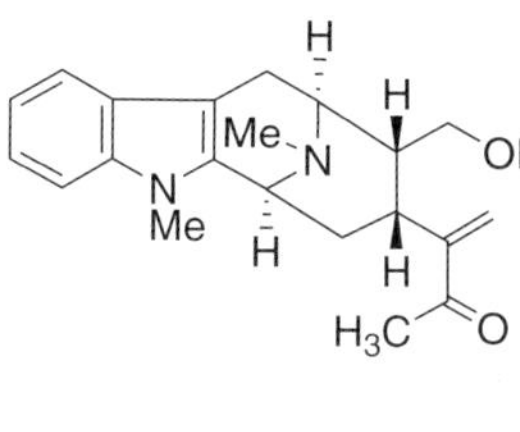

**338** (+)-macroline

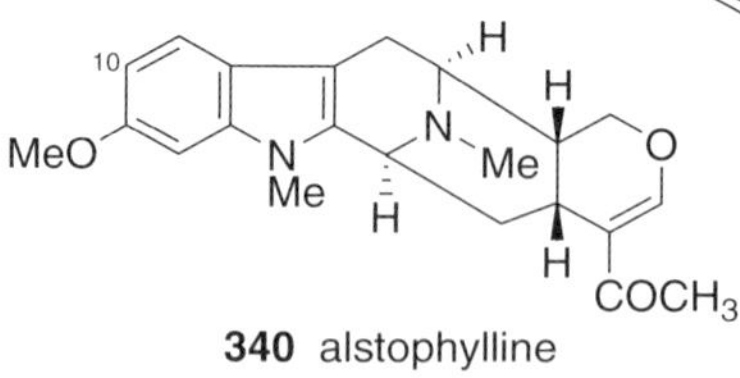

**340** alstophylline

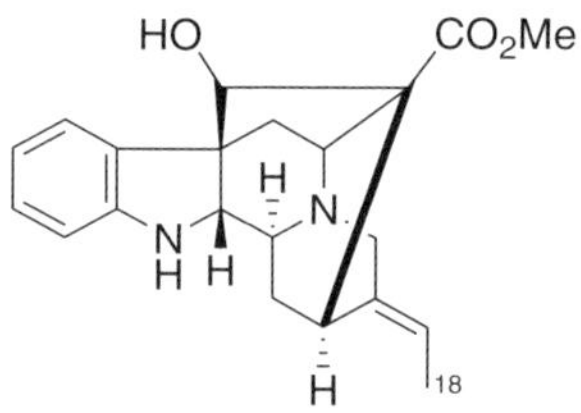

**339** quebrachidine

Alstomacroline (**334**) is constituted from macroline- and ajmaline-type (vincamajine or quebrachidine) halves as revealed by the spectral data. The absence of NH in the NMR as well as IR, the downfield shift of the macroline C(21′) signal to $\delta$ 49.3 compared to that in alstomacrophylline, and the observed NOEs between H(12)/H(21′) and H(2)/H(21′), suggested branching of the bisindole from N(1). The 17-OH signal was shifted unusually upfield to $\delta$ 0.66, which was attributed to shielding by the benzene ring of the quebrachidine unit, while the C(19′) resonance at $\delta$ 97.0 is consistent with the presence of a hemiketal function as part of the macroline ring E. The relative stereochemistries at C(2) and C(20′) were assigned by analogy to villalstonine and alstonisidine as well as from the NOESY spectrum.

The Malaysian *A. macrophylla* provided a new bisindole, perhentinine (**337**), in addition to villalstonine (*222*). The $^{1}$H and $^{13}$C NMR data indicated that the bisindole is constituted from the union of two macroline halves, viz., macroline and alstophylline, with the branching from the nucleophilic C(10′) of alstophylline to C(21) of the macroline unit.

The formation of the above macroline-derived bisindoles can be rationalized based on the biomimetic routes established by LeQuesne (*223,224*), involving coupling of macroline **338** with various electron-rich indole partners. Thus, the three macroline-cabucraline bisindoles can be envisaged as arising through Michael addition of cabucraline (**319**) (or norcabucraline), *via* its nucleophilic C(10′), to C(19) of the $\alpha,\beta$-unsaturated aldehyde of a *seco*-macroline unit, followed by subsequent hemiacetal formation. Similarly, alstomacroline and perhentinine can be considered as arising through coupling of macroline and quebrachidine (**339**) (*via* the indolic nitrogen), and macroline and alstophylline (**340**) moieties, respectively.

A partial synthesis of villalstonine (**322**) has been achieved by Cook, following the biomimetic method of LeQuesne (*223*), by condensation of synthetic (+)-macroline (**338**), or the more stable macroline equivalent (**341**), with natural pleiocarpamine (**342**) in 0.2 N HCl, to furnish villalstonine (Scheme 22). The (+)-macroline was prepared starting from the optically active tetracyclic ketone **343**, prepared from D-(+)-tryptophan by an enantiospecific Pictet–Spengler reaction and stereocontrolled Dieckmann cyclization. The synthesis (Scheme 23) features the use of a stereoselective Claisen rearrangement, followed by stereospecific hydroboration-oxidation of the exocyclic methylene function at C(16), to install the required C(15) and C(16) stereochemistry (*225–227*).

Subsequently, Cook disclosed a similar partial synthesis of macrocarpamine (**323**) by coupling (−)-anhydromacrosalhine-methine (**344**) with natural pleiocarpamine (**342**) in anhydrous 0.2 N HCl in THF (Scheme 24) (*228–230*), providing support for the earlier biogenetic proposal of Hesse (*231*). The required (−)-anhydromacrosalhine-methine was first partially synthesized from (+)-ajmaline (**345**) to provide an authentic sample (Scheme 25). Degradation to the hemiacetal **346** was carried out following the improved procedure of Sakai (*232*). This was followed in succession by dehydration, regioselective oxyselenation, selenoxide elimination, and finally 1,4-elimination to yield anhydromacrosalhine-methine (**344**) in 85% yield.

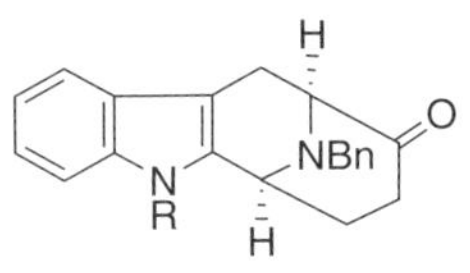


**348a** R = H; **348b** R = Me

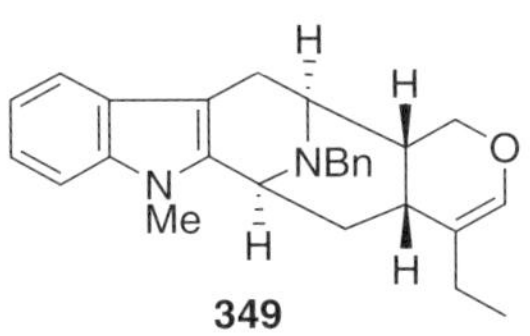


**349**

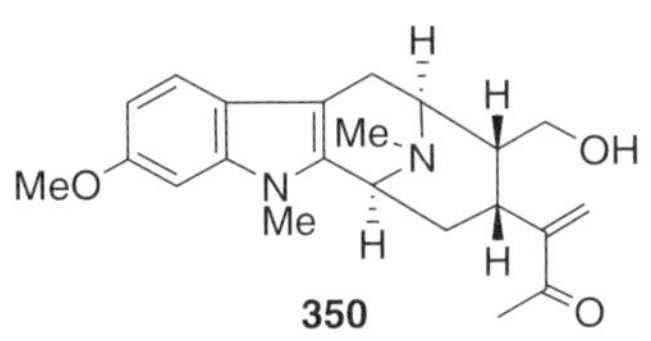


**350**

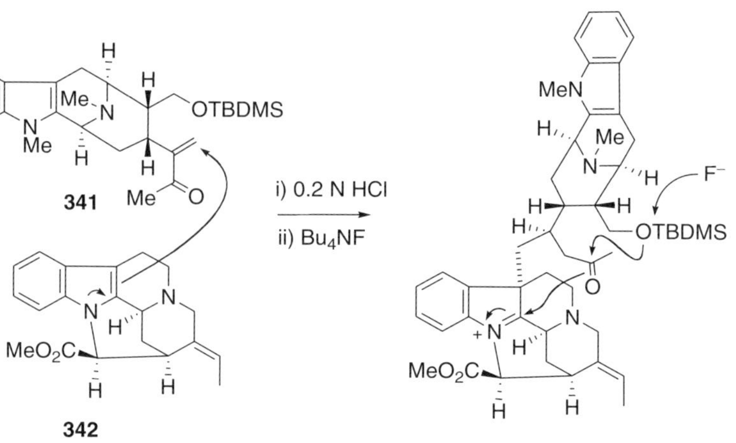


Scheme 22

The enantiospecific synthesis of anhydromacrosalhine-methine (**344**) commenced with the tetracyclic ketone **343** which was converted into alstonerine (**347**) as described previously (*227*). Reduction of alstonerine, followed by dehydration of the resultant allylic alcohol, provided (−)-anhydromacrosalhine-methine (**344**), identical with material prepared from natural (+)-ajmaline (Scheme 25) (*228,230*). In the context of a synthesis of talpinine and talcarpine, an improved synthesis of (−)-anhydromacrosalhine-methine was also achieved from improved routes to the tetracyclic ketones **348** and the enol ether **349**, key intermediates in the talpinine/talcarpine synthesis (*233*).

Cook has also carried out a synthesis of 11-methoxymacroline (**350**), which was then transformed into alstophylline (**340**), which in view of LeQuesne's previous demonstration of biomimetic alstophylline–macroline coupling to macralstonine (**336**) (*234,235*), and Cook's previous synthesis of (+)-macroline (*226,227*), constituted a total synthesis of macralstonine (**336**). Similarly, the synthesis of (+)-*N*-methylsarpargine (**351**) constituted a total synthesis of the bisindole macralstonidine (**352**), in view of LeQuesne's biomimetic macroline-*N*-methylsarpagine coupling (*236*). The synthesis of *N*-methylsarpagine and 11-methoxymacroline, required preparation of the appropriate methoxytetracyclic ketones **353** and **354** (the methoxysubstituted versions of the tetracyclic ketone **348** used above in the Cook synthesis of (+)-macroline), which served as versatile templates for elaboration to alstophylline (**340**) and *N*-methylsarpagine (**351**).

The synthesis of the 10-methoxytetracyclic ketone **353** (Scheme 26) started from the readily available 3-methyl-5-methoxyindole **355**, which after Boc protection, was brominated, and then condensed with the anion of the Schollkopf auxiliary **356** (from L-valine). Removal of the Boc-protecting group, followed in succession by *N*-methylation and hydrolysis, gave the required *N*-methyl-5-methoxy-D-tryptophan ethyl ester **357**, which was then transformed into the key 10-methoxytetracyclic ketone **353**, *via* N-benzylation, Pictet–Spengler condensation, and Dieckmann cyclization. Subsequent N-alkylation by the vinyl iodide **358** followed by Pd-catalyzed (enolate-driven)

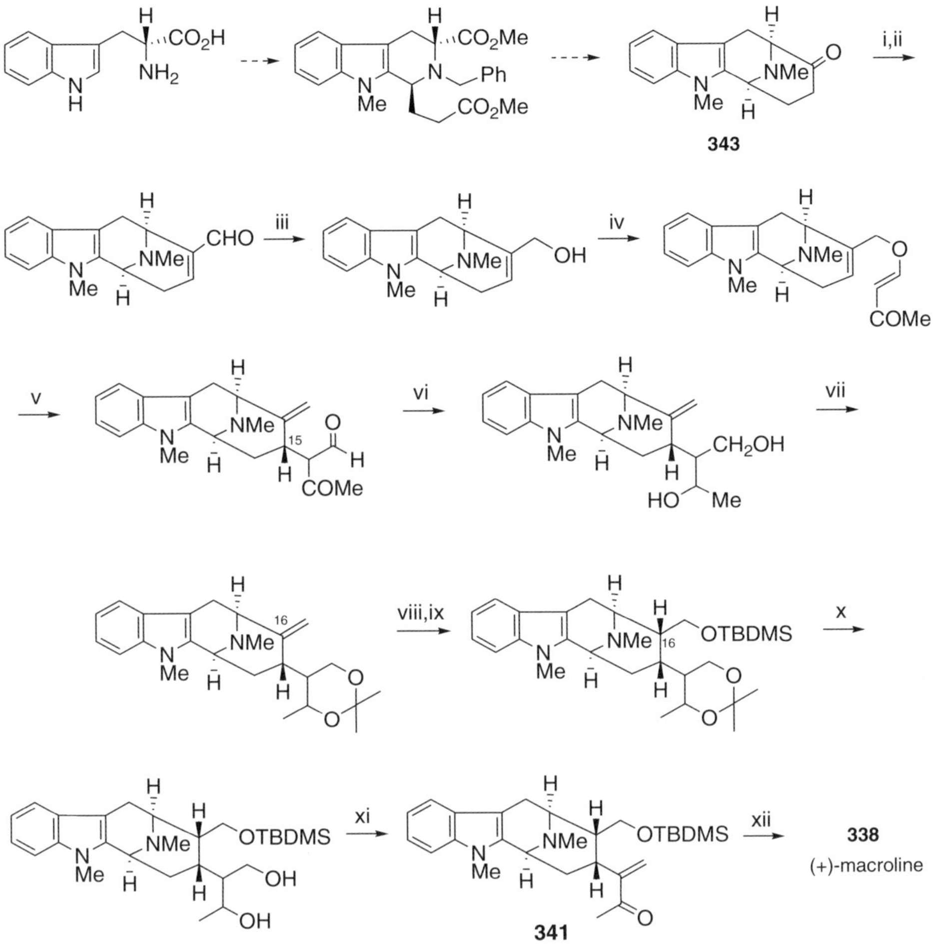

Scheme 23. Reagents: (i) $PhSOCH_2Cl$, LDA, THF; KOH, THF; (ii) $LiClO_4$, $POBu_3$, PhMe; (iii) $LiAlH_4$, $Et_2O$, $-20°C$; (iv) 3-butyn-2-one, dioxane, $Et_3N$; (v) PhH, 145°C; (vi) $NaBH_4$, EtOH; (vii) $Me_2C(OMe)_2$, PTSA; (viii) 9-BBN, THF; NaOH, $H_2O_2$; (ix) TBDMSCl, 4-DMAP, $CH_2Cl_2$; (x) PTSA, MeOH (dry); (xi) $Ac_2O$, py; PDC; (xii) $Bu_4NF$, THF–$H_2O$.

cyclization, forged the essential pentacyclic carbon framework of the sarpagines (*237,238*). The condensation of (+)-macroline (**338**) with *N*-methylsarpagine (**351**) following the method of LeQuesne gave macralstonidine (**352**) (Scheme 27).

The synthesis of 11-methoxymacroline **350** (*239*) required 6-methoxy-D-tryptophan **359**, which was prepared by Larock's Pd-catalyzed heteroannulation of iodoaniline **360** and the propargyl compound **361** (prepared in turn from the Schollkopf chiral auxiliary derived from L-valine), followed by removal of the chiral auxiliary, and N(1)-methylation (Scheme 28). The 6-methoxy-D-tryptophan **359** was then transformed into the pentacyclic sarpagine derivative **362**, *via* the 11-methoxytetracyclic ketone **354**, following the same protocol as that employed in the *N*-methylsarpagine synthesis (*vide*

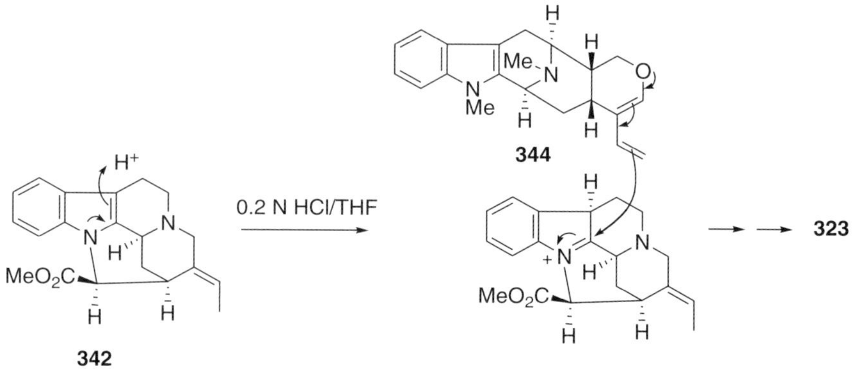


Scheme 24

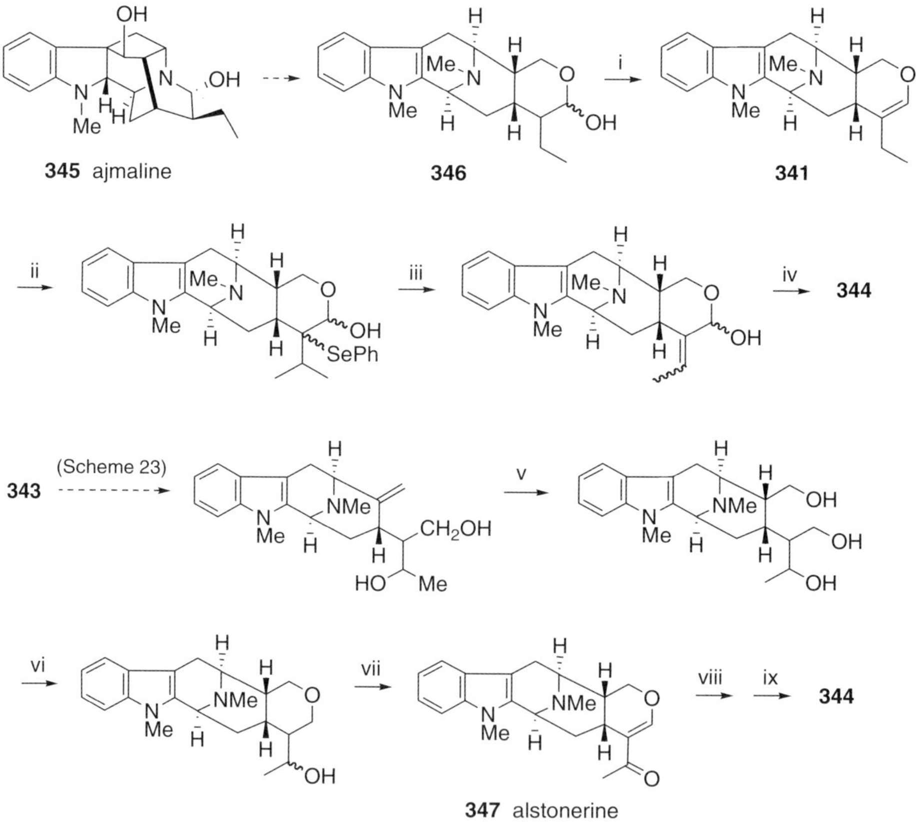


Scheme 25. Reagents: (i) PTSA, PhH, Δ; (ii) *N*-phenylselenophthalimide, PTSA, $CH_2Cl_2$, $H_2O$; (iii) $NaIO_4$, THF–$H_2O$; (iv) PTSA, THF; (v) 9-BBN, THF; NaOH, $H_2O_2$; (vi) TsCl, py; $Et_3N$; (vii) $(COCl)_2$, DMSO, $CH_2Cl_2$, $Et_3N$; (viii) $NaBH_4$, THF; (ix) PTSA, THF.

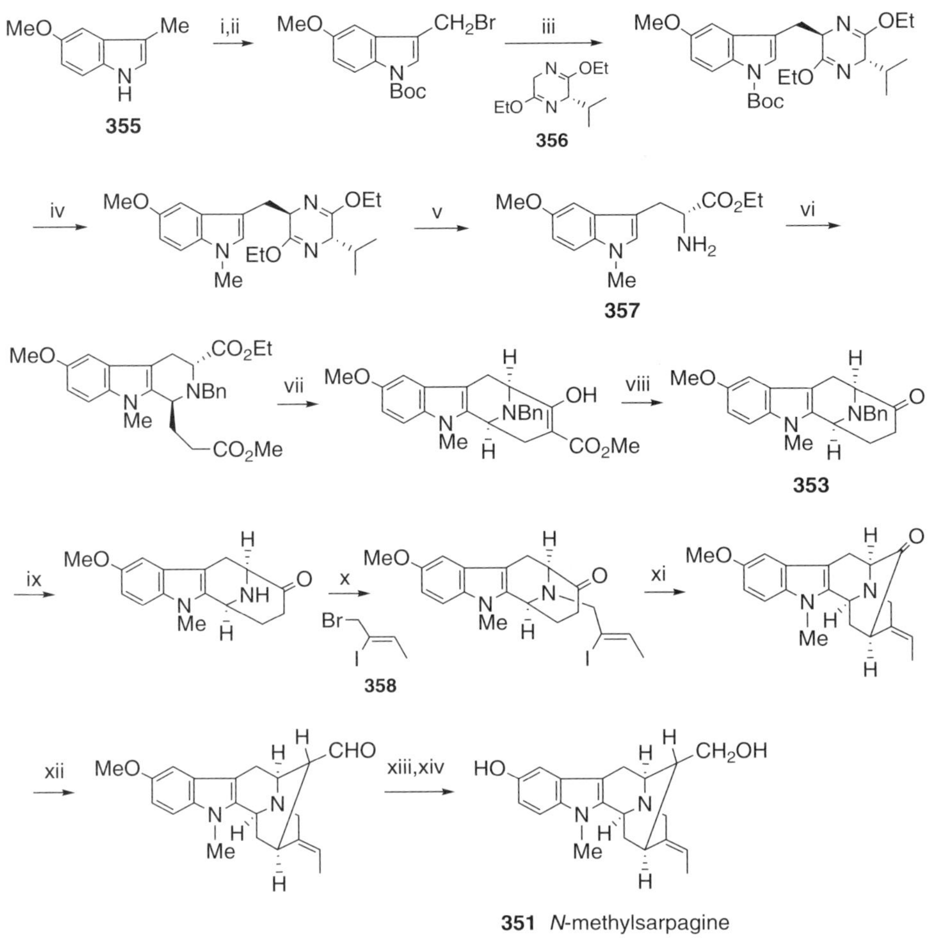

Scheme 26. Reagents: (i) $(Boc)_2O$, DMAP, MeCN; (ii) NBS, AlBN, $C_6H_{12}$, Δ; (iii) **356**, *n*-BuLi, THF, −78°C; (iv) xylenes, Δ; NaH, MeI, DMF; (v) 2 N aq. HCl/THF; (vi) PhCHO, EtOH, 0°C; $NaBH_4$, −5°C; $OHCCH_2CH_2CO_2Me$, HOAc, $CH_2Cl_2$; TFA; (vii) NaH, PhMe, MeOH, Δ; (viii) 40% aq. KOH, 1,4-dioxane, Δ; (ix) Pd/C, $H_2$, 1 atm., EtOH/HCl; (x) **358**, $K_2CO_3$, THF; (xi) $Pd(OAc)_2$, $Ph_3P$, $K_2CO_3$, *n*-$Bu_4NBr$, DMF/$H_2O$; (xii) $CH_3OCH_2PPh_3Cl$, *t*-BuOK, PhH/THF; 2 N HCl/THF; (xiii) 6 equiv. $BBr_3$, $CH_2Cl_2$; (xiv) $NaBH_4$, EtOH.

*supra*). The sarpagine derivative **362** was then processed to alstophylline (**340**) *via* 11-methoxymacroline (**350**) (or 11-methoxy-*N*-methyltrinervine **363**), following which, acid-mediated coupling of macroline **338** with alstophylline (**340**) using the method of LeQuesne, provided macralstonine (**336**) (Scheme 29).

The bisindole alkaloids macrocarpamine (**323**), macralstonine acetate (semi-synthetic), and villalstonine (**322**) from *A. angustifolia* were found to possess significant *in vitro* antiamoebic activity against *E. histolytica* and *P. falciparum*, but were 4–8 times less potent than emetine against *E. histolytica*, and 15–50 times less potent than chloroquine against *P. falciparum* (*240*). Villalstonine and macrocarpamine were

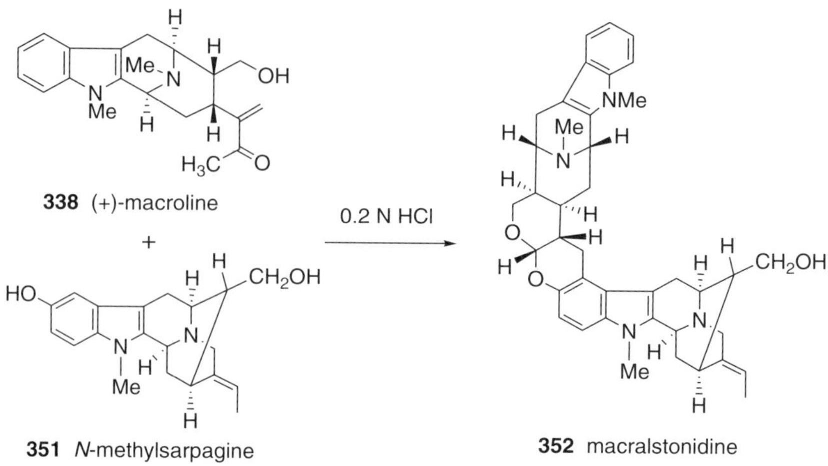


Scheme 27

also found to show pronounced antiplasmodial activity ($IC_{50}$ 0.27 and 0.36 μM, respectively) against the multidrug-resistant K1 strain of *P. falciparum* (*241*). In contrast, the monomeric alkaloids, e.g., echitamine, which is a major alkaloid in several *Alstonia* species, widely used in the treatment of Malaria, was found to display little antiplasmodial activity (*242*). Villalstonine (**322**), macrocarpamine (**323**), and alstomacrophylline (**333**) were found to possess pronounced cytotoxic activity against two human lung cancer cell lines (MOR-P and COR-L23). Villalstonine was by far the most potent with $IC_{50}$ values less than 5 μM, but was about $10^3$ times less potent than vincristine sulphate. Macralstonine (**336**) and *O*-methylmacralstonine (**335**), which were also examined, were considerably less active than villalstonine (*221,243*). Villalstonine, *O*-acetylmacralstonine, and macrocarpamine, also showed pronounced activity when further tested against other human cancer cell lines including melanoma, renal cell carcinoma, breast adenocarcinoma, and colon adenocarcinoma, with $IC_{50}$ values ranging from 2 to 10 μM, with no discernable cell-type selectivity (*243*). Villalstonine and perhentinine (**337**) also showed moderate *in vitro* cytotoxicity toward P388 murine leukemia cell line ($IC_{50}$ 4.4 and 12.3 μg $mL^{-1}$, respectively) (*222*). A significant feature of the bioactivity studies of the *Alstonia* alkaloids is that the bisindole alkaloids were invariably more potent compared to the monomeric alkaloids.

## XII. Akuammiline–Ajmaline Type

*Tonduzia pittieri* (*Alstonia pittieri*) from Guatemala, Central America provided the only bisindole of this group, 10-methoxy-11-[10-(11-methoxyvincamajinyl)]-cathafoline (**364**). The NMR spectral data indicated the presence of methoxycathafoline (cathafoline **365**) and methoxyvincamajinyl moieties (vincamajine **366**), with branching from the aromatic carbons of both units. The presence of only one high field aromatic C(12) at $\delta$ 93.8 due to adjacent C(11) methoxy substitution limited the possibilities to two arrangements, which could be further distinguished by long-range COSY and NOESY experiments. The observed long-range

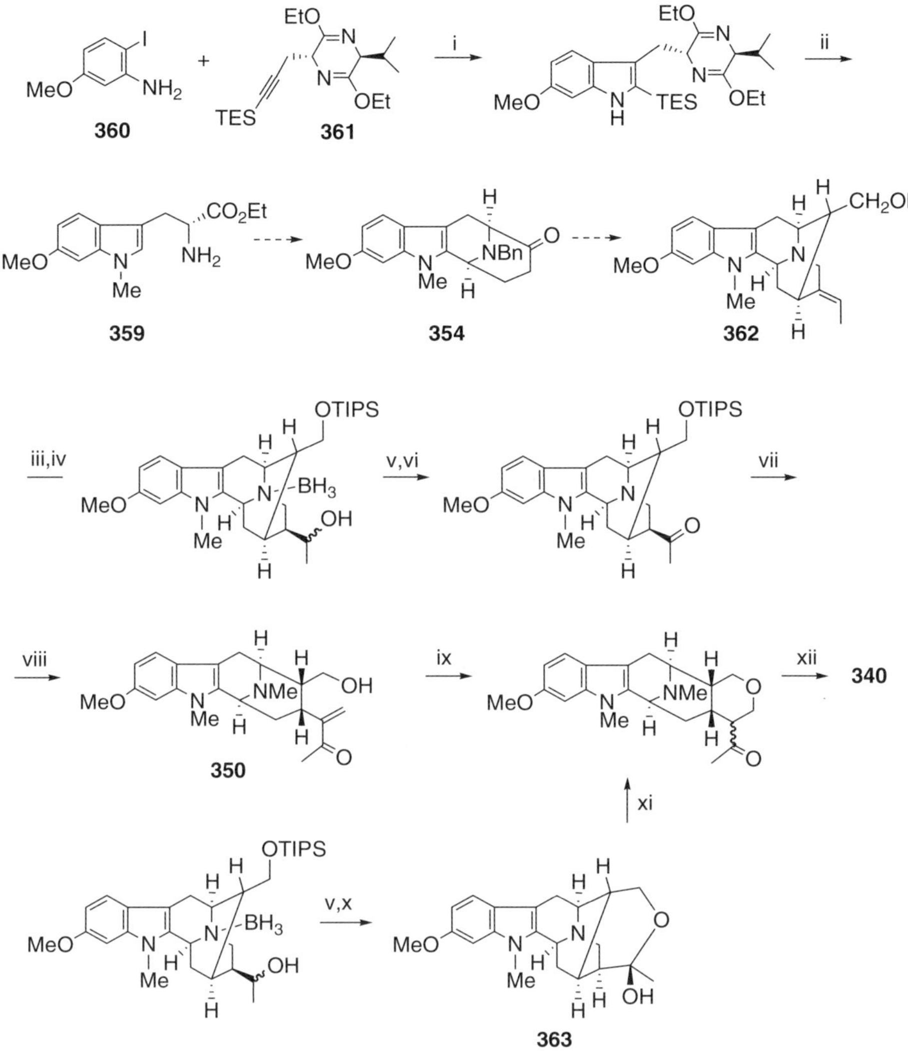

Scheme 28. Reagents: (i) 1% $Pd(OAc)_2$, $Na_2CO_3$, LiCl, DMF, 100°C; (ii) NaH, MeI, DMF; 2 N aq. HCl/THF; (iii) KHMDS, TIPSCl, THF; (iv) 9 equiv. $BH_3$-DMS, THF; $H_2O_2$, NaOH; (v) DMSO, $(COCl)_2$; $Et_3N$; 1.5 equiv. HCl, THF, reflux; (vi) 1.5 equiv. HCl, THF, reflux; (vii) MeI, THF; KOBu-*t*, THF/EtOH, reflux; (viii) TBAF, THF; (ix) NaOEt, THF/EtOH; (x) 10 equiv. HCl (1 N, aq.), THF, reflux; (xi) MeI, THF; 1.5 equiv. KOBu-*t*, THF/EtOH, reflux; (xii) IBX, PTSA, PhMe/DMSO, 70°C.

couplings between the H(12′) at $\delta$ 6.34 to an adjacent 11′-OMe ($\delta$ 3.77) and an NMe′ ($\delta$ 2.68), as well as the observed NOE between the latter and H(2′) of the vincamajinyl moiety, provided firm evidence for the C(11) to C(10′) connection as shown in **364** (*244*).

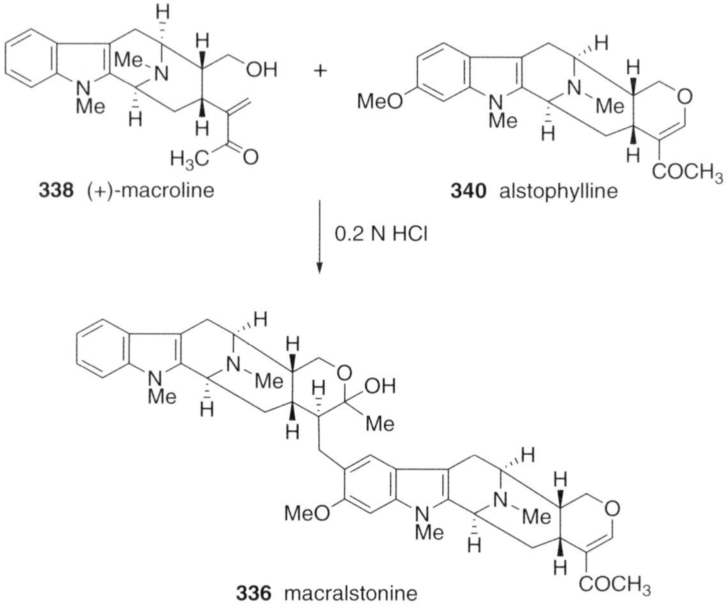


Scheme 29

## XIII. Ajmaline–Vincorine Type

*T. pittieri* (*A. pittieri*) also provided three new bisindoles constituted from 11-methoxyvincamajinyl (**367**) and vincorine (**368**) units, viz., 11-[10-(11-methoxy-17-*epi*-vincamajinyl)]-vincorine (**369**), 11-[10-(11-methoxyvincamajinyl)]-vincorine (**370**), and 11-[10-(11-methoxyvincamedinyl)]-vincorine (**371**). All three bisindoles are branched from C(10′) of the vincamajine half to C(11) of the vincorine unit, which was determined based on long-range COSY and NOESY experiments. The C(12) hydrogens were identified from their coupling with the NMe groups, and only one, H(12′), coupled with an aromatic methoxy. These observations, coupled with the observed NOE between NMe′ and H(2′) of the vincamajinyl unit, allowed assignment of the C(10′) to C(11) bond (*245*).

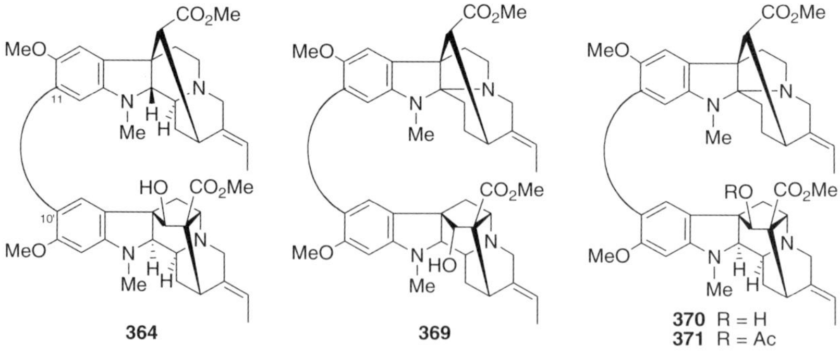

364 369 370 R = H 371 R = Ac

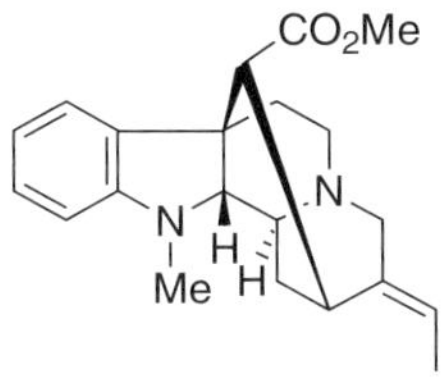

**365** cathafoline

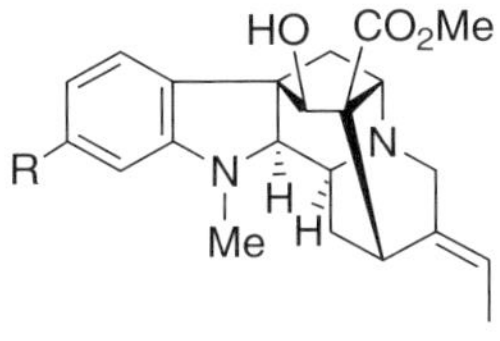

**366** R = H vincamajine
**367** R = OMe

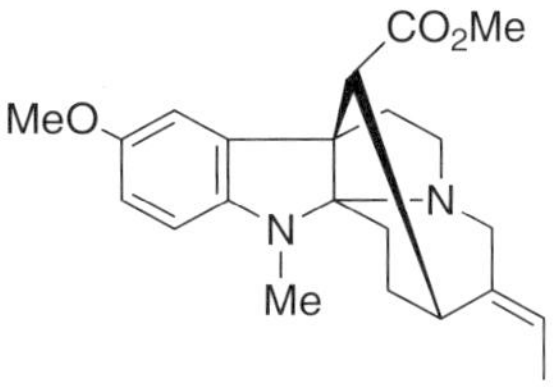

**368** vincorine

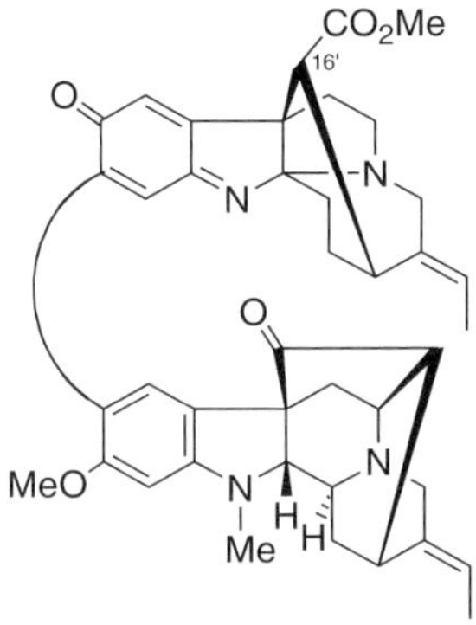

**372** flexicorine

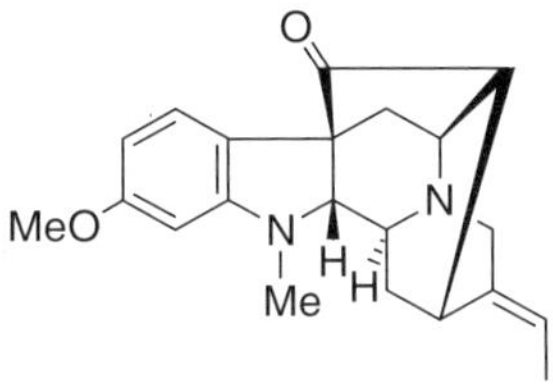

**373** rauflexine

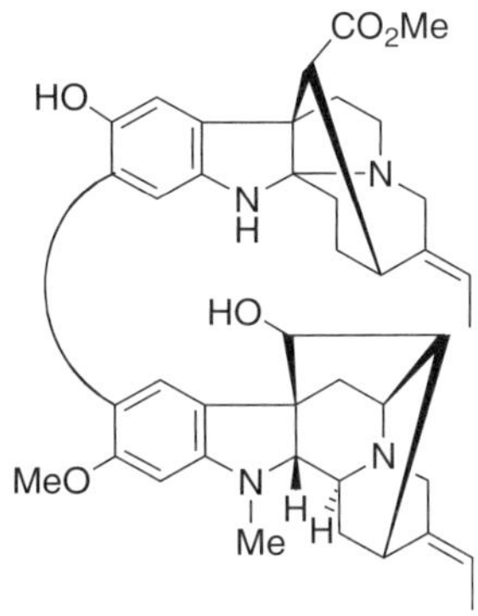

**374**

Flexicorine (**372**) from *Rauvolfia reflexa* is constituted from the union of ajmaline- and vincorine-type halves. The structure elucidation was based on NMR and MS data. The ajmaline half was readily recognized from its correspondence to rauflexine (**373**), which co-occurred with the bisindole in the plant. The vincorine-like unit takes the form of an iminoquinone which was readily demonstrated by its ready reduction to the phenol form (a 10-hydroxyindoline, **374**) by $NaBH_4$. The phenol form was, however, unstable and readily reverted to the iminoquinone on exposure to air. Comparison of the NMR data of flexicorine with those of the reduced form **374** as well as with those of the constituent monomers, rauflexine (**373**) and reflexine, which were both present, greatly facilitated the structure assignment (*246*), although the subsequent revision of some of the $^{13}C$ shift assignments (*vide infra*) required amendment of the C(16′) configuration in **372**.

## XIV. Akuammiline–Vincorine Type

Cabufiline (**375**) (*245,247,248*) and its congeners, deoxycabufiline (**376**), and nordeoxycabufiline (**377**) (*247,249,250*) are found in *Cabucala caudata*, *Rauvolfia sumatrana*, *Alstonia plumosa*, *A. sphaerocapitata*, and *A. pittieri (T. pittieri)*. They are constituted from vincorine and cabucraline (**319**) (or caberine **378**) halves as shown by the spectral data. The connection between the monomeric units is from C(11′) of the vincorine half to C(10) of the cabucraline/caberine half. The occurrence of cabucraline (**319**) and caberine (**378**) in the same plant as well as the acid-induced transformation of cabufiline to the lactone derivative (**379**), facilitated the structure determination. *R. sumatrana* from Thailand provided, in addition to cabufiline (**375**) and flexicorine (**372**), two additional bisindoles, rausutrine (**380**), and rausutranine

(**381**) (*248*). Both alkaloids share a common vincorine unit characterized by an iminoquinone chromophore, identical to that present in flexicorine (**372**).

In common with flexicorine, both alkaloids had a characteristic dark-red color, and were reduced by $NaBH_4$ to the unstable phenol, which reverted to the starting iminoquinone on exposure to air. In rausutrine (**380**), the other half was shown to be a cabucraline-type unit with a 19,20-epoxide function (caberine), similar to that present in cabufiline, while in rausutranine (**381**), it was shown to correspond to caberoline (**382**), the product of intramolecular lactone formation of the 19,20-epoxide precursor, caberine (**378**) (*247*). The monomers are bridged from C(11′) of the vincorine half to C(10) of the caberoline unit, which was determined from the observed H(12′)/C(10) and H(9)/C(11′) correlations in the HMBC spectrum as well as from the observed NOEs between H(12) with 11-OMe and NMe in the PROESY spectrum. The PROESY experiment also allowed assignment of the 19,20-epoxide stereochemistry (19*R*,20*R*) in rausutrine. The $^{13}C$ NMR spectra of cabufiline and flexicorine were also reassigned based on PHSQC and HMBC experiments, which resulted in exchange of some of the previous assignments. The stereochemistry of the epoxide function in cabufiline was not addressed in the earlier report, but may be assumed to be similar to that in rausutrine, based on the similarity of the $^{13}C$ NMR of both compounds with respect to the caberine half. Rausutranine is likely to be derived from rausutrine *via* intramolecular lactone formation (which has been previously demonstrated in the cabufiline to cabufiline lactone conversion), and as such, the configuration at C(19) and C(20) should follow accordingly, an inference that was also supported by the similarity of the Cotton curves for both alkaloids.

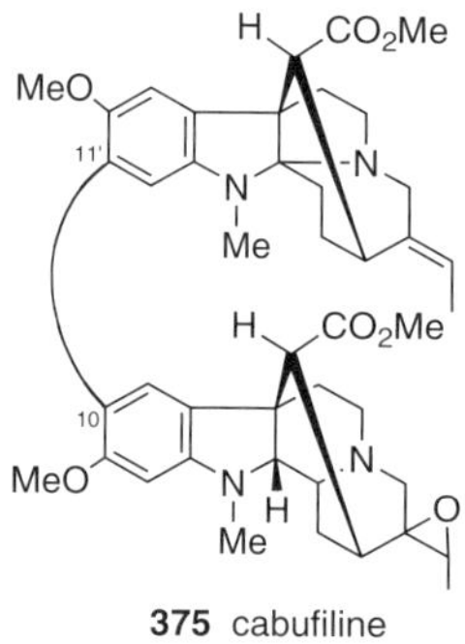

**375** cabufiline

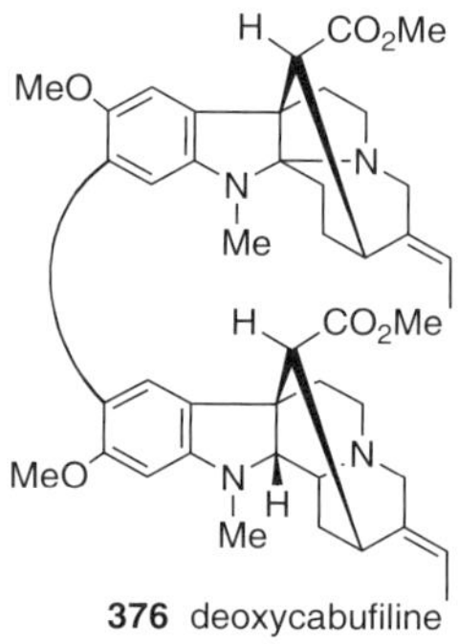

**376** deoxycabufiline

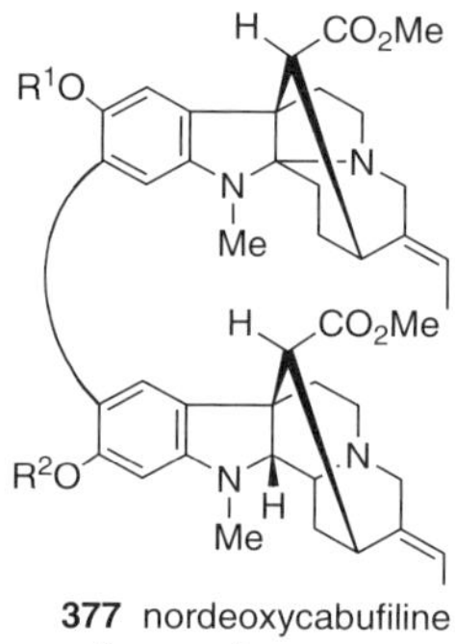

**377** nordeoxycabufiline
($R^1$ = H, $R^2$ = Me
or $R^1$ = Me, $R^2$ = H)

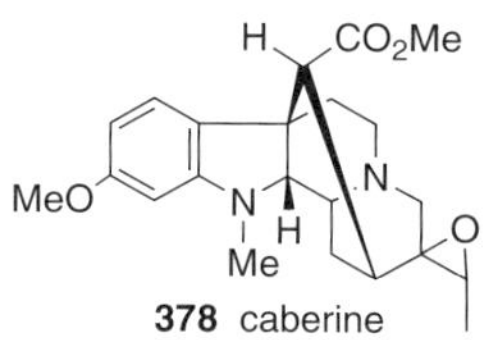

**378** caberine

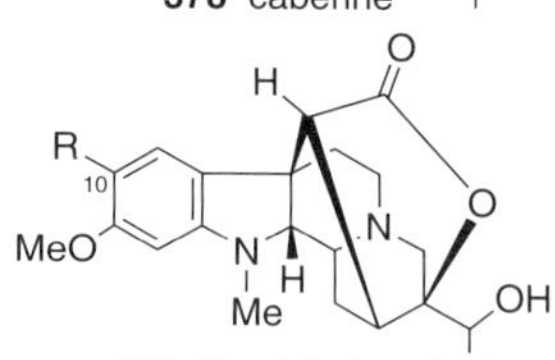

**379** R = 11′-vincorinyl
**382** R = H caberoline

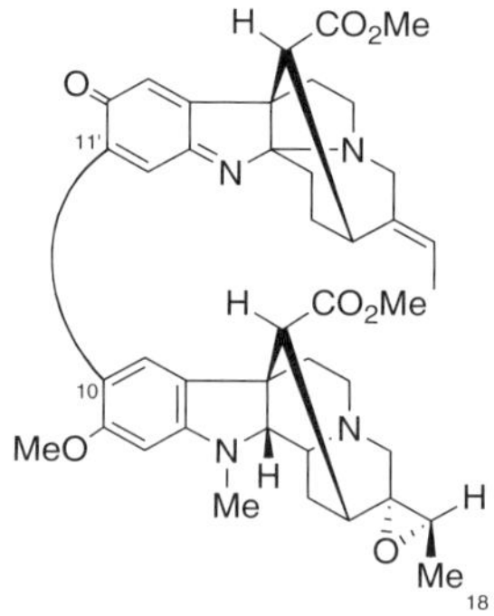

**380** rausutrine

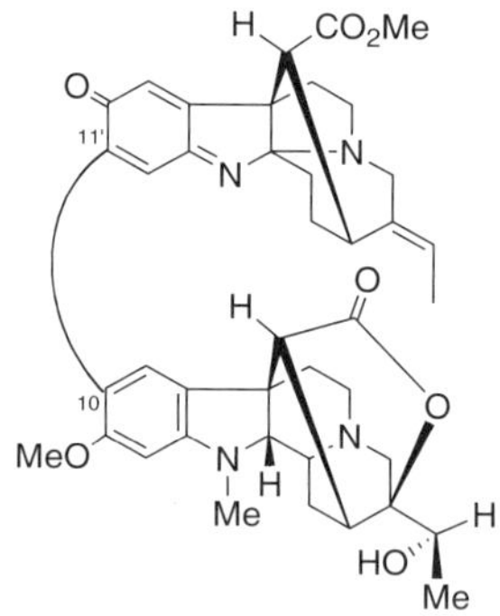

**381** rausutranine

The seeds of the Madagascan *Catharanthus roseus* yielded two alkaloids, vingramine (**383**) and methylvingramine (**384**) (*251*). The structures were elucidated based on HRFABMS and 1- as well as 2-D NMR experiments and revealed a new bisindole skeleton with a vincorine half linked *via* C(11) to C(10′) of the other indole moiety. The latter incorporates several notable features, *inter alia*, the loss of the C(5′)–C(6′) ethylene bridge, the presence of a C(7′)–C(16′) linkage, a 14′-*O*-19′-tetrahydrofuran, and a *N*(4′)-isobutyramide group, while the vincorine moiety is distinguished by the presence of an unusual C(12)-methyl substituent. The $^{1}$H, $^{13}$C, and 2-D NMR spectral data identified the partial structures, which were linked together from examination of the HMBC spectra which were rich in correlations. The branching points from C(11) to C(10′) were deduced based on the observed HMBC correlation between H(9′) and C(11), while the relative configurations of the six methine carbons of the non-vincorine moiety were established based on the observed NOEs. A biogenetic pathway from strictamine (**385**) was proposed, involving a rearrangement and a double fragmentation at C(6′)/C(7′) and N(4′)/C(5′), leading to the non-vincorine half (Scheme 30). The two alkaloids, **383** and **384**, displayed moderate *in vitro* cytotoxicity against KB cells, with $IC_{50}$ values of 5 and 6 μM (4 and 5 μg mL$^{-1}$), respectively. This appears to be the only instance of the occurrence of biphenyl-type bisindoles in *Catharanthus*, although they are found in other Apocynaceae.

## XV. Akuammiline–Pleiocarpamine Type

Plumocraline (**386**) is found in *A. plumosa* (*249,252*) and *A. undulata* (*253*). It is constituted from pleiocarpamine (**342**) and cabucraline (**319**) halves as indicated by $^{13}$C NMR as well as by its ready cleavage under acidic conditions (anhydrous MeOH, HCl) to the two monomeric units, which also occur in both *A. plumosa* and *A. undulata*. Branching from C(2) of the pleiocarpamine half was deduced by comparison of the $^{13}$C NMR shifts with those of macrocarpamine (**323**), another *Alstonia* bisindole with similar C(2) branching from a pleiocarpamine unit. A biomimetic transformation to **386** *via* acid-induced coupling of pleiocarpamine and cabucraline (MeOH, 2% HCl) provided further support for the proposed structure (*252*).

## XVI. Vincorine–Vincorine Type

The Sri Lankan plant, *Petchia ceylanica* (Apocynaceae) provided several interesting bisindoles, including peceyline (**387**), peceylanine (**388**), and pelankine (**389**), which have been discussed in a previous volume of this series (*1,254,255*). Three additional bisindoles have since been reported from the same plant, ceylanine (**390**), ceylanicine (**391**) (*256*), and demethylpeceyline (**392**) (*257*). Ceylanine (**390**) was shown to be constituted from two vincorine units. Comparison of the $^{13}$C NMR spectral data with that of peceylanine (**388**) showed the correspondence of one-half and indicated the presence of a common vincorine moiety. The remaining carbon shifts of the other unit showed features reminiscent of a pleiocarpamine-type alkaloid, such as the absence of NH or NMe, and the downfield shift of C(16′) to $\delta$ 59.7, consistent with this carbon being linked to the indolic nitrogen. In addition, the presence of a quaternary carbon at $\delta$ 96.6, due to bonding to two nitrogen atoms, is attributed to C(2′) which is linked to N(4′) as in vincorine alkaloids, and reminiscent of isodihydropleiocarpamine (**393**), a product of Zn/HCl-induced cleavage of the aspidosperma-pleiocarpamine bisindole, pycnanthine (**394**), from *Pleiocarpa pycnantha* (*258*).

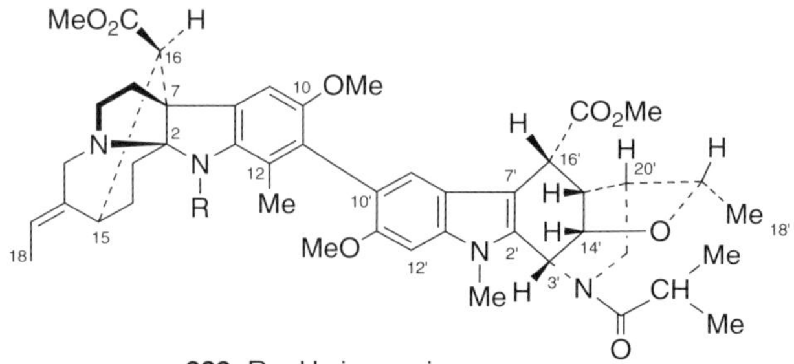

**383** R = H vingramine
**384** R = Me methylvingramine

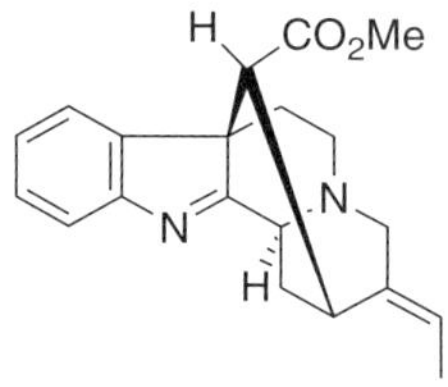

**385** strictamine

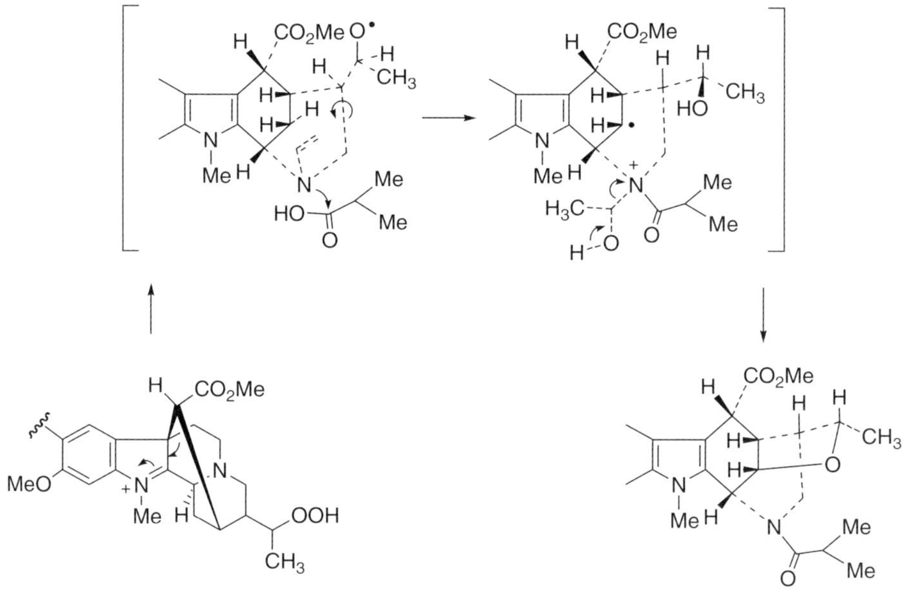

Scheme 30

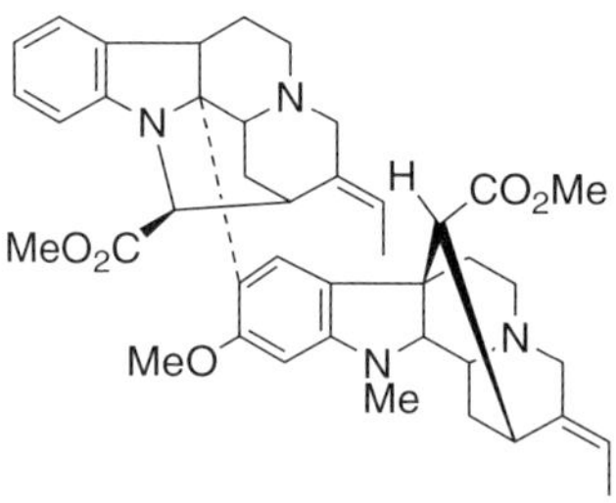

**386** plumocraline

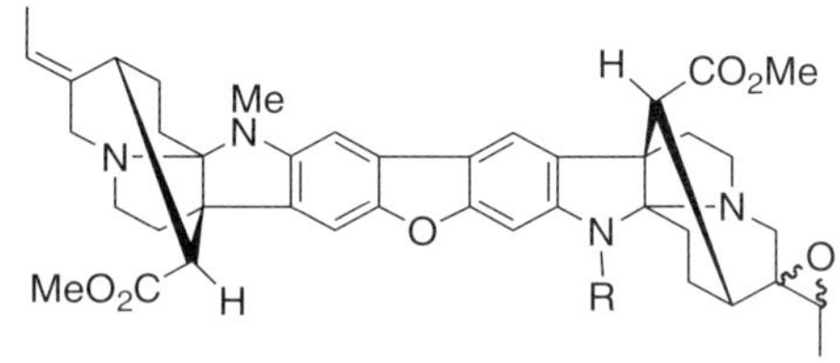

**387** R = Me peceyline
**392** R = H demethylpeceyline

In addition, the downfield shifts of the signals due to C(19′), C(15′), and C(20′), in comparison with those of the corresponding carbons of peceylanine (**388**), are suggestive of formation of a lactone ring. These results are accommodated by formation of a bond between N(1′) and C(16′), following cleavage of the C(7′)–C(16′) bond of peceylanine, followed by intramolecular lactone formation from attack of the carboxylate function on the epoxide moiety. Ceylanicine (**391**) was shown by the spectral data to bear a close similarity to pelankine (**389**), with the monomeric units linked to incorporate a similar central dibenzofuran unit. The main departures concern the cleavage of the C(3′)–N(4′) bond, and the formation of a lactone from reaction of the C(16′)-carboxylate function with the 19′,20′-epoxide moiety. The third bisindole **392** was readily shown to be the *N*′-demethyl derivative of peceyline from the spectral data.

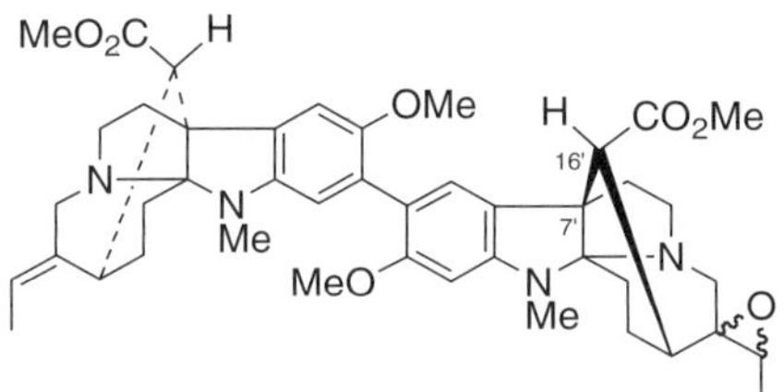

**388** peceylanine

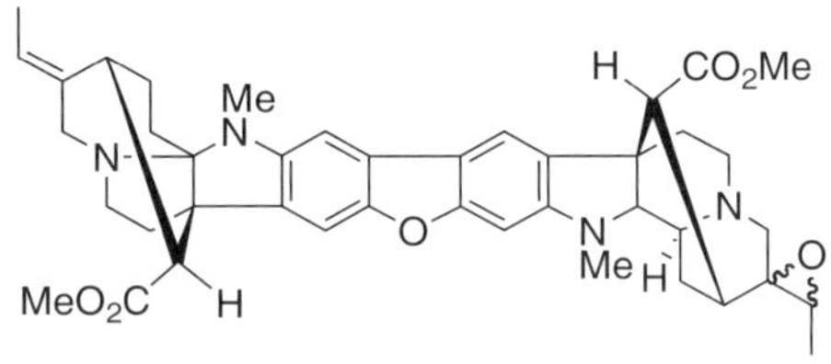

**389** pelankine

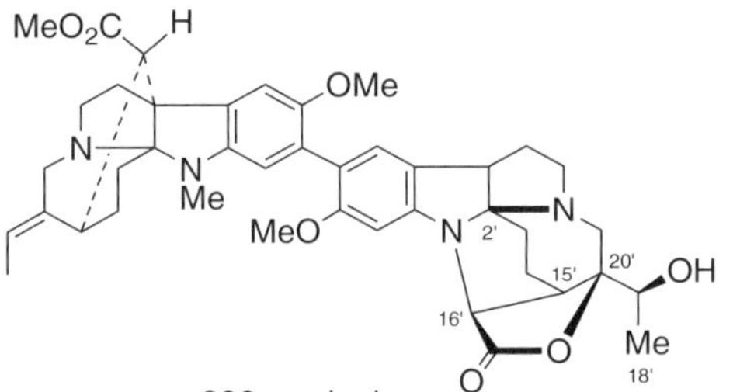

**390** ceylanine

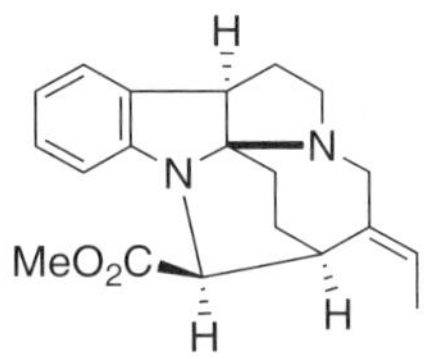

**393**

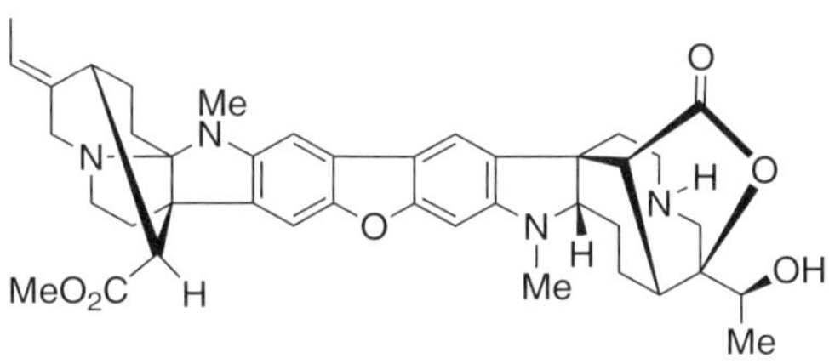

**391** ceylanicine

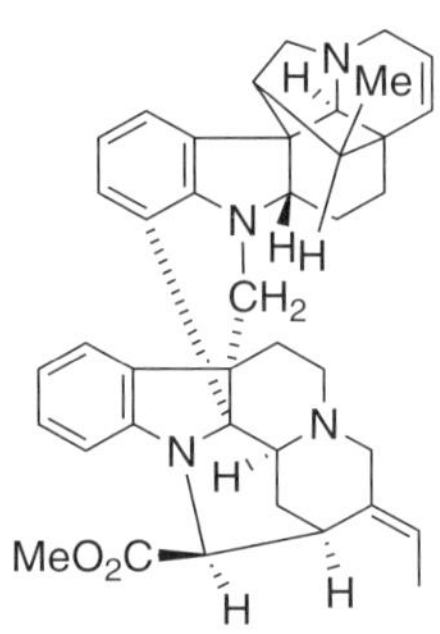

**394**

## XVII. *Strychnos*–Vobasine Type

The roots of the Brazilian *Strychnos* species *S. divaricans* Ducke yielded a polar alkaloid, divaricine (**395**) (*259*). The NMR spectral data indicated the presence of vobasinyl and strychnan halves, which, as all eight aromatic hydrogens were accounted for, were linked from non-aromatic positions. The strychnan half was shown to be associated with an indoline chromophore, while the other half was shown to correspond to vellosimine (**396**). In addition, the characteristic downfield shifts of C(3), C(5), and C(21), associated with the strychnan unit, indicated that it was an N-oxide. Use of long-range heteronuclear correlations from the HMBC spectrum allowed construction of the entire molecule, revealing its constitution from the union of the N(4)-oxide of 18-deoxy Wieland–Gumlich aldehyde (**397**) and vellosimine (**396**), with the two units linked *via* an oxazine ring as in the strychnobilines (*vide infra*). Vobatricine (**398**) was isolated from the Malaysian *T. corymbosa* (*210*). The NMR spectral data revealed the presence of strychnan and vobasinyl moieties. The strychnan half was shown to correspond to norfluorocurarine (**399**) which was also present in the plant, while the vobasinyl unit corresponded to that of a 3-deoxyvobasinol. The bisindole is branched from C(10′) of the strychnan half to C(3) of the vobasinyl half, as indicated by the three-bond correlation from H(9′) and H(11′) of the strychnan unit to C(3) of the vobasine unit.

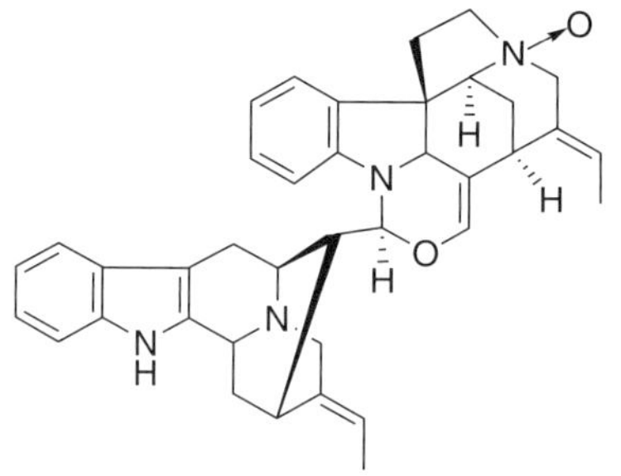

**395** divaricine

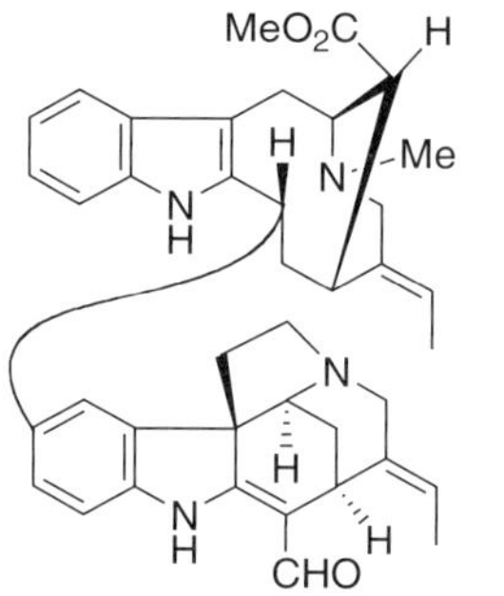

**398** vobatricine

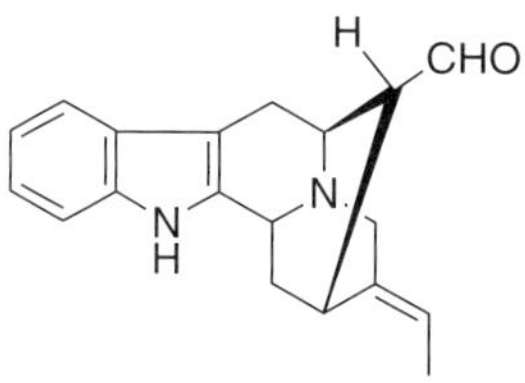

**396** vellosimine

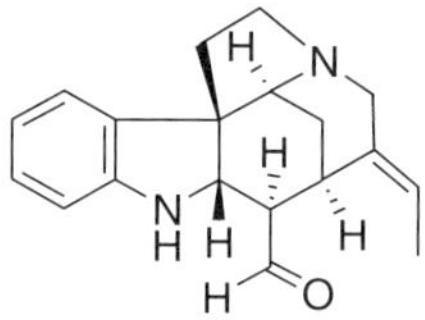

**397**

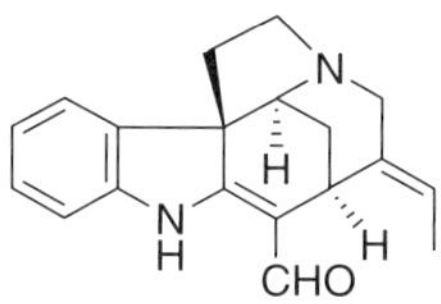

**399** norfluorocurarine

## XVIII. *Strychnos–Strychnos* Type

Matopensine (**400**), a new symmetrical dimeric indole which stains red with Ce(IV) reagents, was isolated from the African *S. matopensis* and *S. kasengaensis* (*188,260,261*). The structure was elucidated by NMR spectroscopy, which revealed a symmetrical dimer, and indicated a similarity with bisnordihydrotoxiferine (**287**), which was also present in the plant. The mass spectrum yielded the formula $C_{38}H_{42}N_4O$, differing from **287** by addition of water. The absence of the 16,17-unsaturation in **400**, coupled with the observation of the C(17)s as oxymethines ($\delta_H$ 5.3, $\delta_C$ 82.7), and the

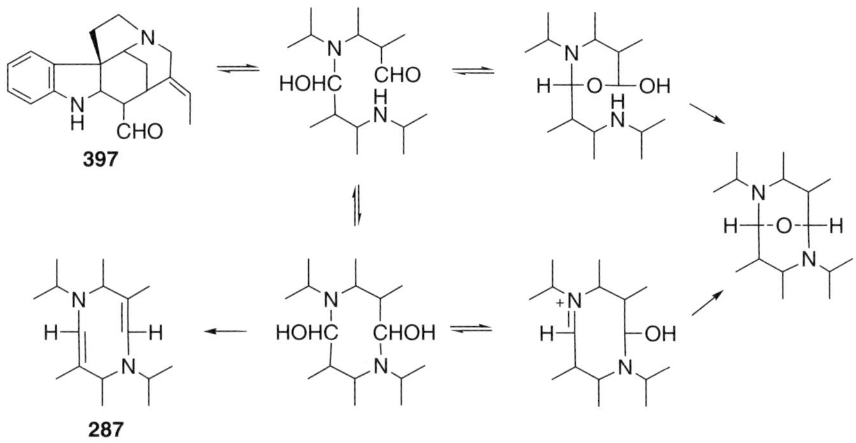

Scheme 31

C(16)s as methines ($\delta_H$ 1.84, $\delta_C$ 41.4), indicated formation of an ether bridge linking the two C(17)s. Analysis of the coupling constants led to assignment of both the H(2) and H(16) stereochemistry as $\beta$ (retuline series; retuline **401**, isoretuline **402**), with the oxazine rings adopting a chair conformation, with H(2) axial, and H(16) and H(17) both equatorial. The structure results in the presence of a $C_2$ axis passing through the oxygen atom. A possible pathway to **400** was proposed from the condensation of two deoxy Wieland–Gumlich aldehyde (**397**) units, *via* initial formation of a carbinolamine, followed by reaction with the aldehyde function to a hemiacetal, and finally dehydration to complete the formation of the second six-membered ring. Formation of an eight-membered ring from the first formed carbinolamine, followed by a double dehydration would lead to **287** (Scheme 31). Several new matopensine derivatives (all staining red with Ce(IV)) have been subsequently isolated from *S. matopensis*, including the N-oxide (**403**), 16-methoxyisomatopensine (**404**), 16-ethoxyisomatopensine (**405**), 18-hydroxymatopensine (**406**), and 18,18′-bishydroxymatopensine (**407**) (*188*).

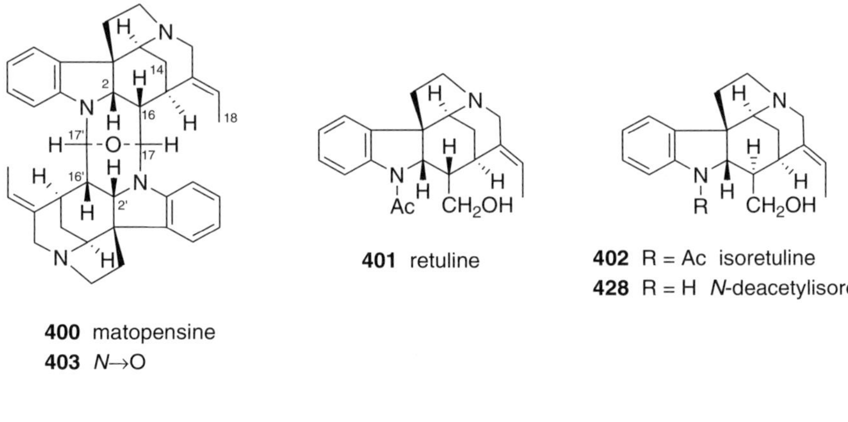

**401** retuline

**402** R = Ac isoretuline
**428** R = H *N*-deacetylisoretuline

**400** matopensine
**403** *N*→O

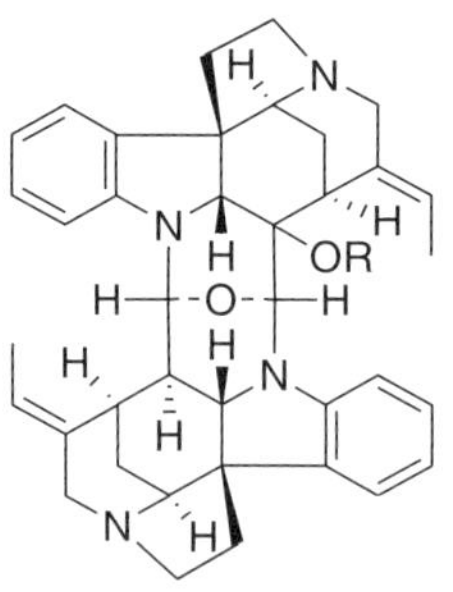

**404** R = Me
**405** R = Et

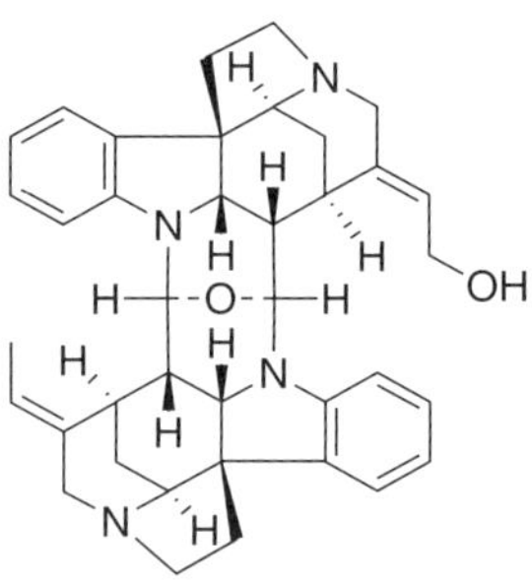

**406**

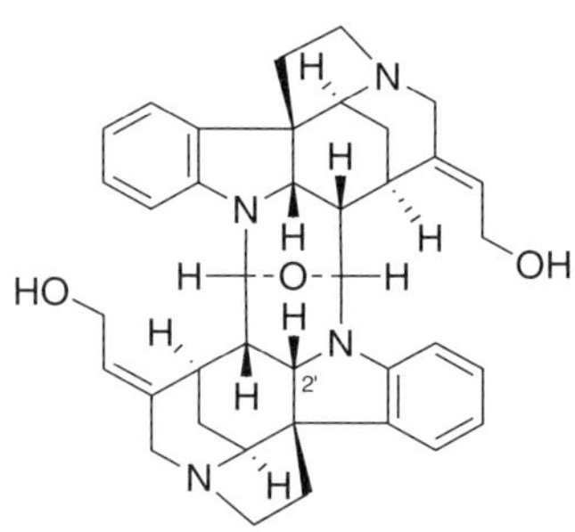

**407**

The $^1$H NMR spectral data of **404** indicated a matopensine derivative in which the symmetry has been lifted through substitution by a methoxy group. This was supported by the observed molecular ion, detected by FABMS, which showed an $MH^+$ ion at $m/z$ 601, differing from matopensine by addition of 30 mass units, as well as by the $^1$H and $^{13}$C NMR spectral data, which indicated the presence of an additional methoxy group. The downfield shifts for one of the C(17) and C(2), compared with matopensine, coupled with the observation of the corresponding H(17) and H(2) as singlets instead of doublets, indicated C(16) as the site of methoxy substitution. In addition, H(2′) of the unsubstituted part of the molecule was observed as a doublet with $J$ 10.5 Hz, compared with 5.5 Hz in matopensine, indicating that this unit belongs to the isoretuline series (H(16$\alpha$)). Another notable observation was the large line shape differences in the $^1$H NMR signals of the two ethylidenes, which were attributed to the two halves having different C(16) configurations. Compound **404** therefore possesses a C(16) configuration similar to that of matopensine in the methoxylated half, but different in the unsubstituted half. The same considerations apply in the case of 16-ethoxyisomatopensine (**405**), i.e., isoretuline configuration in the unsubstituted half, and retuline in the ethoxylated half.

Alkaloid **406** showed an $M^+$ at $m/z$ 586, differing from matopensine by replacement of H with OH. This was also reflected in the NMR spectral data which showed the presence of ethylidene and hydroxyethylidene side chains, suggesting **406** to be a 18-hydroxy matopensine derivative. The observed coupling constants for H(2), H(16), and H(17) were similar to those of matopensine (**400**), indicating the same configurations at these centers in the two halves. The bishydroxy derivative **407**, had $M^+$ at $m/z$ 602, indicating substitution by two OH groups compared to matopensine, while the NMR data indicated a symmetrical dimer with hydroxyethylidene side chains in place of ethylidene, as well as similar configurations of the asymmetric centers as those in matopensine. Bisnor-C-alkaloid D (**408**), characterized by the unusual 16–16′ linkage, showed the same red color with Ce(IV) as the matopensine alkaloids. It was also found in *S. dolichothyrsa* (*262*), and had been previously obtained as an unidentified product from the treatment of bisnordihydrotoxiferine (**287**) with dilute HCl (*263*). The mass spectrum showed the characteristic retuline fragments, while the NMR spectra indicated a symmetrical dimeric alkaloid. The key features of the $^1$H NMR spectrum which led to the structure assignment were the observation of two one-H singlets at $\delta$ 4.95 and 4.55, attributed to H(17) and H(2), respectively, and the unprecedented upfield shift of the 18-methyls at $\delta$ 0.58, which was attributed to their spatial proximity to the C(17) oxygens. In addition to **287**, two other bisindoles possessing the dihydrotoxiferine skeleton were also isolated, viz., bisnor-C-alkaloid H (**284**, *vide supra*) and bisnor-C-curarine (*188*).

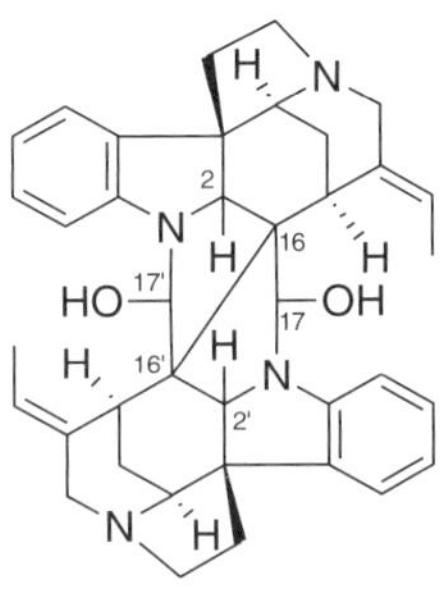

**408** bisnor-C-alkaloid D

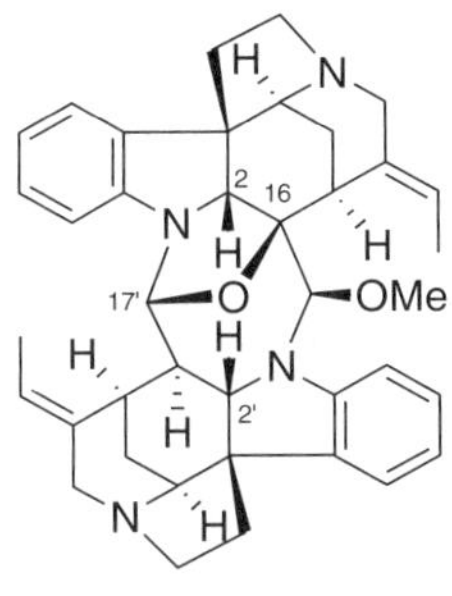

**409a** divarine

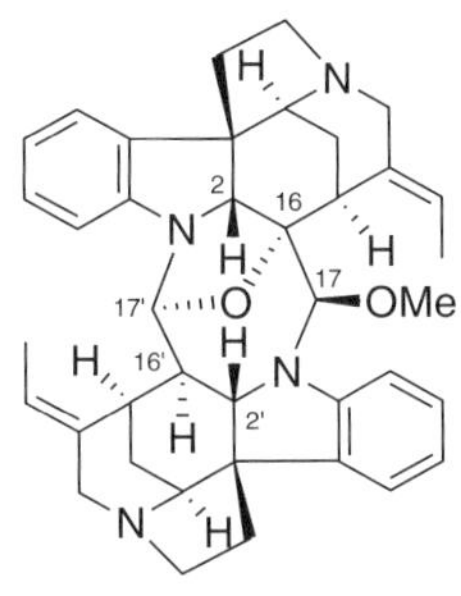

**409b** divarine

*S. divaricans* also provided in addition to divaricine (**395**) (*vide supra*), another new bisindole, divarine (**409**), derived from the condensation of two 18-deoxy Wieland–Gumlich aldehyde moieties (*264*). Divarine (**409**) is isomeric with 16-methoxyisomatopensine (**404**) and, in addition, showed a close similarity in its NMR spectral data with that of **404**. The purported difference between the two structures concerns the site of methoxy substitution. In the case of divarine (**409**), an ether bridge intervenes between C(16) and C(17′), while C(17) is substituted by a methoxy group. The structure was established based on the extensive application of various 1- and 2-D NMR methods. The possibility that divarine (**409**) and 16-methoxyisomatopensine (**404**) are the same alkaloid cannot be ruled out as the NMR data showed very close correspondence, although the more extensive NMR data which were available in the case of divarine (**409**) provided a sound basis for the structure assignment. The lack of an authentic sample of 16-methoxyisomatopensine, however, precluded a direct comparison between the two alkaloids. In addition, due in all probability to an error in the drawing of the structure of divarine during the preparation of the manuscript, the stereochemistry of the ether bridge linking C(16) to C(17′) has been drawn as $\beta$ as depicted in **409a**, which raises several problems. In addition to the considerable strain incurred, the stereochemistry as depicted is entirely incompatible with the observed NOE data (e.g., the observed NOE between H(2) and H(2′) is impossible in structure **409a**). These problems would vanish if the oxygen bridge is $\alpha$ instead of $\beta$, as drawn. The NOE data are now in perfect harmony. It is therefore likely that the correct structure of divarine is **409b**, instead of **409a** as originally given.

Strychnobiline (**410**), isostrychnobiline (**411**), and 12′-hydroxyisostrychnobiline (**412**) were previously isolated from *S. variabilis* (*265,266*). A new derivative, 12′-hydroxystrychnobiline (**413**), has been subsequently isolated from the same plant (*267*), while dehydroisostrychnobiline (**414**) was isolated from *S. kasengaensis* (*261*). The NMR spectral data of the latter indicated similarity with the isostrychnobiline series when compared with those of 12′-hydroxyisostrychnobiline (**412**) and isostrychnobiline

(**411**). The observation of a vinylic singlet at $\delta$ 6.2 associated with an enol ether function, and the appearance of H(2) as a doublet ($\delta$ 3.7) and H(15) as a triplet, placed the site of unsaturation at C(16)–C(17). The configuration of the oxazinic C(17′), however, could not be ascertained with certainty.

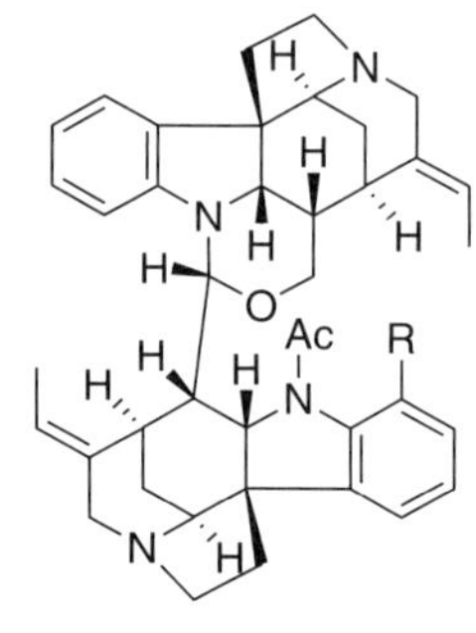

**410** R = H strychnobiline
**413** R = OH

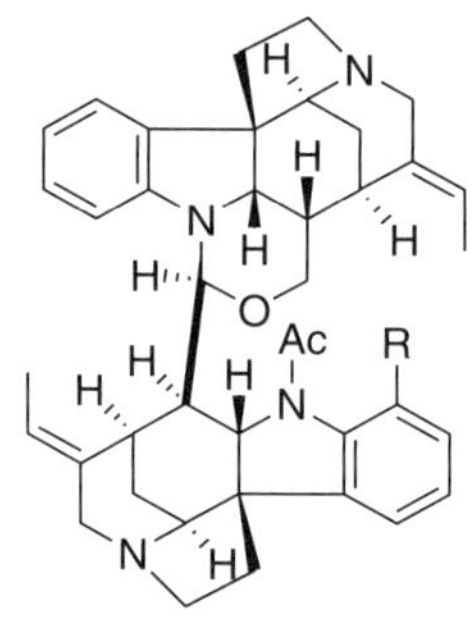

**411** R = H isostrychnobiline
**412** R = OH

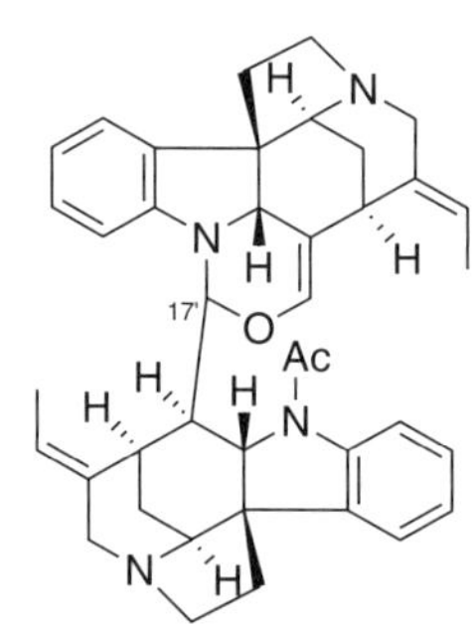

**414** dehydroisostrychnobiline

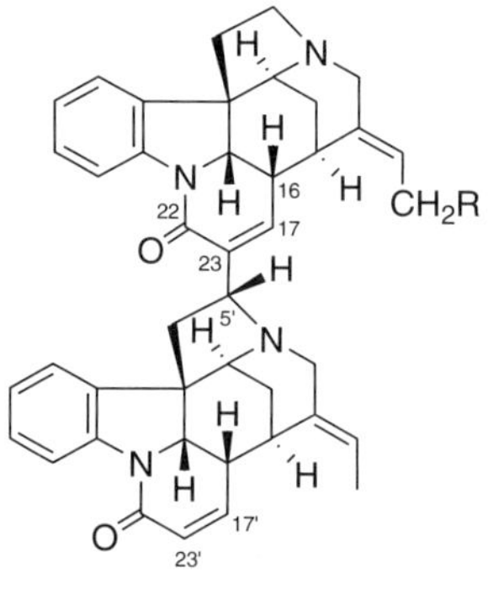

**415** R = H sungucine
**417** R = OH

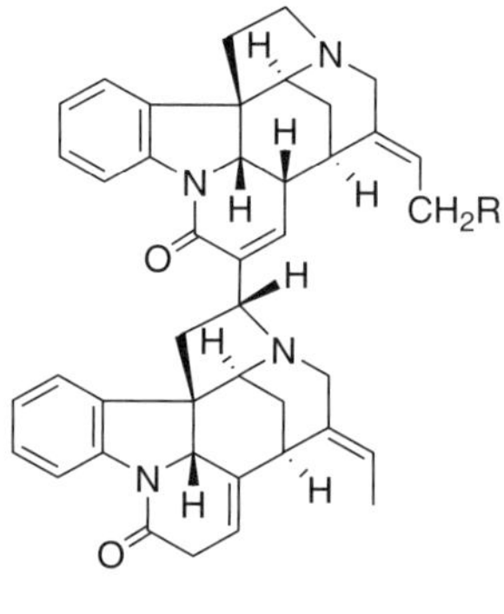

**416** R = H isosungucine
**418** R = OH

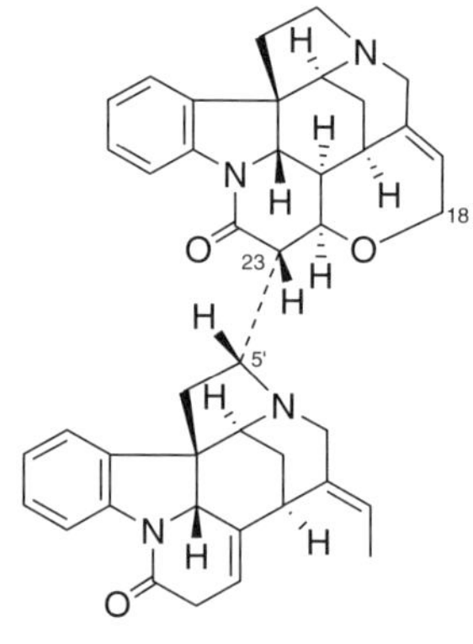

**419** strychnogucine A

*S. icaja*, the only African *Strychnos* containing strychnine, also provided sungucine (**415**), a *Strychnos–Strychnos* bisindole with the unprecedented C(5′)–C(23) link, whose structure was solved by X-ray analysis (*268,269*). Three new sungucine derivatives, isosungucine (**416**), 18-hydroxysungucine (**417**), and 18-hydroxyisosungucine (**418**) were subsequently isolated from the same plant (*270*). Isosungucine (**416**) showed NMR spectral data consistent with the change of the site of unsaturation from 17′-23′ to 16′-17′ for the lower half of the molecule, viz., the absence of the H(16′) signal ($\delta_{C16'}$ 140.9), and H(2′) observed as a singlet instead of a doublet in sungucine. The structures of 18-hydroxysungucine (**417**) and 18-hydroxyisosungucine (**418**), were readily assigned as the 18-hydroxy derivatives of sungucine and isosuguncine, respectively, from the NMR spectral data. The same report also provided $^{1}$H and $^{13}$C NMR data for sungucine. In subsequent reinvestigations of the same plant, additional bisindoles with antiplasmodial activity, viz., strychnogucines A (**419**), B (**420**), and C (**421**), were further isolated (*271,272*). All three bisindoles are characterized by the same C(5′)–C(23) link present in sungucine and its derivatives. The NMR data of **419** indicated a similarity with strychnine (**285**) for its upper half, while the lower unit was identical with that of the lower half of isosungucine. The C(5′)-C(23) link was supported by the observed H(5′)/C(17),C(23) and H(17)/C(5′) correlations in the HMBC spectrum. Similarly, the NMR data revealed **420** to be constituted from isostrychnine II (17,23-double bond, upper half) and strychnine (lower half), and **421** to be constituted from strychnine (upper half) and deoxyisostrychnine II (lower half). *S. icaja* also provided a new trimeric indole, strychnohexamine (**422**), incorporating three strychnan units, as indicated by the molecular formula ($C_{59}H_{60}N_6O$) established by HRESIMS (*273*). The NMR data readily revealed the presence of a bisnordihydrotoxiferine moiety for the lower portion linked *via* C(5) to C(23″) of the upper deoxyisostrychnine II unit (similar to the upper unit of sunguсine). Strychnohexamine was also found to be active against the chloroquine-sensitive FCA strain of *P. falciparum* (*272*).

*S. panganensis* from East Africa provided four new bisindole alkaloids, panganensines R (**423**), S (**424**), X (**425**), and Y (**426**), in addition to matopensine (*274*).

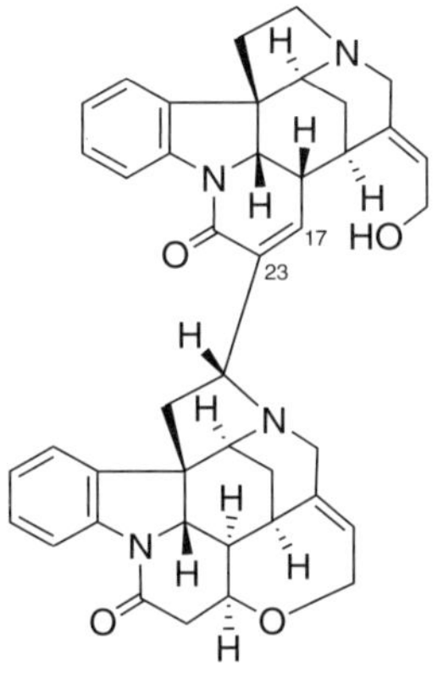

**420** strychnogucine B

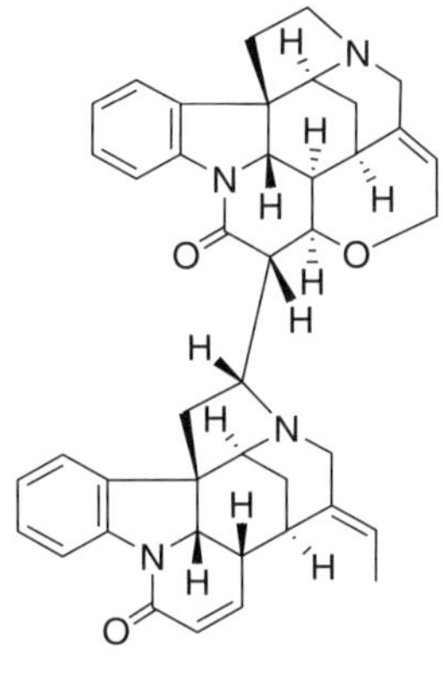

**421** strychnogucine C

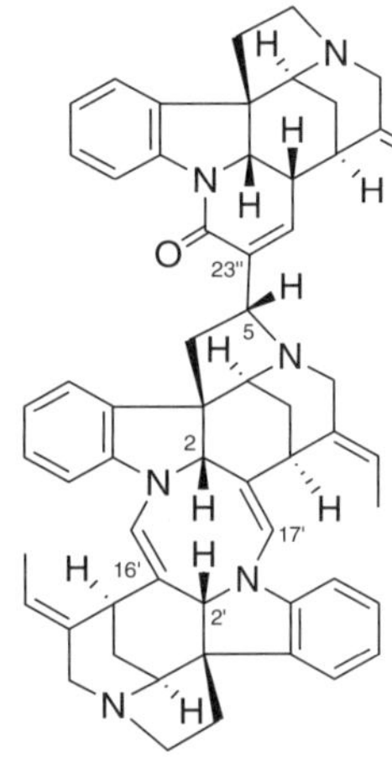

**422** strychnohexamine

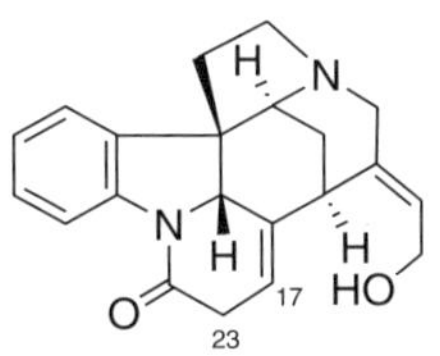

isostrychnine

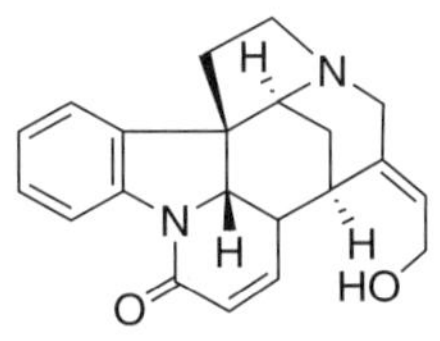

isostrychnine II

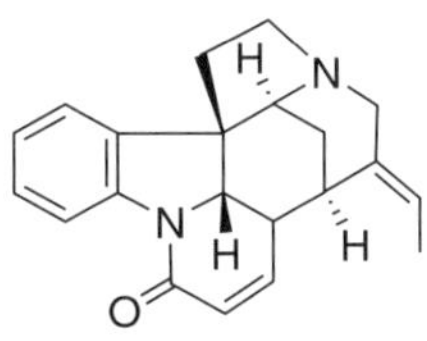

18-deoxyisostrychnine II

Panganensines R (**423**) and S (**424**) showed almost identical spectral features. The NMR spectral data revealed constitution from *N*-deacetylspermostrychnine (**427**) and *N*-deacetylisoretuline (**428**), both of which were also present in the plant. The dimer is branched from C(18) of **428** to C(21′) of **427**, as inferred from the observed H(21′)/C(19) and H(18)/C(20′) long-range couplings in the HMBC spectrum. The only difference between the two alkaloids lies in the configuration at C(19′) which was resolved from the ROESY spectrum. The observation of a H(21′)/H(18′) correlation in the ROESY spectrum indicated that the H(19′) stereochemistry is α (or 19′*R*) in panganensine R, while conversely, the stereochemistry of H(19′) in panganensine S is β (or 19′*S*) from the observed H(21′)/H(19′) interaction. Panganensines X and Y were shown to be constituted from the union of *N*-deacetylretuline (upper half) and *N*-deacetylisoretuline (lower half) units. The bisindoles are linked at C(19) of the retuline and C(10′) of the isoretuline half, as indicated by the HMBC data (H(9′)/C(19), H(11′)/C(19), H(19)/C(10′), H(18)/C(10′)). As in the previous pair, panganensines X and Y differ only in the configurations at C(19), which, however, could not be determined based on the available data.

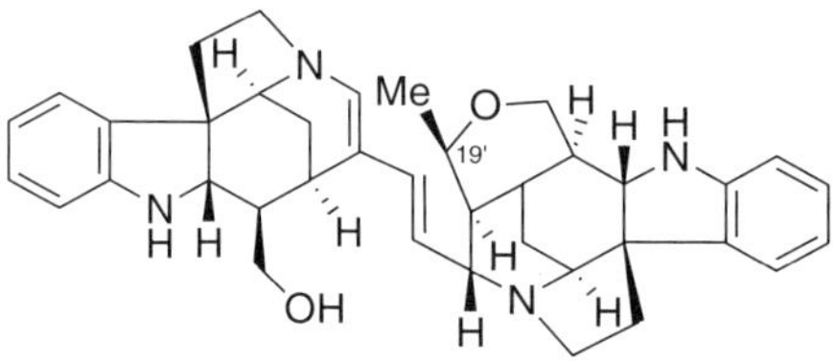

**423** panganensine R

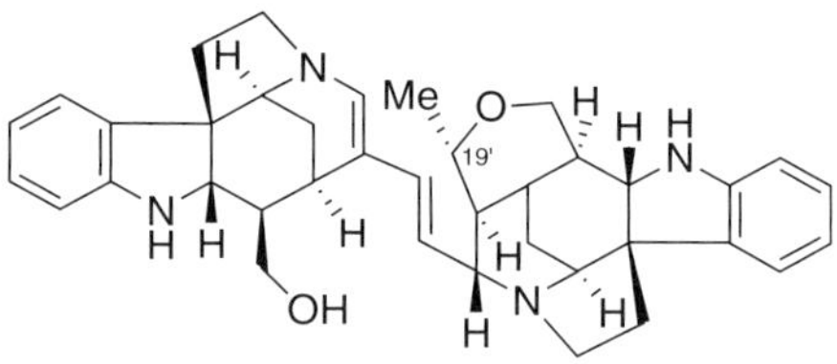

**424** panganensine S

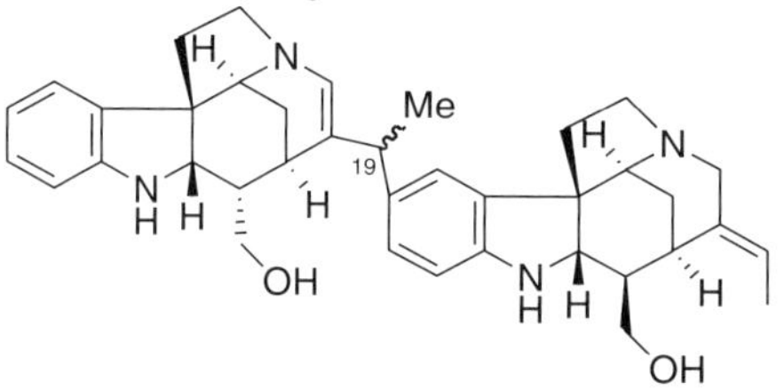

**425**, **426** panganensines X and Y

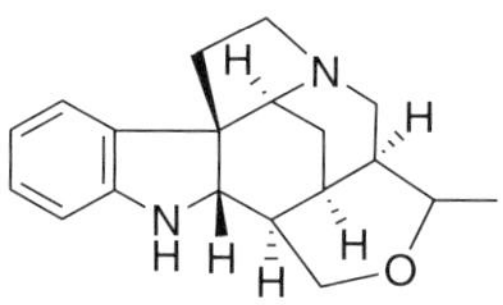

**427** *N*-deacetylspermostrychnine

A number of these *Strychnos* bisindoles were found to possess significant *in vitro* antiplasmodial activity against *P. falciparum*. Matopensine (**400**), 18-hydroxymatopensine (**406**), 16-methoxyisomatopensine (**404**), isosungucine (**416**), 18-hydroxyisosungucine (**418**), and strychnogucine B (**420**) were found to be effective against both chloroquine-sensitive, as well as chloroquine-resistant, strains of *P. falciparum*. Strychnogucine B (**420**) was the most potent alkaloid of this group, and all isolates displayed higher potency against the chloroquine-resistant strain (W2) compared to the sensitive FCA strain of *P. falciparum* (*172,270,271*). Sungucine (**415**), isosungucine (**416**), and 18-hydroxyisosungucine (**418**) also showed cytotoxic activity against the human cancer cell lines, HeLa and KB, as well as against the human fibroblasts W138. The cytotoxic effects of sungucine (**415**) and isosungucine (**416**) were further investigated using human leukemia HL-60 cells. The results indicated that these alkaloids induced apoptosis in these cells, with evidence of binding to DNA and inhibition of nucleic acid synthesis (*275*).

The biological activity of bisnordihydrotoxiferine (**287**) from *S. trinervis* and *S. divaricans* has been investigated in some detail. It was found to show a wide spectrum of antimicrobial action against Gram-positive, Gram-negative, and acid-fast microorganisms, filamentous and yeast-like fungi, and certain phytopathogenic microorganisms. The antimicrobial activity though broad, was relatively weak (*276*). Bisnordihydrotoxiferine also inhibited *E. coli*-induced diarrhoea and cholera toxin-stimulated intestinal fluid accumulation in mice (*277*). It was also found to antagonize in a non-specific manner, oxytocin- and acetylcholine-induced contractions in the rat uterus, and acetylcholine and histamine responses in the guinea pig ileum (*278*).

## XIX. Secodine–Secodine Type

*Aspidosperma excelsum* provided in addition to tetrahydrosecamine (**429**) and 16-demethoxycarbonyltetrahydrosecamine (**430**), two new secamine alkaloids, 16-hydroxytetrahydrosecamine (**431**) and 16-hydroxy-16-demethoxycarbonyltetrahydrosecamine (**432**) (*279*). All four alkaloids showed antimicrobial activity and **429** and **430** were previously identified from *A. marcgravianum* (*280*). Alkaloids **431** and **432** both showed the typical secamine fragment ion at $m/z$ 126, as well as similar UV spectra. The absence of the H(16) signal at $\delta$ 4.3 in the $^1$H NMR spectrum of **431** compared to that of **429**, indicated C(16) as the site of OH substitution ($\delta_{C16}$ 70.1). The $^1$H NMR spectrum of **432** was essentially similar to that of **431**, except that only one methoxy signal was observed, and another signal, a broad one-H singlet, absent in the spectrum of **431**, was observed at $\delta$ 5.04 suggesting hydroxy substitution at C(16). Unfortunately, paucity of material prevented recording of the $^{13}$C NMR spectrum. Didemethoxycarbonyltetrahydrosecamine (**433**), the product of demethoxycarbonylation of **429** and **430**, was also found to be antimicrobially active. It has also been subsequently isolated as a natural product from *Rhazya stricta* (*281*).

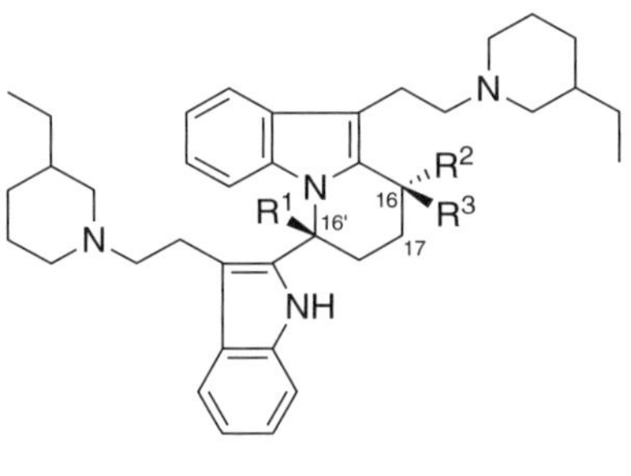


**429** $R^1 = R^2 = CO_2Me$, $R^3 = H$
**430** $R^1 = CO_2Me$, $R^2 = R^3 = H$
**434** $R^1 = R^3 = H$, $R^2 = CO_2Me$
**435** $R^1 = R^2 = H$, $R^3 = CO_2Me$

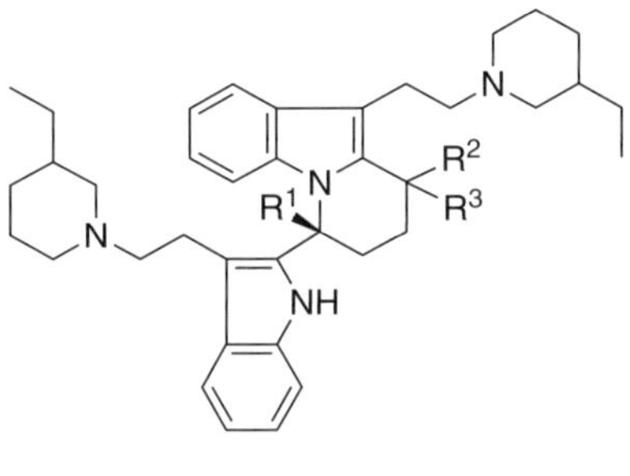


**431** $R^1 = R^2 = CO_2Me$, $R^3 = OH$
**432** $R^1 = CO_2Me$, $R^2 = H$, $R^3 = OH$
**433** $R^1 = R^2 = R^3 = H$

Both the epimeric pairs of 16′-demethoxycarbonyltetrahydrosecamine, **434** and **435**, have also been isolated from *R. stricta* (*282,283*). The two alkaloids are prone to interconversion on TLC. The configuration at C(16) was determined based on the chemical shifts of the 16- and 16′- ester methyl hydrogens. Tetrahydrosecamine and its derivatives were found to display *in vitro* cytotoxicity toward KB cells (*284*).

## XX. Apparicine–Vobasine Type

Vobparicine (**436**) (*285*) and its N-oxide (**437**) (*286*), were obtained in minute amounts from *Tabernaemontana chippii*. The $^1$H NMR spectrum of **436** indicated the presence of vobasinyl and apparicine halves, as evident from comparison with the spectra of vobasinol (**438**), conodurine (vobasinyl half), and apparicine (**439**). The spectrum also showed the presence of eight aromatic hydrogens, indicating branching from non-aromatic positions of the constituent moieties. Notable differences were the presence of a doublet at $\delta$ 5.96 ($J$ 10.1 Hz), absence of the two characteristic apparicine H(22) singlets at ca. $\delta$ 5.3, and the additional coupling of 10.1 Hz for the vobasinyl H(3′). These observations suggested branching from C(22) of the apparicine half to C(3′) of the vobasinyl half, an inference which was supported by the $^{13}$C NMR data, which showed a downfield shift for C(22) and an upfield shift for C(16) when compared with the spectra of apparicine and voacamine (vobasinyl half). The geometry of the 16, 22-double bond was deduced to be *Z* based on NOE experiments (NOE between H(22) and NH′). Finally, a biomimetic conversion by reaction of vobasinol with excess apparicine in 1.5% methanolic HCl, provided further confirmation of the structure.

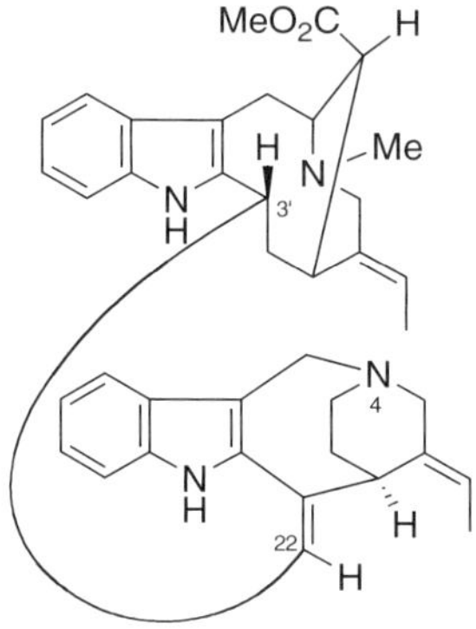

**436** vobparicine
**437** *N*(4)→O

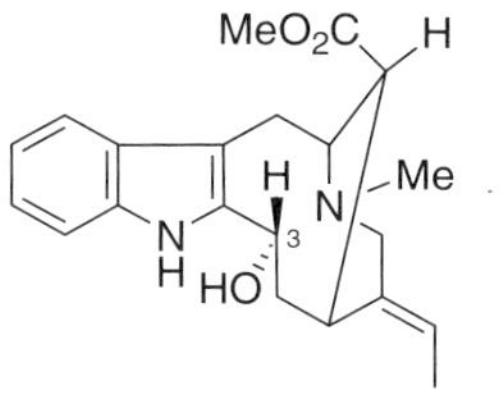

**438** vobasinol

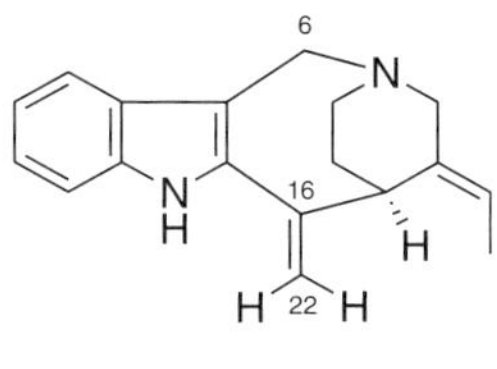

**439** apparicine

A third bisindole constituted from vobasinyl and apparicine moieties, and isomeric with **436**, pseudovobparicine (**440**), was subsequently isolated from another *Tabernaemontana, T. divaricata* (*287*). The $^{1}$H NMR data showed the presence of seven aromatic hydrogens, two indolic NH, two intact apparicine H(22) singlets at $\delta$ 5.27 and 5.39, and only one H(3′) signal for the vobasinyl part which showed the usual coupling to the two C(14′) hydrogens. From these observations, as well as from a comparison of the aromatic hydrogen shifts of the apparicine half with those of tabernamine, it was deduced that the branching points in pseudovobparicine are from C(3′) of the vobasinyl half to C(10) of the apparicine half.

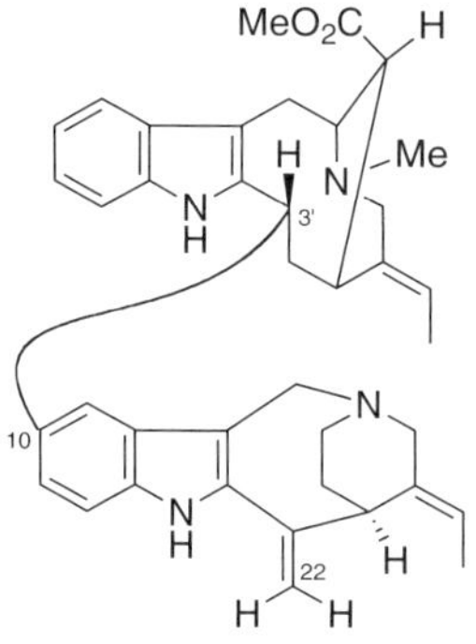

**440** pseudovobparicine

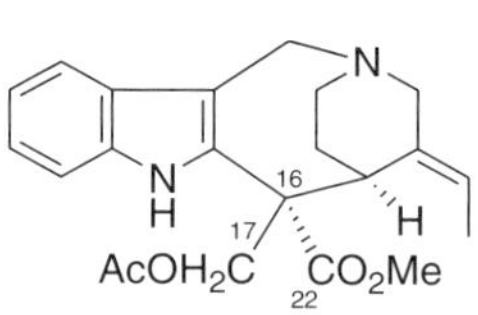

**442** *O*-acetylvallesamine

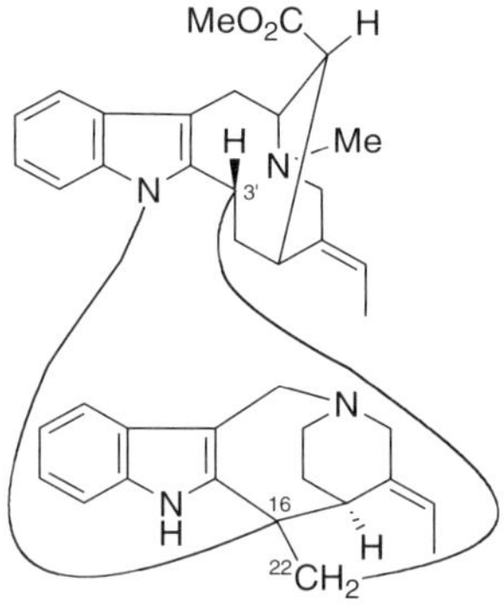

**441** monogagaine

Another member of this group is monogagaine (**441**), which was isolated in small amounts from *T. chippii* and *T. dichotoma*. It is also constituted from the union of vobasinyl and apparicine halves, and is also isomeric with **436** and **440**. The eventual structure elucidation required recourse to 500 MHz NMR analysis, in addition to HREIMS and FDMS. The structure is characterized by the formation of a bond from N(1′) of the vobasinyl unit to C(16) of the apparicine unit, and another bond from C(22) to C(3′), resulting in C(16) being a spirocyclic centre (*288*). Although the EIMS of **441** showed fragments ($m/z$ 182, 180, and 122) characteristic of a 3′-vobasinyl half, the $^1$H NMR spectrum indicated that the characteristic vobasinyl signals, viz., NMe, $CO_2Me$, H(16), H(21), and H(14) were not observed at their usual positions, suggesting that a vobasinyl unit, if present, must be in a different conformation. The NMR spectrum further showed the presence of two unsubstituted aromatic rings, only one indolic NH, two ethylidene side chains, two AB doublets ($J$ 17.7 Hz) at $\delta$ 4.80 and 4.45, a $CO_2Me$ at $\delta$ 3.47, and an NMe at $\delta$ 2.45, suggesting the presence of two bonds connecting the two halves. The AB doublets are reminiscent of the geminal H(6) of the apparicine half of vobparicine, and based on the COSY spectrum, which indicated the presence of a 3′-vobasinyl half with an additional methylene attached to C(3′), as well as a $CH_2CH_2CH$ fragment strongly reminiscent of the C(3)–C(14)–C(15) spin system in *O*-acetylvallesamine (**442**), the gross structure of monogagaine could be assembled.

The proposed structure is consistent with the $^{13}$C NMR spectral data, and bears a resemblance to vobparicine, except for the absence of the 16–22 double bond, and the formation of a new bond between the vobasinyl nitrogen and the apparicine C(16). The observed $J_{5'-16'}$ coupling of ca. 7 Hz indicated that the configuration of C(16′) is the same as in vobasinol. The difference in the chemical shifts and coupling constants of the vobasinyl piperidine ring hydrogens in monogagaine and vobparicine, compared with other vobasinyl-containing bisindoles, is rationalized on the grounds that the piperidine ring of the vobasinyl half adopts the boat instead of the chair conformation. Likewise, examination of the coupling constants indicated that the configuration of C(3′) is also the same as in vobparicine, i.e. with H(3′$\beta$). The configuration of the spirocyclic C(16) was established to be *S* from extensive NOE experiments, in particular the observation of strong NOE between H(12′) and H($14_R$).

### XXI. Ellipticine–Ellipticine Type

Strellidimine (**443**), isolated from *S. dinklagei*, is the first, and to date the only example of a natural bisellipticine alkaloid (*289*). The UV spectrum displayed characteristic absorptions associated with a highly conjugated polyaromatic system involving at least one pyridine ring. The molecular ion could not be detected by EIMS, which, however, showed a strong fragment ion at $m/z$ 195, suggesting the presence of a dimethylcarbazole-derived unit. The molecular ion ($MH^+$ 509) was detected by CIMS, which also showed the two ions at $m/z$ 263 and 247, characteristic of hydroxyellipticine and dihydroellipticine units, respectively.

Except for a few differences, the $^1$H NMR spectrum showed a correspondence to most of the signals of 10-hydroxyellipticine (**444**) and 3,14-dihydroellipticine (**445**), which were also present in the plant. The characteristic signals due to H(9) of **444**, and H(21) of **445**, were absent from the spectrum of **443**, while the H(11) and H(12)

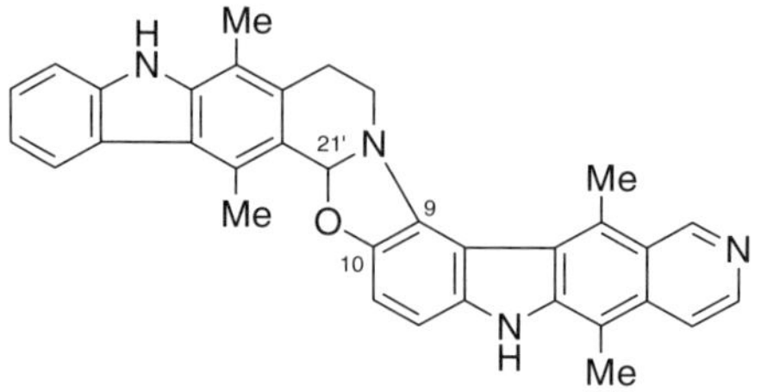

**443** strellidimine

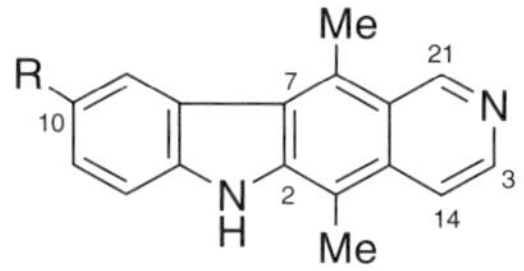

**444** R = OH
**445** R = H, 3,14-dihydro

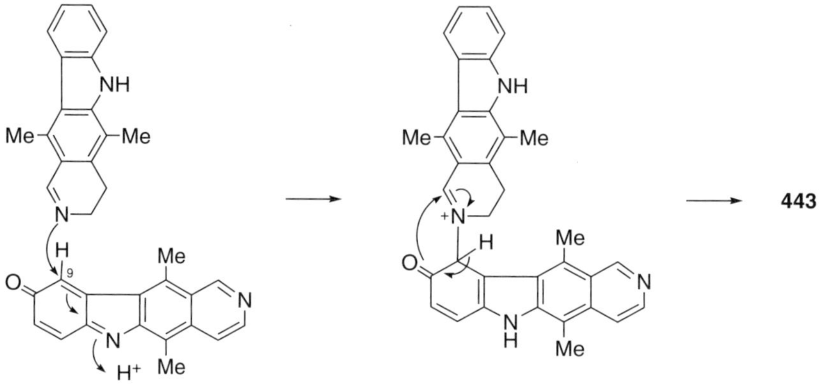

Scheme 32

signals of **443** appeared as an AB system, and the isolated H(21′) of a dihydrooxazole unit was seen as a singlet at $\delta$ 6.55. A biogenetic pathway was proposed, involving condensation of the two monomeric units *via* addition of **445** through its nucleophilic dihydopyridine nitrogen to C(9) of the quinone form of **444**, followed by cyclization to **443** (Scheme 32). Support for the proposal was obtained from the oxidation (horseradish peroxidase-$H_2O_2$) of **444** in the presence of **445** at pH 7.4 to provide a quantitative yield of strellidimine (**443**).

## XXII. Iboga–Vobasine Type

Bisindoles of the iboga–vobasine type abound in the genus *Tabernaemontana* (*Ervatamia*), due in large measure to the propensity of vobasinyl cations to react with the nucleophilic aromatic carbons of iboga alkaloids. *E. hainanensis* (*T. bufalina*) from Hainan, China, provided three new bisindoles, ervahanines A (**446**), B (**447**), and C (**448**), branched from C(3) of a vobasine unit to C(11′), C(10′), and C(12′), respectively, of a ibogan moiety, which was shown by the spectral data to correspond to coronaridine (*290*). The mass spectra indicated that all three were isomers ($M^+$ $m/z$ 674), and displayed voacamine-like fragmentation. The determination of the branching point on the ibogan part in ervahanine A was based on comparison of the $^1H$ and $^{13}C$ NMR spectral data with those of capuvosine and tabernamine. Additional support for the structure was provided by a semisynthetic conversion, through the reaction of vobasinol (**438**) with coronaridine (**449**) in HCl/

MeOH to give ervahanine A (**446**) as the major product, accompanied by some ervahanine B (**447**).

Four additional bisindoles were subsequently obtained from the same plant, ervahaimines A (**450**) and B (**451**), and ervahainamidines A (**452**) and B (**453**), differing from the ervahanines by modification at C(3′) of the ibogan moiety (*291*). Thus ervahaimine B (**451**) is the 3′-oxo derivative of ervahanine B (**447**), as duly reflected in the changes in the spectral data of **451** in comparison with those of **447**. The IR spectrum showed the presence of a second carbonyl function (1660 $cm^{-1}$) in addition to the methyl ester absorption at 1730 $cm^{-1}$. The molecular ion was observed at *m/z* 688, 14 mass units higher than that of **447**, while the characteristic fragment ions of the vobasinyl unit (*m/z* 122, 180, 181, 194) remained intact, indicating that the carbonyl function should reside in the iboga moiety, an inference further supported by observation of an intense voacamine type fragment ion at *m/z* 493. The location of the carbonyl group at C(3′) was indicated by the $^1$H NMR spectrum which showed a complex 2H multiplet at δ 4.5, which was interpreted as arising from superposition of a 1H singlet on another 1H multiplet, assigned to H(14′) and H(21′), respectively, by analogy with 3-oxocoronaridine. Comparison of the aromatic shifts with those of ervahanine B confirmed similar C(10′) substitution in ervahaimine B. In a similar manner, ervahaimine A (**450**) was deduced to be the 3′-oxo derivative of ervahanine A (**446**). A partial synthesis involving condensation of vobasinol (**438**) with 3-oxocoronaridine (**454**) under acidic conditions to yield a mixture of ervahaimines A and B (major product), provided additional confirmation for the proposed structures. Ervahainamidines A (**452**) and B (**453**) had the same molecular formula as shown by HREIMS. Comparison of the mass spectral data with those of the ervahaimines and ervahanines, indicated an intact vobasinyl part with changes in the iboga moiety in the ervahainamidines (intense ion at *m/z* 523). The IR spectra showed the presence of NH, OH, and ester carbonyl functions, but absence of the carbonyl peak at 1660 $cm^{-1}$. In the $^1$H NMR spectra, the 2H multiplet at δ 4.5 associated with 3-oxo iboga derivatives was absent, instead, signals due to a hydroxyethyl ($CHOHCH_3$) group were observed (δ 1.15, *d*, *J* 7 Hz, 3H; 3.55, *m*, 1H). The hydroxyethyl group was deduced to be at C(3′) from comparison of the $^1$H NMR data with that of 3-hydroxyethylcoronaridine (**455**) which was also present in the plant.

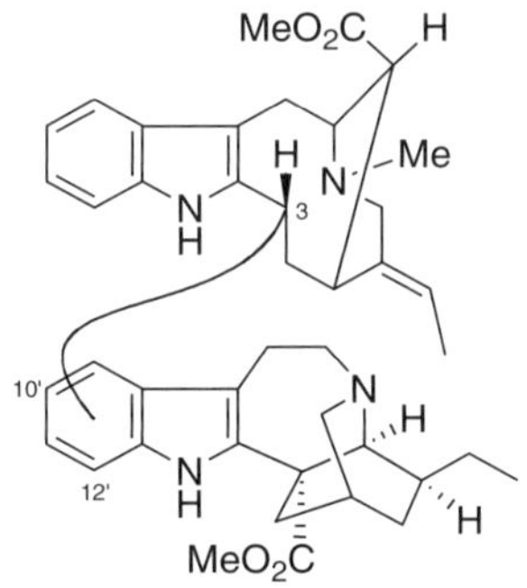


**446** C(3)-C(11′) bond ervahanine A
**447** C(3)-C(10′) bond ervahanine B
**448** C(3)-C(12′) bond ervahanine C

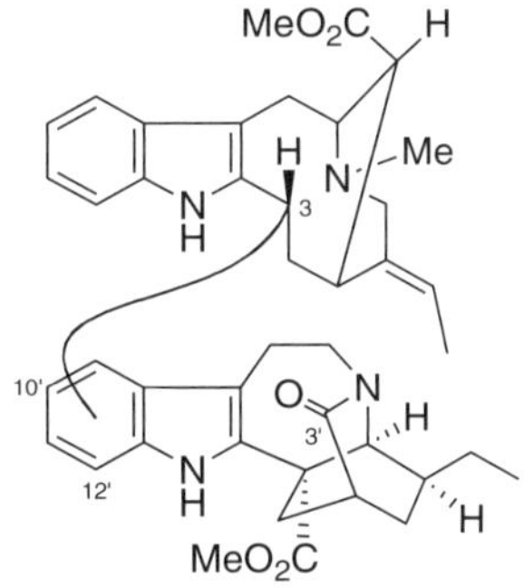


**450** C(3)-C(11′) bond ervahaimine A
**451** C(3)-C(10′) bond ervahaimine B

Another ervahanine congener, 19,20-dihydroervahanine A (**456**), was isolated from *E. coronaria* (*292*). The mass spectrum showed peaks at $m/z$ 196, 182, and 124, and at $m/z$ 136 and 122, which also occur in the mass spectra of 19,20-dihydrovobasinols and coronaridine, respectively, suggesting the presence of 19,20-dihydrovobasinyl and coronaridine moieties in **456**. Further confirmation was provided by the $^1$H NMR spectrum which displayed a close similarity with that of ervahanine A (**446**), except for replacement of signals due to ethylidene with signals of an ethyl group. The signals for the aromatic region resembled those of tabernamine and ervahanine A, indicating a similar 3-11′ bond in **456**. Another ervahanine derivative, 19′(*S*)-hydroxyervahanine A (**457**) was obtained from *T. corymbosa* from Malaysia (*293*). The FABMS showed an $MH^+$ ion at $m/z$ 691, 16 mass units higher than that of ervahanine A (**446**), which was also obtained from the plant. In addition, a fragment peak due to loss of $H_2O$ was also observed in the mass spectrum. The location of the hydroxyl susbstituent was deduced to be at C(19′) of the iboga unit from a comparison of the NMR spectral data of **457** with those of ervahanine A (**446**). The spectral data of the two alkaloids showed a general similarity, except for replacement of the iboga C(20′) ethyl side chain in **446**, by a hydroxyethyl in **457**. The chemical shifts of C(15′) and C(21′) at $\delta$ 22.9 and 59.5 respectively, allowed assignment of the C(19′) configuration as *S*, by analogy with heyneanine (**458**) (*294*).

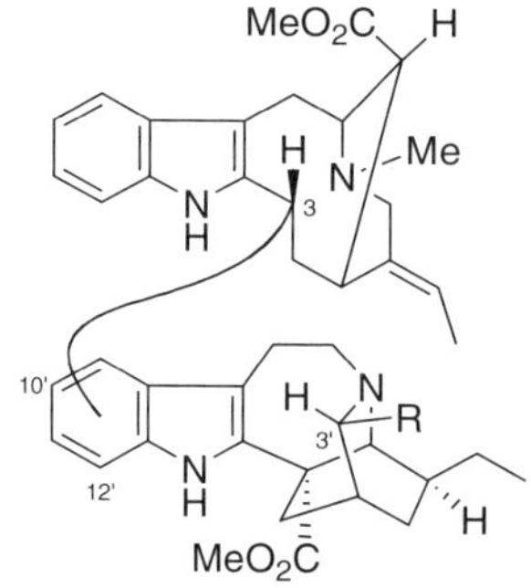

**452** R = CH(OH)Me, C(3)-C(11') bond
**453** R = CH(OH)Me, C(3)-C(10') bond

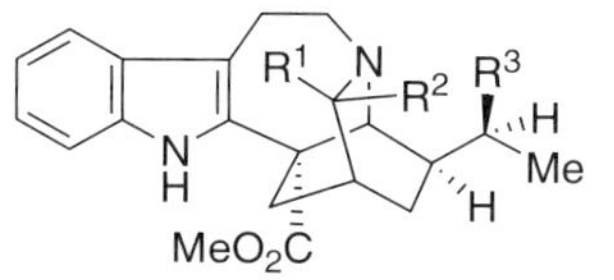

**449** R¹ = R² = R³ = H
**454** R¹ = R² = O, R³ = H
**455** R¹ = R³ = H, R² = CH(OH)Me
**458** R¹ = R² = H, R³ = OH

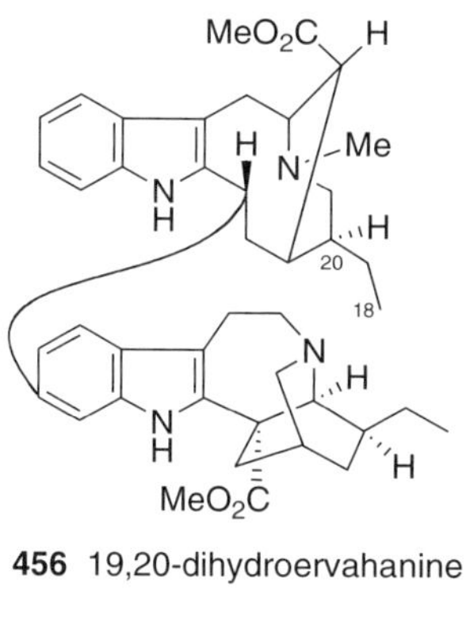

**456** 19,20-dihydroervahanine A

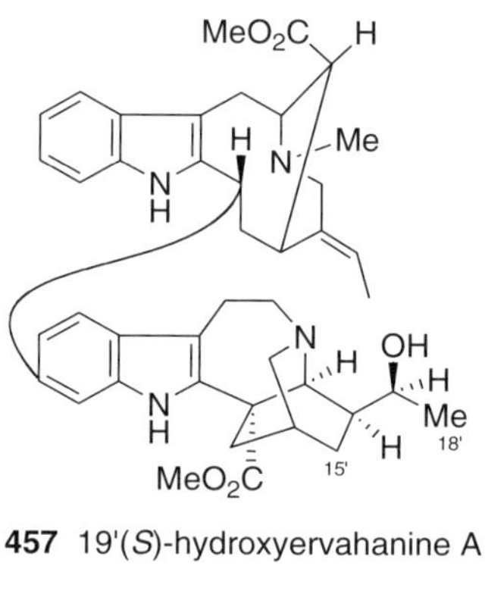

**457** 19'(*S*)-hydroxyervahanine A

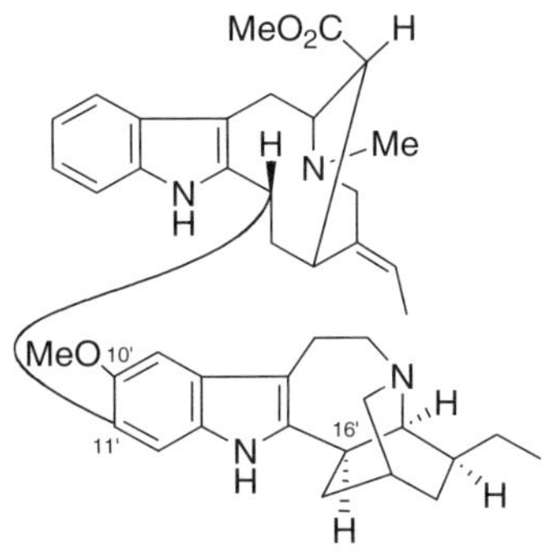

459

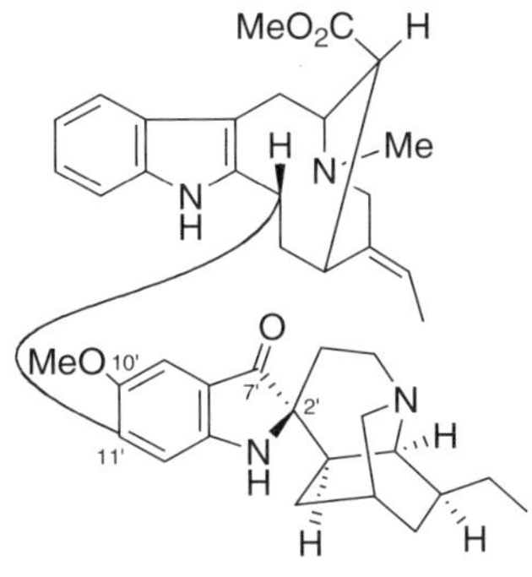

460

The Malaysian *E. hirta* (*T. hirta*) gave in addition to 16′-decarbomethoxyvoacamine (**459**) and 19,20-dihydro-16′-decarbomethoxyvoacamine, a new bisindole, 16′-decarbomethoxyvoacaminepseudoindoxyl (**460**) (*295*). Alkaloid **460** in solution showed the intense yellow-green fluorescent coloration characteristic of a pseudoindoxyl chromophore. This was readily confirmed by the UV spectrum (228, 372, 294, and 417 nm) as well as the observation of an IR band at 1660 $cm^{-1}$. The mass spectrum showed an $M^+$ at $m/z$ 662 analyzing for $C_{41}H_{50}N_4O_4$, 16 mass units more than 16′-decarbomethoxyvoacamine. The mass spectrum also showed characteristic fragments of vobasinyl as well as iboluteinyl moieties. The $^1H$ NMR of **460** was similar to that of **459**, except for the upfield shift of the iboluteinyl NH, in agreement with a pseudoindoxyl structure. The $^{13}C$ NMR spectrum provided further support for this from the characteristic shifts for C(2′) and C(7′), at $\delta$ 67.5 and 204.8, respectively. The C(11′) substitution was inferred by comparison of the aromatic carbon shifts with those of iboluteine. The substitution at C(3) was deduced to be $\alpha$ for steric reasons, in common with other vobasinyl–iboga bisindoles. Two voacamine derivatives, ervadivaricatines A and B (20-$\beta$H-voacamine **461** and 20-$\alpha$H-voacamine **462**, respectively) were obtained from the Chinese plant, *E. divaricata* (*T. divaricata*) (*296*).

Another voacamine derivative, existing as an unresolvable mixture of epimers, 3′(*R*/*S*)-hydroxyvoacamine (**463**), was isolated from *T. chippii* (*286*) and *T. dichotoma* (*297*). The structure was deduced based on the mass spectrum, the $^1H$ NMR spectrum, and from chemical correlation ($NaBH_4$ reduction) with voacamine (**464**). Other similar 3-hydroxy bisindoles isolated from *T. chippii* include 3′(*R*/*S*)-hydroxyconoduramine (**465**), 3′(*R*/*S*)-hydroxyconodurine (**466**), and 3′(*R*/*S*)-hydroxy-16′-decarbomethoxyconodurine (**467**) (*286*). The mass spectra of these alkaloids were characterized by weak or absent molecular ions, and stronger M-2, M-16, M-17, and M-18 fragment ions as well as doubling of many of the signals of the iboga half in the $^1H$ and $^{13}C$ NMR spectra. The 3′*R* and 3′*S* isomers can be distinguished by the C(3) carbinolamine shift (for example, $\delta$ 95.9 for the 3′*R*-epimer and 86.1 for the 3′*S*-epimer for **465**). As in the case of **463**, $NaBH_4$ reduction of **465** gave conoduramine (**468**), while similar reduction of **466** gave conodurine (**469**). The *R*- and *S*-epimers could not be resolved due probably to rapid interconversion catalyzed by the $SiO_2$ layer.

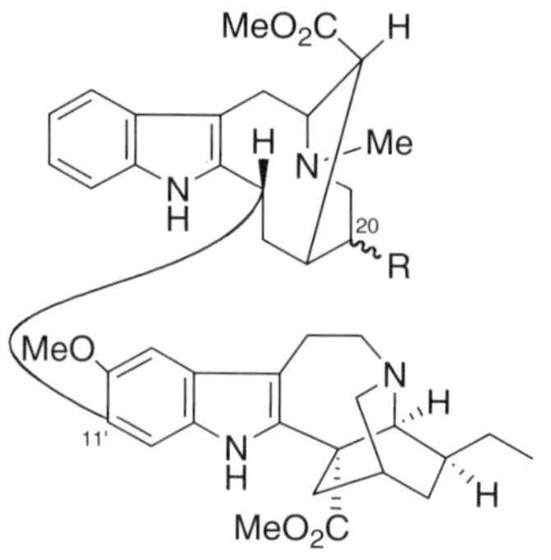

**461** R = α-Et
**462** R = β-Et

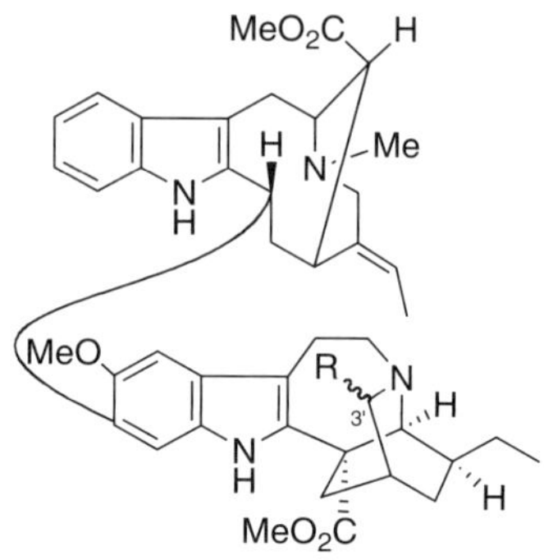

**463** R = OH 3'(*R/S*)-hydroxyvoacamine
**464** R = H voacamine

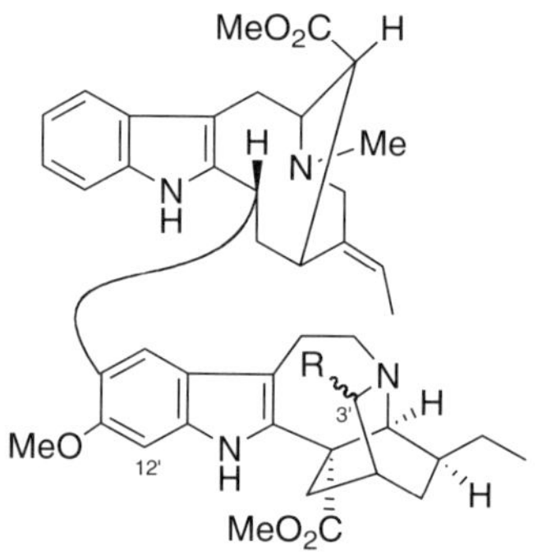

**465** R = OH 3'(*R/S*)-hydroxyconoduramine
**468** R = H conoduramine

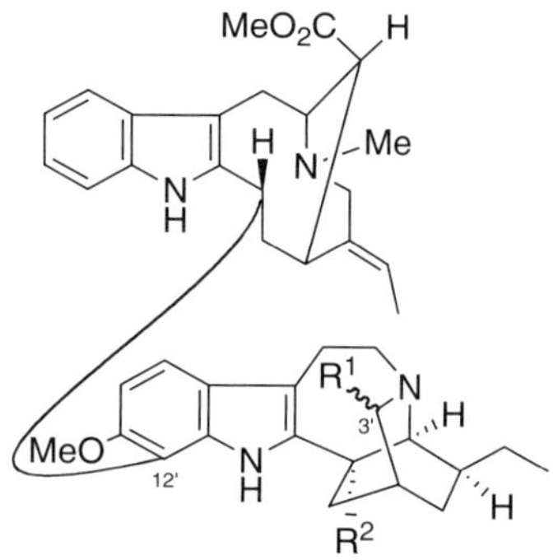

**466** $R^1$ = OH, $R^2$ = $CO_2Me$
**467** $R^1$ = OH, $R^2$ = H
**469** $R^1$ = H, $R^2$ = $CO_2Me$ conodurine

*T. dichotoma* was another source of 3′-hydroxysubstituted bisindoles, such as 3′(*R*/*S*)-hydroxytabernamine (**470**), 3′(*R*/*S*)-hydroxy-*N*(4)-demethyltabernamine (**471**), 3′(*R*/*S*)-hydroxy-*N*(4)-demethylervahanines A (**472**) and B (**473**) (*297*). As in the previous cases, $NaBH_4$ reduction of 3′(*R*/*S*)-hydroxytabernamine (**470**) and 3′(*R*/*S*)-hydroxy-*N*(4)-demethyltabernamine (**471**), gave tabernamine (**474**) and *N*(4)-demethyltabernamine (**475**), respectively. The latter alkaloid **475**, whose $^1$H NMR spectrum was identical to that of tabernamine, except for the absence of the N(4)-methyl group ($\delta$ 2.62), was also present in the plant.

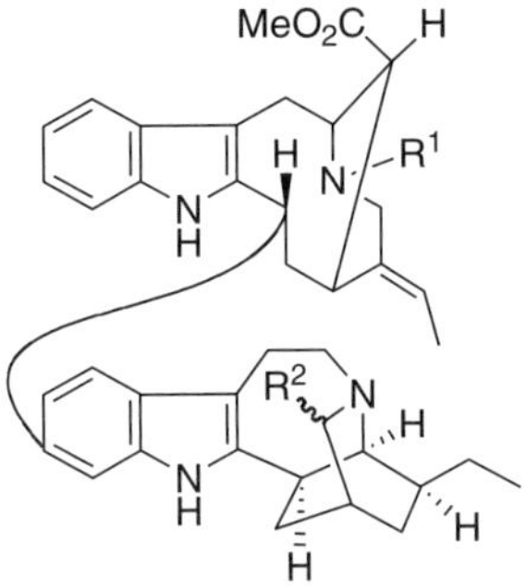

**470** $R^1$ = Me, $R^2$ = OH
**471** $R^1$ = H, $R^2$ = OH
**474** $R^1$ = Me, $R^2$ = H
**475** $R^1$ = $R^2$ = H

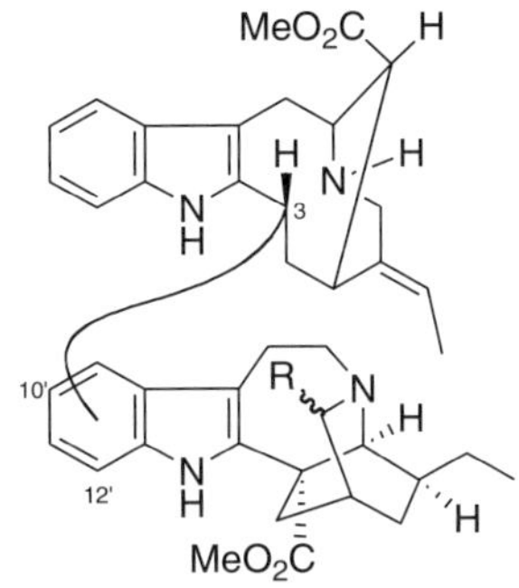

**472** R = OH, C(3)-C(11') bond
**473** R = OH, C(3)-C(10') bond

*T. pachysiphon* yielded 11′-demethylconoduramine (**476**) ($M^+$ *m/z* 690). The UV spectrum showed a bathochromic shift on basification (additional maximum at 323 nm), indicating the presence of a phenolic group. The $^1H$ NMR spectrum was identical to that of conoduramine (**468**) in all respects, except for the signal of the 11′-methoxy group at $\delta$ 3.97, which was absent in the spectrum of **476**. Attempts to correlate **476** with **468** by prolonged reaction of **476** with $CH_2N_2/BF_3$ led only to recovery of starting material, the lack of reactivity presumably being due to steric hindrance exerted by the vobasinyl moiety (*298*).

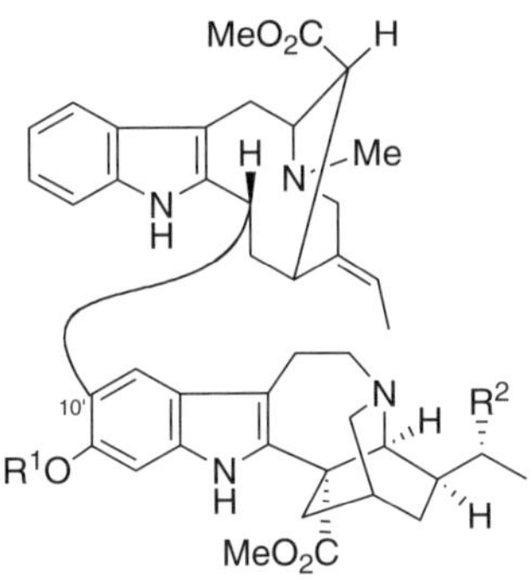

**476** $R^1$ = $R^2$ = H
**478** $R^1$ = Me, $R^2$ = OH

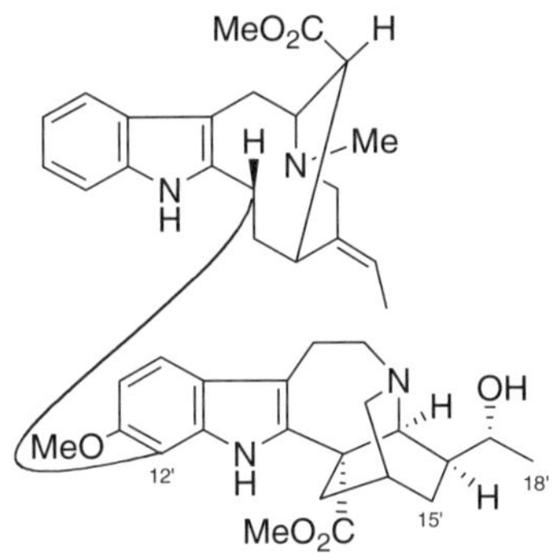

**477** 19'(*R*)-hydroxyconodurine

*Tabernaemontana subglobosa*, native to Taiwan, gave 19′(*R*)-hydroxyconodurine (**477**) and 19′(*R*)-hydroxyconoduramine (**478**). The UV and mass spectral data suggested that **477** is a derivative of conodurine (**469**), which was also isolated. The molecular formula from HREIMS indicated the addition of one oxygen atom compared to conodurine. The $^1H$ NMR spectrum showed a similarity to that of conodurine, except for the presencc of a hydroxyethyl group in place of an ethyl. The $^{13}C$ NMR spectrum was assigned with the use of 2-D methods. Examination of the C(19′) and C(15′) resonances and comparison with heyneanine and 19-*epi*-heyneanine, allowed the assignment of the C(19′) configuration in **477** as *R*.

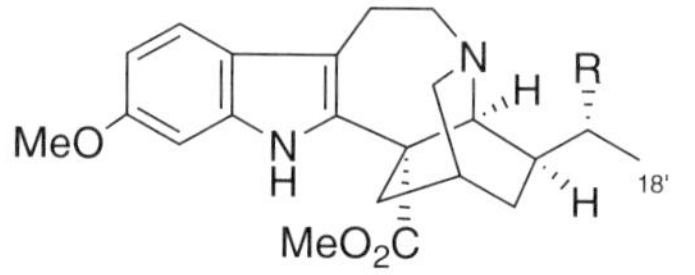


**479** R = H isovoacangine
**480** R = OH 19-*epi*-isovoacristine

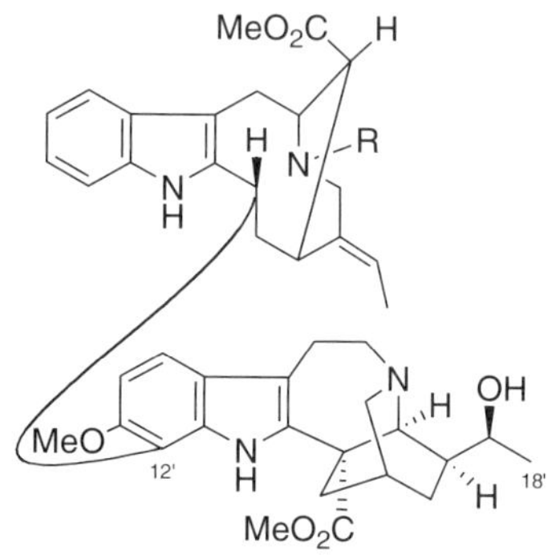


**481** R = Me 19′(*S*)-hydroxyconodurine
**482** R = H conodurinine

The structure of **478** was similarly deduced to be the 19′-hydroxy derivative of conoduramine (**468**), based on the mass spectrum and comparison of the NMR spectral data with that of conoduramine. Treatment of conodurine (**469**) with concentrated HCl in methanol gave isovoacangine (**479**) and conoduramine (**468**), together with unreacted conodurine. Similar treatment of **477** gave 19-*epi*-isovoacristine (**480**) and a small amount of the rearranged product, 19′(*R*)-hydroxyconoduramine (**478**), together with unreacted starting material. The formation of these products was rationalized based on a pathway involving acid-induced rearrangement and fragmentation (*299*).

*T. corymbosa* also provided several new conodurine and conoduramine derivatives, 19′(*S*)-hydroxyconodurine (**481**), conodurinine (**482**), and 19′(*S*)-hydroxyconoduramine (**483**) (*293*). The structure assignment of **481** was based on comparison of the MS and NMR spectral data with those of conodurine (**469**), while that of **483** was based on comparison with the spectral data of conoduramine (**468**) and 19′(*R*)-hydroxyconoduramine (**478**). The FABMS of conodurinine (**482**) showed an $MH^+$ ion at *m/z* 707, differing from **481** by 14 mass units, and suggesting the replacement of a methyl group by H. This was confirmed by the NMR spectral data which showed that absence of the vobasinyl NMe was the main difference in the spectra of the two compounds. The absence of the *N*-methyl substituent caused the resonances of C(5) and C(21) to be shifted upfield from $\delta$ 59.5 and 52.0, respectively, in **481**, to $\delta$ 53.3 and 44.1, respectively, in **482**, a feature that has been also observed previously in the related bisindoles, conodiparines E and F (*vide infra*). *T. corymbosa* also provided, in addition to tabernamine (**474**), three new tabernamine derivatives, the epimeric 19′(*S*)-hydroxytabernamine (**484**) and 19′(*R*)-hydroxytabernamine (**485**), and 19′-oxotabernamine (**486**) (*300*). The structures were readily assigned by comparison of the spectral data (MS and NMR) with those of tabernamine (**474**).

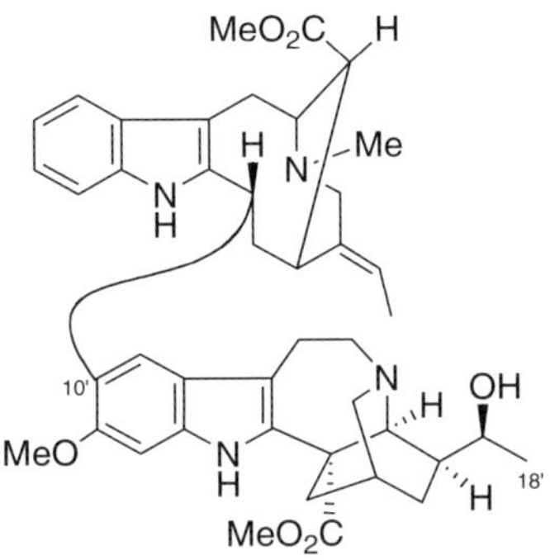

**483** 19'(*S*)-hydroxyconoduramine

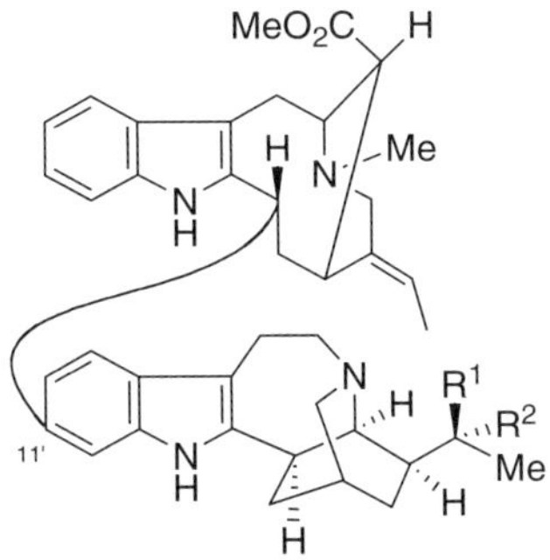

**484** $R^1$ = OH, $R^2$ = H
**485** $R^1$ = H, $R^2$ = OH
**486** $R^1$, $R^2$ = O

*H. modesta* (*Tabernaemontana coffeoides*) from Madagascar gave, in addition to the known tabernaelegantine A (**487**) (*301*), two additional bisindoles, 19′(*R*)-hydroxytabernaelegantine A (**488**) and 19,20-dihydro-20(*S*)-tabernamine (**489**) (*302*). The mass spectra of both alkaloids indicated the presence of dihydrovobasinyl and ibogan units. The $M^+$ of **488** at $m/z$ 722 differed from that of **487** by addition 16 mass units, while the $^1$H NMR showed similarity with that of **487**, except for replacement of one ethyl side chain (triplet at $\delta$ 0.82) with a hydroxyethyl (doublet at $\delta$ 1.18). The presence of an OH was also indicated by a M-18 fragment in the mass spectrum and the 3450 $cm^{-1}$ absorption in the IR spectrum, while the observation of ibogan-containing fragments at $m/z$ 522 and 509, 2 mass units less compared to the corresponding fragments in the mass spectrum of **487**, indicated that OH substitution was in the iboga moiety. This was confirmed by the $^{13}$C NMR spectral data which showed a correspondence for the dihydrovobasinyl half. The carbon spectrum also allowed the assignment of the C(19′) configuration as *R*, by comparison of the C(19′) and C(15′) shifts with those of heyneanine and 19-*epi*-heyneanine as before. The stereochemistry of the C(20′) ethyl side chain ($\beta$), and the branching from C(12′) of the iboga unit, were also based on the $^{13}$C NMR spectral data, through comparison with tabernamontanine (**490**) (C(14) and (C(16)) and tabernaelegantine A (**487**).

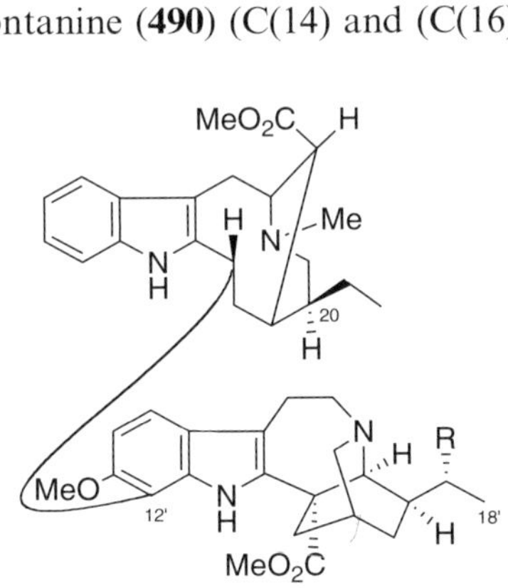

**487** R = H tabernaelegantine A
**488** R = OH

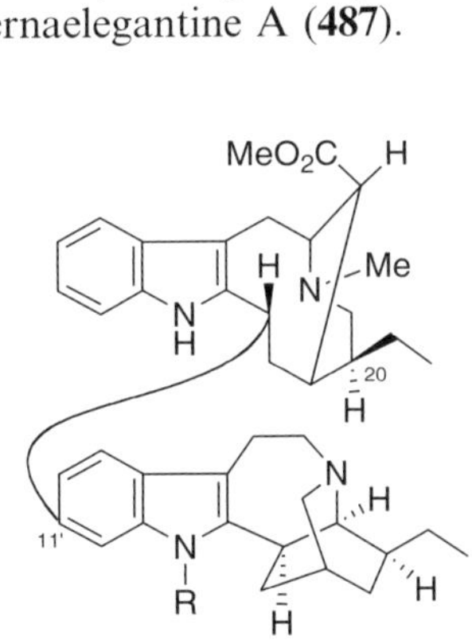

**489** R = H
**494** R = Me

Alkaloid **489** showed an $M^+$ at $m/z$ 618. The $^1$H NMR spectrum differed from that of **487** by the absence of aromatic methoxy and carbomethoxy functions, suggesting the presence of ibogamine (**491**) in place of isovoacangine (**479**) for the iboga half. This was supported by the $^{13}$C NMR spectral data, which, however, did not allow differentiation between C(11′) or C(10′) as the branching point on the iboga half. This was ascertained in an indirect manner by condensation of tabernaemontaninol (**492**) with *N*(1)-methylibogamine (**493**) (prepared by methylation of ibogamine) using a modified Büchi procedure, to give the *N*(1′)-methylated bisindole (**494**) (condensation of **492** with **491** under the same conditions gave **490**). The observed NOE between H(12′) ($\delta$ 7.01, br d, 2 Hz) and N(1′)-Me allowed unambiguous assignment of the aromatic hydrogens, which led to the assignment of a C(11′) to C(3) bond connecting the monomer units.

Further examination of *T. elegans* (whole plant extract) yielded two new derivatives of the tabernaelegantine series, 3′(*R*/*S*)-hydroxytabernaelegantine B (**495**) and 3′(*R*/*S*)-methoxytabernaelegantine C (**496**), in addition to the known bisindoles, 3′(*R*/*S*)-hydroxyconodurine and tabernaelegantines A–D (*303*). Reduction of **495** and **496** with $NaBH_4$ gave tabernaelegantines B and C, respectively. Alkaloid **496** is likely an artifact derived from 3′(*R*/*S*)-hydroxytabernaelegantine C. Interestingly, the tabernaelegantines were not detected in callus culture extracts of *T. elegans*, which provided only two bisindoles, 3′(*R*/*S*)-hydroxyconodurine (**466**) and monogagaine (**441**), in addition to the monomeric alkaloids (*304*).

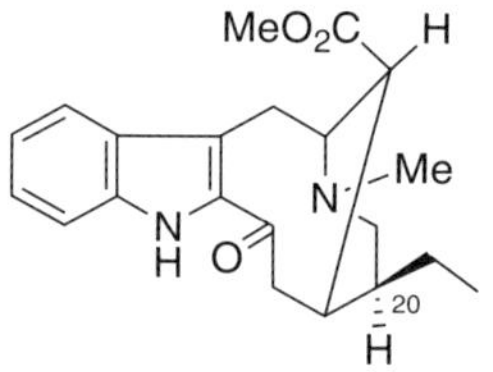

**490** tabernaemontanine

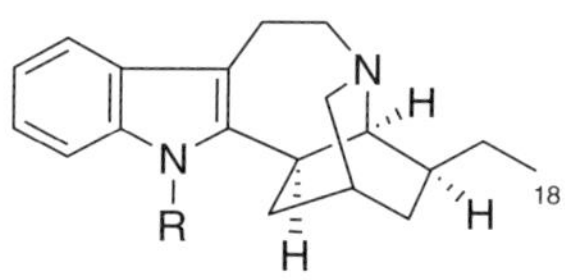

**491** R = H ibogamine
**493** R = Me

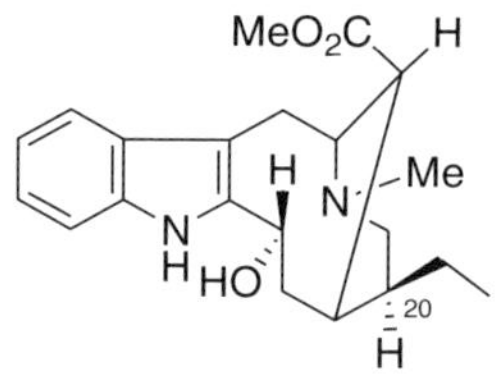

**492** tabernaemontaninol

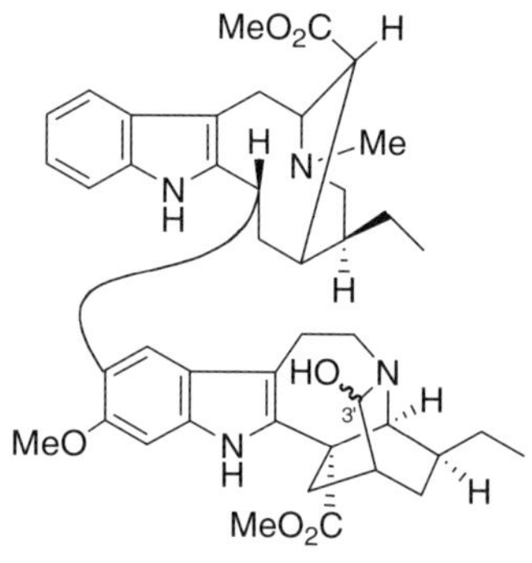

**495**

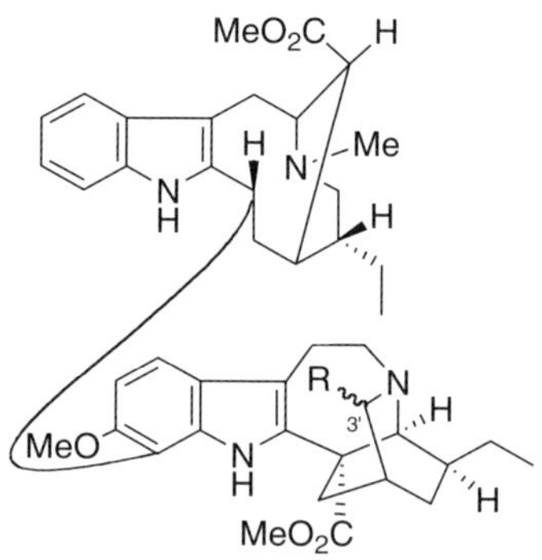

**496** R = OMe

The Malaysian *T. corymbosa* provided, in addition to vobasonidine (**313**), vobatricine (**398**) (*210*), conodurine, conoduramine, ervahanine (*293*), and tabernamine derivatives (*300*), a series of new vobasinyl–iboga alkaloids, **497**–**506**, characterized by the presence of hydroxymethyl group at C(16) of the vobasinyl moiety (*211*,*305*).

All 10 bisindoles were isolated from the leaves, of which conodiparine A (**497**) was the most abundant. The configuration of C(16) in all the 10 alkaloids can be ascertained from the unusual upfield shift of the C(16) ester methyl signal ($\delta$ 2.4), due to its placement within the anisotropic influence of the aromatic ring. The EI-mass spectra of conodiparines A–F (**497–502**) showed fragment ions (*m*/*z* 180, 136, 124, and 122) which are characteristic of vobasine–iboga bisindoles. In addition, the fragment at *m*/*z* 367 due to an intact vobasinyl fragment was common in the mass spectra of conodiparines A–D (**497–500**), suggesting that they share a common vobasinyl moiety, while the N(4)-demethyl compounds, conodiparines E and F (**501**,**502**), showed the corresponding fragment at *m*/*z* 353. Only three alkaloids, conodiparines A, C, and E, (**497**,**499** and **501**) are characterized by a C(10′) to C(3) bond linking the iboga and vobasinyl moieties, while the rest are characterized by a C(12′) to C(3) linkage, with the different branching clearly indicated in the aromatic region of the $^{1}$H NMR spectra. In addition to two one-H singlets observed for the aromatic H(9′) and H(12′) in the 10′-3 linked bisindoles, the H(12′) signal (confirmed by NOE between H(12′) and NH′) was shifted upfield, which was a characteristic of adjacent oxygenation. The 12′-3 linked bisindoles were readily recognized by the presence of an AB doublet due to the iboga H(9′) and H(10′). Another distinction between the two types of branching was afforded by the HMBC data. The 10′-3 linked bisindoles showed long range correlations from C(3) to H(9′), while the 12′-3 linked bisindoles showed correlations from C(12′) to H(3). The vobasinyl H(3) resonance appeared as a broad doublet with $J_{3-14}$ value of 13 Hz. This observation (H(3) and H(14) *trans*-diaxial), coupled with the observed NOE between H(3) and NH, provided proof of the C(3$\alpha$) substitution in these alkaloids. The NMR spectral data for conodiparines E and F (**501**,**502**) are distinguished by the conspicuous absence of the signals due to the N(4)-Me of the vobasinyl unit, while conodutarines A and B (**503**,**504**) are notable for the lack of the iboga C(16′) carbomethoxy group.

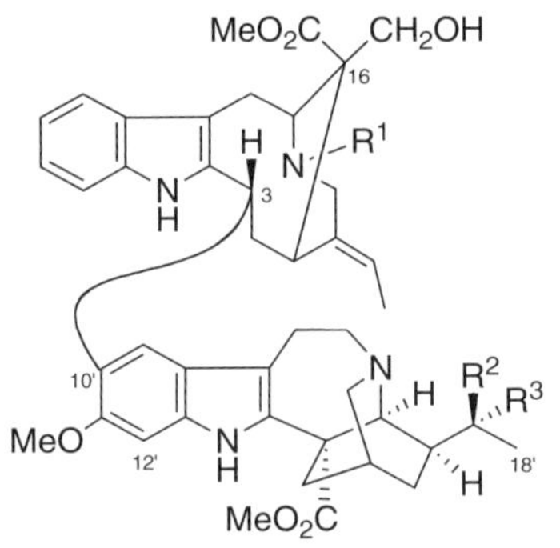

**497** R$^1$ = Me, R$^2$ = OH, R$^3$ = H
**499** R$^1$ = Me, R$^2$, R$^3$ = O
**501** R$^1$ = H, R$^2$ = OH, R$^3$ = H

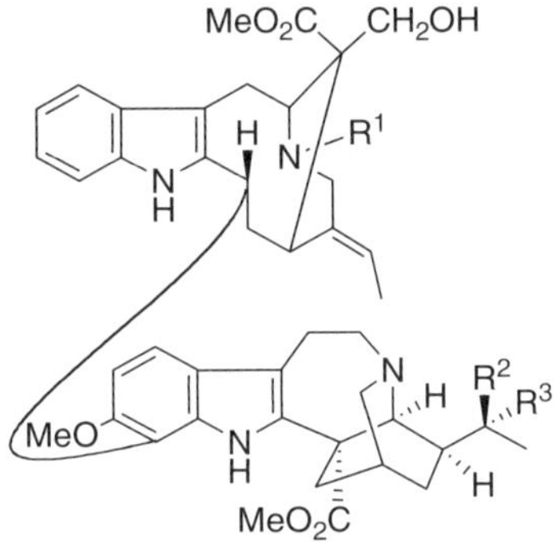

**498** R$^1$ = Me, R$^2$ = OH, R$^3$ = H
**500** R$^1$ = Me, R$^2$, R$^3$ = O
**502** R$^1$ = H, R$^2$ = OH, R$^3$ = H

Cononitarines A and B (**505**,**506**) are characterized by the presence of a phenolic substituent at C(11′) instead of OMe. In **506**, the side chain at the iboga C(20′) is an ethyl group in place of a hydroxyethyl in **505**. Two additional bisindoles of the same type, conodirinines A and B (**507**,**508**), in which the vobasinyl unit has incorporated an additional tetrahydrooxazine ring, were also obtained from the same plant (*306*). Notable differences in the NMR spectral data of **507** and **508** compared to the other vobasinyl–iboga bisindoles (e.g., **497–500**), were the absence of the vobasinyl NMe signal, and the appearance of a low field AB doublet at ca. $\delta$ 4.6 and 4.7, corresponding to the oxymethylene hydrogens at C(22). In the $^{13}$C NMR spectra, two low field signals were observed at ca. $\delta$ 76 and 88. The former was due to C(17) while the latter was attributed to C(22), which was $\alpha$ to both an oxygen and a nitrogen atom. The observed long-range correlations from H(22) to C(5), C(17), and C(21), in the HMBC spectrum, provided further support for the incorporation of the tetrahydro-1,3-oxazine ring on the vobasinyl unit. From a biogenetic viewpoint, these bisindoles can be considered as arising from the appropriate tetracyclic precursors *via* N(4)-condensation with formaldehyde followed by intramolecular cyclization. Such derivatives are known for the iboga alkaloids (e.g., chippiine (*286*) and the dippinines (*307*)), and have been encountered only once in the vobasine series (pagicerine (*308*)).

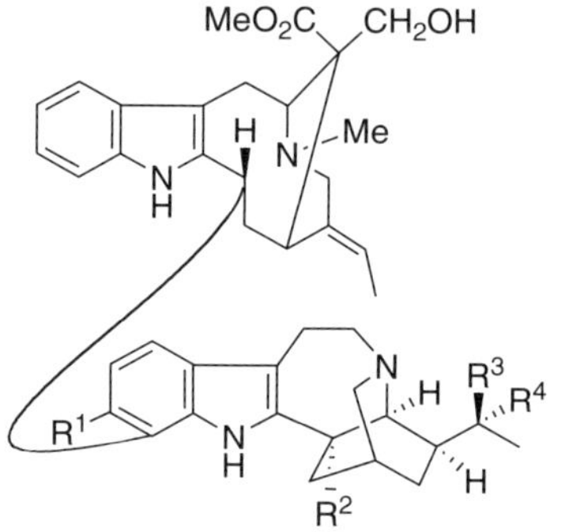

**503** R$^1$ = OMe, R$^2$ = H, R$^3$ = OH, R$^4$ = H
**504** R$^1$ = OMe, R$^2$ = H, R$^3$, R$^4$ = O
**505** R$^1$ = OH, R$^2$ = CO$_2$Me, R$^3$ = OH, R$^4$ = H
**506** R$^1$ = OH, R$^2$ = CO$_2$Me, R$^3$ = R$^4$ = H

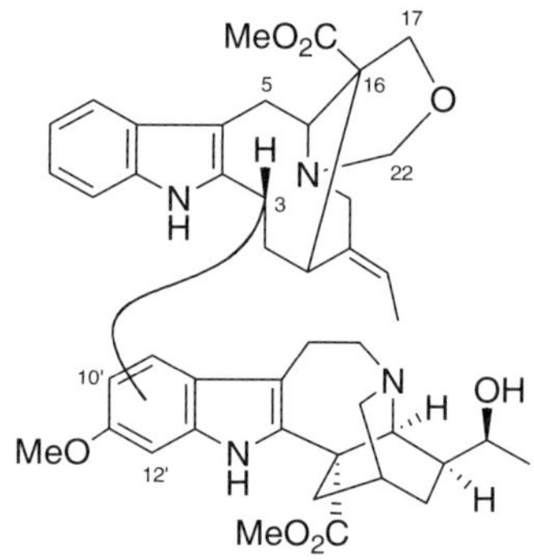

**507** C(3)-C(10') bond
**508** C(3)-C(12') bond

In contrast to the many *Aspidosperma–Aspidosperma* type bisindoles found in the leaf extract of the Malaysian *T. divaricata* (*vide infra*), the stem extract provided only one bisindole, conodusarine (**509**), which was of the iboga–vobasine type (*309*). The spectral data indicated that conodusarine (**509**) is constituted from the union of 3-vobasinyl and 3-oxovoacangine moieties *via* a 3-11′ bond. The presence of the latter alkaloid in the same plant provided additional support for the assignment.

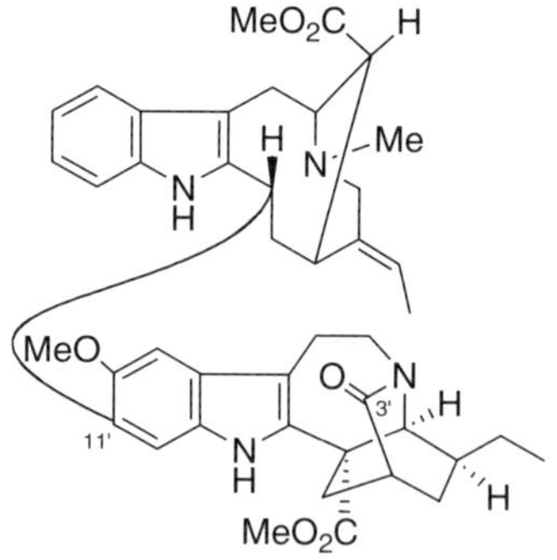

**509** conodusarine

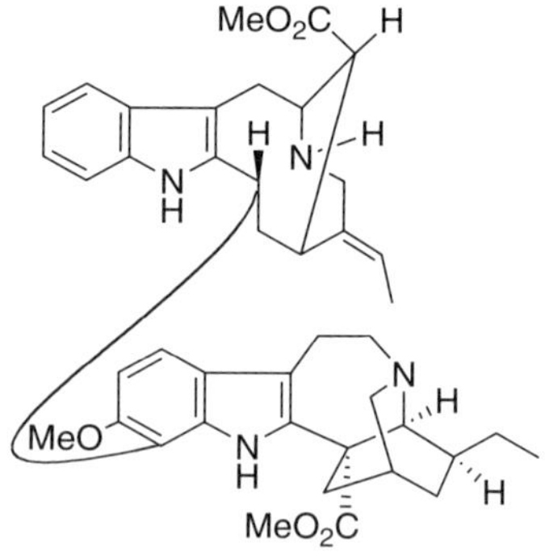

**510**

Conodiparines A–D (**497–500**) from *T. corymbosa* were found to be effective in reversing multidrug resistance in vincristine-resistant KB cells (*305*), while conoduramine (**468**) and voacamine (**464**) from *Peschiera laeta* (*T. laeta*), were found to reverse multidrug resistance in vinblastine-resistant KB cells (*310*). The bisindole alkaloids from *T. chippii* and *T. pachysiphon* (conoduramine, conodurine, voacamine, 3-hydroxyconoduramine, 3-hydroxyconodurine) displayed moderately strong antimicrobial activity against Gram-positive bacteria, and to a lesser extent against Gram-negative bacteria (*286,298*). Leishmanicidal and antibacterial activities were also reported for conodurine (**469**), *N*-demethylconodurine (gabunine **510**), and conoduramine (**468**), isolated from *Peschiera van heurkii* (*212*). Voacamine (**464**) from *P. fuchsiaefolia* was shown to be effective against the $D_6$ strain, and the chloroquine-resistant $W_2$ strain of *P. falciparum* (*311*). In a subsequent study, it was shown to exhibit *in vivo* antimalarial activity, but with a potency which was less pronounced compared with chloroquine (*312*).

## XXIII. Iboga–Iboga Type

Bisiboga-type alkaloids are rare. The first member, the symmetrical dimer, bis(11-hydroxycoronaridin-12-yl) (**511**) isolated from *Bonafousia tetrastachya* (*T. siphilitica*) (*313*), has been discussed previously (*1*). In the intervening years, only one other example has been encountered, obovatine (**512**), from *Stemmadenia obovata*, which is a derivative of **511**, in which one of the aromatic hydroxy substituents has been replaced by methoxy (*314*). This is evident on comparison of the mass spectra and NMR spectral data with those of **512**. The loss of symmetry has resulted in a near doubling of the signals in both the $^{1}$H and $^{13}$C NMR spectra, with some of the signals coincident. The C(12) to C(12′) linkage was confirmed by a HMBC experiment, while methylation of **511** ($CH_2N_2/Et_2O$) yielded both the monomethyl derivative **512** as well as the dimethyl derivative **513**. Alkaloid **511** and the monomer alkaloid, 11-hydroxycoronaridine were also present in this plant.

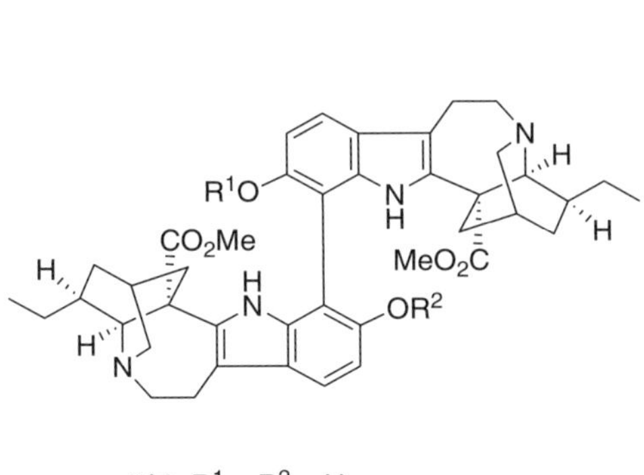


**511** R¹ = R² = H
**512** R¹ = H, R² = Me
**513** R¹ = R² = Me

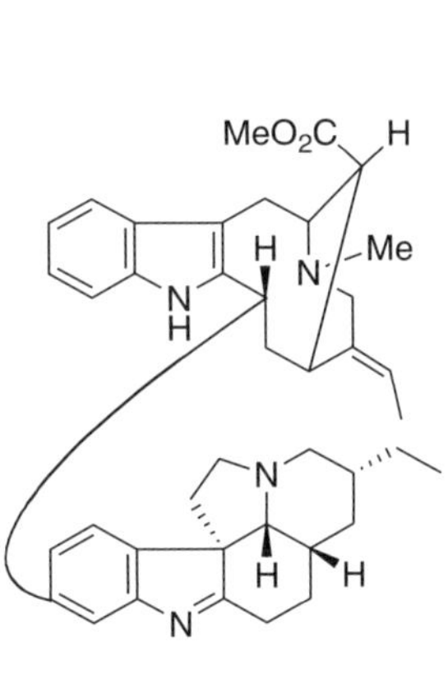


**514** capuvosidine

## XXIV. *Aspidosperma*–Vobasine Type

Bisindoles of this group appear to be rare. Capuvosidine (**514**), a pseudo-aspidosperma–vobasine-type bisindole was isolated from *Capuronetta elegans* (*1,315*). In the intervening period, only two bisindoles of the *Aspidosperma*–vobasine type, viz., tabernaemontabovine and tabernaemontavine, have been reported from *Tabernaemontana bovina*, occurring in Vietnam, in addition to two *Aspidosperma–Aspidosperma* bisindoles, methylenebismehranine and tabernaebovine, which were also isolated from the plant (*vide infra*). The structures which were initially attributed to tabernaemontabovine and tabernaemontavine (*316*) were subsequently revised to **515** and **516**, respectively, based on additional NMR data obtained (*317*). Inspection of the $^{13}$C NMR data revealed that a common mehranine unit is present in both alkaloids, while the vobasinyl parts in **515** and **516** were deduced by analogy to the vobasinyl parts of conoduramine and conodiparine A (*vide supra*), respectively.

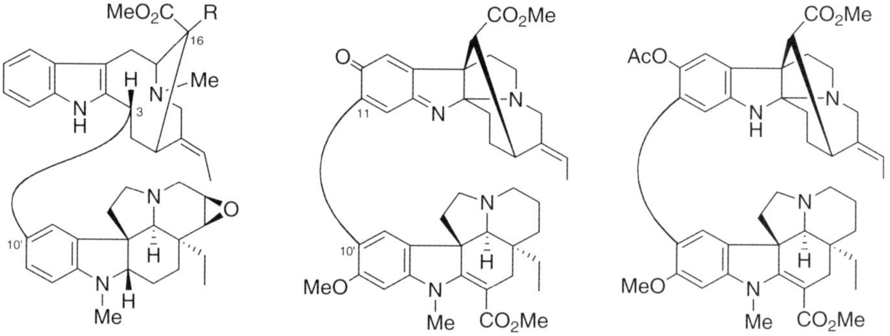


**515** R = H
**516** R = $CH_2OH$

**517** vincarubine

**518**

## XXV. *Aspidosperma*–Vincorine Type

Vincarubine (**517**), characterized by a dark-red color reminiscent of flexicorine (**372**), was isolated from *Vinca minor* (*318,319*). As with flexicorine (**372**), vincarubine (**517**) on reduction with $NaBH_4$ in methanol afforded a yellow compound, presumably the phenol, which reverted spontaneously to the iminoquinone form **517** on exposure to air. Reduction of **517** with sodium dithionate followed by direct *in situ* acetylation of the product, allowed the phenol to be isolated as the stable acetate **518**. Addition of acid resulted in a hypsochromic shift to 343 nm, and the appearance of a shoulder at 320 nm, compared to the original absorption maxima at 259 and 347 nm in the UV spectrum of **518**. This behavior is diagnostic of vincorane-type compounds possessing a $C_6H_5$–N–C–N function and is consistent with the observed shift of the C(2′) resonance, from 104.0 in **517** to 94.7 in **518**, the latter value being close to that in vincorine (**268**) and *N*-demethylvincorine (*320*). Comparison of the $^{13}$C NMR spectral data of **517** and **518** with those of flexicorine (**372**) and dihydroflexicorine indicated a similarity of the vincorine moiety between these two groups of alkaloids. The other half of vincarubine (**517**) was readily identified from the remaining signals as belonging to 11-methoxy-*N*(1)-methylvincadifformine, by comparison with the $^{13}C$ NMR spectrum of vincadifformine (**529**). The assignment of the $^{13}C$ NMR data of **517** was facilitated by the application of 2-D NMR methods. Vincarubine inhibited incorporation of precursors of protein and nucleic acid synthesis into murine leukemia P388 cells. The effect was less compared to vincadifformine, but greater than that of vinblastine (*319*).

## XXVI. *Aspidosperma*–Macroline Type

Pandicine (**519**) was isolated from *Pandacastrum saccharatum* as a brown amorphous solid. HREIMS yielded the formula $C_{44}H_{50}N_4O_7$, indicating a highly oxygenated molecule, while characteristic peaks at *m/z* 170 and 190 are typical of the N(1)-Me and N(4)-Me macroline skeleton. In alkaline medium, the UV spectrum (231, 234, 297, 307, 342 nm) underwent a bathochromic shift above 350 nm with no appreciable change below 300 nm. Treatment of **519** with $MnO_2$ in $CHCl_3$ resulted in quantitaive conversion into the iminoquinone **520**, which did not show the OH/NH band at 3400 cm$^{-1}$ present in the IR spectrum of **519**. The UV spectrum of **520** was also markedly different from that of **519**, with absorption maxima at 233, 260, 288 (sh), and 396 nm.

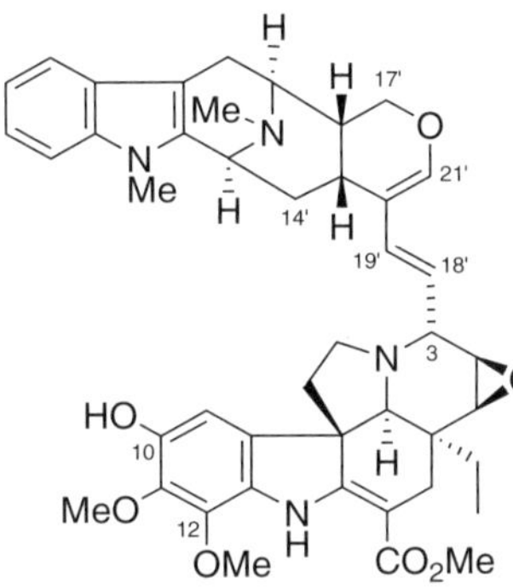

**519** pandicine

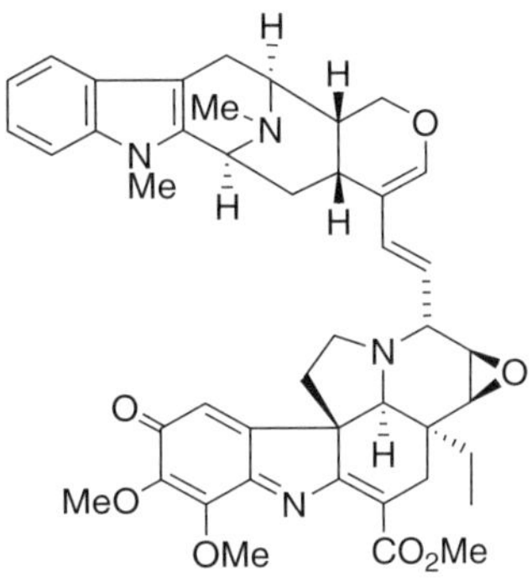

**520**

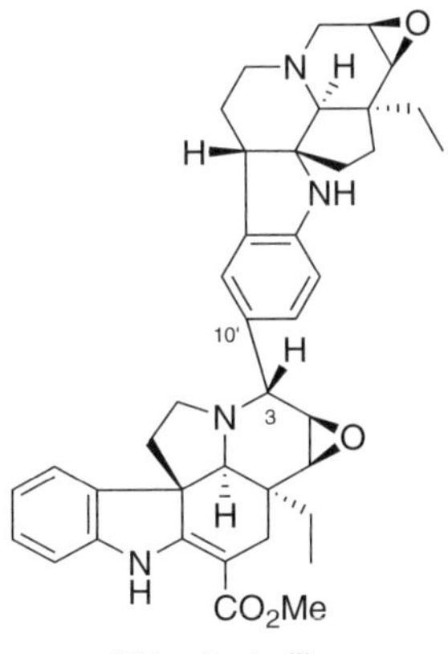

**521** criophylline

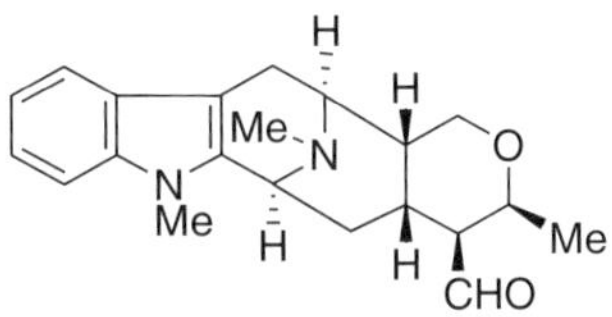

**522**

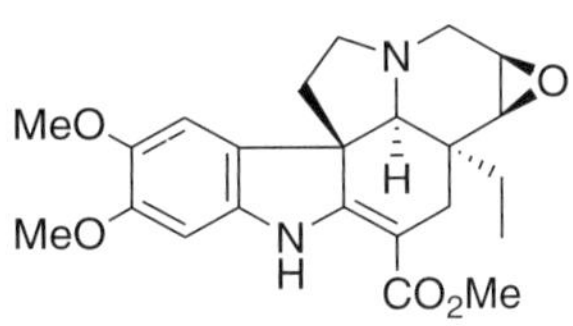

**523**

The complete structure elucidation was based on analysis of the $^{13}C$ NMR spectral data by comparison with the known bisindoles, criophylline (**521**) and villalstonine (**322**). The former contains a 3-substituted 14,15-tabersonine-$\beta$-epoxide moiety, while the latter contains a macroline unit. The $^{13}C$ NMR spectrum of **519** showed correspondence of all the non-aromatic $^{13}C$ shifts with those of criophylline, with the exception of C(3) which is shifted slightly upfield. Similarly, comparison of the remaining $^{13}C$ NMR shifts of **519** revealed analogy with the macroline half of villalstonine, with the exception of signals due to C(18′), C(19′), C(20′), and C(21′), reflecting the structural changes in this part of the molecule compared with villalstonine. The $^{1}H$ signals for H(18′), H(19′), and H(21′) at $\delta$ 5.05 (dd, $J$ 16, 8 Hz), 5.85 (d, $J$ 16 Hz), and 6.41 (s), are in agreement with an exocyclic *trans*-disubstituted 18′, 19′ double bond as part of a diene system with the 20′, 21′ olefinic bond, as in **519**. The excellent correlation of the ring E carbons of the tabersonine epoxide unit in **519** with those of criophylline, is also consistent with the stereochemistry of the epoxide ring and the 3$\alpha$-substitution, while the substitution pattern on the aromatic moiety is consistent with the observed aromatic carbon shift values, the facile oxidation to the iminoquinone **520** as well as the change of the H(9) shift from $\delta$ 6.34 in **519** to 5.63 in **520**. The proposed relative stereochemistry of the constituent moieties in **519** was predicated on a presumed correspondence with the configurations of talcarpine (**522**) and hazuntinine (**523**) (*321*).

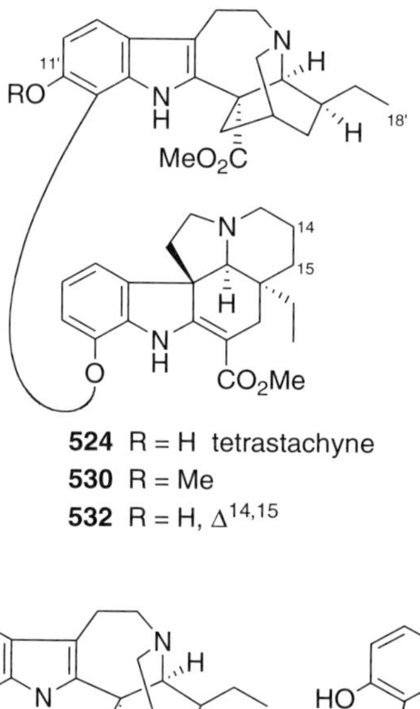

**524** R = H tetrastachyne
**530** R = Me
**532** R = H, $\Delta^{14,15}$

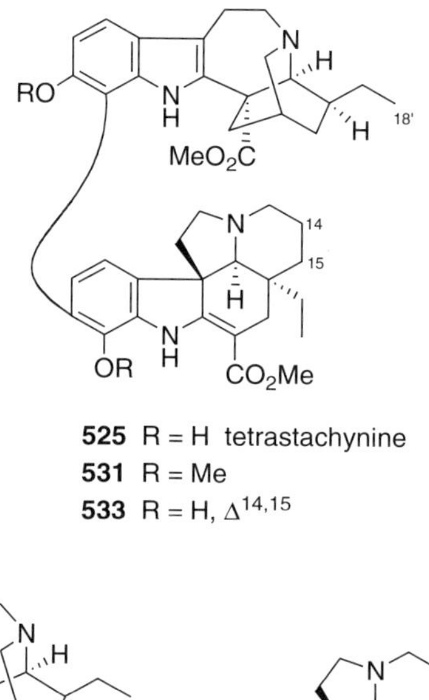

**525** R = H tetrastachynine
**531** R = Me
**533** R = H, $\Delta^{14,15}$

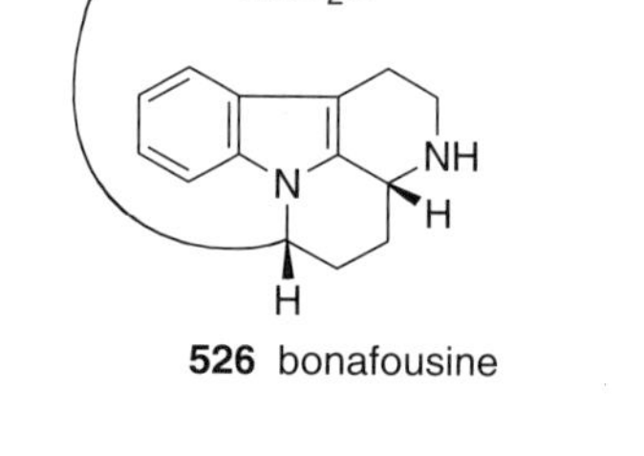

**526** bonafousine

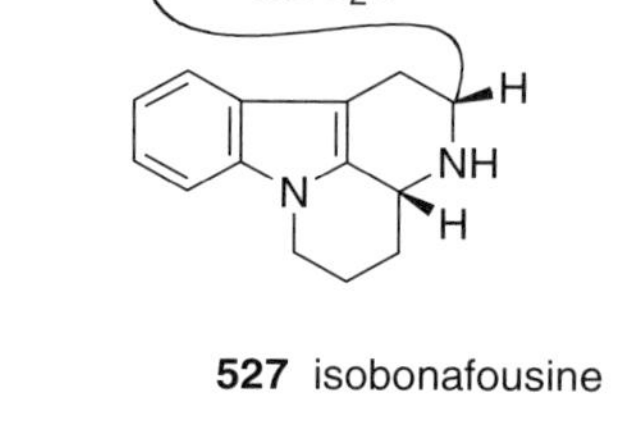

**527** isobonafousine

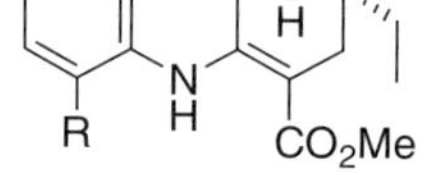

**528** R = OH
**529** R = H
**535** R = H, $\Delta^{14,15}$

## XXVII. *Aspidosperma*–Iboga Type

The isomeric tetrastachyne (**524**) and tetrastachynine (**525**) were isolated from *Bonafousia tetrastachya* (*T. siphilitica*), in addition to the previously known iboga–canthinone alkaloids, bonafousine (**526**) and isobonafousine (**527**) (*322*). Both alkaloids showed a molecular ion at *m/z* 704, together with a higher mass fragment at *m/z* 720, attributed to intramolecular transmethylation. The base peak at *m/z* 124 was a characteristic of the mass spectrum of 12-hydroxyvincadifformine (**528**), which was also present in the plant, and indicated the presence of a similar aspidospermane moiety. The molecular formula of **524** ($C_{42}H_{50}N_4O_6$) was confirmed from microanalytical data. The UV spectra of both alkaloids were similar to that of **528** (hydroxyindole and $\beta$-anilinoacrylate chromophores) and showed characteristic bathochromic shifts on addition of alkali. Both **524** and **525** were shown by $^{13}C$ NMR to be constituted from the union of a 12-hydroxyvincadifformine unit and a 11-hydroxycoronaridine unit, by comparison with 12-hydroxyvincadifformine (**528**) and bis(11-hydroxycoronaridin-12-yl) (**511**). Both alkaloids were also isolated from the plant. The $^{13}C$ shifts of the non-aromatic part of both compounds were virtually identical, with differences confined to the aromatic carbons, suggesting a variation in the structures involving aromatic substitution as the main difference between **524** and **525**. The $^{1}H$ NMR of **524** showed two sets of signals in the aromatic region, a pair of AB doublets corresponding to two adjacent *ortho*-coupled hydrogens, and another set corresponding to three contiguous aromatic hydrogens. Methylation ($CH_2N_2$) of **524** resulted in a monomethyl derivative **530** ($M^+$ *m/z* 720), with the appearance of a methoxy singlet at $\delta$ 3.79 in the $^{1}H$ NMR spectrum, indicating that the other oxygen must be part of an ether function, consistent with the proposed structure and the $^{13}C$ NMR shifts ($\delta_{C12}$ 143.6, $\delta_{C12'}$ 127.9 vs. $\delta_{C12}$ 140.8 and 100.0 in **528** and **511**, respectively). The $^{1}H$ NMR of tetrastachynine on the other hand showed two sets of aromatic AB doublets corresponding to two pairs of *ortho*-coupled aromatic hydrogens. Furthermore, methylation afforded the dimethoxy derivative **531** ($M^+$ *m/z* 734), in agreement with the proposed structure **525** for tetrastachynine ($\delta_{C12}$ 139.0, $\delta_{C12'}$ 100.5 vs. $\delta_{C12}$ 140.8 and 100.0 in **528** and **511**, respectively).

The dehydroderivatives of **524** and **525**, 14,15-dehydrotetrastachyne (**532**) and 14,15-dehydrotetrastachynine (**533**), respectively, in addition to bis(11-hydroxycoronaridin-12-yl) (**511**), were later isolated from *T. citrifolia* (*323*). The former alkaloid **532** was also isolated from *Peschiera echinata* (*T. echinata*) (*324*), while the latter alkaloid **533** was also isolated from *Stemmadenia grandiflora* (*325*). The change in the $^{13}C$ shifts for C(14) and C(15), to $\delta$ 125.0 and 133.2, respectively, coupled with the appearance of the corresponding olefinic hydrogen shifts at $\delta$ 5.85 and 5.78, respectively, when compared to the NMR spectra of **524**, indicated the location of the unsaturation in the aspidosperma half in **532**. The same considerations apply to **533**, except that the olefinic hydrogens were observed as a multiplet at $\delta$ 5.78. In any case, catalytic hydrogenation of **533** gave the parent alkaloid **525**. Doubling of the signals was observed in the $^{1}H$ NMR spectrum of **533** due to atropisomerism, as a result of the highly and non-symmetrically substituted biphenyl unit, with coalescence observed at 60°C in $CD_3OD$ (*325*).

## XXVIII. *Aspidosperma–Aspidosperma* Type

*Voacanga grandifolia* (Apocynaceae) from India, provided a new bisindole, voacinol (**534**), in addition to vobtusine, deoxyvobtusine, and amataine (*326*). The UV spectrum indicated the presence of a $\beta$-anilinoacrylate chromophore as in tabersonine (**535**). The molecular ion was detected by FABMS, showing an $MH^+$ at $m/z$ 717 ($C_{43}H_{48}N_4O_6$). The $^{13}C$ NMR spectral data showed a close correspondence for all the carbon shifts with those of tabersonine, except for some of the ring D carbons. The downfield shift of C(14) to $\delta$ 129.7 from 124.8 in **535**, is consistent with it being a quaternary center in **534**, while the signal at $\delta$ 39.0 was attributed to the methylene C(22) linking the two tabersonine moieties. This is also in agreement with the observation of only one olefinic H at $\delta$ 5.5 in the $^1H$ NMR spectrum of **534**, which was a broadened singlet due to allylic coupling with the two H(22). The ethyl side chain has also been modified to 18-hydroxyethyl, as indicated by the shifts for C(18) at $\delta$ 58.8 in **534** compared to 7.3 in **535**.

Another bisindole with a methylene bridge linking two *Aspidosperma*-type units is methylenebismehranine (**536**) which was isolated, together with tabernaebovine (**537**) from *T. bovina* from Vietnam (*327*). Alkaloid **536** is constituted from the linking of two (−)-mehranine units by a methylene bridge ($\delta_C$ 41.1; $\delta_H$ 3.79). The monomeric unit (−)-mehranine (**538**), was also isolated from the same plant. The other bisindole, tabernaebovine (**537**) is also consitituted from the union of two (−)-mehranine moieties, but in a different manner from **536**. In **537**, the bisindole is branched from the aromatic C(10) of one, to C(2′) of the other mehranine unit.

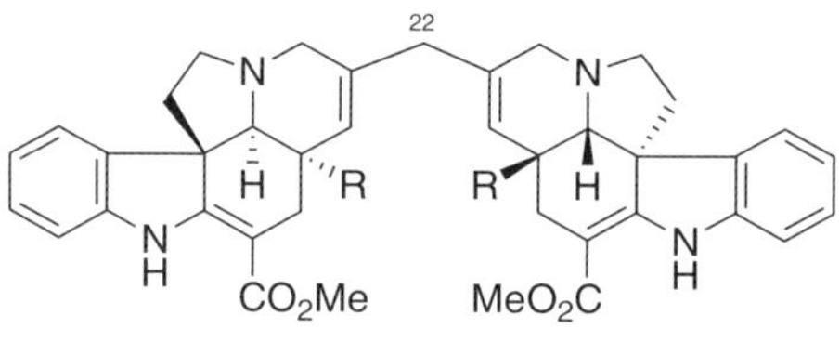

**534** R = $CH_2CH_2OH$ voacinol

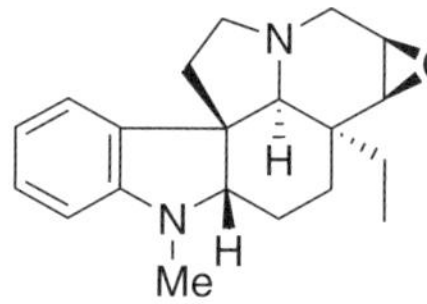

**538** (−)-mehranine

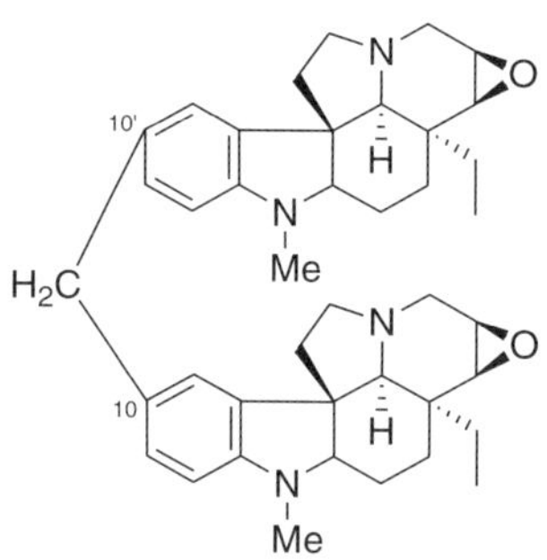

**536** methylenebismehranine

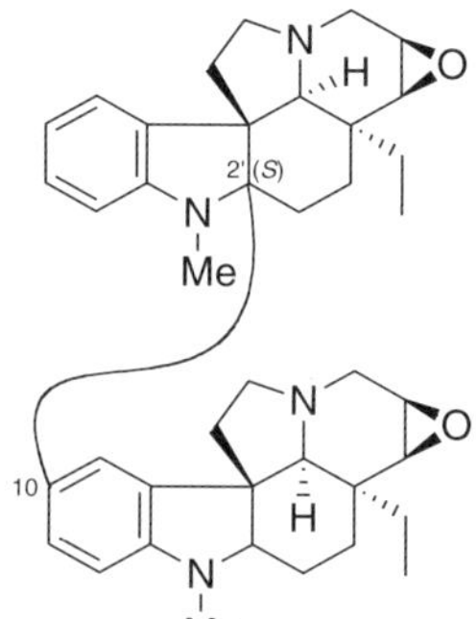

**537** tabernaebovine

*Pterotaberna inconspicua* yielded kisantine (**539**), an unstable bisindole constituted from the union of two highly oxidized tabersonine units, as shown by its $^{13}C$

NMR spectrum. The UV spectrum showed the presence of β-anilinoacrylate chromophores, while the molecular formula, $C_{44}H_{50}N_4O_{11}$, was established from HRFABMS. The $^1H$ NMR spectrum (analyzed by COSY and delayed COSY experiments) showed two aromatic hydrogens as singlets at δ 6.67 and 5.53, two sets of two coupled olefinic hydrogens, one ethyl side chain linked to C(20), and four methoxy groups. The other side chain takes the form of an 18-hydroxy-substituted ethyl, linked to the other C(20′). One of the olefinic pair showed coupling to the aminomethylene hydrogens on C(3′) at δ 3.4, while the other olefinic pair showed coupling to a single allylic H (δ 4.16) attached to the other C(3), which is an oxymethine from its observed shift at δ 84.8. The presence of two β-anilinoacrylate moieties was confirmed from the characteristic $^{13}C$ shifts, and taken with the other data, indicated the involvement of two tabersonine type halves.

The aromatic shifts of one of the tabersonine units showed correspondence to the 10-hydroxy-11,12-dimethoxy tabersonine unit in pandicine (**519**), while the other aromatic moiety required substitution by three oxygens to accommodate the molecular formula and the presence of only one aromatic hydrogen. Since all four methoxyls have been accounted for, this left two OH and an ether as likely subsituents on the second aromatic moiety. In the mass spectrum, the observation of the $m/z$ 390 fragment from *retro*-Diels–Alder cleavage is consistent with the aromatic substitution of the pandicine-like tabersonine unit, while the $m/z$ 138 ion, derived from fragmentation of the other tabersonine unit, indicated that the 18′-hydroxyethyl side chain was attached to this moiety. The branching from the first tabersonine moiety is from C(3) since it is an oxymethine as shown by its $^1H$ and $^{13}C$ shifts, with the substitution likely to be α, from the low value of 3.8 Hz observed for $J_{3-14}$. The lone aromatic hydrogen at δ 5.53 has to be placed at C(9′) of the second aromatic moiety to account for the observed N(1′)-H to H(9′) long range coupling. Its substantial shielding compared to the other H(9) at δ 6.67 was attributed to proximity of the second tabersonine moiety. These considerations led to structure **539** for kisantine (*328*).

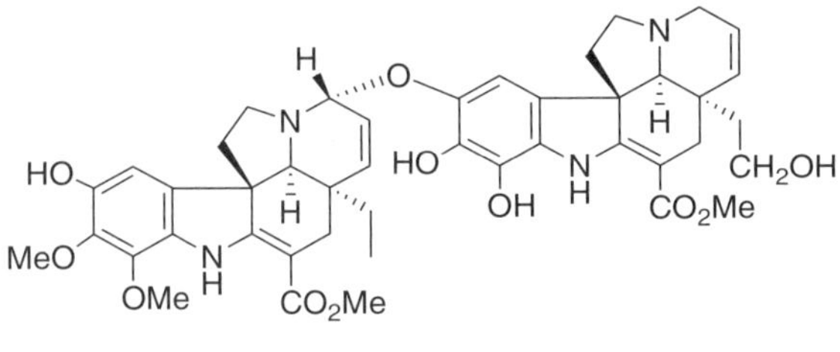

**539** kisantine

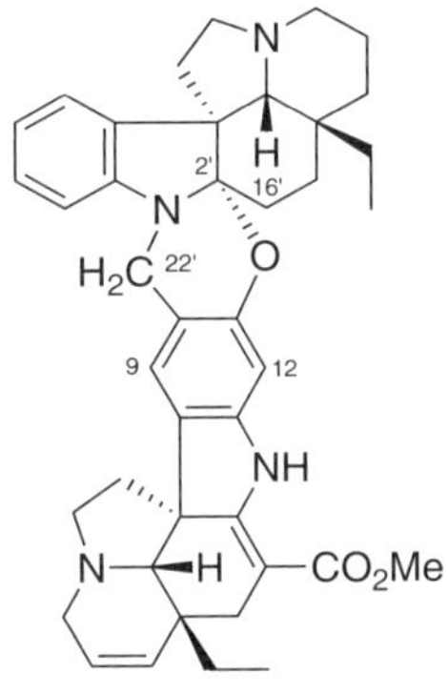

**540** melomorsine

*Melodinus morsei* from China provided melomorsine (**540**), which analyzed for $C_{41}H_{48}N_4O_3$ by HREIMS. The UV spectrum (206, 253, 315, 327 nm) could be interpreted as a composite of the chromophores of aspidospermidine and tabersonine. The $^1H$ NMR spectrum indicated the presence of one unsubstituted aromatic ring, a 10,11-disubstituted aromatic ring, and a methylene linked to an aromatic carbon and an indoline nitrogen. Comparison of the $^{13}$C NMR spectral data of melomorsine with those of aspidospermidine and tabersonine showed a close correspondence of most of the signals, except for changes in the shifts and multiplicity of C(10) and C(11) of the tabersonine unit, and C(2′) of the aspidospermidine unit, indicating these positions as points of attachment of the monomeric moieties. The substantial downfield shift of C(2′) to $\delta$ 96.9 in melomorsine is indicative of an oxygenated quaternary carbon, and is consistent with the proposed structure of melomorsine in which N(1′) and C(2′) of the aspidospermidine unit are linked to C(10) and C(11) of the tabersonine unit, respectively, with the former linkage mediated by a methylene, and the latter by an oxygen. The proposed structure as well as the relative configuration of **540**, received additional support from the results of NOE experiments. Irradiation of H(22′α) caused enhancement of the H(12′) and H(9) signals, while irradiation of H(16′α) resulted in enhancement of the H(22′β) signal, indicating bridging of N(1′) and C(11) by the 22′-methylene group as well as α-substitution at C(2′) (*329*).

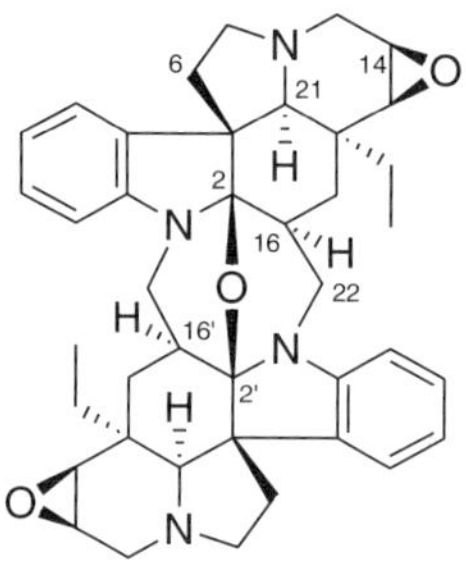

**541** hazuntiphylline

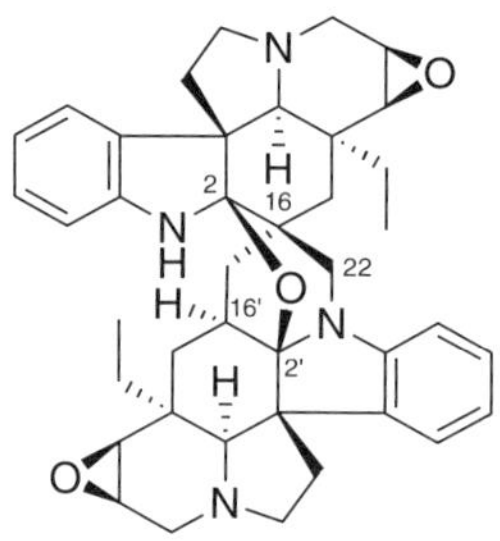

**542** hazuntiphyllidine

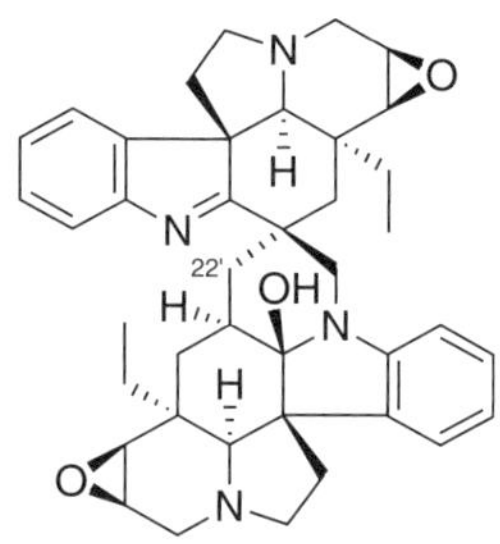

**543** hazuntiphyllidine

*H. modesta* (*T. coffeoides*) from Madagascar provided three bisindoles, hazuntiphylline (**541**), hazuntiphyllidine (**542** or **543**), and anhydrohazuntiphyllidine (**544**). Hazuntiphylline (**541**) presented an indoline chromophore in the UV spectrum, while the NMR spectral data indicated a symmetrical bisindole. In addition to the $M^+$ at $m/z$ 630 ($C_{40}H_{46}N_4O_3$), the mass spectrum showed fragments ($m/z$ 138, 108, and M–138), which indicated the presence of a 14,15-epoxy substituted *Aspidosperma*-type moiety as found in lochnericine, pachysiphine, and hazuntine. The absence of NH, OH or C=O absorptions in the IR spectrum suggested involvement of the indolic nitrogens in the linking of the two monomeric halves, as well as the presence of an ether linkage to account for the third oxygen. These features led to the structure **541** for hazuntiphylline which is in agreement with the $^{13}$C NMR spectral data, the assignment of which was facilitated by comparison with the spectra of folicangine (**545**) and ervafoline (*vide infra*). The observation of a quaternary carbon signal at $\delta$ 94.2 indicating a linkage to two heteroatoms, can be attributed to C(2) of an *Aspidosperma*-type unit, while the correspondence of the piperidine ring D carbon signals with those of pachysiphine rather than lochnericine pointed to a $\beta$ stereochemistry of the epoxide function. Analysis of the coupling constants for H(16) indicated that it is axially oriented ($J_{16-17}$ 13.8 Hz), while strong NOEs between H(16) and H(21) are indicative of their *syn* relationship. In addition, examination of models required the oxygen bridge linking C(2) and C(2′) to be $\beta$, to be consistent with the NOE data (*330*). Hazuntiphyllidine (**542**,**543**) was isomeric with **541**. The NMR spectrum in

$CDCl_3$ was complicated by the existence of two equilibrating forms resulting in apparent doubling of the signals. These two forms were shown to comprise the indoline form **542**, which predominates in $C_6D_6$, or the ring-opened, indolenine form **543**, which predominates in DMSO-$d_6$ solution. In either solvent, the signals of all the 40 carbons were distinguishable, although the shifts were slightly, but significantly different in the two solvents. In particular, the C(2) shift at $\delta$ 190.7 in DMSO-$d_6$ due to an indolenine, was seen at $\delta$ 102.2 in $C_6D_6$. The $^1H$ NMR spectrum in $C_6D_6$ showed features due to two 14,15-epoxy substituted *Aspidosperma*-type moieties, in common with **541**. The main departure was the loss of symmetry indicating a change in the manner of branching of the constituent moieties. Other notable differences included the upfield shift of H(22′) to $\delta$ 1.68, and the presence of an isolated aminomethylene at $\delta$ 2.97 attributed to C(22). These features are accommodated in **542**, which is also consistent with the $^{13}C$ NMR data. It is to be noted that in a freshly prepared solution of hazuntiphyllidine in $C_6D_6$, the NMR spectrum corresponded with the indolenine form **543**, and only after several hours did the equilibrium shift in favor of the indoline form **542**, suggesting that the open indolenine-form probably represents the initial structure of hazuntiphyllidine. Anhydrohazuntiphyllidine (**544**) analyzed for $C_{40}H_{44}N_4O_2$, differing from hazuntiphyllidine by loss of $H_2O$. The structure was readily deduced from examination of the spectral data, as well as by its identity with the product obtained from hazuntiphyllidine by acid-induced dehydration. The depicted configuration of the spirocarbon C(16) in these compounds (**542**–**544**) was based on the observed NOE between H(22) and H(6) (*331*).

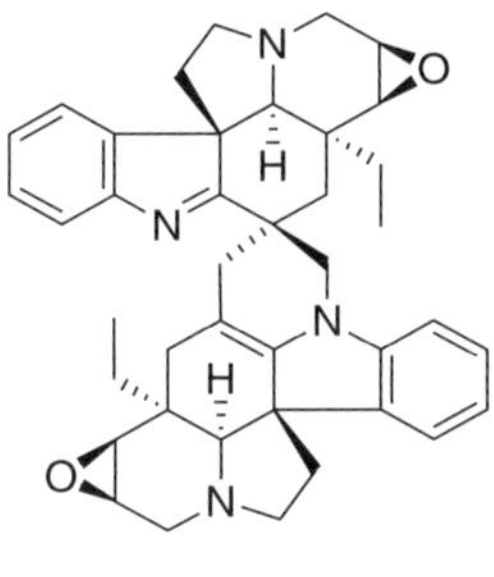

**544** anhydrohazuntiphyllidine

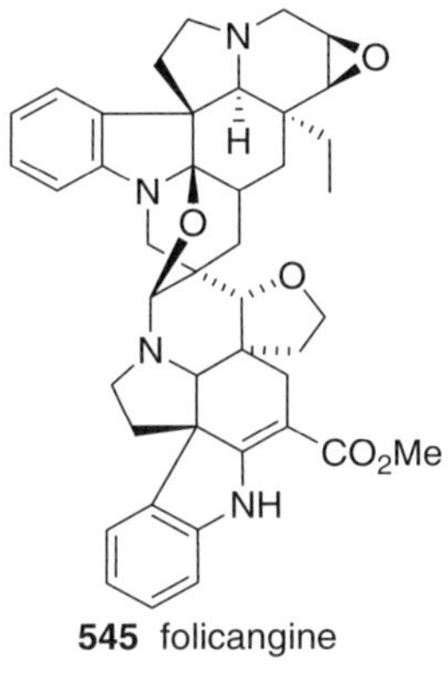

**545** folicangine

The Malaysian *T. divaricata* provided a series of bisindoles, constituted from the union of tabersonine type units, but linked in a manner different from those considered above. Conophylline (**546**) (*332–334*) showed an $MH^+$ in the FABMS at $m/z$ 795 ($C_{44}H_{50}N_4O_{10}$). The UV spectrum was characteristic of alkaloids with $\beta$-anilinoacrylate chromophores and the IR spectrum showed the presence of NH, OH, and conjugated carbonyl functions. The $^1H$ and $^{13}C$ NMR spectral data indicated a bisindole constituted from highly oxygenated vincadifformine–tabersonine epoxide moieties. The $^1H$ NMR spectrum showed the presence of two indole NH, three isolated aromatic hydrogens one of which was significantly shielded ($\delta$ 5.55), an OH function, four methoxy groups, of which two were associated with the presence of two ester carbomethoxy functions, and two ethyl groups. The presence of only three aromatic singlets indicated highly substituted indole rings, where one indole ring was substituted at the 10, 11, and 12 positions, while the other was substituted at the 10′ and 11′ positions. This was confirmed by the NOE interactions observed between one of the indole NH and the aromatic MeO at C(12), and between the other indole NH′ and the aromatic H(12′). The unusually low-field aromatic MeO carbon absorptions ($\delta_C$ 60.5, 61.0) suggested that they were in an *ortho* arrangement and indicated that the other aromatic MeO group was at C(11) with C(10) substituted by a OH group ($\delta_C$ 138.7), an arrangement reminiscent of that in the bisindole pandicine (**519**) (*321*).

Comparison of the spectral data with those of pandicine (**519**), in fact revealed generally good agreement of the aromatic $^{13}$C shifts in particular as well as those of the non-aromatic carbons, with the exception of the piperidine ring D carbons. The other unit of the bisindole was clearly shown to be a 10-alkyl-11-oxy-tabersonine-$\beta$-epoxide from the excellent correlation of the non-aromatic $^{13}$C shifts with those of pachysiphine (**547**), and the aromatic $^{13}$C shifts with those of vandrikine and the bisindole alkaloid vincarubine (**517**).

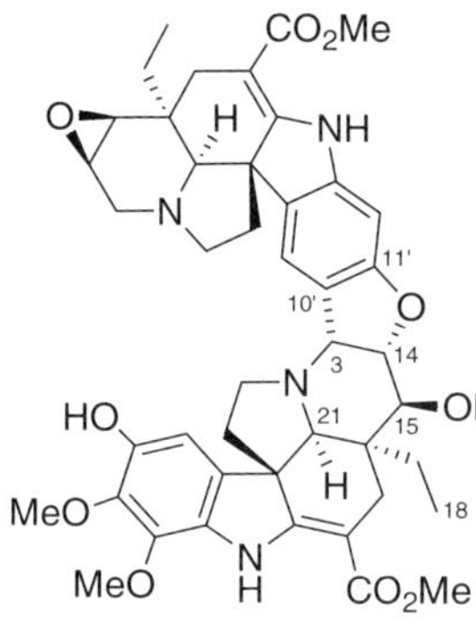

**546** conophylline

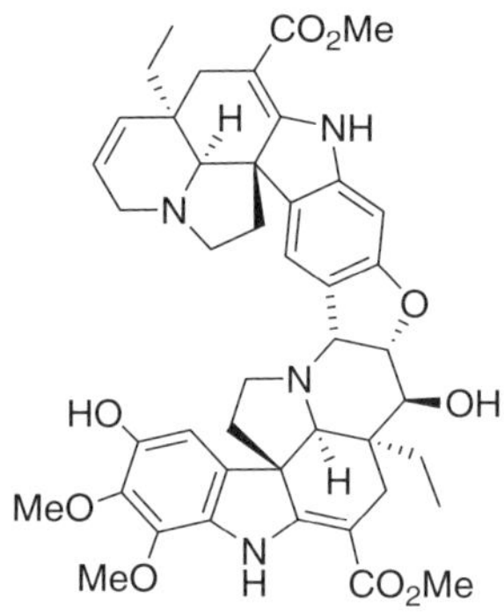

**548** conophyllidine

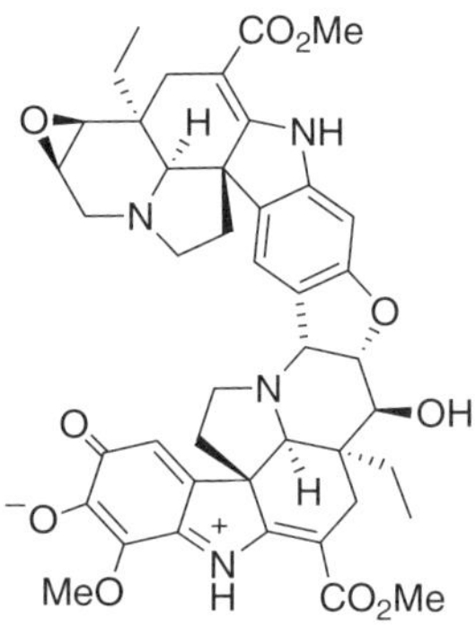

**549** polyervinine

The mode of attachment of the monomer units was deduced from examination of the H(3), H(14), and H(15) resonances of the piperidine ring of the vincadifformine unit, which were clear and well resolved in the region from ca. $\delta$ 4 to 5. The OH function was placed on C(15), since, when exchanged with deuterium, the H(15) doublet of doublets ($J$ 11.0, 3.7 Hz) collapsed to a doublet ($J$ 3.7 Hz). The mode of attachment of the monomeric units is thus *via* C(3) and C(14) of the vincadifformine unit to C(10′) and C(11′) of the pachysiphine unit, respectively, the C(14) to C(11′) connection being mediated by an ether oxygen. The substitution at C(3) and C(14) has to be *cis*, this being dictated by the fact that carbons 3 and 14 form part of a dihydrofuran unit. The similarity of the C(3) shift to those in the dimers criophylline (**521**) (*335*) and pandicine (**519**) (*321*) suggested that conophylline also has a 3$\alpha$ substitution. This was further confirmed by the observation that H(9) of conophylline ($\delta$ 5.55) was significantly shielded compared with H(9) of pandicine ($\delta$ 6.34) in the $^1$H NMR spectrum, owing to its being affected by the anisotropy of the aromatic ring of the tabersonine–epoxide unit, a feature which would only be possible if the tabersonine epoxide unit was attached on the $\alpha$-face at C(3) and C(14). The configuration of the remaining stereocenter, viz., that of the C(15)–OH, was readily deduced to be $\beta$ from the NOE interaction observed between H(15) and the C(18) hydrogens of the C(20) $\alpha$-ethyl substituent. The structure deduced from spectroscopic data has been subsequently confirmed by X-ray analysis (*333*).

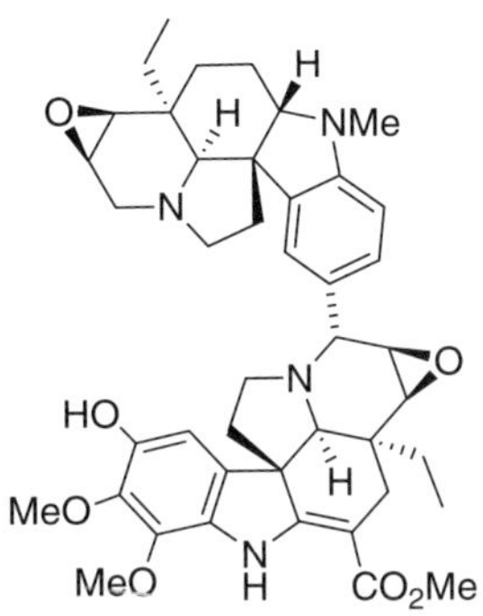

**550** conofoline

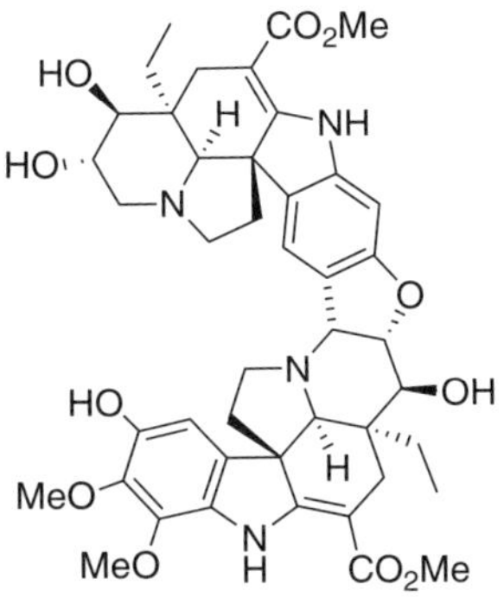

**551** conophyllinine

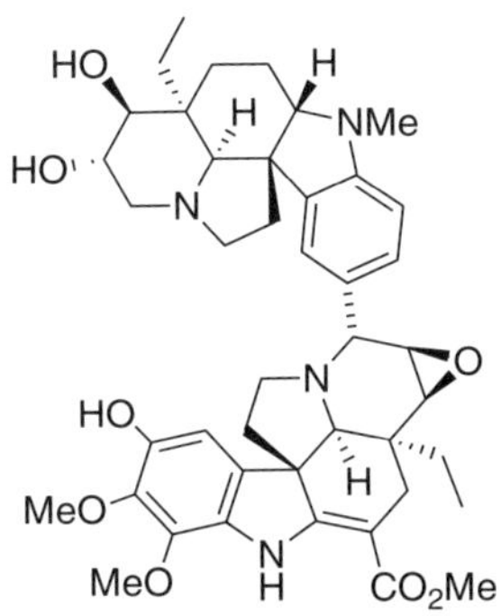

**552** peduncularidine

Conophyllidine (**548**) is identical in all respects with conophylline except for replacement of the epoxide function at positions 14′ and 15′ of the tabersonine epoxide unit by a double bond (*333*). Conophylline has also been subsequently isolated from the South American species, *Tabernaemontana glandulosa* (*336*), and from the Thai species, *Ervatamia microphylla* (in all probability *T. divaricata*) (*337*). It has also been reported from another Malaysian species, *Ervatamia polyneura* (*T. dichotoma*), under the name polyervine (*338*). The leaf extract of *E. polyneura* also yielded the related bisindole, polyervinine (**549**), which is similar to conophylline, except for the aromatic portion of the vincadifformine unit, which in the case of **549** incorporates an indoline dione chromophore, existing predominantly in its zwitterionic quinoniminium form.

Conofoline (**550**), which occurs only in the double flower variety of *T. divaricata*, has the same 10-hydroxy-11,12-dimethoxy-tabersonine-$\beta$-epoxide unit as in conophylline, but differs in the identity of the other monomeric unit as well as in the mode of attachment of the monomers (*339*). This was shown by the $^{1}$H and $^{13}$C NMR data which also revealed the other unit to be a 10-alkylmehranine moiety from the excellent agreement of the non-aromatic $^{13}$C NMR resonances with those of (−)-mehranine (**538**), which also occurs in the plant (*vide supra*). The NMR data also showed that the dimer was branched from C(3) of the oxygenated tabersonine-$\beta$-epoxide moiety to the aromatic C(10′) of the mehranine unit. The stereochemistry at C(3) was readily ascertained from the NMR signal attributed to H(3) ($\delta$ 4.52) which was a singlet, requiring the H(3)/H(14) dihedral angle to be ca. 90°, an arrangement possible only if H(3) is $\beta$. The structure has also been subsequently confirmed by X-ray analysis (*340*). Another bisindole from the same plant is conophyllinine (**551**) which shares the same pandicine-like tabersonine unit with **546**, **548**, and **550**, as well as a similar mode of linking *via* a central dihydrofuran ring (*334*). The difference is in the other monomeric unit which is now an opened form of tabersonine-$\beta$-epoxide, with a *trans* diol functionality at positions 14′ and 15′ instead of a $\beta$-epoxide function. The epoxide ring opening by a water molecule is anticipated to occur preferentially at the less hindered C(14′) resulting in a H(14′$\beta$), H(15′$\alpha$) configuration. This proposal was supported by the $J_{14'-15'}$ value of 9 Hz and the observed NOEs for H(15′)/H(18′), H(19′), H(21′).

Conofoline has also been isolated from another Malaysian *Tabernaemontana* species (*T. peduncularis*) under the name peduncuIine (*341*) as well as from *T. bovina* (*342*). The former plant also furnished another bisindole alkaloid, pedunculaTidine (**552**), which shares the common oxygenated tabersonine-β-epoxide moiety of conofoline, but differs in the second unit which is now an opened form of (−)-mehranine, with a *trans* diol functionality at positions 14′ and 15′. The stereochemistry at C(14′) and C(15′) was determined from the ROESY spectrum which showed correlations between the axial H(2′) and H(17′) with H(14′), indicating that the latter has a β stereochemistry.

The common, highly oxygenated tabersonine/pachysiphine unit which occurs in conophylline (**546**), conophyllidine (**548**), conofoline (**550**), conophyllinine (**551**), pedunculaTidine (**552**), kisantine (**539**), and pandicine (**519**), has been finally isolated from *T. divaricata*, and named taberhanine (*334*). The co-occurrence of taberhanine and conophylline in the same plant suggested a biogenetic pathway involving electrophilic attack of the taberhanine iminium ion (**553**) on the activated position 10′ of the hypothetical 11′-hydroxypachysiphine acceptor unit (**554**), followed by intramolecular epoxide ring opening to form the central dihydrofuran unit (Scheme 33) (*334*).

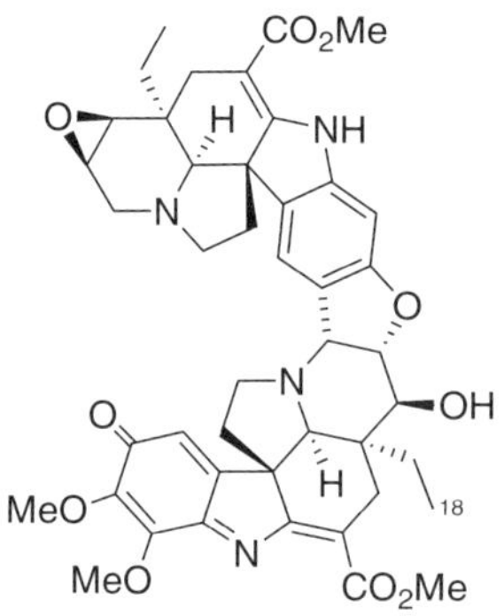

**555**

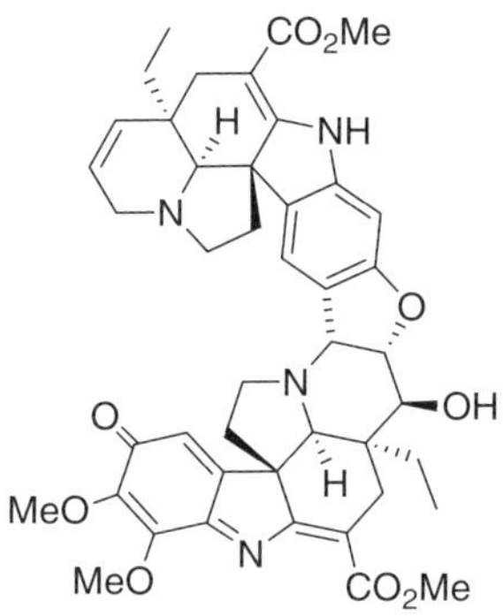

**556**

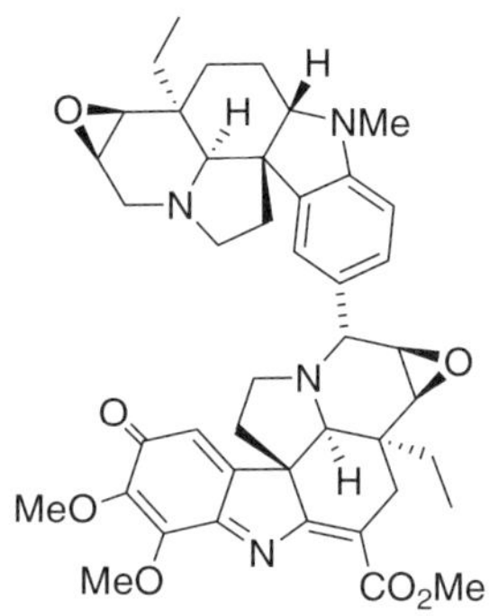

**557**

Conophylline (**546**) has been shown to possess important biological activity. It has been shown to induce morphological normalization in *ras*-overexpressing cell lines such as K-*ras*-NRK and K-*ras*-NIH3T3 (*337*). In both *in vitro* as well as *in vivo* experiments using K-*ras*-NRK cells, conophylline was shown to inhibit tumor infiltration and metastasis (*343,344*). Its antitumor effects were later demonstrated in human cancer cells when it was shown to inhibit attachment and chemotactic invasion of human uterine endometrial cancer cells (*345*). Other biological effects of conophylline include the suppression of TGF-β signaling (*346*) and the downregulation of TNF-α receptors in human T-cell leukemia cells (*347*). Recently, conophylline has also been shown to induce morphological change, as well as insulin production, in rat pancreatic acinar carcinoma cells (*348,349*). In a structure-activity study it was shown that the highest effect was shown by conophylline (**546**), followed by conophyllidine (**548**). Conofoline (**550**), which has a single bond connecting the monomeric moieties was ineffective, indicating that an intact central dihydrofuran ring is essential for manifestation of the biological activity. The putative monomeric precursors of the bisindoles, taberhanine, pachysiphine, and mehranine were themselves ineffective. The respective semisynthetic iminoquinone derivatives of the bisindoles (e.g., **555**–**557**) were also examined where it was found that of a pair of active bisindoles, the iminoquinone derivative invariably showed the higher potency (*348*).

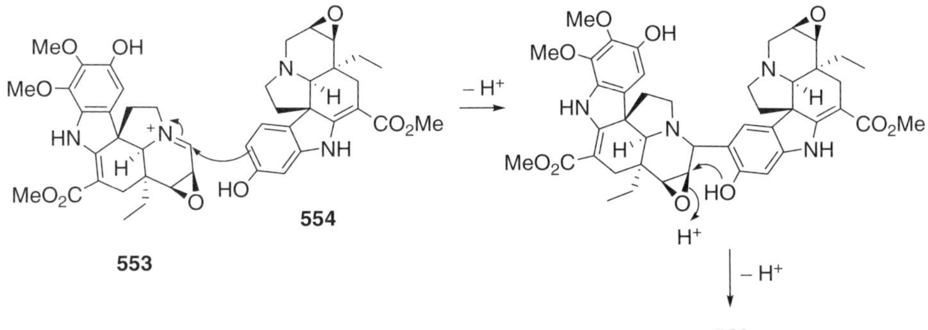


Scheme 33

In a subsequent study of the effect of conophylline on the differentiation of pancreatic precursor cells, it was found that conophylline inhibited the formation of cystic structure and increased the number of insulin-positive cells in the rat pancreatic rudiment in organ culture. Conophylline was also found to increase the expression of mRNA for insulin and the number of homeobox-1-positive cells. *In vivo* studies using neonatal streptozotocin-induced diabetic rats showed that conophylline-treatment resulted in significant reduction of the plasma glucose concentration and improved glucose tolerance on glucose loading. In addition, the insulin content, the $\beta$-cell mass, the number of islet-like cell clusters, and pancreatic duodenal homeobox-1-positive ductal cells, were all significantly increased in conophylline-treated rats. These results suggest that conophylline induces differentiation of pancreatic precursor cells and increases the formation of $\beta$-cells (*350*).

The structures of the alkaloids of *Melodinus scandens*, viz., scandomeline, scandomelonine, and their C(19)-epimers have been reported previously (*351*). From the concentrated mother liquors of scandomeline, another bisindole scandomelidine (**558**) has been obtained, which was shown by the spectral data to be constituted from tabersonine- and aspidofractinine-type halves (*352*). Examination of the $^{13}C$ NMR spectral data and comparison with those of venalstonine (**559**), pachysiphine (**547**), and scandomelonine (**560**) led to structure **558** for scandomelidine, in which venalstonine is linked *via* C(10) to C(3′) of pachysiphine, in the same manner as in scandomelonine (3′α-substitution).

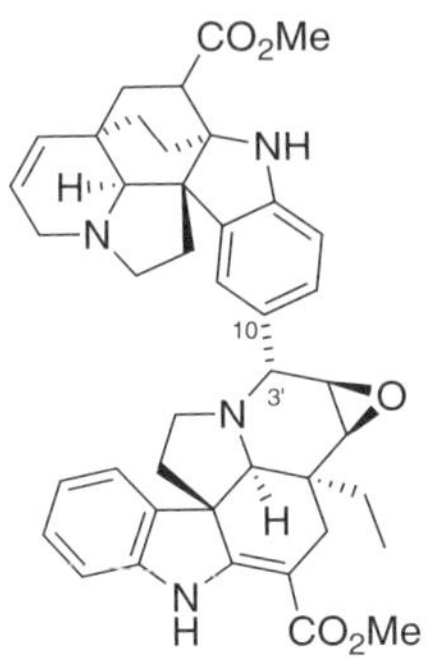


**558** scandomelidine

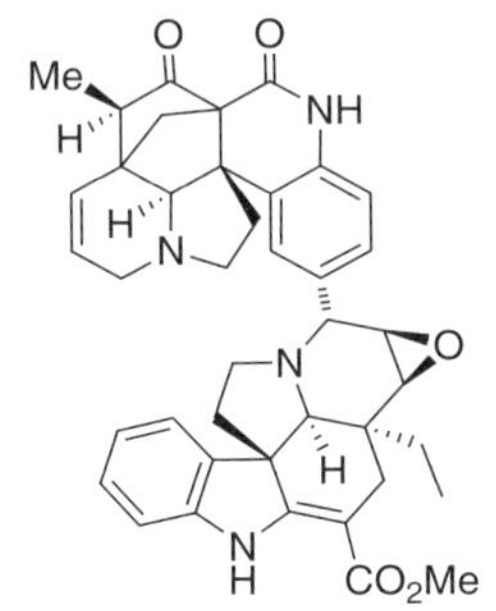


**560** scandomelonine

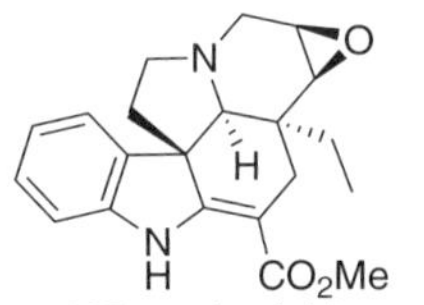


**547** pachysiphine

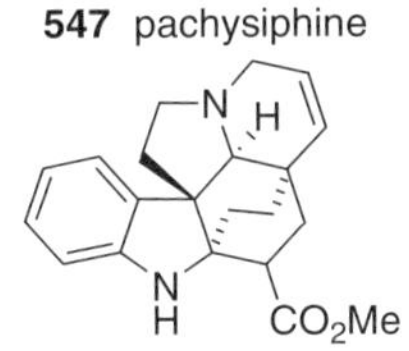


**559** venalstonine

Voafrines A (**561**) and B (**562**), isolated from cell suspension cultures of *Voacanga africana*, represent the first instance of fully characterized bisindole alkaloids produced by plant cell cultures (*353*). HRMS showed that **561** and **562** are isomers with molecular formula $C_{42}H_{46}N_4O_4$, while the UV spectra indicated the presence of $\beta$-anilinoacrylate chromophores. The fragment ions observed at $m/z$ 214 and 228, which are typical of tabersonine (**535**), and the fact that the molecular formula corresponds to a dehydrodimer of tabersonine, suggested strongly that both alkaloids are constituted from the union of two tabersonine units. Interpretation of the NMR spectral data was facilitated by comparison with that of tabersonine as well as application of 2-D techniques. The $^1$H NMR spectrum showed the presence of eight aromatic hydrogens and two NH signals, excluding these positions as possible sites of attachment of the monomeric units. However, only three olefinic hydrogens, two diastereotopic aminomethylene hydrogens, and a single aminomethine H, were observed, indicating branching from an olefinic carbon of one unit to C(3′) of the other tabersonine unit. The attachment point on the olefinic carbon was deduced to be C(14), since H(3) in both compounds showed only small allylic coupling with H(15). The other two olefinic hydrogens form part of an ABX system with the lone H(3′). Analysis of the vicinal coupling constants of these hydrogens indicated an arrangement in which H(3′) is nearly orthogonal to the plane of the double bond in voafrine A (H(3′α)), and nearly coplanar to the plane of the double bond in voafrine B (H(3′β)).

These conclusions led to the structures depicted in **561** and **562** for voafrines A and B, respectively, which also represent their absolute stereochemistry, in view of the similarity of the chiroptical characteristics (CD spectra) with those of (−)-tabersonine (**535**).

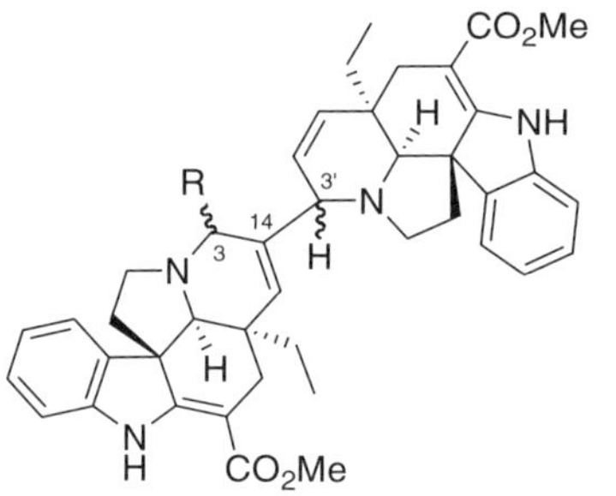


**561** R = H, H(3′α) voafrine A

**562** R = H, H(3′β) voafrine B

**563** R = OH, H(3′α)

**564** R = OH, H(3′β)

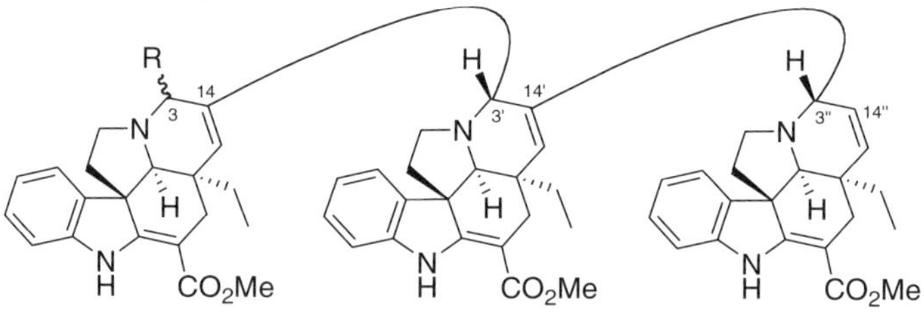


**565** R = OH

**566** R = H

Dimerization of tabersonine (**535**) in the presence of oxygen by a crude enzyme preparation from *C. roseus* yielded 3-hydroxyvoafrines A (**563**) and B (**564**), which were reduced enzymatically or by $NaBH_4$ to voafrines A and B, respectively

(*354*). The $^1H$ NMR spectrum of 3-hydroxyvoafrine B (**564**) in DMSO-$d_6$ or acetone-$d_6$ was complicated by the existence of equilibrating epimers of **564** ($3\alpha$-hydroxyvoafrine B and $3\beta$-hydroxyvoafrine B). Addition of water shifted the equilibrium toward the $3\alpha$-hydroxy epimer, which was the sole species observed in DMSO-$d_6$-$H_2O$ or acetone-$d_6$-$H_2O$ mixed solvents. The involvement of the tabersonine iminium ion as an intermediate was suggested by the isolation of $3\alpha$-acetonyltabersonine and $3\alpha$-acetonylvoafrine B when the enzyme-mediated conversion was carried out in the presence of acetone. A third polar product which was also isolated was subsequently identified to be a trimeric alkaloid (**565**), constituted from three tabersonine units, and, like the hydroxyvoafrines, was readily reduced by $NaBH_4$ to **566** (*355*). The dimerization of tabersonine (**535**) was also probed by anodic oxidation (Pt/$LiClO_4$/MeCN) which produced a different result compared to the enzyme-mediated process, yielding instead the 16,10′-linked dimeric product **567** as the major product accompanied by minor amounts of the 10,10′-linked dimer **568**, and the trimers **569** and **570** (*356*). Anodic oxidation of 3-oxotabersonine gave the corresponding 16,10′-linked didehydrodimer **571** as the major product (*356*) as did reaction of 3-oxotabersonine with nitrosonium tetrafluoroborate ($NOBF_4$) (*357*).

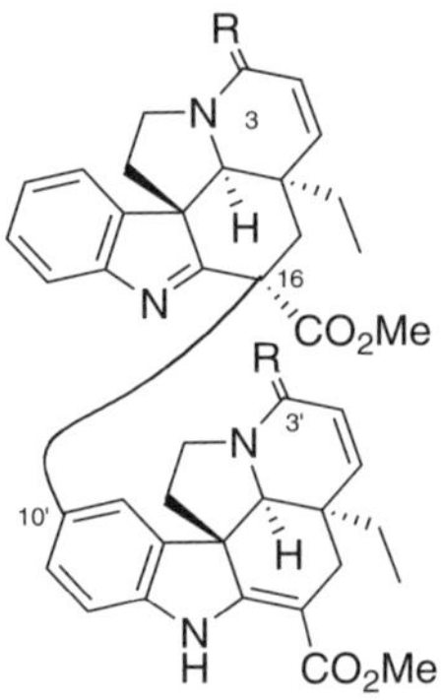

**567** R = $H_2$
**571** R = O

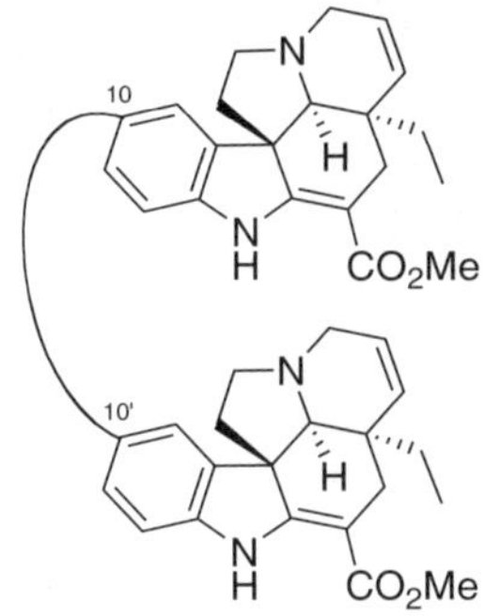

**568**

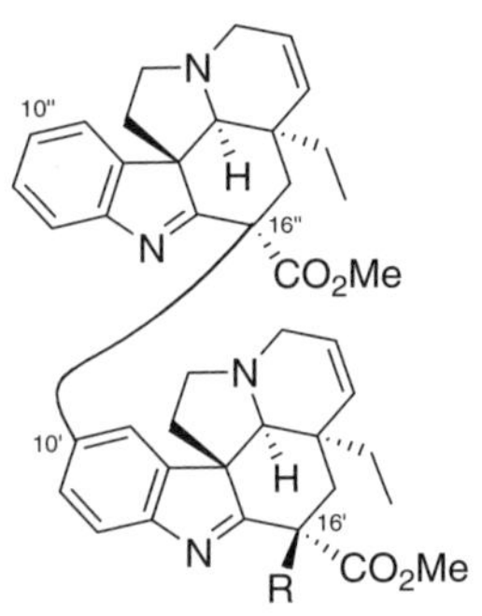

**569** R = 10-tabersonine

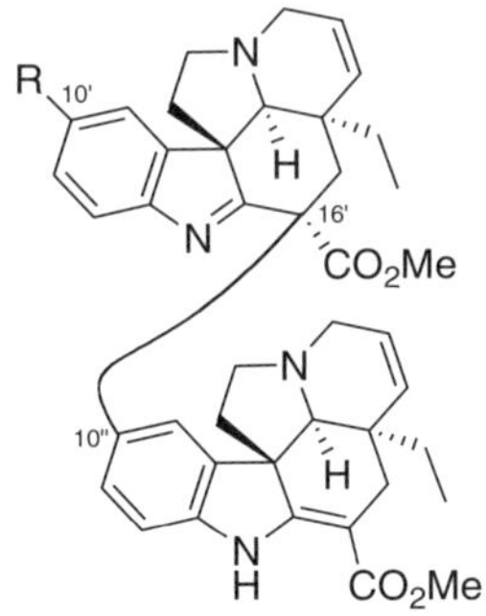

**570** R = 10-tabersonine

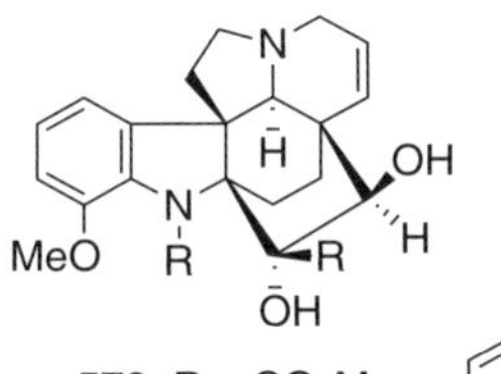

**573** R = $CO_2Me$

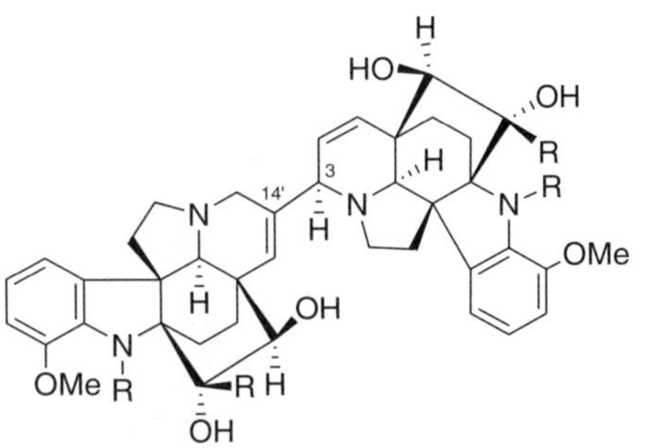

**572** R = $CO_2Me$ nitaphylline

Nitaphylline (**572**), was the only bisindole isolated from *Kopsia teoi* which otherwise provided a large number of new indole alkaloids. The mass spectrum showed a molecular ion at *m/z* 910 with a significant peak at *m/z* 455 suggesting symmetrical cleavage of the parent ion along the bond bridging the monomeric moieties. This, together with the observation that the UV spectrum of nitaphylline was virtually superimposable with that of kopsingine (**573**) provided an early indication that nitaphylline is constituted from union of two kopsingine units. This was further reinforced by the NMR spectral data. The $^{13}C$ NMR spectrum accounted for only 33 peaks, indicating overlap of 15 carbon resonances. In addition, four pairs of signals although distinguishable were only just so, differing in chemical shift by only 0.1 or 0.2 ppm. The bulk of the overlapped signals were readily assigned to the aromatic portions of the alkaloid as well as the methoxy, carbamate, and ester groups which were similar to those of kopsingine. The remaining signals could also be assigned based on 2-D NMR data and by comparison with kopsingine. The observation that the resonances of the piperidine ring carbons showed greater departure from kopsingine suggested that attachment of the monomeric entities involved the piperidine ring carbons.

The $^{1}H$ NMR spectrum showed six aromatic hydrogens, three olefinic hydrogens, two geminal hydrogens of an aminomethylene group and only one H due to an aminomethine, indicating branching of the dimer from C(3) of one kopsingine unit to the olefinic C(14′) or C(15′) of the other kopsingine unit. Since the geminal hydrogens of C(3′) were doublets (*J* 16 Hz) with no evidence of coupling to any adjacent olefinic hydrogen, the dimer was therefore deduced to be bridged from C(3) of one kopsingine unit to C(14′) of the other. The stereochemistry of the point of branching could not be ascertained directly from the NMR spectrum due to overlap of the aminomethylene signal and poor resolution of the olefinic signals, but could be assigned by analogy to the bistabersonine alkaloids voafrine A (**561**) ($\beta$-branching) and voafrine B (**562**) ($\alpha$-branching), which possess the same mode of attachment of the monomeric units (*353*). In nitaphylline (**572**), the carbon shifts of C(21) and C(21′) are identical (overlapped) and are similar to the value in the monomer indicating that the substitution is $\beta$, with the pseudoequatorially oriented second kopsingine unit pointing away from the first by analogy to voafrine A. In the $\alpha$-branched voafrine B, the resulting greater spatial proximity between the two monomeric units is reflected in the very different shifts observed for C(21). Nitaphylline (**572**) represents the first example of an aspidofractinine–aspidofractinine-type bisindole alkaloid (*358,359*).

Tenuiphylline (**574**) was isolated in minute amounts, together with the tenuisines (*vide infra*) from *Kopsia tenuis* occurring in Malaysian Borneo (*360*). The FABMS showed a $MH^+$ peak at *m/z* 717 and high-resolution measurements provided the formula $C_{42}H_{44}N_4O_7$. The UV spectrum indicated the presence of dihydroindole chromophores and the IR spectrum indicated the presence of hydroxyl, urethane, ester, and lactone functions. In view of the small amount obtained, definitive structure elucidation of tenuiphylline required resort to 600 MHz NMR spectra since extensive overlap was encountered with lower field instruments.

The NMR spectral data showed the presence of one carbamate and one methyl ester function. The aromatic region integrated for seven hydrogens indicating branching of the dimer from an aromatic carbon of one monomeric unit. Absence of an NH signal and the presence of only one urethane function indicated the dihydroindole nitrogen of the other monomeric moiety as the other site of attachment. The NMR spectral data also showed the presence of four olefinic hydrogens, two associated with a six-membered ring and two with a five-membered ring. One unit of the bisindole was readily deduced to be identical to lapidilectine B (**575**) from the excellent correlation of the $^{13}C$ shifts, in particular the non-aromatic shifts, with those of lapidilectine B. The other monomeric unit can be considered as having a novel carbon skeleton or a rearranged venalstonine. The carbon shifts of this second unit generally resembled those of venalstonine (**559**), except for changes involving carbons (2), (12), (16), (17), (18), (20), and (21). The COSY spectrum indicated the presence of the same groups as in venalstonine and the downfield shifts of C(20) and C(21) were reminiscent of those of the vindolinine derivatives (e.g., *N*-methylvindolinine **576**), suggesting a change in the connectivity involving the C(18)–C(19) fragment. The presence of a hydroxyl group on C(2) was supported by its downfield shift at $\delta$ 99.8 due to its being linked to an oxygen and a nitrogen and by the observed three-bond correlations from C(2) to H(6) and H(21) in HMBC. The presence of an OH on C(2) required that C(18) be now linked to C(16) which was consistent with the downfield shift of C(16) when compared with venalstonine as well as with the observed correlations from both C(18) and C(19) to H(17) in the HMBC spectrum. From the $^{1}H$ NMR spectrum (COSY, NOE), it could be established that the aromatic portion associated with this unit was unsubstituted. Furthermore, the upfield shift of H(12), when compared with venalstonine, was consistent with the change from NH in venalstonine to N–C(11′) in tenuiphylline.

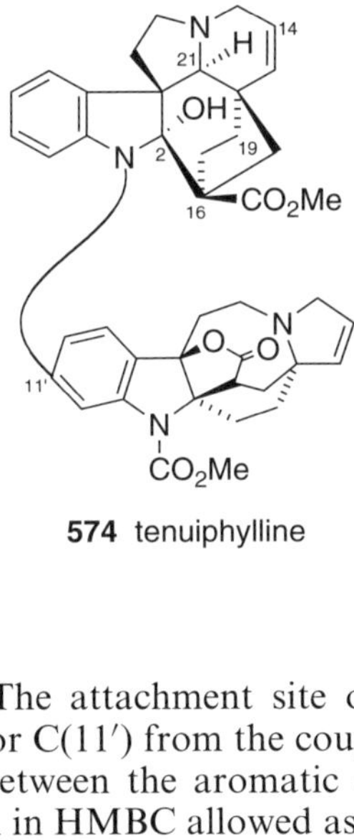


**574** tenuiphylline

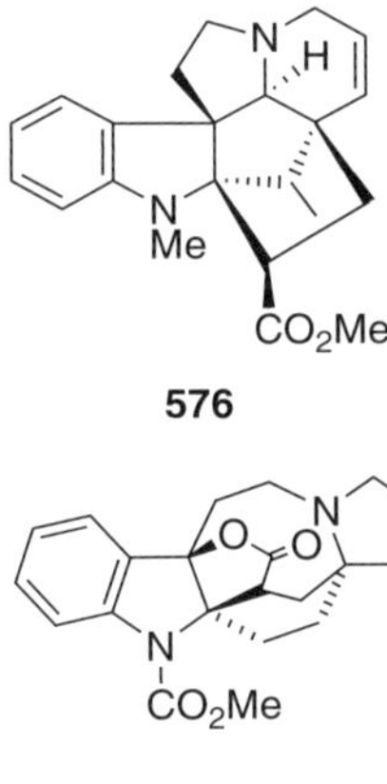


**576**

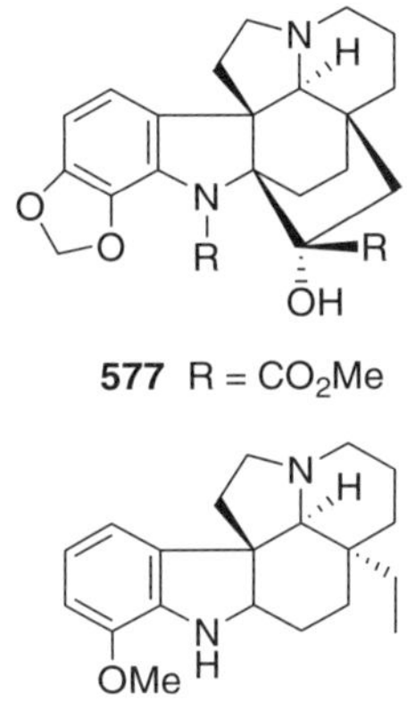


**577** R = $CO_2Me$

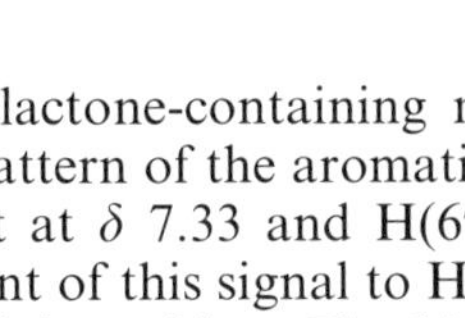


**575**

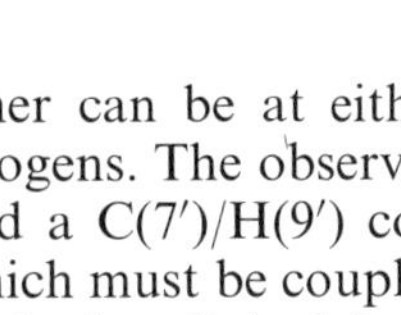


**578**

The attachment site on the lactone-containing monomer can be at either C(10′) or C(11′) from the coupling pattern of the aromatic hydrogens. The observed NOE between the aromatic doublet at $\delta$ 7.33 and H(6′$\beta$), and a C(7′)/H(9′) correlation in HMBC allowed assignment of this signal to H(9′) which must be coupled to H(10′), furnishing proof of C(11′) branching. The bisindole is thus linked from N(1) of one unit to C(11′) of the other. The stereochemistry of the C(2)-OH was assigned the $\alpha$ configuration since, had it been $\beta$, the ester function would have been too far removed to experience NOE with H(9′) and H(10′) as well as the observed anisotropy from the other aromatic ring.

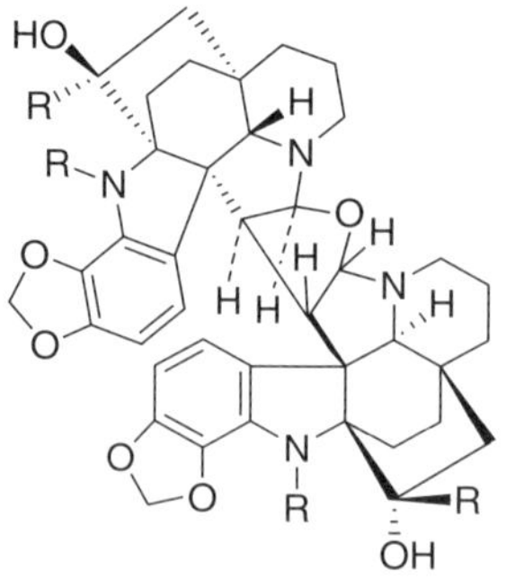


**579** R = $CO_2Me$

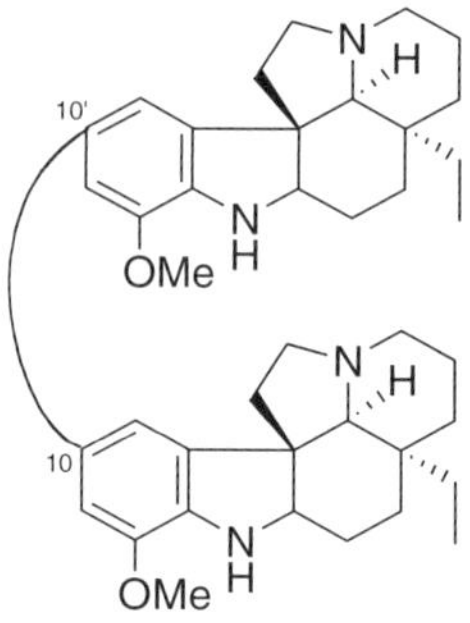


**580**

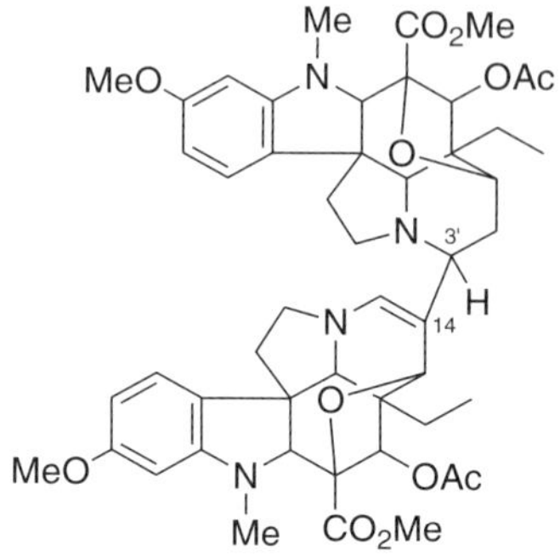


**581**

Six alkaloids (the kopsifolines), possessing the carbon skeleton corresponding to the novel venalstonine-like monomer in tenuiphylline, and constituting a new structural subclass of the monoterpenoid indole alkaloids, have since been obtained from another Malaysian *Kopsia* (*361,362*).

Anodic oxidation (C/$LiClO_4$/MeOH) of the aspidofractinine-type alkaloid, kopsamine (**577**) yielded a dimerization product **579**, characterized by the presence of a central tetrahydrofuran ring and a $C_2$ axis of symmetry (*363*). Photochemically induced coupling of deacetylaspidospermine (**578**) gave the dimer **580**, characterized by a 10, 10′-bond (*364*). Oxidation of vindoline by horseradish peroxidase gave the 14,3′-linked enamine dimer **581**, while similar oxidation of 16-*O*-acetylvindoline led to the stable conjugated iminium salt, without formation of any dimerization product (*365*).

## XXIX. *Aspidosperma*–Pseudoaspidosperma Type

The alkaloids of the ervafoline group isolated from *Stenosolen heterophyllus* (Apocynaceae), exemplify an additional dimension in the structural complexity of the bisindoles, in which the constituent units are linked by three bonds. Ervafoline (**582**) has been mentioned in a previous volume of this series (*1*). The structure was solved by X-ray analysis (*366*) and augmented by a full $^{1}H$ NMR analysis to facilitate structure elucidation of analogues (*367,368*). Subsequent to the initial disclosures on the structure of ervafoline, seven other related bisindoles were described in a full report (*369*). These include 19′-hydroxyervafoline (**583**), ervafolene (**584**), and 19′-hydroxyervafolene (**585**), which, with **582**, constitute the four members of the ervafoline series, while four others, ervafolidine (**586**), 3-*epi*-ervafolidine (**587**), 19′(*R*)-hydroxyervafolidine (**588**), and 19′-hydroxyepiervafolidine (**589**), comprise the evafolidine series.

Ervafoline (**582**) analyzed for $C_{40}H_{44}N_4O_4$. The UV spectrum indicated the presence of dihydroindole and $\beta$-anilinoacrylate chromophores, while the $^{1}H$ NMR spectrum indicated the presence of a methyl ester function ($\delta$ 3.74). When the $^{1}H$ NMR was obtained in $CDCl_3$ containing a small amount of $CF_3COOH$, the 6H triplet at $\delta$ 0.94 was split into two three-H triplets at $\delta$ 0.97 and 1.06, indicating the presence of two ethyl side chains. Resistance to acetylation ($Ac_2O$/pyridine) indicated that the remaining two oxygens were part of ether linkages. This was supported by the $^{13}C$ NMR spectrum which showed the presence of an N-C-O linkage ($\delta$ 103.4) as well as a 14′,15′-epoxide unit ($\delta$ 51.7, 57.0) of an *Aspidosperma* moiety. The presence of only one indolic NH ($\delta$ 9.20) suggested the possibility that the other indolic nitrogen could be involved in the linking of the two halves. Assignment of the $^1$H and $^{13}$C NMR spectral data was made possible after unravelling of the structure and relative stereochemistry by X-ray analysis. Of particular note was the H(3) singlet at $\delta$ 3.86, and the unusually shielded multiplet at $\delta$ 5.64 for H(12′), possibly as a result of the anisotropy exerted by the other aromatic ring. These features, as well as the mass spectral fragment at $m/z$ 333 due to the pseudoaspidosperma (lower) half, constitute valuable diagnostic tools for the recognition of related bisindoles belonging to the ervafoline series. Thus, the observation of the $m/z$ 333 fragment in the mass spectrum of 19′-hydroxyervafoline (**583**) indicated that the additional oxygen was located in the *Aspidosperma* half of the molecule. Formation of a monoacetate derivative on acetylation, and the appearance of the $CH_3$-18′ signal as a doublet in the $^1$H NMR spectrum, placed the OH substituent on C(19′). The $^1$H NMR spectral data of ervafolene (**584**) indicated replacement of the 14′,15′-epoxide function by a double bond in the upper *Aspidosperma* unit (absence of signals at $\delta$ 3.25 and 3.03, and appearance of olefinic signals at $\delta$ 5.92 and 5.75) as the main difference between **584** and **582**.

Similarly, **585** was readily shown by the NMR spectral data to be 19′-hydroxyervafolene, from the presence of the 14′,15′-unsaturation, and replacement of an ethyl with a hydroxyethyl side chain at C(20′). Formation of a monoacetate derivative on acetylation and catalytic hydrogenation of the 14′,15′-double bond provided additional support for the assignment.

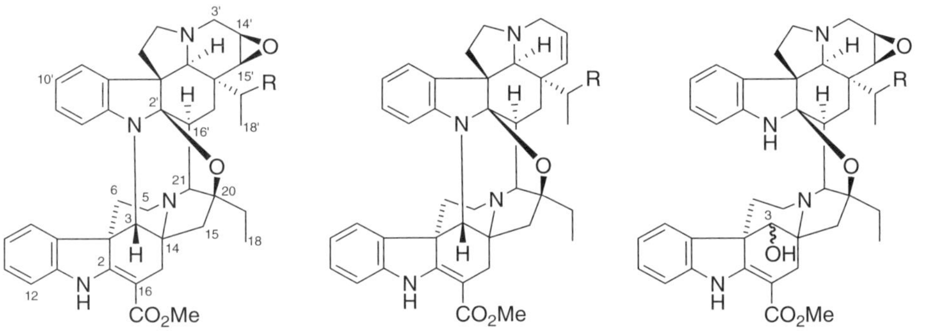

**582** R = H ervafoline
**583** R = OH

**584** R = H ervafolene
**585** R = OH

**586** 3(*R*), R = H ervafolidine
**587** 3(*S*), R = H
**588** 3(*R*), 19′(*R*), R = OH
**589** 3(*S*), R = OH

The ervafolidine alkaloids **586**–**589** are characterized by loss of one ring due to hydrolytic cleavage and scission of the N(1′)-C(3) bond. This was confirmed by an X-ray analysis of 3-*epi*-ervafolidine (**587**), which allowed full NMR assignments of **587**, as well as ervafolidine (**586**). HREIMS indicated that **586** and **587** were isomeric, and differed from **582** by the addition of $H_2O$, while the notable absence of the $m/z$ 333 fragment in the mass spectrum indicated a change in the pseudoaspidosperma half. The $^1$H NMR spectral data showed an intact 14′,15′-epoxide moiety and the presence of two ethyl side chains. Acetylation to a monoacetate derivative resulted in a significant downfield shift of the 1H singlet at $\delta$ 3.87 to $\delta$ 5.20, suggesting the presence of a secondary hydroxyl group linked to two quaternary centers. The main difference in the $^1$H NMR spectrum of **586** (3*R*) with that of its C(3) epimer **587** (3*S*), concerns the signal of H(3), which was changed from a singlet ($\delta$ 3.87) in **586**, to a doublet ($\delta$ 4.14, *J* 5 Hz) in **587**, due to coupling with OH, as demonstrated by deuterium exchange. As with the ervafoline alkaloids, the identification of **588** and **589** as the 19′-hydroxy substituted derivatives of **586** and **587**, respectively, followed from a comparison of the NMR spectral data. The configuration at C(19′) of **588** was determined from X-ray analysis, which also confirmed the proposed structure.

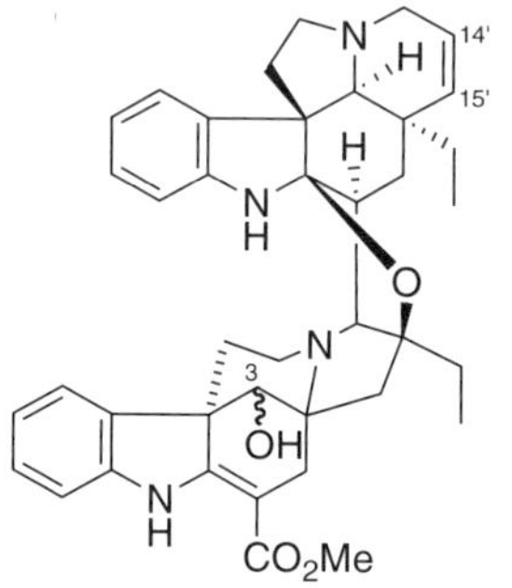

**590** 3(*S*) ervafolidene
**591** 3(*R*) *epi*ervafolidene

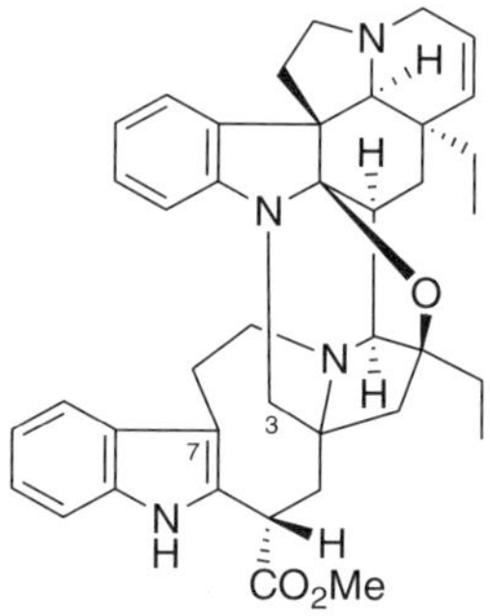

**592**

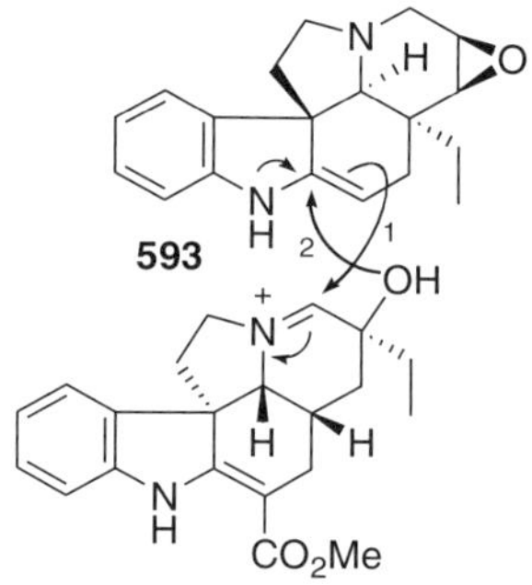

**593**

**594**

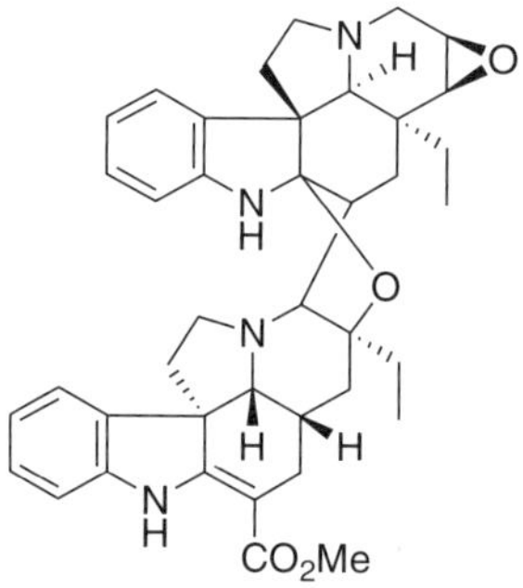

**595**

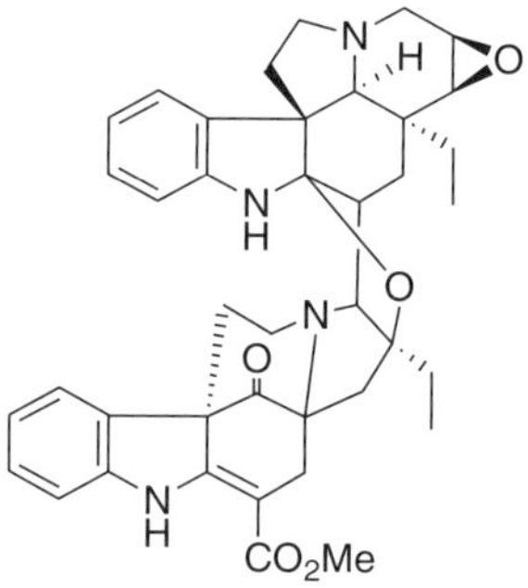

**596**

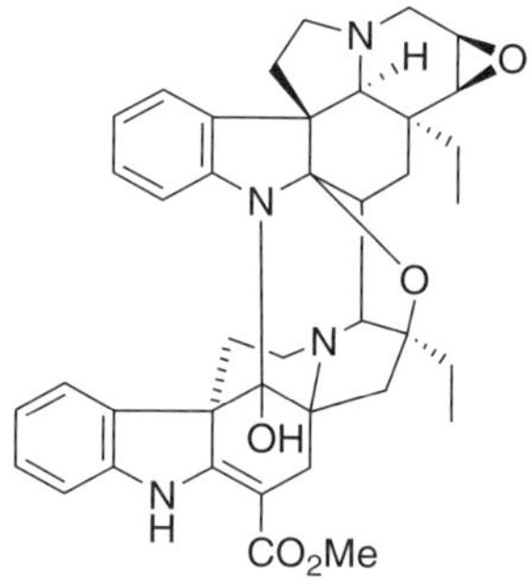

**597**

Two additional bisindoles of the ervafolidine series, ervafolidene (**590**) and the isomeric *epi*-ervafolidene (**591**), were isolated from *Pandaca caducifolia* (Apocynaceae) (*370*). The structures were established based on comparison of the NMR spectral data with ervafolene (**584**), which was also isolated from the plant. Ervafolidene (**590**) and 3-*epi*-ervafolidene (**591**) are similar to 3-*epi*-ervafolidine (**587**) and ervafolidine (**586**), respectively, except for replacement of the 14′,15′-epoxide function with a double bond. Reduction of ervafolene (**584**) with $NaBH_3CN$ afforded the unexpected ring-opened alkaloid **592** as the major product, *via* formation of the 1,2-indoleninium ion, followed in succession by N(1′)-assisted fragmentation leading to cleavage of the C(3)–C(7) bond, and reductive trapping of the N(1′)–C(3) iminium ion by $NaBH_3CN$.

A possible biogenetic pathway has been proposed to the ervafoline alkaloids (*369*), involving coupling of the enamine **593** (from decarboxylation of a tabersonine

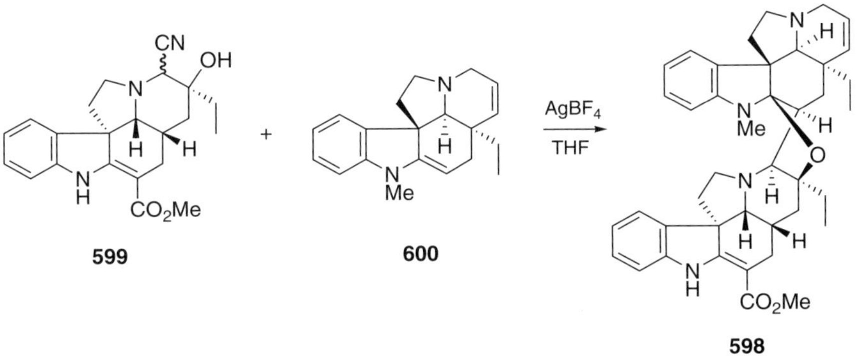


Scheme 34

type precursor) with the iminium ion **594** (from a modified pandoline-type precursor) to give **595**, which then rearranges to the spiroketone **596**. The ervafolidines can then be formed from **596** by reduction. Alternatively, cyclization to the carbinol amine **597**, followed by reduction would lead to the ervafolines. A synthesis of the bisindole **598** along the lines of the proposed biogenetic pathway, through coupling of the α-cyanopiperidine **599** (derived from 20-*epi*-pandoline) as a synthetic equivalent of the N(4)–C(3) iminium ion, with the enamine **600** (derived from tabersonine), in the presence of $AgBF_4$ in THF (Scheme 34), provided support for the proposed biogenesis of the ervafolines (*369,371*).

## XXX. *Aspidosperma*–Eburnea Type

Earlier studies of the structure of callichiline from the African plant, *Callichilia subsessilis* were inconclusive, although the available evidence appeared to favor structure **601** (*1*). The structure has been solved by an X-ray crystallographic analysis of the alkaloid which revealed the structure and relative configuration of callichiline as shown in **602** (*372*).

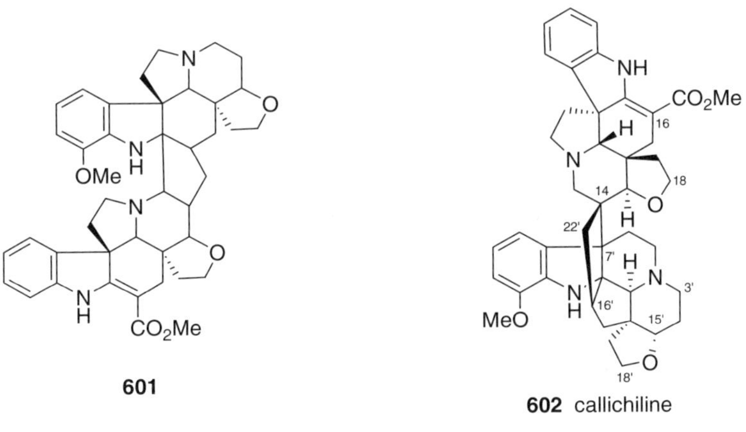


**602** callichiline

Vobtusine (**603**), the spirofused bisindole alkaloid isolated from *Voacanga chalotiana* has been discussed in a previous volume of this series (*1*). In the course of purifying large quantities of **603**, a small amount of another bisindole, vobtusamine (**604**), was isolated (*373*). The mass spectrum of **604** showed an $M^+$ at $m/z$ 734 ($C_{43}H_{50}N_4O_7$), 16 mass units higher than vobtusine, indicating the presence of an additional oxygen. The $^{13}C$ NMR spectrum of **604** showed that 20 carbons were identical in chemical shift and multiplicity with the upper (12-methoxyaspidosperma) half of vobtusine, except for the methylenes C(22′) and C(23′) ($\delta$ 27.6 and 45.8, respectively) which are adjacent to the spirocyclic C(14). The remaining carbons suggested a cuanzine-like moiety devoid of aromatic methoxy substitution and having disubstitution at C(14). These observations led to the proposed structure **604**, which was supported by the observation of the $m/z$ 224 ion in the mass spectrum which is diagnostic of vincamine-type alkaloids. The proposed structure was in agreement with the $^1H$ and $^{13}C$ NMR data. The shielding of the 12′-OMe and H(23′) signals in **604** compared to those in **603**, was attributed to diamagnetic anisotropy of the indole ring of the cuanzine-like half.

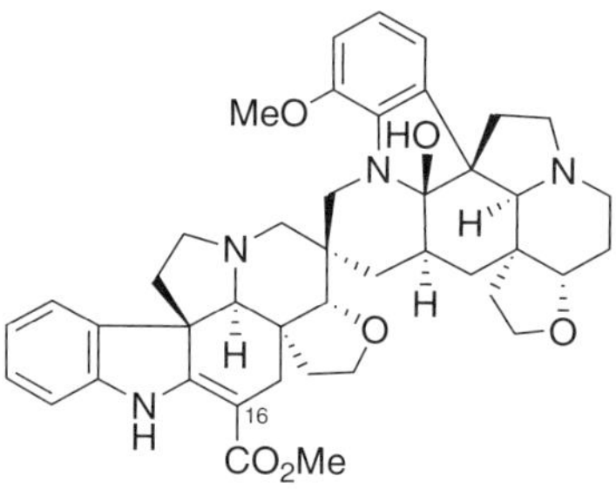

**603** vobtusine

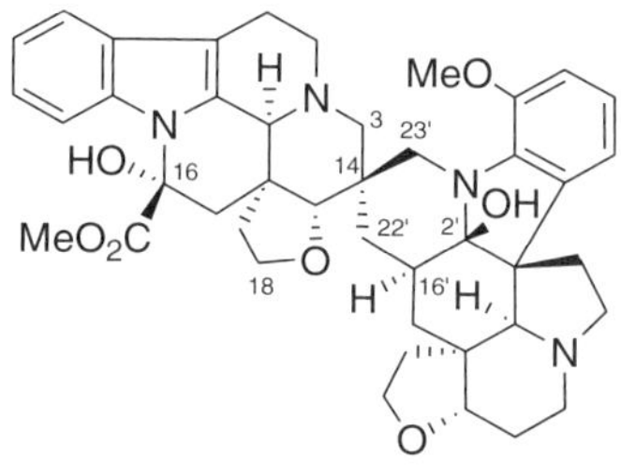

**604** vobtusamine

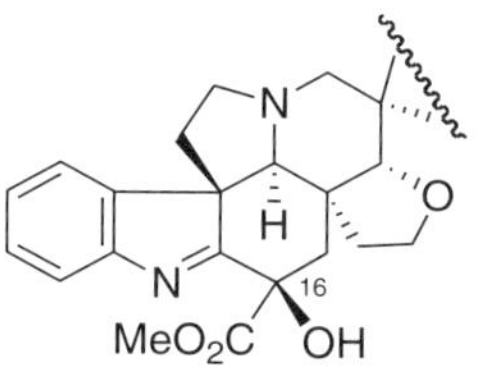

**605** 16-hydroxyvobtusine

Additional confirmation was obtained by a biomimetic conversion from vobtusine (**603**) based on the oxidative *Aspidosperma*→eburnea rearrangement proposed by Wenkert. This was achieved by ozonization of vobtusine (**603**) to the 16-hydroxyderivative (**605**), followed by an acid-induced rearrangement to give **604** accompanied by the C(16)-epimer as a minor product. Since the structure and absolute configuration of vobtusine (**603**) have been previously established by X-ray analysis (*374*), the above conversion confirmed the absolute configuration of vobtusamine as depicted in **604**.

(−)-Norpleoimutine (**606**) was isolated from *H. zeylanica* collected in Kenya, in addition to pleiomutinine and 19′-*epi*-pleiomutinine (*vide infra*). The $M^+$ was 14 mass units less compared to (−)-pleiomutine (**607**), and the missing methylene can be attributed to the kopsinine half as its fragments were shifted by 14 mass units from those of pleiomutine. This was confirmed on examination of the NMR data, and by methylation (HCHO, HOAc, $NaBH_3CN$) of norpleiomutine to pleiomutine (*375*).

Magnus reported a total synthesis of norpleiomutine (**606**) by the acid-induced coupling of kopsinine (**608**) (synthesized from the tetracyclic amine **609**) and (−)-eburnamine (**611**) (from $LiAlH_4$ reduction of (+)-eburnamonine **612**, which was in turn synthesized starting from the Diels–Alder adduct **610**) (*376*).

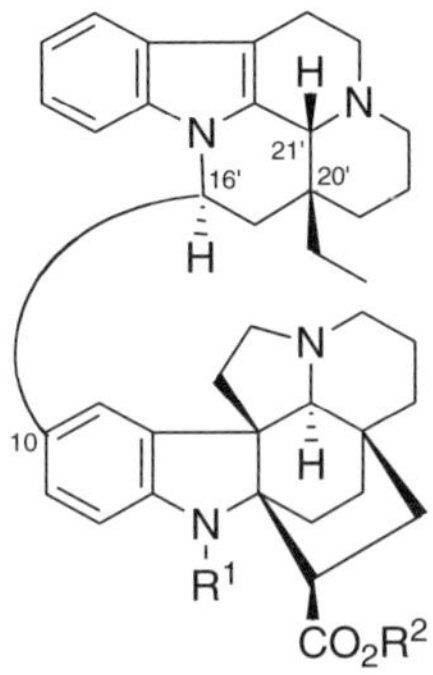


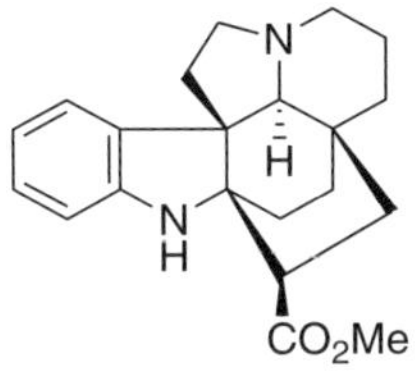

**608** kopsinine

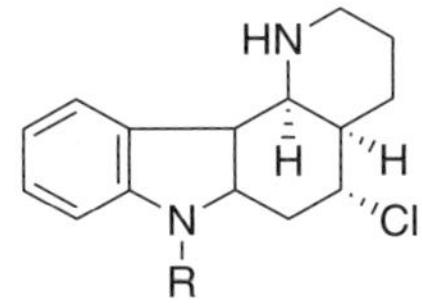

**609** R = $-SO_2C_6H_4OMe$-*p*

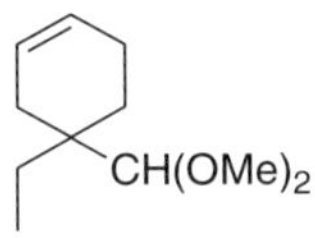

**606** $R^1$ = H, $R^2$ = Me (−)-norpleiomutine
**607** $R^1$ = Me, $R^2$ = Me (−)-pleiomutine
**616** $R^1$ = $R^2$ = H (−)-demethylnorpleiomutine

**610**

*Kopsia officinalis*, endemic to the Yunnan province of China, gave the bisindole alkaloid, (+)-kopsoffine (**613**) (*377*). As in the case of norpleiomutine (**606**), the mass spectral fragmentation was similar to that of (−)-pleiomutine (**607**), except that the $M^+$ of **613** ($m/z$ 616) is 14 mass units less than that of **607**. The $^1H$ NMR spectrum confirmed the absence of NMe in the kopsinine half, while the H(16) coupling constants of 11 and 5 Hz indicated that H(16′) is axial or $\beta$, and that the C(16′) substitution is $\alpha$. Comparison of the $^{13}C$ NMR spectral data with those of (−)-kopsinine (**608**) and (+)-vincamine provided further support for the proposed structure in which the eburnane half is the optical antipode of the corresponding half in (−)-norpleiomutine (**606**). Confirmation of this was provided by semisynthesis, from the condensation of (+)-eburnamine (**614**) (from $LiAlH_4$ reduction of (−)-eburnamonine **615**) with (−)-kopsinine (**608**) under acidic conditions, which led to a product identical with (+)-kopsoffine.

*Kopsia pauciflora* from Malaysian Borneo provided, in addition to kopsoffine (**613**) and norpleiomutine (**606**), two related bisindole alkaloids, (−)-demethylnorpleiomutine (**616**), and (+)-kopsoffinol (**617**) (*378*). These alkaloids are also constituted from the union of eburnane and kopsinine units and are remarkable in that while norpleiomutine (**606**) and demethylnorpleiomutine (**616**) incorporate eburnane units having the 20$\beta$,21$\beta$ configuration, kopsoffine (**613**) and kopsoffinol (**617**) apparently incorporate eburnane units having the opposite (enantiomeric) configuration (20$\alpha$,21$\alpha$). The eburnane alkaloids present in the plant (including 19-hydroxyeburnamine **619** which corresponds to the eburnane half of kopsoffinol), however, belonged to the (−)-eburnamine series, with the 20$\beta$,21$\beta$

configuration (*379*). It would thus appear from these results that *K. pauciflora* is exceptional in that both enantiomers of the eburnane moiety are involved in different bisindoles, even though the eburnane monomers existing in the plant belonged exclusively to one enantiomeric group. In another study of the same plant, only one bisindole, norpleiomutine (**606**) was isolated, while the eburnane alkaloids which were present were also all of the 20$\beta$,21$\beta$ configuration (*380*). Norpleiomutine (**606**), demethylnorpleiomutine (**616**), and kopsoffinol (formulated as **618**) have been subsequently isolated from another Malaysian *Kopsia* from Borneo, *K. dasyrachis* and complete high field NMR spectral data were provided (*381*). In addition, (+)-19(*R*)-hydroxyeburnamine (**619**), which constitutes the eburnane half in kopsoffinol was also isolated and its structure was established by X-ray analysis (*381,382*).

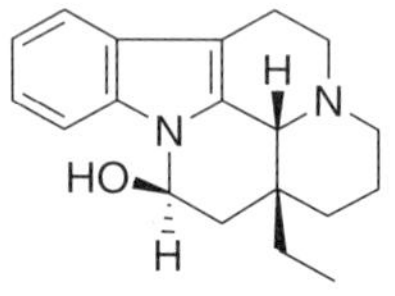

**611** (–)-eburnamine

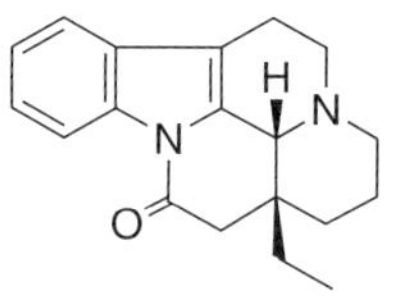

**612** (+)-eburnamonine

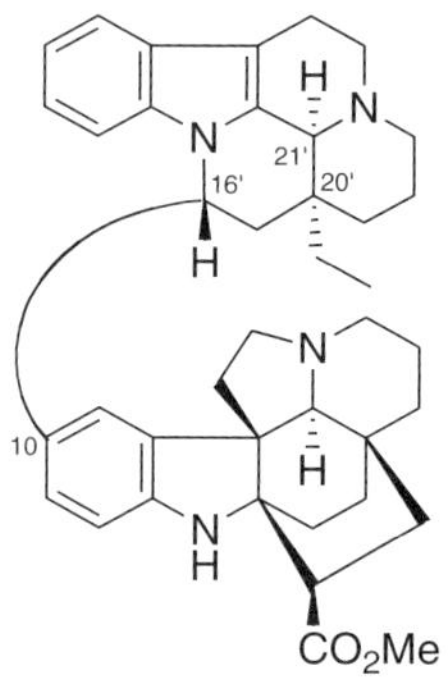

**613** (+)-kopsoffine

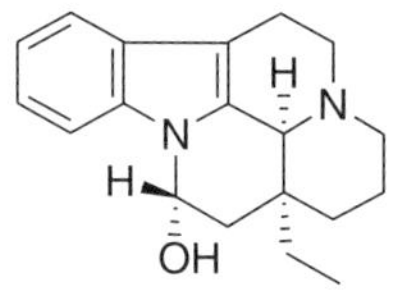

**614** (+)-eburnamine

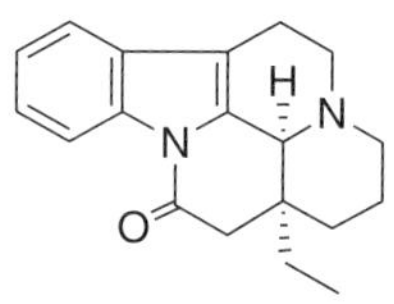

**615** (–)-eburnamonine

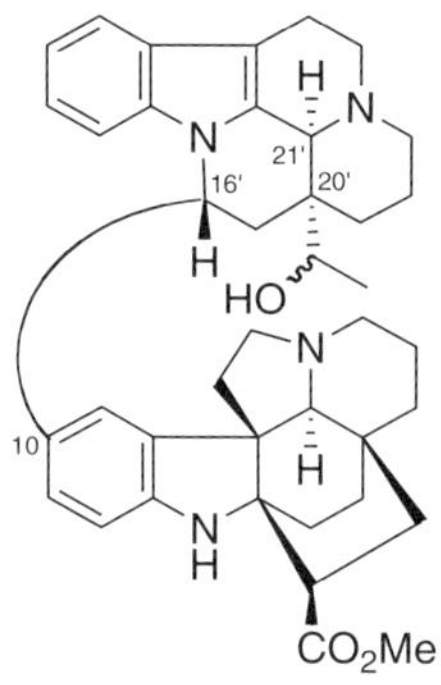

**617** (+)-kopsoffinol

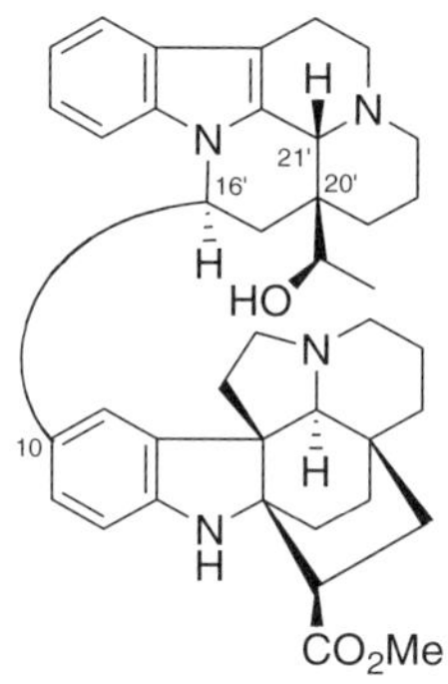

**618** (+)-kopsoffinol

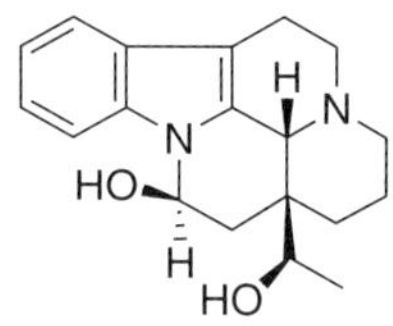

**619**

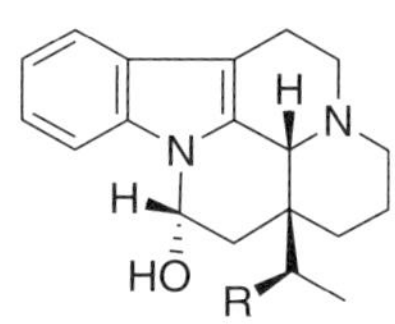

**620** R = H
**621** R = OH

It was contended that all three bisindoles (**606**, **616** and **618**) from *K. dasyrachis* incorporate in common, eburnane units having the 20$\beta$,21$\beta$ configuration, for the following reasons (*381*). First, since the other two bisindoles present (**606** and **616**) have eburnane units with the 20$\beta$,21$\beta$ configuration, it would be extremely unlikely that a third, odd bisindole should have a eburnane unit of the opposite configuration. Moreover, since all four monomeric eburnane compounds present in the plant, viz., (+)-eburnamonine (**612**), (+)-19(*R*)-hydroxyeburnamine (**619**), (+)-isoeburnamine (**620**),

(−)-19(*R*)-hydroxyisoeburnamine (**621**), were of the 20*β*,21*β* configuration, it would be singularly unusual for kopsoffinol to incorporate a eburnane unit of the opposite configuration. Lastly, considering that the likely precursor of the eburnane half in kopsoffinol, (+)-19(*R*)-hydroxyeburnamine (**619**), was also obtained from this plant, and its structure has been established by X-ray analysis, the most likely structure of kopsoffinol is **618**, with the eburnane half corresponding to that of the monomeric precursor, **619**. The attachment of the kopsinine unit at C(16′) is now *β* (H(16′) *α*), as required by the observed coupling constants for the H(16′) signal of 11 and 5 Hz for kopsoffinol (*383*).

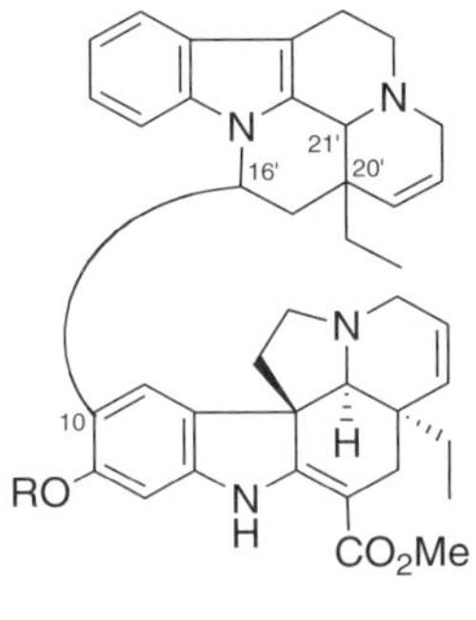

**622** R = OMe
**623** R = H

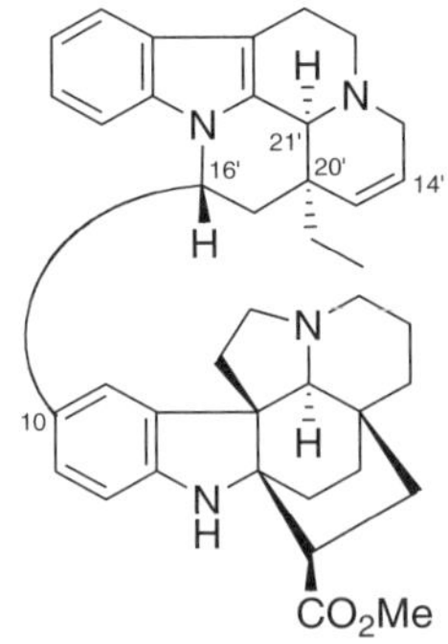

**624** insulopinine

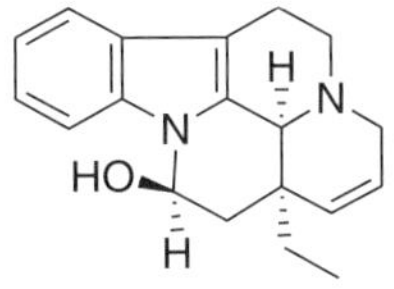

**625**

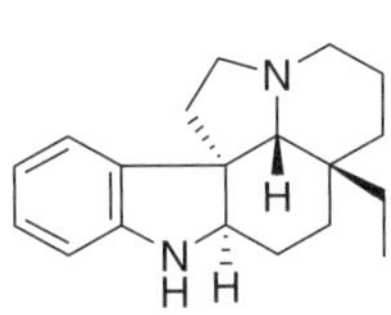

**627**

Tenuicausine (**622**) (*384*) and demethyltenuicausine (**623**) (*385*) were isolated from *Melodinus tenuicaudatus* collected from the Yunnan province of China. The UV (232, 331 nm) and IR (1680, 1620 cm$^{-1}$) data indicated the presence of *β*-anilinoacrylate chromophores, while the $^1$H NMR spectral data indicated the presence of two methoxy groups, four olefinic hydrogens, an unsubstituted aromatic ring, and another 10,11-disubstituted aromatic moiety. The mass spectrum showed diagnostic fragments of tabersonine (*m/z* 107, 121, 122, 135) and 14,15-dehydroeburnamine (*m/z* 249, 250), suggesting that tenuicausine may be constituted from the union of these units. Examination of the $^{13}$C NMR spectral data led to the proposed structure **622** for tenuicausine (the configurations at C(16′), C(20′), and C(21′) were not addressed), which was confirmed by partial synthesis from the condensation of 11-methoxytabersonine and 14,15-dehydroeburnamine. A similar condensation of 11-hydroxytabersonine and isoeburnamine gave demethyltenuicausine (**623**).

Insulopinine (**624**) was isolated from *Melodinus insulae-pinorum* from New Caledonia (*386*). It is constituted from the linkage of kopsinine *via* C(10) to C(16′) of a (−)-14,15-dehydrovincane unit. The occurrence of the monomer, (−)-14,15-dehydrovincanol ((−)-14,15-dehydroisoeburnamine) (**625**), provided support for the structure assignment. The observed coupling constants of 10 and 5 Hz for H(16′) were invoked to support *α*-substitution at C(16′) of **624**, for which the optical rotation was not reported.

Strempeliopidine (**626**) was obtained from *Strempeliopsis strempelioides*, a plant indigenous to Cuba (*387*). Acid hydrolysis gave aspidospermidine (**627**), while reductive cleavage (Sn/HCl) yielded dihydroeburnamenine, from which it was deduced that strempeliopidine was constituted from the union of aspidospermidine and eburnane moieties. This was supported by partial synthesis through acid-induced condensation of eburnamine (not stated which enantiomer was used) with aspidospermidine. Since

(+)-eburnamonine (**612**) was also isolated from the plant, it is likely that the eburnane unit in **626** has the 20′β,21′β configuration. The coupling constants of 11.5 and 4.6 Hz for H(16′) indicated that the substitution at C(16′) is β. The structure as originally presented (**626a**) required amendment to that depicted in **626b**.

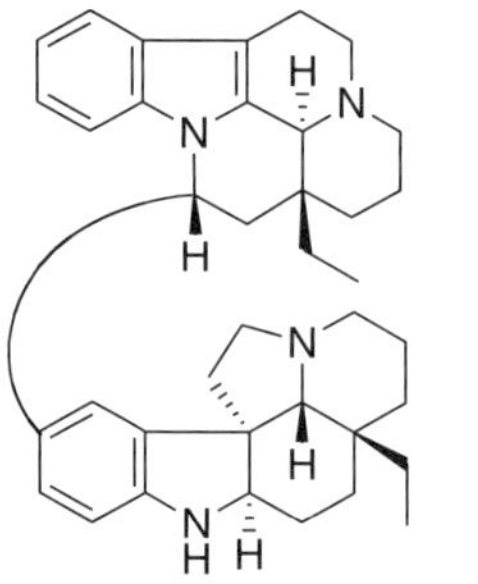

**626a** strempeliopidine

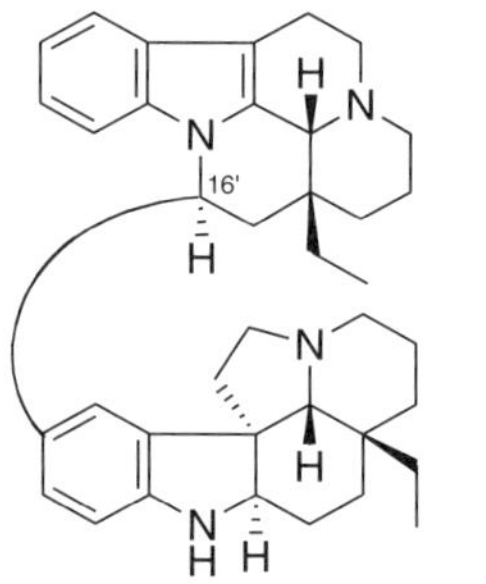

**626b** strempeliopidine

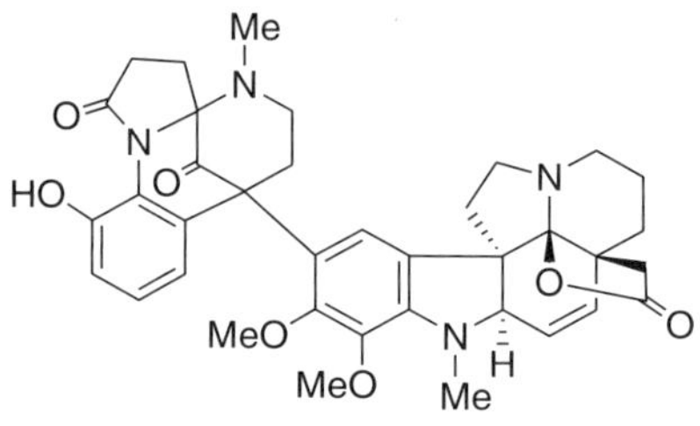

**628** haplophytine

## XXXI. *Aspidosperma*–Canthinone Type

Bisindole alkaloids of this group, exemplified by haplophytine are found exclusively in the genus *Haplophyton* (Apocynaceae). Haplophytine (**628**), the major alkaloid of *H. cimicidum*, was isolated in 1973 by Cava *et al.* (*388*). Subsequently, several new bisindole alkaloids belonging to this group have been isolated, by subjecting the alkali-soluble base residues from the isolation of **628** to repeated pH-gradient countercurrent distribution. These include cimiciphytine (**629**) and norcimiciphytine (**630**) (*1,389*), cimilophytine (**631**), haplocidiphytine (**632**), norisohaplophytine (**633**), and cimiciduphytine (**634**). *H. crooksii* was another source of such alkaloids, providing the new bisindole crooksiine (**635**), in addition to haplophytine.

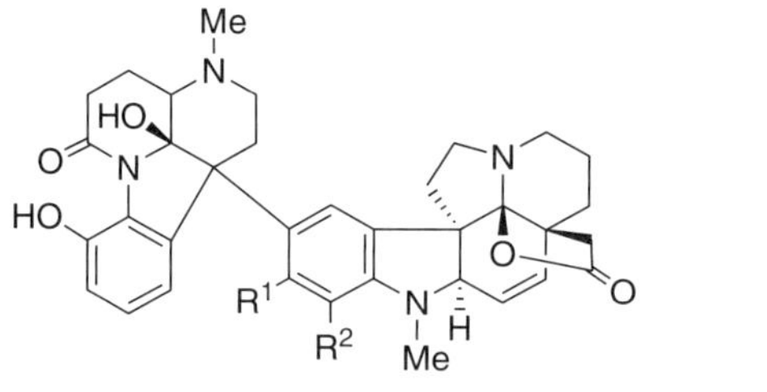

**629** $R^1 = R^2 =$ OMe cimiciphytine
**630** $R^1 =$ OH, $R^2 =$ OMe norcimiciphytine
**635** $R^1 =$ OMe, $R^2 =$ OH crooksiine

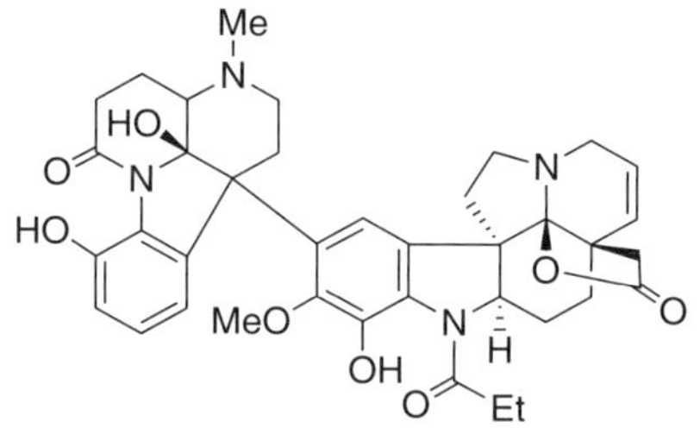

**631** cimilophytine

Cimilophytine (**631**) showed a weak $M^+$ at $m/z$ 682 which analyzed for $C_{38}H_{42}N_4O_8$. The $^1H$ NMR spectrum showed an *N*-propionyl group, an NMe, a shielded methoxyl (δ 3.24), two olefinic hydrogens, four aromatic hydrogens, and three $D_2O$ exchangeable hydrogens (δ 7.08, 10.16, 11.49). The IR spectrum indicated the presence of OH (3425 cm$^{-1}$), lactone (1751 cm$^{-1}$), and amide (1621 cm$^{-1}$) functions, while the UV spectrum (228, 266, 300 nm) was that of a dihydroindole chromophore, with the 266 nm peak shifted to 315 nm on addition of alkali, indicating the presence of a phenolic function. The presence of the same canthinone unit as that in **629** and **630** was indicated by the common prominent ion at *m/z* 255 ($C_{15}H_{15}N_2O_2$) in the mass spectra of all three alkaloids (**629–631**) due to the canthinone moiety, after the loss of the tertiary hydroxyl as water.

The NMR spectrum of **631** showed several important similarities to those of **629** and **630**, such as the similar pattern of aromatic substitution, the presence of the H-bonded hydrogen at $\delta$ 11.49, due to the phenolic OH bonded to the $\delta$-lactam carbonyl of the canthinone unit. The second H-bonded phenolic hydrogen at $\delta$ 10.16 was attributed to the 12-OH bonded to the carbonyl of the *N*(1)-propionyl group, by analogy with the monomeric alkaloid, cimicidine (**636**). The lone methoxyl was placed at C(11) to account for its shielding, which has been noted previously in **628** and **629**. The two 16,17-olefinic hydrogens in **628**, **629**, and **630** displayed an AB system with allylic coupling. The olefinic hydrogens of **631** on the other hand were observed as a singlet at $\delta$ 5.66, analogous to the single sharp peak at $\delta$ 5.7 for the 14,15-olefinic hydrogens of obscurinervine (**637**) and obscurinervidine (**638**), two lactone-containing *Aspidosperma* alkaloids. Based on such a comparison, the unsaturation in the *Aspidosperma* half was placed at C(14)/C(15). Catalytic hydrogenation of **631** ($H_2/PtO_2$), followed by methylation ($CH_2N_2$), gave the tetrahydro derivative **639**, characterized by a new OMe at $\delta$ 3.57, and absence of the olefinic signals at $\delta$ 5.66, in contrast to catalytic hydrogenation of obscurinervine (**637**), which resulted only in saturation of the double bond, leaving the lactone moiety intact. The proposed structure was also corroborated by the $^{13}C$ NMR spectral data, by comparison with the spectra of **628**, **629**, and **636** (*390*).

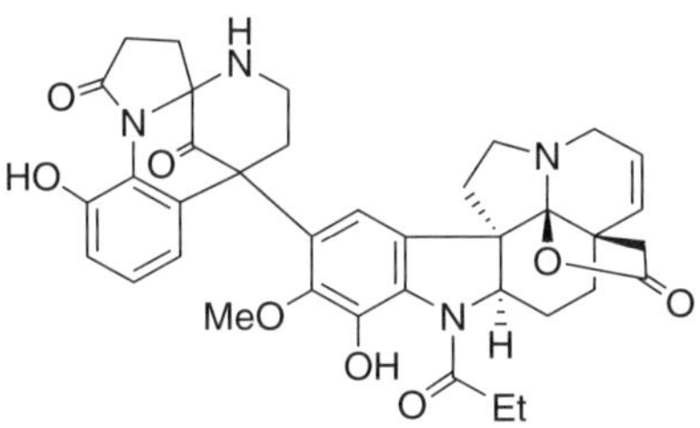


**632** haplocidiphytine

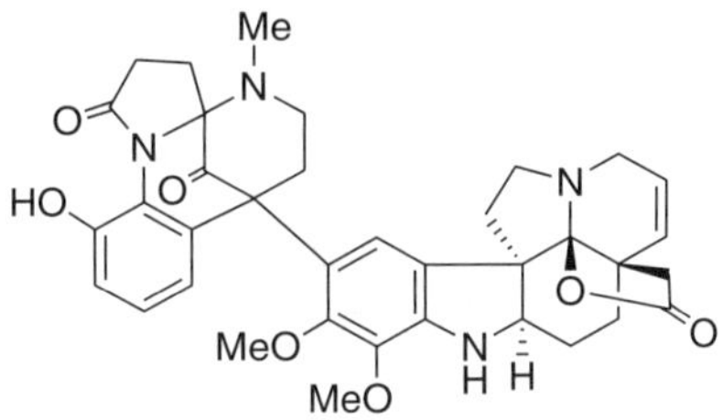


**633** norisohaplophytine

Examination of the NMR spectral data and use of the same hydrogenation/methylation sequence facilitated the structure elucidation of the other members of this group. Haplocidiphytine (**632**) was shown to retain the same *Aspidosperma* moiety as in cimilophytine (**631**), but the canthinone portion was that of haplophytine (**628**), except for replacement of NMe′ by NH′. The presence of the second downfield phenolic hydrogen at $\delta$ 8.97, instead of at about $\delta$ 11.4, indicated that **632** had a rearranged canthiphytine unit as in **628**, and not an unrearranged canthiphytine unit as in **629** and **631**. The C(16′) carbonyl was observed at $\delta$ 197.0 (*c.f.* $\delta$ 197.2 in **628**), while no signal due to the canthinone NMe′ was observed. Norisohaplophytine (**633**) had the same canthinone portion as **628**, and a similar *Aspidosperma* unit, except for the location of the unsaturation at C(14)/C(15), and replacement of NMe with NH (*391*).

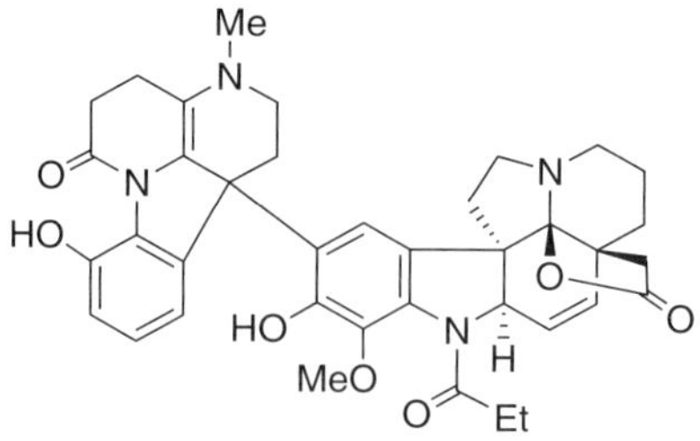

**634** cimiciduphytine

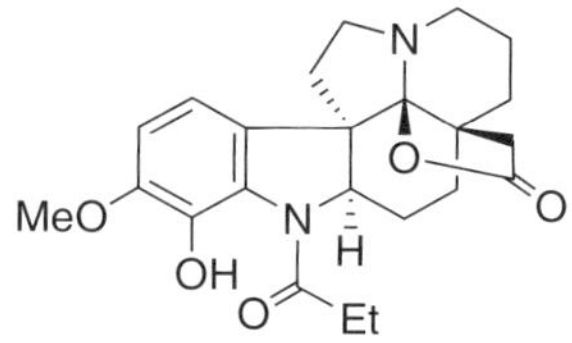

**636** cimicidine

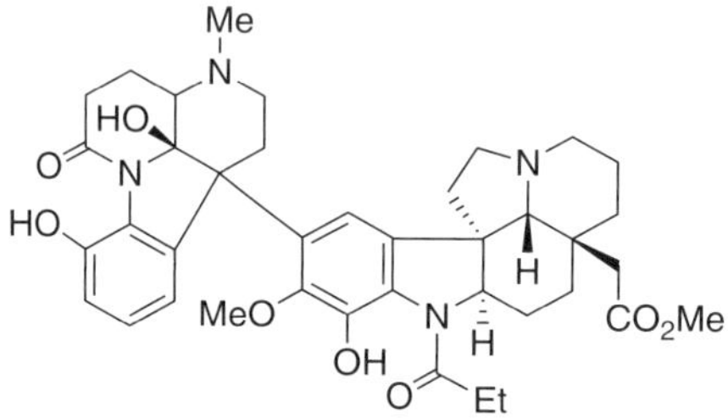

**639**

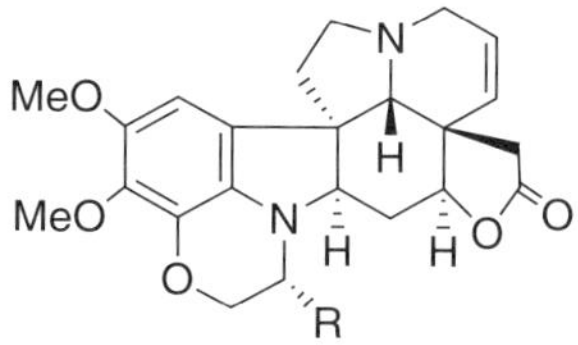

**637** R = Et obscurinervine
**638** R = Me obscurinervidine

Cimiciduphytine (**634**) showed a number of similarities with cimilophytine (**631**), with some important departures (*392*). First, the aromatic substitution in the *Aspidosperma* unit has undergone an inversion to 11-OH and 12-OMe, as the downfield signal at ca. $\delta$ 10.2 characteristic of 12-OH bonded to the carbonyl of the *N*(1)-propionyl group was absent, and was replaced by an OH signal at $\delta$ 7.20 attributed to the OH at C(11). The other downfield signal characteristic of the phenolic hydrogen, H-bonded to the $\delta$-lactam carbonyl of the canthinone unit, was observed at $\delta$ 11.49. The location of the double bond has also changed to 16,17- from 14,15- in cimilophytine (**631**), from the AB pattern with allylic coupling shown by the olefinic hydrogens, which was similar to that in **628**, **629**, and **630**. The $^1$H NMR, as well as the mass spectral data indicated loss of the tertiary OH associated with the canthinone half, suggesting dehydration leading to a 5′,16′-double bond in the canthinone moiety. Cimiciduphytine is therefore derived from coupling of an unrearranged canthiphytine unit with a dehydroisocimicidine unit.

Crooksiine (**635**) from *H. crooksii* was isomeric with **630**, and showed very similar NMR spectral data with those of **629** and **630**. The spectral data indicated that **635** is a regioisomer of **630** in which the aromatic substitution in the *Aspidosperma* unit has undergone an inversion to 11-OMe and 12-OH. Evidence for this inference was provided by NOE experiments where irradiation of H(11′) resulted in greater enhancement of the aromatic OMe ($\delta$ 3.57) when compared to the enhancement caused by irradiation of H(12′). Based on an assumed preferred conformation of crooksiine, in which the two halves are mutually orthogonal, the NOE results favored the aromatic substitution depicted in **635**. Crooksiine showed complete inhibition of acetylcholinesterase activity at a concentration of 3.06 μg mL$^{-1}$ (4.78 μM) (*393*).

## XXXII. Eburnea–Eburnea Type

Eburnea–eburnea-type bisindoles are rare. Only one, designated as methylene 10,10′-bis-[(+)-*N*(1)-norvallesamidine] (**640**) has been isolated from *Melodinus celastroides* from New Caledonia, together with three other quaternary alkaloids (**641–643**) which are probably artifacts arising from the use of dichloromethane in the extraction procedure (*394*). The molecular ion of **640** was detected at *m/z* 576 ($C_{39}H_{54}N_4$), while the UV spectrum showed the presence of dihydroindole moieties. The $^{13}C$ NMR spectrum showed only 20 signals, which, coupled with the observation of fragment ions at *m/z* 296 and 282 in the mass spectrum, suggested a molecule with two identical indoline units connected by a methylene bridge. Examination of the $^{13}C$ NMR spectrum, and comparison with the spectrum of *N*(1)-norvallesamidine (**644**), revealed a virtual congruence of the non-aromatic carbon signals. The two main differences noted were the presence of the additional signal due to the methylene bridge at $\delta$ 41.1 (connected to two quaternary centers from the observed 2H singlet at $\delta$ 3.85 in the $^1H$ NMR spectrum), and the deshielding of C(10), which identified these carbons as the points of branching in the dimeric alkaloid. Compound **641** was deduced to be methylene *N*(4),*N*(4′)-bis-[(+)-*N*(1)-melonimium] dichloride from comparison of the $^{13}C$ NMR spectral data ($\delta_{CH2}$ 71.5) with that of melonine (**645**), as well as from its cleavage to **645** by treatment with hot KOH. Treatment of **645** with $CH_2I_2$/MeOH, followed by HCl/MeOH, regenerated **641**. Similar treatment of **642** and **643** with ethanolamine gave the tertiary alkaloids celastromeline (**646**) and celastromelidine (**647**), respectively, the structures of which were established by comparison of the $^{13}C$ NMR data with those of 14,15-dehydrovincanol (**625**) and melonine (**645**).

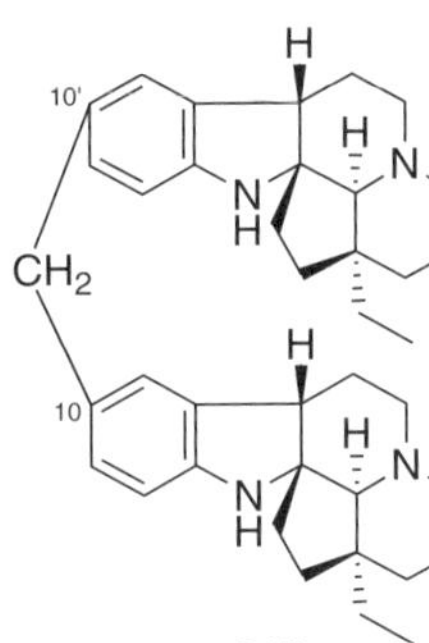

**640**

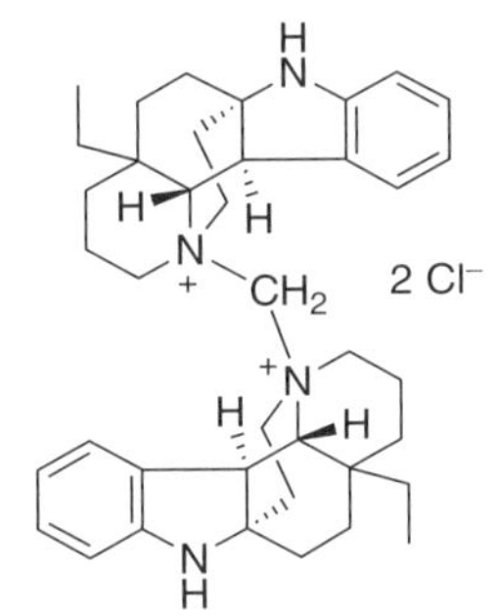

**641**

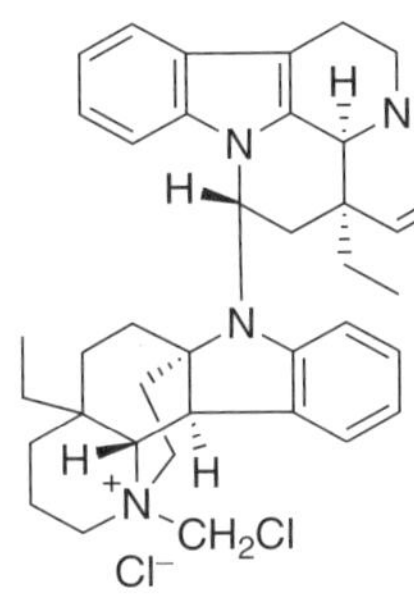

**642**

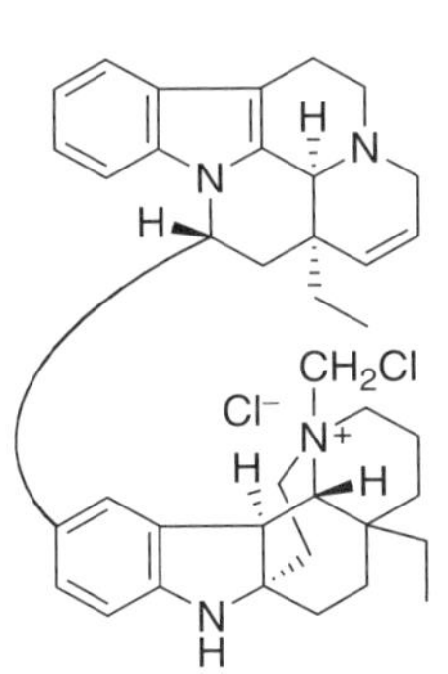

**643**

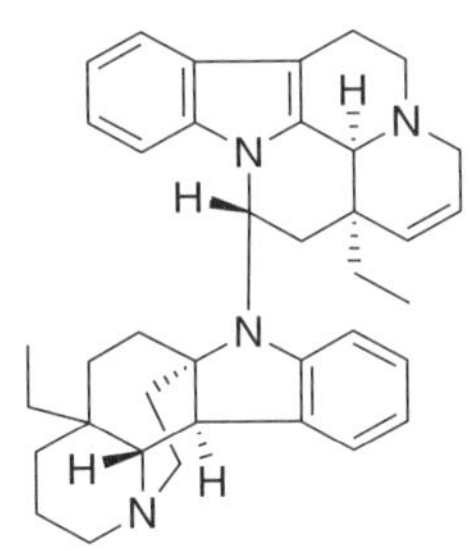

**646**

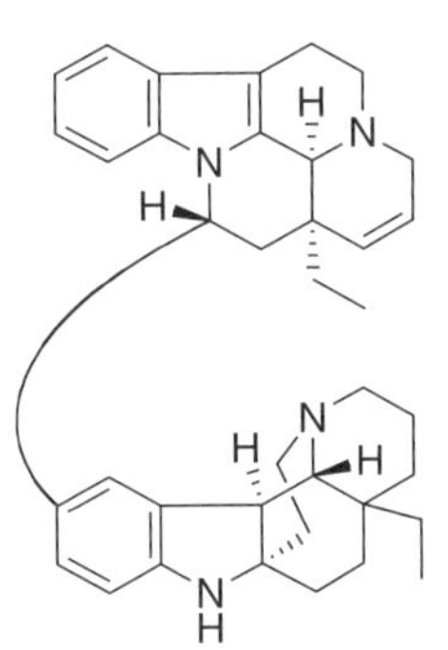

**647**

Several eburnea–eburnea bisindoles have been obtained through coupling of iminium and enamine species which are in equilibrium under the reaction conditions. For instance, treatment of dehydrovincamone (**648**) with acetic acid gave the dimeric compound **649**. The same dimer was formed from the reaction of vincamone-*N*-oxide with acetic anhydride. Extension of the reaction to criocerine (**650**) and its derivatives resulted in formation of the dimer **651** (*395,396*). The phenazine derivative **652** was formed in 30% yield from the reaction of (+)-11-aminovincamine **653** with iodine in $CHCl_3/NaHCO_3$ (*397*).

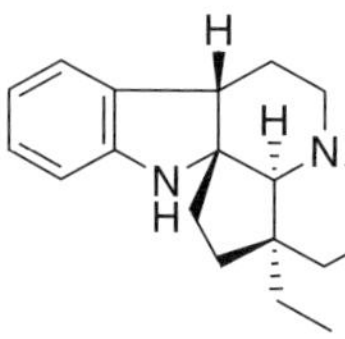

**644**

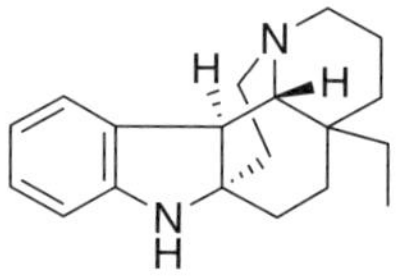

**645** melonine

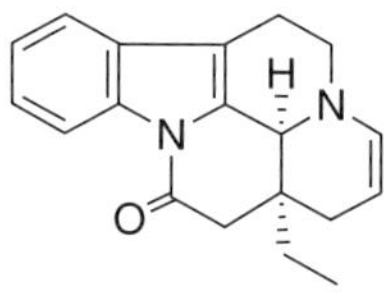

**648**

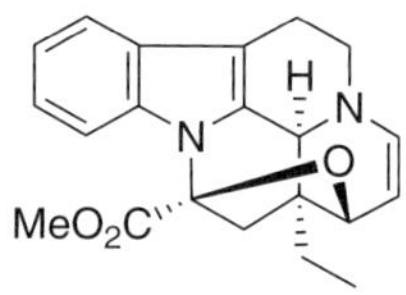

**650**

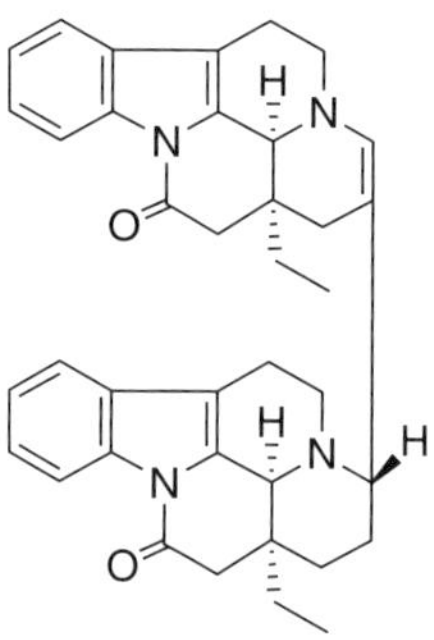

**649**

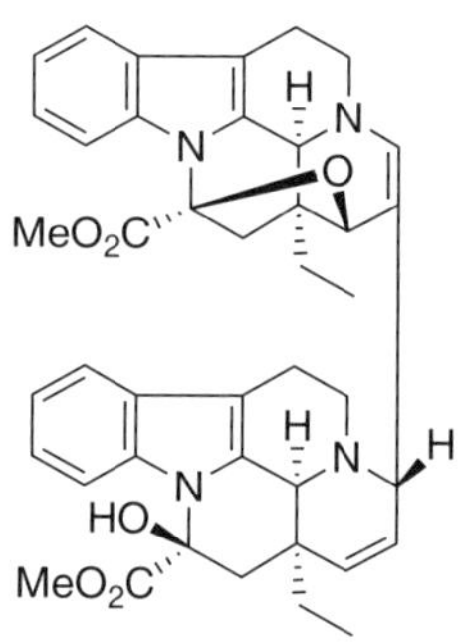

**651**

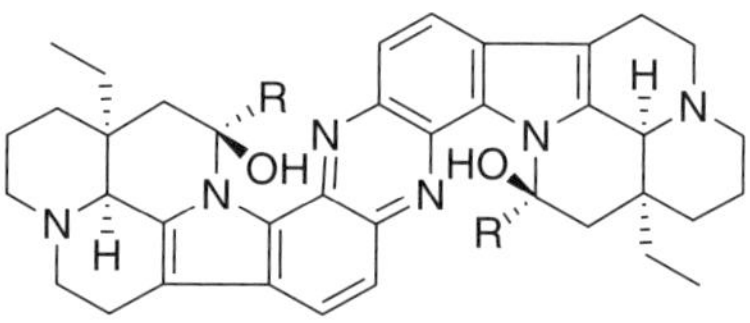

**652** R = $CO_2Me$

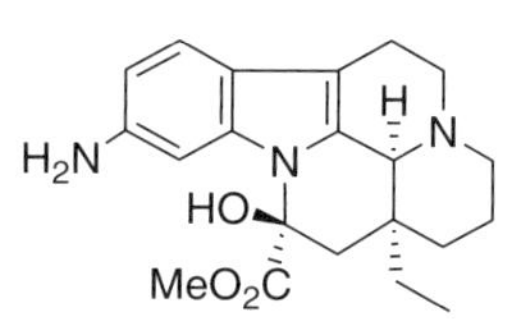

**653**

## XXXIII. Miscellaneous Bisindole Alkaloids

*Hunteria congolana* from Zaire, gave 19′-epipleiomutinine (**654**, 19′*R*-epimer of pleiomutinine), in addition to the previously known pleiomutinine (**655**, 14′,15′-dihydropycnanthine) (*398*). Both alkaloids also occur in *H. zeylanica* collected in Kenya (*vide supra*).

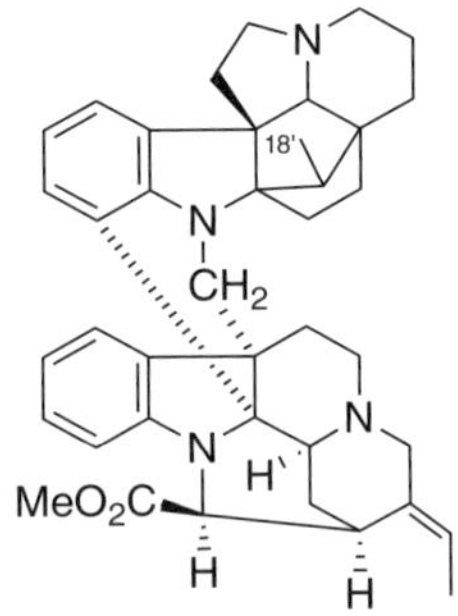

**654** 19′(*R*) 19′-epipleiomutinine
**655** 19′(*S*) pleiomutinine

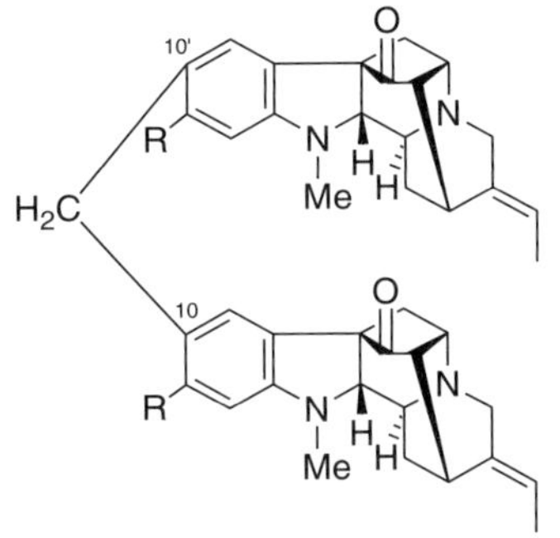

**656** R = OMe raureflexine

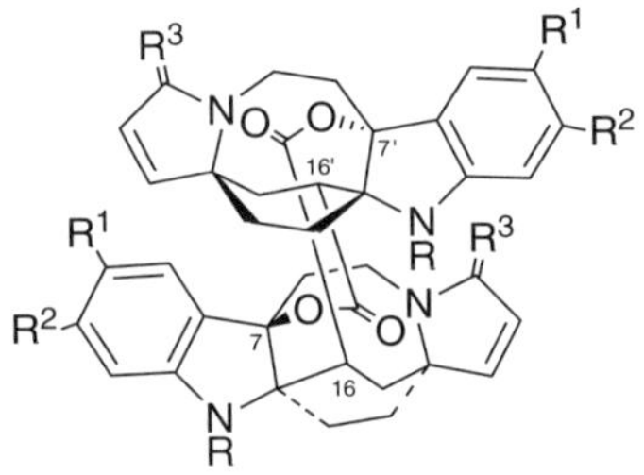

**657** R = $CO_2Me$, $R^1$ = OMe, $R^2$ = H, $R^3$ = $H_2$
**658** R = $CO_2Me$, $R^1$ = OMe, $R^2$ = OMe, $R^3$ = $H_2$
**659** R = $CO_2Me$, $R^1$ = OMe, $R^2$ = H, $R^3$ = O

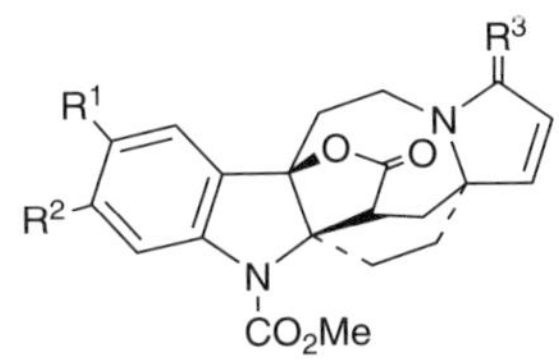

**660** $R^1$ = OMe, $R^2$ = H, $R^3$ = $H_2$
**661** $R^1$ = OMe, $R^2$ = OMe, $R^3$ = $H_2$
**662** $R^1$ = OMe, $R^2$ = H, $R^3$ = O

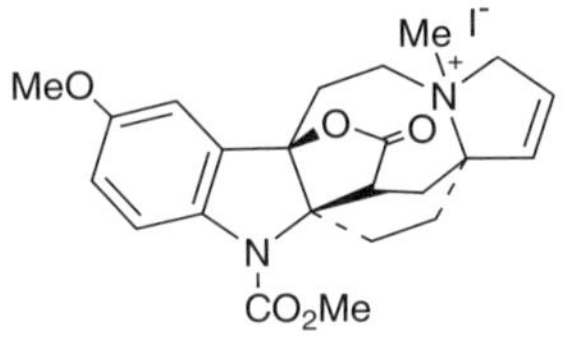

**663**

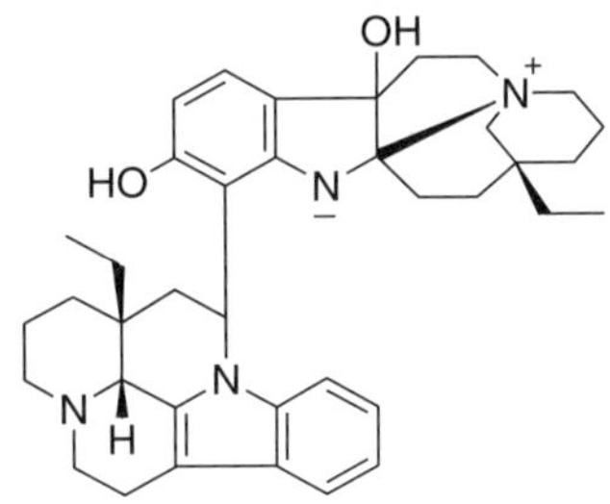

**664**

Further investigation of the leaf alkaloids of *R. reflexa* yielded, in addition to flexicorine (**372**), the symmetric bisindole, raureflexine (**656**), constituted from two rauflexine units linked *via* a methylene bridge from the aromatic C(10) and C(10′) of the monomer halves (*399*). The EIMS gave an $M^+$ at $m/z$ 684, while the $^1H$ NMR spectrum was essentially similar to that of rauflexine (**373**), except for the absence of the C(10) hydrogen signals, and the appearance of the methylene signal as a singlet at $\delta$ 3.54. Similarly, the $^{13}C$ NMR spectrum was essentially similar to that of the monomer alkaloid except for the substituted C(10) and the appearance of the methylene carbon signal at $\delta$ 32.9. The structures of tenuisines A, B, and C, isolated from *K. tenuis* occurring in Malaysian Borneo, have been revised from the dimeric structure with an axis of symmetry (**657–659**) (*360,400*), to the monomeric structure (**660–662**, respectively), based on new HRLSIMS data, as well as LSIMS and NMR analysis of the methyl iodide salt of tenuisine A (**663**) (*401*). Melaxillarinine (**664**) was isolated from *Melodinus axillaris* and its structure established based on an analysis of the spectral data (*402*).

## XXXIV. *Aspidosperma*–Cleavamine Type

Alkaloids of this group are almost entirely dominated by the bisindole alkaloids of *C. roseus*, in particular the antitumor alkaloids, vinblastine (**665**, or vincaleukoblastine, VLB) and vincristine (**666**, or leurocristine, VCR), due to their important role as prized antineoplastic agents in cancer chemotherapy. In addition to the extensive coverage in Cordell and Saxton's previous review on the bisindole alkaloids (*1*), a recent volume in this series (*4*) has been entirely devoted to various aspects of these antitumor bisindole alkaloids. Other than the akuammiline–vincorine alkaloids, vingramine (**383**) and methylvingramine (**384**) discussed above (*vide supra*) (*251*), there have been no new bisindole alkaloids isolated from this plant in the intervening period. The efficacy of vinblastine and vincristine has not been surpassed by any recent natural congener, and they remain the two clinically most important bisindoles of this group. As such they have been the focus of intensive research which has in turn generated a voluminous literature, detailed coverage of which would be beyond the scope of the present chapter. The reader is referred to the previous dedicated volume in this series for more comprehensive coverage. In the context of this chapter, other than brief mention to set the background, only selected aspects of structure, synthesis, biosynthesis, and biological activity, which have appeared in the literature since the 1990 volume (*4*), will be discussed.

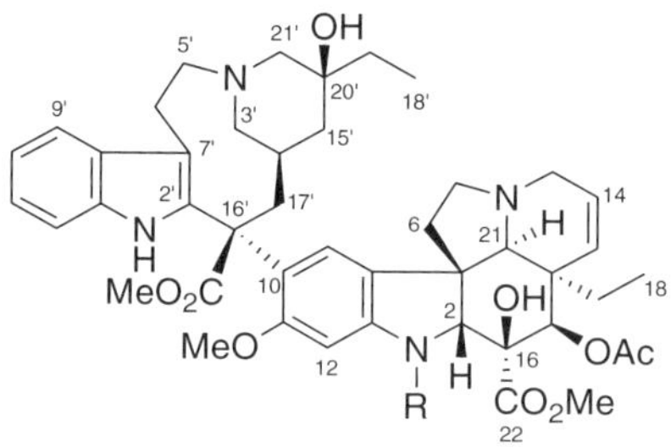


**665** R = Me vinblastine
**666** R = CHO vincristine

## A. STRUCTURE AND BIOSYNTHESIS

The absolute configuration of vincristine had been established previously by X-ray analysis of its methiodide derivative (*403,404*), from which the configurations of the stereocenters in vinblastine were inferred from the known relationship between the two compounds. Recently, the structure of vinblastine has been determined by X-ray crystallographic analysis of its sulfate, which was successfully crystallized by the hanging drop vapor diffusion method against polyethylene glycol and lithium sulfate. The gross molecular geometry was found to be rather similar to that of vincristine, except that the C(16′) methyl ester group is oriented in an opposite direction from that in vincristine. Half of the nine-membered azacyclononane ring in the catharanthine portion is coplanar with the indole ring, while the other half is in a boat conformation, and fused with the piperidine ring in a chair conformation (*405*). The crystal structure of 16′-epivinblastine, obtained by Magnus in the course of his vinblastine synthesis (*vide infra*), has also been reported (*406*). A preliminary theoretical analysis of the CD of vinblastine has been carried out (*407*). It was shown that the intense bisignate curve at 228 and 214 nm in vinblastine CD arises from exciton coupling of the indole and indoline transitions, and furthermore, the sign of the couplet is directly related to the C(16′) configuration, providing vindication for Kutney and Scott's earlier demonstration that CD could be used for the interpretation of the vinblastine C(16′) stereochemistry (*408*). Subsequently, the analysis of a series of vinblastine analogues by circular dichroism has been described (*409*). Exciton coupling of the indole and indoline chromophores of these derivatives provided a general method for the assignment of the C(16′) configuration with the bioactive C(16′*S*) and the inactive C(16′*R*) analogues giving rise to positive and negative couplets, respectively. In addition, analysis of the non-coupled transitions of the CDs indicated that all bioactive vinblastine analogues showed a positive Cotton effect at 305 nm, which, but for one exception, correlates with a C(14′*R*) configuration. Complete $^{1}H$ and $^{13}C$ NMR assignments for Navelbine (5′-nor-anhydrovinblastine) (*410*), its analogues (*411*), and anhydrovinblastine (*412*), have been carried out with the extensive application of 2-D techniques.

The biosynthesis of the *Catharanthus* bisindole alkaloids has been extensively reviewed in previous volumes of this series (*413,414*). It has also been partially addressed in several other recent reviews (*415,416*). The bisindoles are derived from condensation of catharanthine (**667**) and vindoline (**668**) (Scheme 35). Catharanthine (**667**) is thought to be formed from strictosidine (**669**) *via* 4,21-dehydrogeissoschizine, stemmadenine, and dehydrosecodine, with only a few details known (*413,414*). Vindoline (**668**) is derived from strictosidine *via* stemmadenine and tabersonine (**535**). Little is known concerning the post-strictosidine pathway to tabersonine. The transformation of tabersonine into vindoline, however, has received considerable attention and a substantial amount of detail has become available. The process involves six steps, comprising, aromatic hydroxylation, *O*-methylation, hydration of the 2,16-double bond, N(1)-methylation, C(17)-hydroxylation, and 17-*O*-acetylation. Except for the double bond hydration step, the enzymes involved in the other steps are now known (*413,414,416–418*). The coupling of catharanthine and vindoline, and the subsequent transformation to the vinblastine-type bisindoles have been intensively investigated using plant enzymes and the results have been reviewed (*3,413–415,419,420*). The exact pathway to the bisindoles has not been firmly established. While some consider the iminium ion to be a key intermediate, others assign this role to anhydrovinblastine (**670**), although there is common agreement that both are interconvertible (*415*).

Recently however, a basic peroxidase, anhydrovinblastine synthase, which couples catharanthine and vindoline to yield anhydrovinblastine, the putative precursor to vinblastine and vincristine, has been purified and characterized from *C. roseus* leaves (*421,422*). The enzyme showed a specific anhydrovinblastine synthase activity of 1.8 nkat $mg^{-1}$, and a molecular weight of 45.40 kDa. It was shown to be a high-spin ferric heme protein belonging to the plant peroxidases superfamily (class III peroxidases), and cytochemical studies showed that the enzyme is localized in the mesophyll vacuoles, in individual spots at the inner surface of the tonoplast. On the basis of the ability of the monomeric alkaloid substrates to reduce the *C. roseus* basic

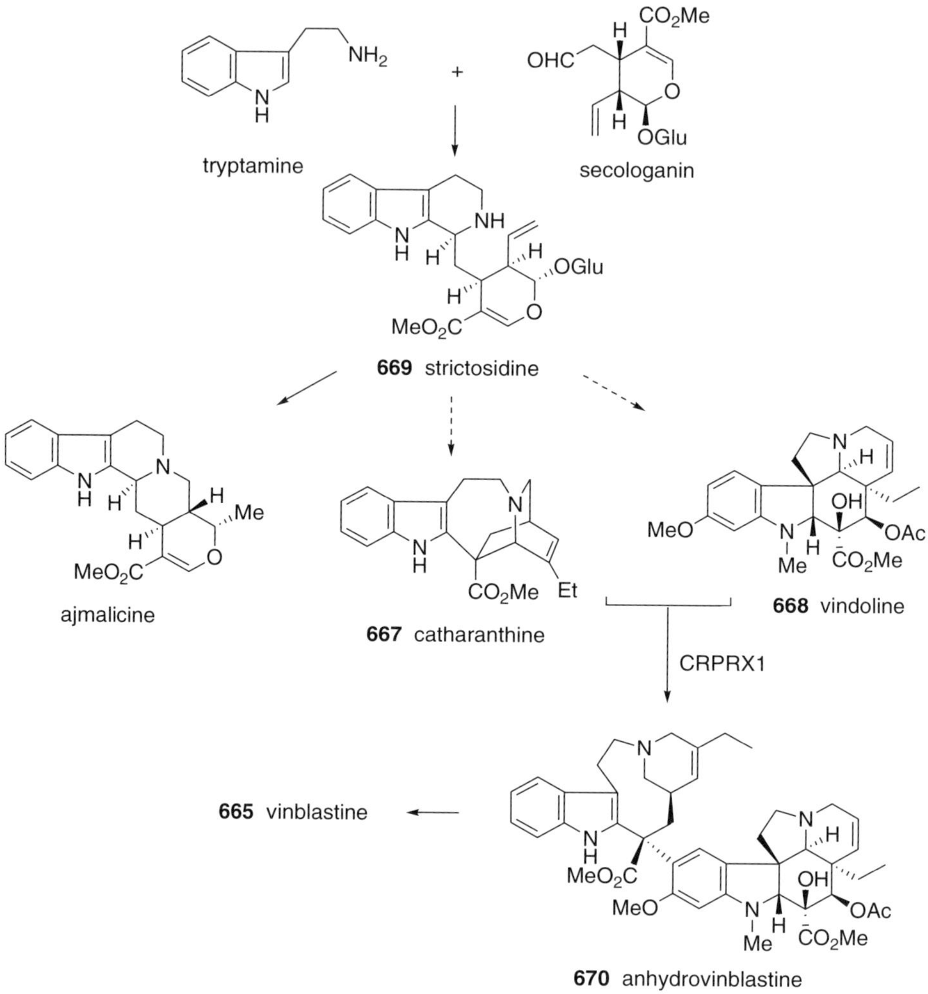


Scheme 35

peroxidase catalytic intermediates, the substrate specificity of the enzyme, and its localization in mesophyll cells, the authors contend that the enzyme fulfills all the requirements to be considered as an anhydrovinblastine synthase (subsequently designated CRPRX1 or *C. roseus* peroxidase 1) (*415*), and proposed a channelling mechanism for the peroxidase-mediated synthesis of anhydrovinblastine in plants (*415*,*422*).

## B. SYNTHESIS OF VINBLASTINE

The Volume 37 account by Kuehne and Marko (*423*) has considered the history of the various approaches to the synthesis of vinblastine and related analogues. The subject has also been addressed in several other recent reviews (*3*,*424*).

The significant breakthrough of the earlier period was the biomimetic approach of Potier (*425–427*) and Kutney (*428–431*), based on coupling of the bisiminium ion **671** (generated from Polonovski oxidation of catharanthine *N*-oxide, **672**), with vindoline (**668**), followed by borohydride reduction to deliver the coupled product (anhydrovinblastine, **670**), *en route* to vinblastine (**665**) (Scheme 36). The stereochemical outcome of this reaction was found to be strongly temperature-dependent, reaction at −50°C provided the desired C(16′*S*) configuration, while reaction at 0°C or higher, led to the undesired C(16′*R*) epimer. The observed temperature dependence has been rationalized based on rapid equilibration to another iminium ion conformer at the higher temperatures, the capture of which proceeds preferentially from the *β*-face, resulting in formation of the coupled product with the unnatural C(16′) configuration. A recent study of this reaction under conditions which allowed observation of intermediates by low temperature NMR, has confirmed the general features of the originally proposed mechanism, with the addition of some new details (*432*).

The chloroindolenine approach (Scheme 37) on the other hand, has consistently failed to deliver coupling products with the correct C(16′*S*) stereochemistry, which is an essential prerequisite for the biological activity (*423*). Kuehne was the first to attempt a modified approach based on the use of a chloroimine precursor with inverted stereochemistry at C(3) and C(7), such as the *D*/*E-trans*-deethyl-*ψ*-vincadifformine **673** (*433*). The desired C(3)–C(7) cleavage required to deliver the cleavamine-type structures, however, did not occur due to unfavorable stereoelectronic factors, which led Kuehne to consider the D-seco compounds, such as **674**, as substrates for coupling with vindoline, based on the expectation that such substrates would not suffer from the stereochemical restraints to reductive cleavage of the C(3)–C(7) bond. In the event, reactions of vindoline with the racemic chloroimines **675** derived from 14-*epi*-D-secodeethylvincadifformine **676** resulted in high yields of the correct C(16′*S*), C(14′*R*)-type coupling products which could then be processed to 20′-deethyl-20′-deoxyvinblastine (**677**) and 20′-deethyl-20′-deoxyvincovaline (**678**) (*434*). Use of the same approach delivered the 20′-deoxy derivatives of vinblastine (**679**), leurosidine (**680**), vincovaline (**681**,**682**), and vincristine (Scheme 38) (*435*).

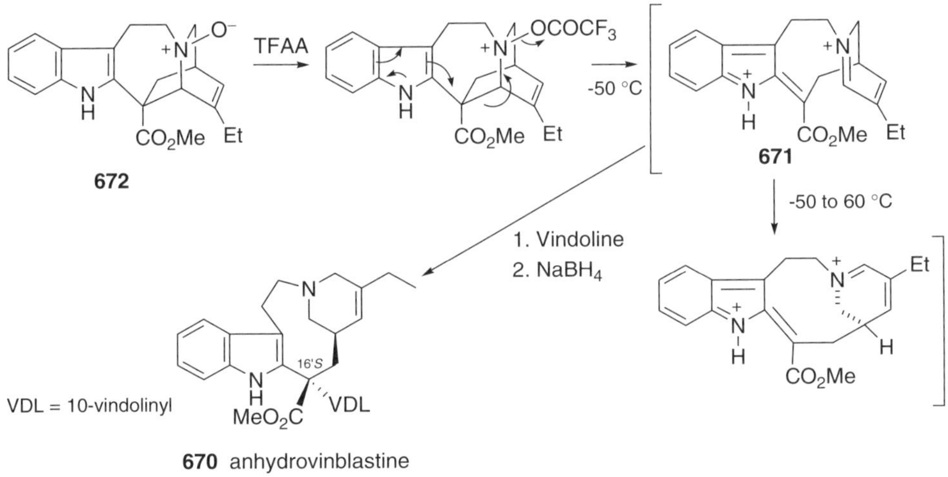


Scheme 36

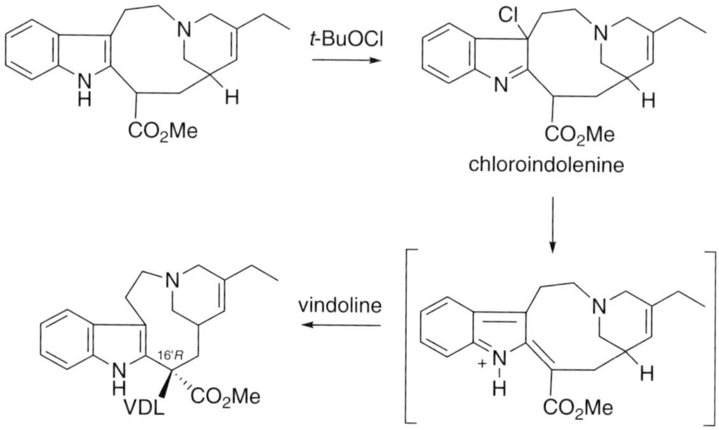

Scheme 37

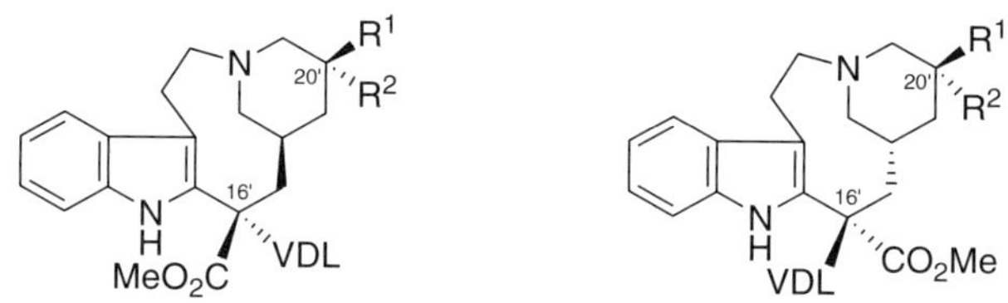

**677** $R^1 = R^2 = H$
**679** $R^1 = H, R^2 = Et$
**680** $R^1 = Et, R^2 = H$
**683** $R^1 = Et, R^2 = OH$

**678** $R^1 = R^2 = H$
**681** $R^1 = H, R^2 = Et$
**682** $R^1 = Et, R^2 = H$
**684** $R^1 = OH, R^2 = Et$
**685** $R^1 = Et, R^2 = OH$

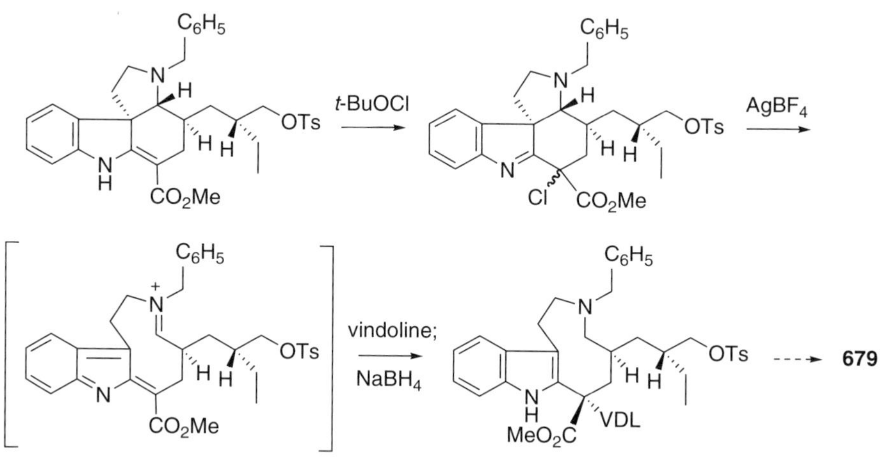

Scheme 38

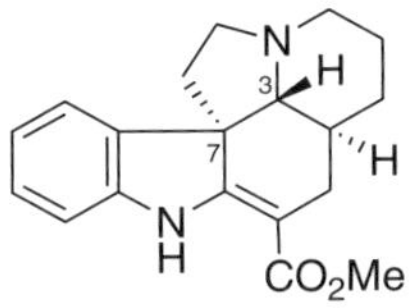

673

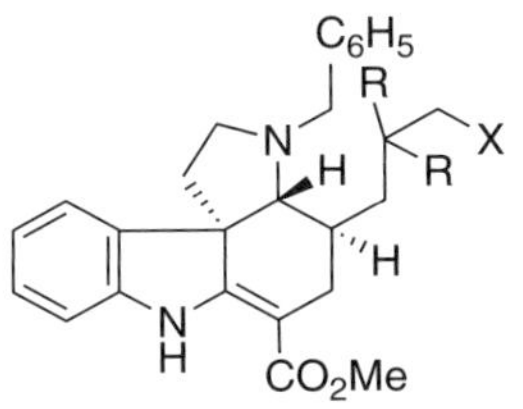

674
676 R = H, X = OTs

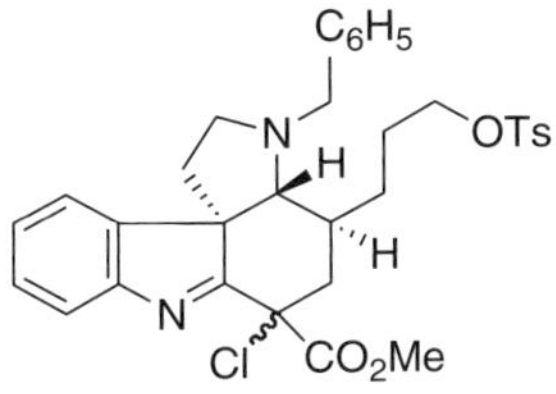

675

The culmination of this approach was the enantioselecive synthesis of vinblastine (**665**) (Scheme 39), as well as leurosidine (**683**), vincovaline (**684**), and 20′-*epi*-vincovaline (**685**) (*436*). Establishment of the C(20′) configuration in the tetracyclic coupling partners was through a first-step Sharpless epoxidation of 2-ethyl-2-propenol (leading to the chiral aldehyde **686**), followed by a later diastereomeric separation (**688** and diastereomer). Additionally, use of the naphthylethyl-substituted indoloazepine **689** for condensation with the chiral aldehyde **686**, resulted in improved diastereoselectivity of the tetracyclic amine product **690**, which was then processed to eventually yield the higher energy atropisomer of vinblastine, from which the natural conformer (**665**) was obtained on heating in toluene (Scheme 40). Kuehne has subsequently reported an improved route to the tetracyclic *N*-benzylamine **687**, *via* the use of α-ferrocenylethyl N-substituents on the indoloazepine, to improve the diastereoselectivity in the secodine-type cyclization step (*437*).

Another modification of the chloroindolenine approach was that developed by Schill who generated the racemic D-seco chloroindolenine **692** from the corresponding racemic indoloazacycloundecane **691**, which on $BF_3 \cdot Et_2O$-mediated coupling with vindoline did lead to coupled products of the correct stereochemistry (**693** and **694**), which could then be processed to 20′-deethyl-20′-deoxyvinblastine (**677**) and 20′-deethyl-20′-deoxyvincovaline (**678**), respectively (*438–441*). Application of the same approach, but using the azecinoindole derivative **695** to generate the chloroindolenine intermediate, followed by coupling with vindoline and subsequent cyclization, led to four diastereomeric products, from which 20′-deethyl-20′-deoxy-C′-homovinblastine and 20′-deethyl-20′-deoxy-C′-homovincovaline could be separated by repeated chromatography. The yield of the former compound was 27% (*442,443*).

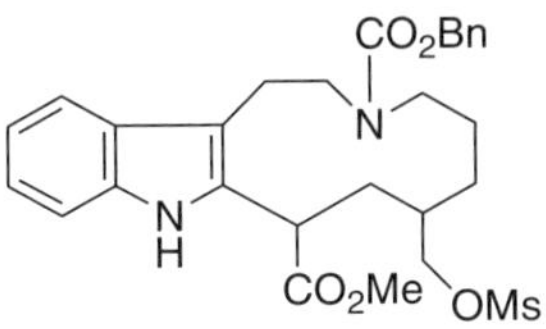

691

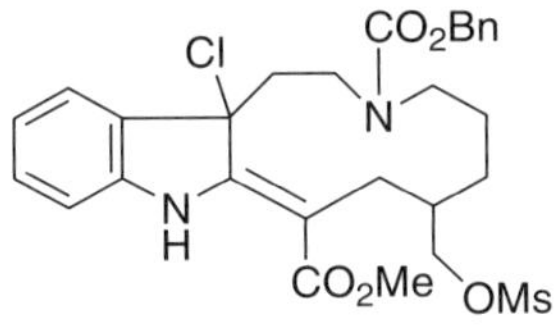

692

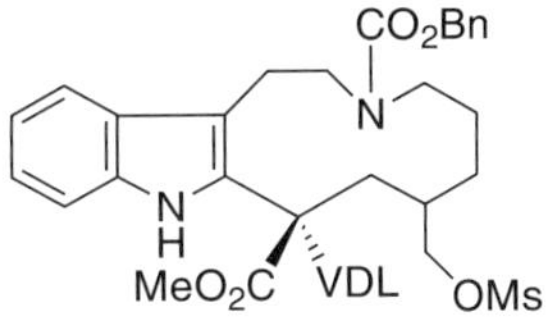

693

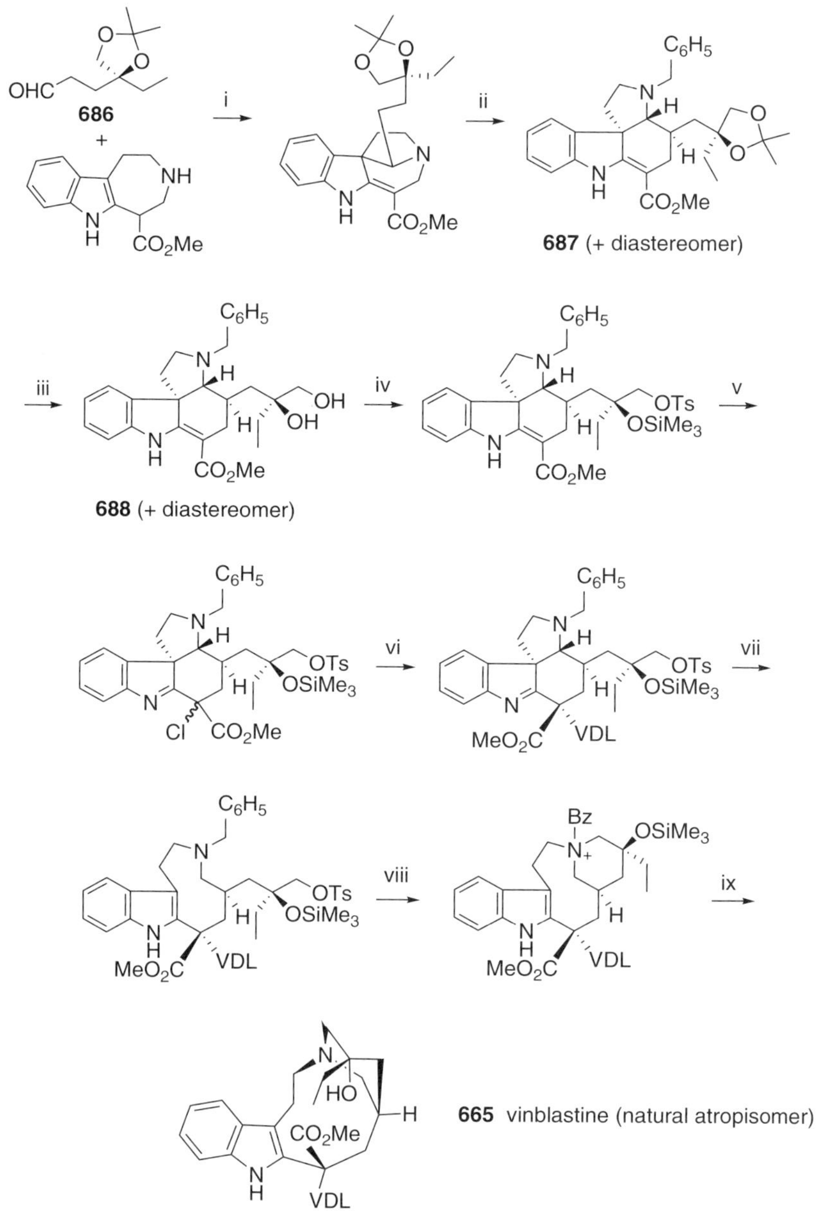

Scheme 39. Reagents: (i) Δ, THF; (ii) BnBr; $Et_3N$, THF; (iii) HCl aq.; (iv) $Ts_2O$, $Et_3N$; TMSOTF, *i*-$Pr_2NEt$; (v) *t*-BuOCl; (vi) vindoline, $AgBF_4$; (vii) $KBH_4$, AcOH; (viii) Δ, MeOH; (ix) $H_2$/Pd; $Bu_4NF$.

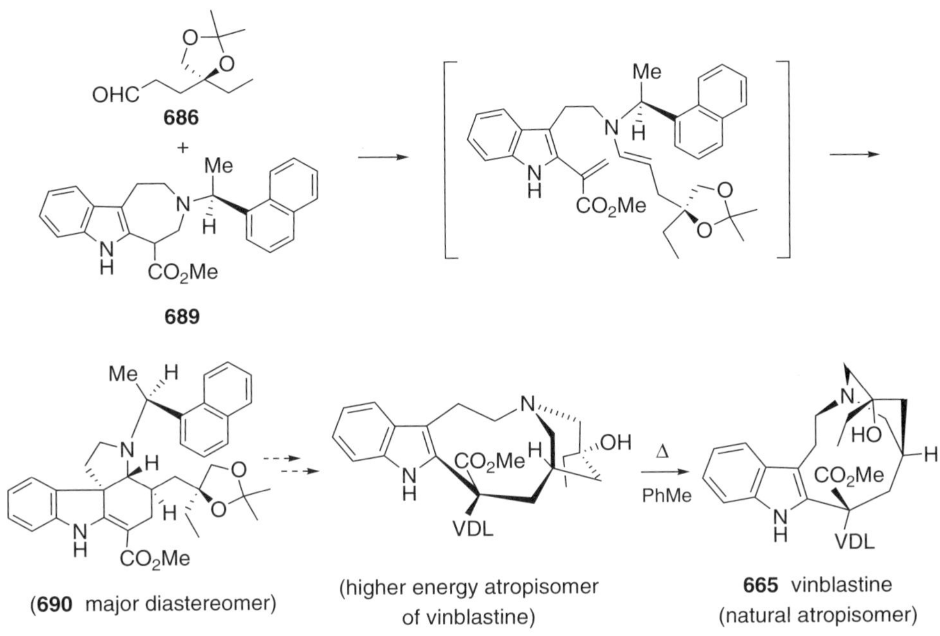

Scheme 40

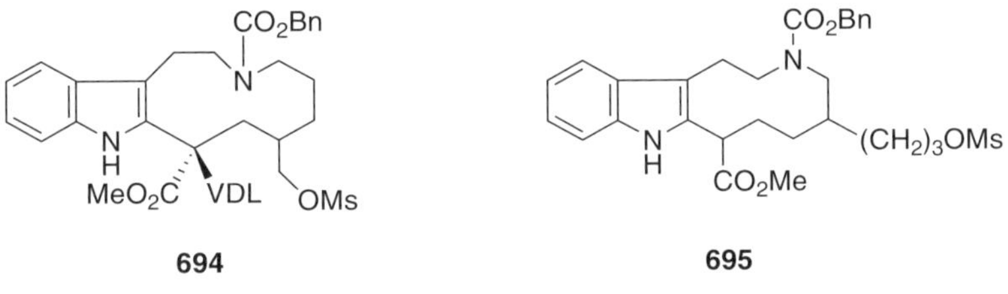

The first successful coupling reaction to deliver bisindoles of the vinblastine type was one developed by Harley-Mason and Atta-ur-Rahman (*444,445*), and involved a non-oxidative coupling, leading to the bisindole product **696** as a mixture of stereoisomers, lacking however, the C(16′) carbomethoxy and C(20′) hydroxyl groups, as well as possessing the wrong C(16′)-C(14′) relative stereochemistry (Scheme 41). Recently, Magnus has also developed a non-oxidative approach, which has culminated in a synthesis of vinblastine (*446–449*). The overall strategy adopted was based on the non-oxidative cleavage of the tertiary amine (+)-**697** (prepared from L-tryptophan), to generate the intermediate delocalized cation, **698**, which could then be trapped by electron-rich aromatic nucleophiles at C(16′) to give the bisindole, **699**. Treatment of (+)-**697** with *p*-nitrobenzyl chloroformate in $CH_2Cl_2$ at 25°C for 48 h in the presence of 3-$MeOC_6H_4NMe_2$ gave **700** (72%, 58% *ee*), indicating that the coupling reaction proceeded with predominant retention of configuration at C(16′). With more reactive nucleophiles, such as 3,5-$(MeO)_2C_6H_3NMe_2$, the enantiomeric excess was in excess of 90%, while treatment of (+)-**697** with $ClCO_2CH_2C_6H_4NO_2$-*p*/$CH_2Cl_2$ in the presence

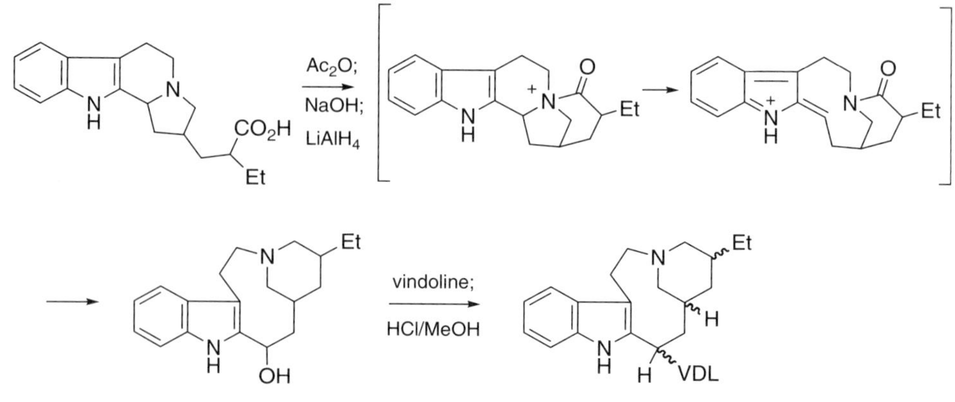


**696**

Scheme 41

of vindoline (a less reactive nucleophile) at 25°C, gave **701** as 1:1 mixture of C(16′) epimers, on account of the reversible nature of the coupling reaction. The results from these model studies indicated that for a coupling which will lead to vinblastine, a tetracyclic amine substrate is required with a side chain at C(14′) that can eventually become the piperidine ring C(15′), C(20′), and C(21′). Such an appendage may also slow down the conformational inversion of the nine-membered ring, to allow coupling with vindoline to proceed with retention of configuration. These requirements were met in the tetracyclic amine **702**.

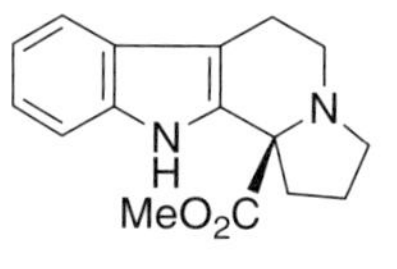

(+)-**697**

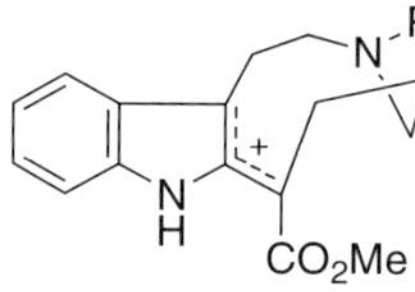

**698**

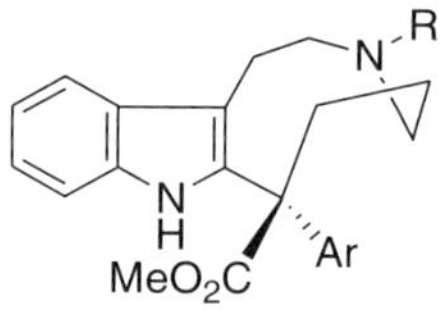

**699**

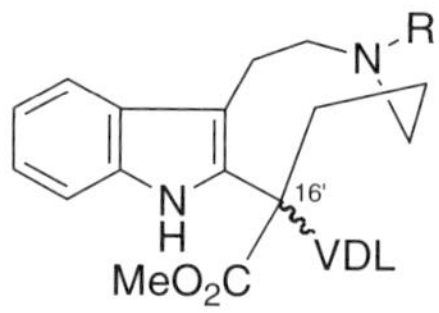

**701**

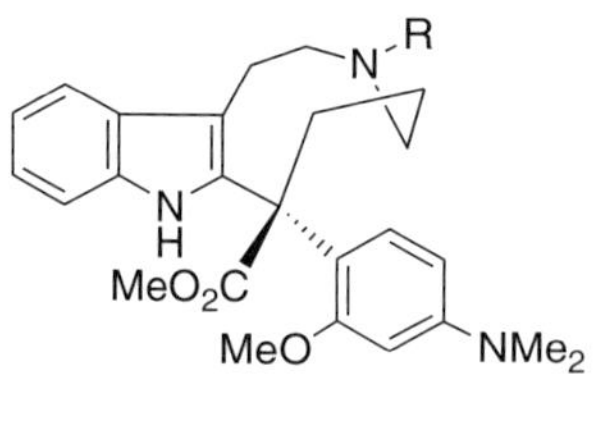

**700**

R = $CO_2CH_2C_6H_4NO_2$-*p*

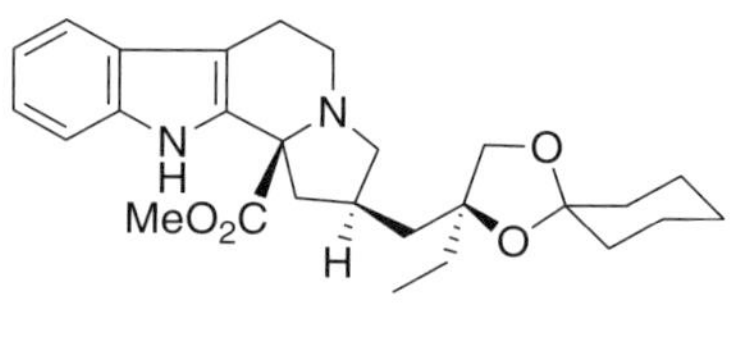

**702**

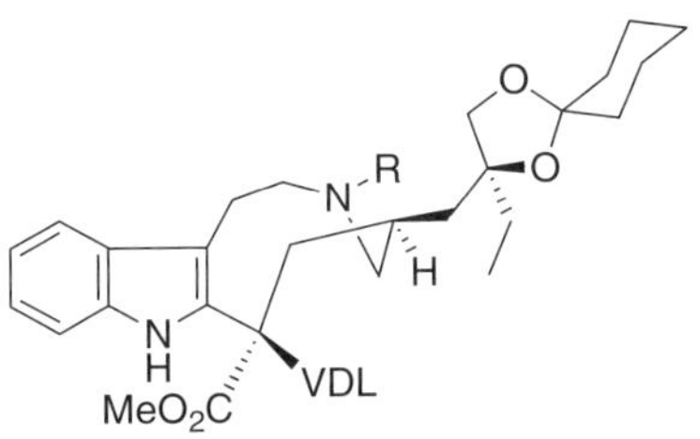

**703**

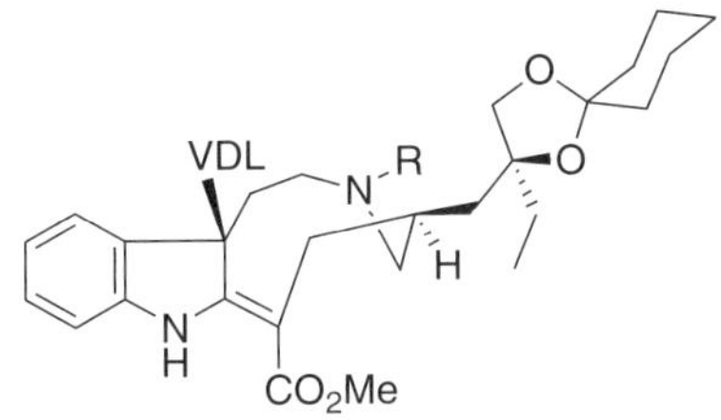

**704**

Treatment of (−)-**702** with $ClCO_2CH_2C_6H_4NO_2$-*p*/vindoline/$CH_2Cl_2$/25°C, gave **703** and **704**, but none of the adduct **705** with the correct C(16′) stereochemistry, whereas changing the solvent to nitromethane and lowering of the reaction temperature to −20°C, resulted in formation of the desired **705** (46%), along with **703** (33%) and traces of **704**. The adduct **705** can be processed to vinblastine, following hydrolysis to **706**, oxidation to **707**, and hydrogenation to vinblastine (**665**), as shown in Scheme 42. The pronounced favorable effect of the solvent ($CH_3NO_2$ or $CH_3CN$ vs. $CH_2Cl_2$) in reversing the stereochemistry at C(16′) was attributed to preferential solvation of the 'closed' iminium ion **708** (favored in $CH_3NO_2$ and $CH_3CN$), vs. the more delocalized 'open' iminium ion **709** (favored in $CH_2Cl_2$). The closed ion adopts a conformation that allowed the coupling with vindoline to take place from the least hindered α-face to establish the natural C(16′) configuration. Since of all the various stereoisomeric precursors of the top half of vinblastine tested, only the (−)-**702** diastereomer gave the coupled product of the correct C(16′) stereochemistry, a shorter and more stereoselective synthesis of (−)-**702** from 2-ethyl-2-propenol was also described (*446,450*).

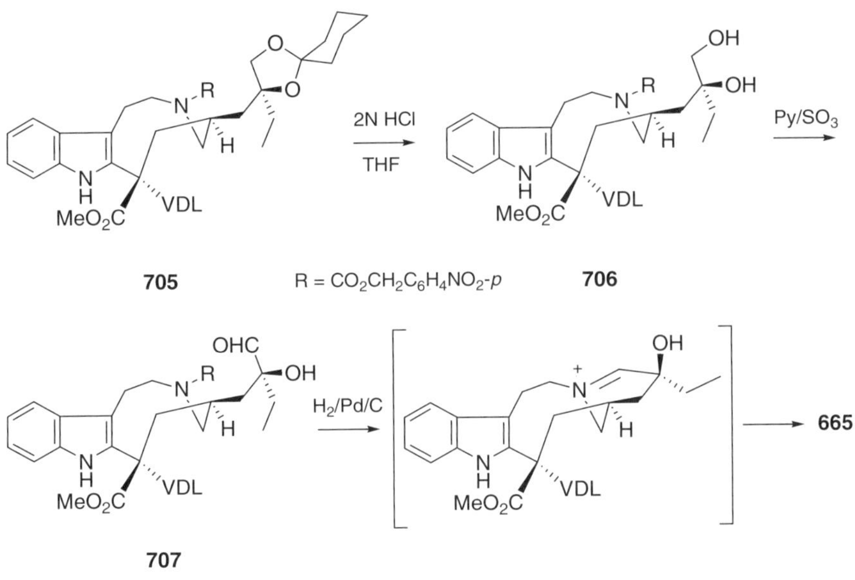


**705** R = $CO_2CH_2C_6H_4NO_2$-*p* **706**

**707**

Scheme 42

708 709

The previous syntheses of vinblastine (Potier, Kutney, Kuehne, and Magnus) have relied on coupling of various top half-precursor moieties to naturally occurring vindoline. The first total synthesis of vinblastine by coupling of an appropriate top half unit to a wholly synthetic vindoline was reported by Fukuyama who took advantage of the previous results of Schill, who had earlier demonstrated the feasibility of coupling vindoline with a chloroindolenine generated from an indoloazacycloundecane precursor, to deliver coupling products with the correct C(16′)-C(14′) stereochemistry (*vide supra*). On the basis of these results as well as his own molecular modeling studies, Fukuyama's synthetic plan was predicated on the construction of the two coupling moieties, viz., the 11-membered intermediate **734**, and (−)-vindoline (**668**), the syntheses of which were elegantly executed by Fukuyama (*451*).

The vindoline synthesis required prior preparation of the amine coupling partner, the 2,4-dinitrobenzenesulfonamide **713**, which was prepared from the pentanal **710**, as shown in Scheme 43. A notable feature of this route was the enzyme-mediated resolution of the cyanohydrin acetate **711**, *via* enzymatic hydrolysis to selectively afford a diastereomeric mixture of only the (*S*)-cyanohydrins **712**.

The indole-coupling partner was prepared by the newly developed Fukuyama indole synthesis (*452*) which provides 2,3-disubstituted indoles from quinolines, and features a $Bu_3SnH$ initiated 5-*exo* radical cyclization (Scheme 43). The starting 7-mesyloxyquinoline **714** (prepared by a modified Skraup quinoline synthesis from 3-hydroxyaniline in four steps), on treatment with thiophosgene followed by $NaBH_4$ reduction, provided isothiocyanate **715**, which was processed to the thioanilide **716**. The ensuing radical cyclization initiated by $Bu_3SnH$/AIBN proceeded smoothly to furnish the 2,3-disubstituted indole **717**, which was then transformed to the indole fragment **718**. Mitsunobu reaction of **718** with the previously prepared amine **713** afforded **719**. Treatment of **719** with TFA in $Me_2S/CH_2Cl_2$, resulted in Boc cleavage and hydration of the enol ether to the lactol, which, without purification, was subjected to sulfonamide deprotection by treatment with pyrrolidine in MeOH/MeCN. This yielded the amine **720**, which on reflux, initiated a cascade, culminating in an intramolecular inverse electron demand Diels–Alder reaction of the enamine **721**, to deliver the pentacyclic indole **722**, *en route* to (−)-vindoline (**668**). This procedure represented a marked improvement over Fukuyama's previous vindoline synthesis (*453*) and provided (−)-vindoline in several hundred-milligram quantities.

The construction of the top half coupling partner of vinblastine, the macrocycle **734**, required preparation of the ester **727** (Scheme 44). Installation of a single stereocenter at the outset was achieved through application of the Evan's

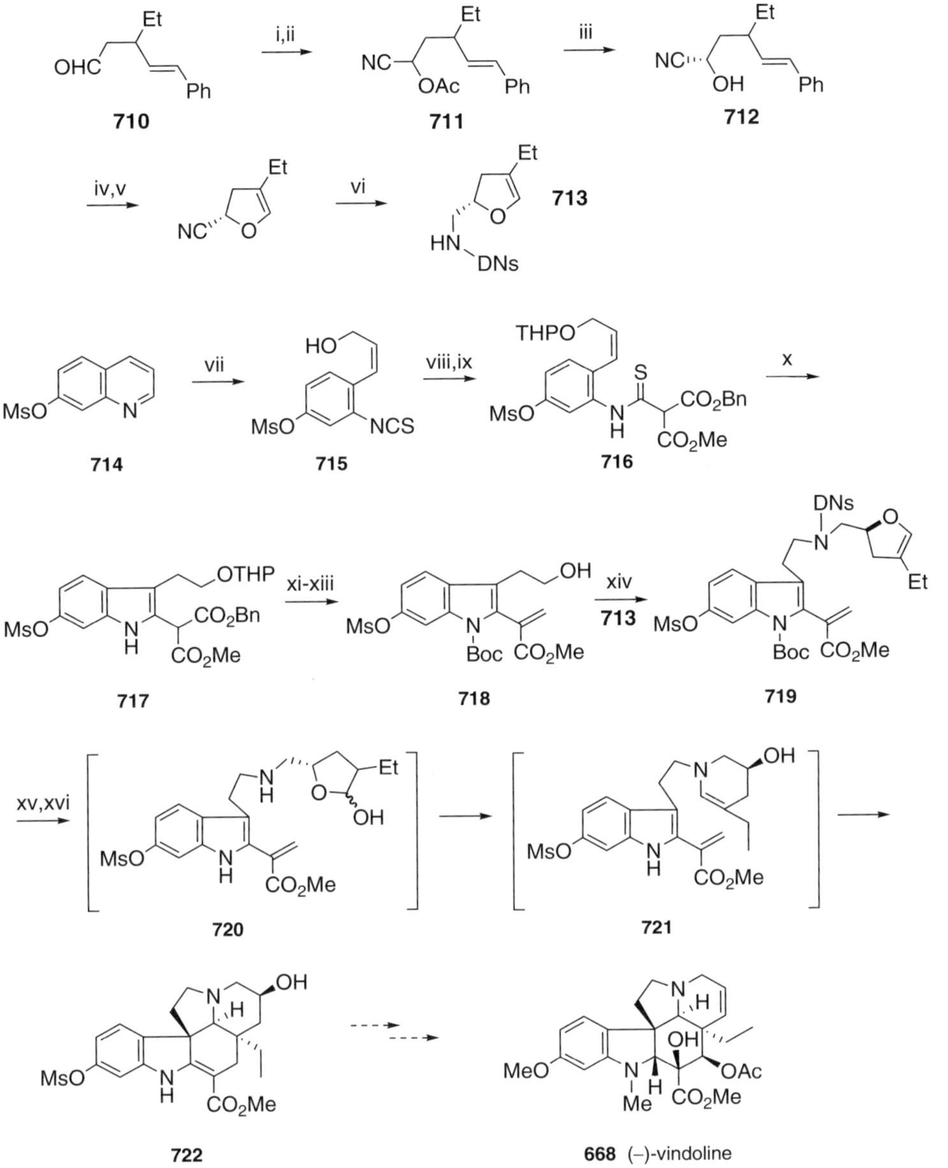

Scheme 43. Reagents: (i) NaCN, AcOH; (ii) $Ac_2O$, pyr.; (iii) Lipase PS, THF–$H_2O$; (iv) $O_3$, MeOH, $CH_2Cl_2$; $Me_2S$; (v) MsCl, $Et_3N$, PhMe; (vi) $LiAlH_4$, THF; $H_2O$, NaOH; DNsCl, $CH_2Cl_2$ (DNs = 2,4-dinitrobenzenesulfonyl); (vii) thiophosgene, $Na_2CO_3$, THF–$H_2O$; $NaBH_4$, MeOH; (viii) DHP, CSA, $CH_2Cl_2$; (ix) benzyl methyl malonate, NaH, THF; (x) AIBN, $Bu_3SnH$, PhMe; (xi) $Boc_2O$, $Et_3N$, DMAP, $CH_2Cl_2$; (xii) $H_2$, Pd/C, EtOH; $Me_2NH \cdot HCl$, HCHO, AcONa, AcOH-EtOH; (xiii) CSA, MeOH; (xiv) **713**, DEAD, $Ph_3P$, PhH; (xv) TFA, $Me_2S$, $CH_2Cl_2$; (xvi) pyrrolidine, MeOH–MeCN.

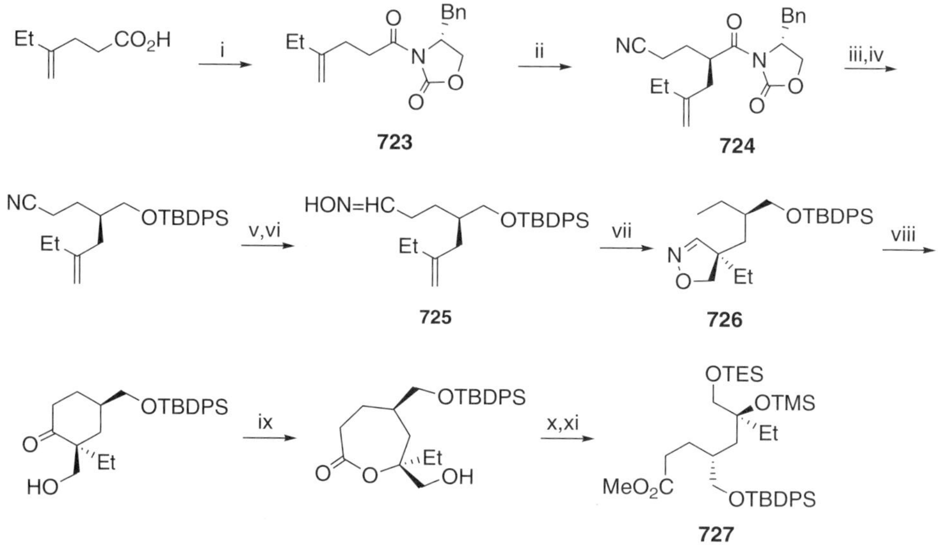

Scheme 44. Reagents: (i) PivCl, $Et_3N$, $Et_2O$, 0°C; *n*-BuLi, (*R*)-4-benzyl-2-oxazolidinone, THF, −78°C; (ii) (*i*-PrO)$TiCl_3$, *i*-$Pr_2NEt$, acrylonitrile, $CH_2Cl_2$, 0°C; (iii) $NaBH_4$, THF–$H_2O$; (iv) TBDPSCl, imid., DMF; (v) DIBAL-H, $CH_2Cl_2$, −78°C; (vi) $H_2NOH \cdot HCl$, NaOAc, EtOH; (vii) NaOCl aq., $CH_2Cl_2$; (viii) Zn, AcOH; (ix) *m*CPBA, AcOH; (x) $K_2CO_3$, MeOH; (xi) TESCl, imid., DMF; TMSCl.

chiral auxiliary method (*454*). Thus, addition of the deprotonated form of (*R*)-4-benzyloxazolidinone to the mixed anhydride of 4-ethylpent-4-enoic acid gave the imide **723** which underwent diastereoselective cyanoethylation to give **724**, which was then processed in several steps to the oxime **725**. The oxime on exposure to sodium hypochlorite, underwent facile intramolecular 1,3-dipolar cycloaddition *via* the nitrile oxide, to afford isoxazoline **726** as a single diastereomer. Isoxazoline **726** was then processed in several steps to the desired ester **727** (Scheme 44).

Addition of the enolate of **727** to isothiocynate **728** afforded a mixture of thioamides **729**, which was then subjected the Fukuyama radical cyclization conditions leading to indole **730** (Scheme 45). Selective protection of the primary alcohol of the 1,2-diol unit in the triol **731** was achieved with TsCl/$Et_3N$ in the presence of dibutyltin oxide. This was then followed by epoxide formation, after which a Mitsunobu reaction on the remaining primary alcohol provided the nosyl-protected amine **732** to set the stage for the crucial macrocyclization, which was achieved by heating **732** in the presence of $K_2CO_3$ in DMF. The resultant macrocycle **733** was then processed in several steps to the desired key coupling partner, **734**. The coupling of **734** with vindoline was achieved in the usual manner by conversion to the chloroindolenine **735**, which upon exposure to TFA in the presence of vindoline underwent oxidative coupling *via* the intermediate iminium ion **736** to provide the coupled adduct as the sole isomer in 97% yield. Deprotection of the tertiary alcohol, and removal of the nosyl group, which triggered cyclization to the piperidine ring, completed the synthesis of vinblastine (**665**) (Scheme 45). This first

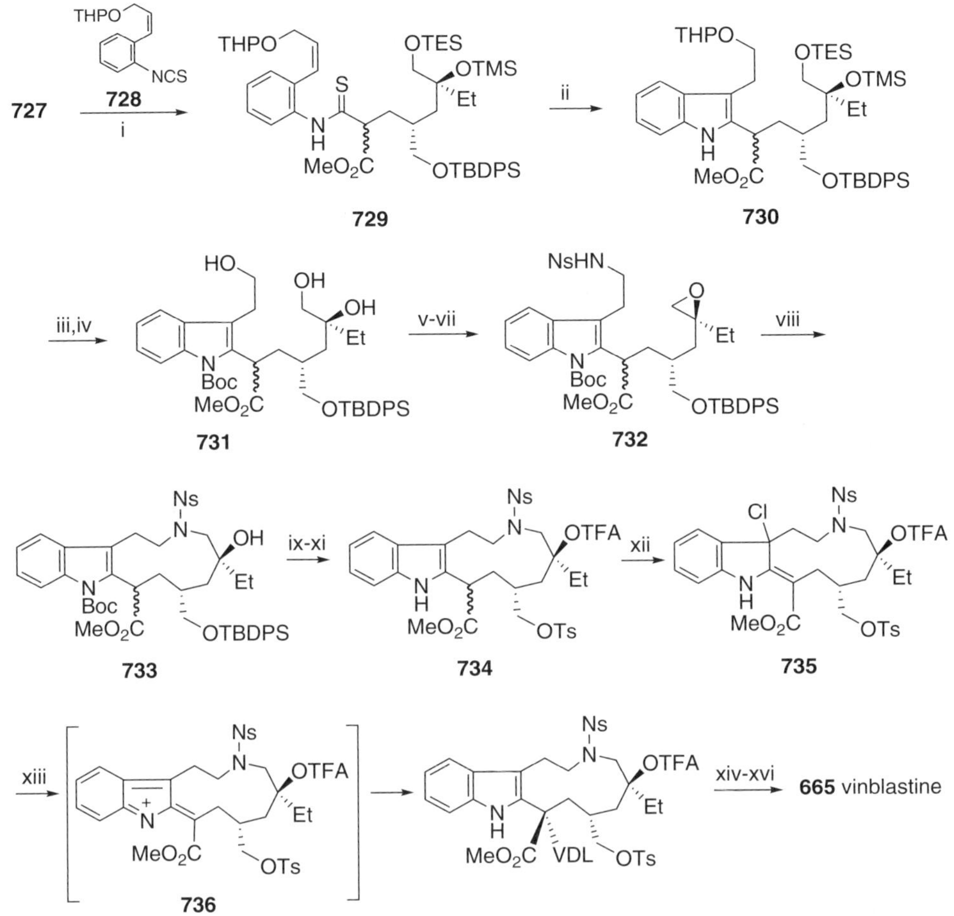

Scheme 45. Reagents: (i) LDA, THF, −78°C; **728**; (ii) $Bu_3SnH$, $Et_3B$, THF; (iii) $Boc_2O$, $Et_3N$, DMAP, $CH_2Cl_2$; (iv) AcOH–$H_2O$; (v) TsCl, $Bu_2SnO$, $Et_3N$, $CH_2Cl_2$; (vi) $NaHCO_3$, DMF, 80°C; (vii) $NsNH_2$, DEAD, $Ph_3P$, PhMe; (viii) $K_2CO_3$, DMF, 90°C; (ix) TFA, $CH_2Cl_2$; (x) TsCl, $Me_2N(CH_2)_3NMe_2$, MeCN–PhMe; (xi) TFAA, pyr., $CH_2Cl_2$; (xii) *t*-BuOCl, $CH_2Cl_2$, 0°C; (xiii) (−)-vindoline(**668**), TFA, $CH_2Cl_2$; (xiv) $Et_3N$, MeOH; (xv) $HSCH_2CH_2OH$, DBU, MeCN; (xvi) $NaHCO_3$, *i*-PrOH-$H_2O$ (Ns = 2-$NO_2C_6H_4SO_2$).

and efficient total synthesis of vinblastine opens the possibility to the elaboration of better and more potent analogues for applications in chemotherapy.

## C. SYNTHESIS OF ANALOGUES AND BIOLOGICAL ACTIVITY

The synthesis of analogues and the various structural modifications of vinblastine, as well as the attendant evaluation of their biological activities, covering the period up to 1990, have been discussed in Volume 37 (*455*). Other recent reviews have also addressed these issues (*424,456–458*). Despite a very large number of analogues prepared and evaluated, vinorelbine (Navelbine) (**737**) and vindesine

(**738**) remain the only two that have entered clinical use. Several others held initial promise, but further clinical evaluations were discontinued for various reasons, including toxicity (e.g., vinepidine, vinzolidine, vinglycinate, vinfosiltine).

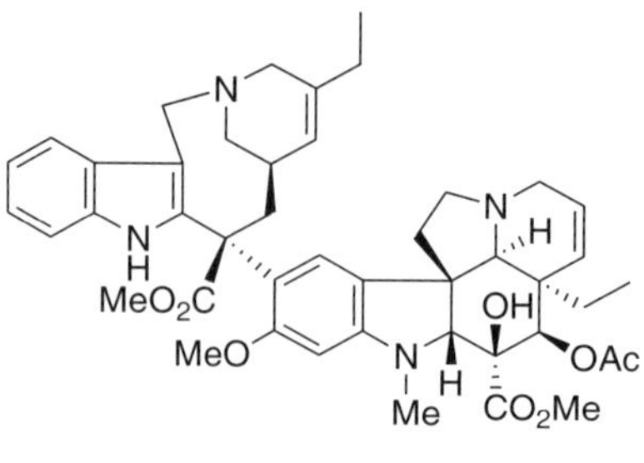

**737** vinorelbine

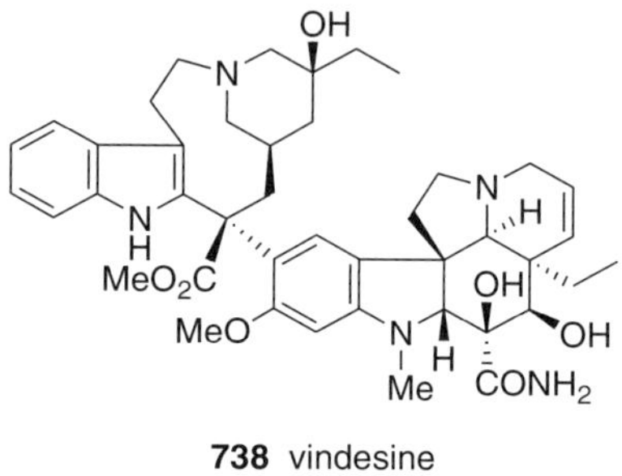

**738** vindesine

There have been several recent syntheses of anhydrovinblastine, stimulated in part by its role as a key intermediate in the synthesis of the anticancer drug, Navelbine (vinorelbine). Anhydrovinblastine has been prepared *via* an electrochemically mediated coupling of catharanthine (**667**) and vindoline (**668**). The oxidation of catharanthine on a platinum anode in MeCN-$Et_4NClO_4$ at controlled potential (0.6 V vs. SCE) in the presence of vindoline, gave, after *in situ* reduction with $NaBH_4$, (16′*S*)- and (16′*R*)-anhydrovinblastine, in yields of 52 and 12%, respectively (Scheme 46) (*459,460*).

Anhydrovinblastine (**670**) has been prepared in 50% yield by coupling of catharanthine and vindoline under near-UV light irradiation (370 nm) in the presence of flavin mononucleotide, followed by $NaBH_4$ reduction of the first-formed dihydropyridinium intermediate (**739**) (*461*). A one-pot procedure *via* irradiation of

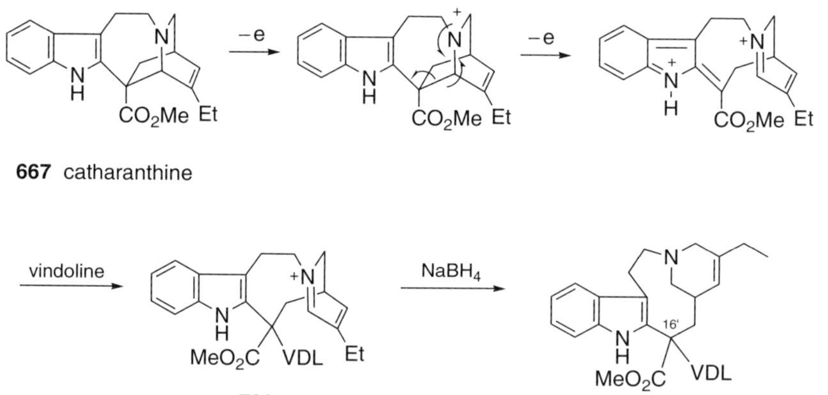

**667** catharanthine

**739**

16′(*S*)-anhydrovinblastine (52%)
16′(*R*)-anhydrovinblastine (12%)

Scheme 46

the two monomers at 370 nm, in the presence of flavin mononucleotide, followed by NADPH reduction however, resulted in low yields of anhydrovinblastine (*462*). Anhydrovinblastine has also been synthesized in 70% yield by a one-step, $Cp_2TiCl$-mediated deoxygenation of leurosine (**740**), a major bisindole alkaloid occurring in the leaves of *C. roseus*. The low valent titanium complex was generated by the *in situ* reduction of $Cp_2TiCl_2$ by powdered zinc, while the trapping of the putative radical intermediate by a hydrogen-atom donor to yield the reduced alkaloid, provided support for the proposed radical mechanism (Scheme 47) (*463,464*). Another transformation of leurosine (**740**) to anhydrovinblastine was by treatment with $Me_3SiCl$/powdered zinc in dry $CH_2Cl_2$ at −23°C, which gave the product in 95% yield (*465*). There has been recent renewal of interest in anhydrovinblastine for potential applications in anticancer therapy (*466,467*).

The dihydropyridinium ion (**739**) generated in the Potier-Polonovski coupling of catharanthine and vindoline can be trapped by organozinc reagents, prior to borohydride reduction, to give either C(15′) or C(21′) alkylated adducts, depending on the reaction conditions. The former were unstable enamines, but treatment with borohydride allowed the isolation of the corresponding C(20′)-deoxyvinblastine analogues. Several compounds, viz., 21′-phenylanhydrovinblastine (**741**), 15′-ethyl-, 15′-*t*-butyl-, and 15′-phenyl-20′-deoxyvinblastine (**742**–**744**), showed activity as tubulin assembly inhibitors in the $\mu$M range (*468*).

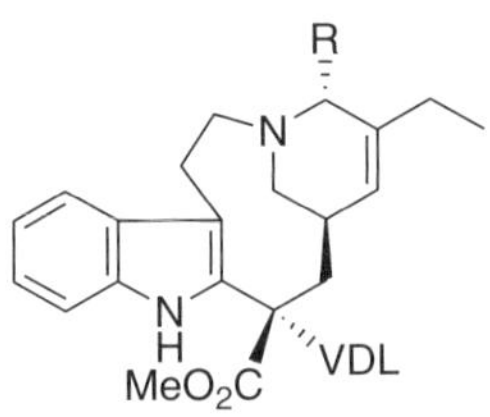


**741** R = Ph

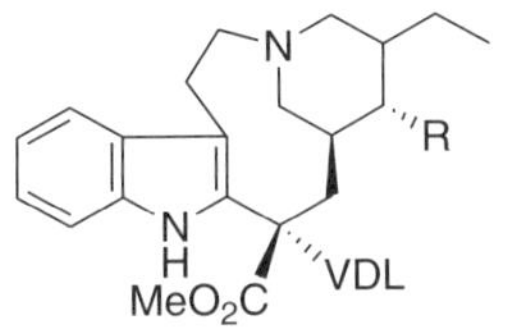


**742** R = Et
**743** R = *t*-Bu
**744** R = Ph

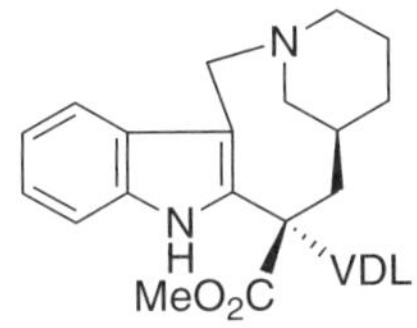


**745** deethyldihydronavelbine

Magnus has reported a synthesis of the Navelbine (7′-noranhydrovinblastine) analogue, deethyldihydronavelbine (**745**), starting from (*R*)-(−)-ethyl nipecotate (**746**), as shown in Scheme 48 (*469*). The reaction of the Grignard reagent **748** derived from the *N*-allyl bromide **747**, with the N-protected, 2-(methoxyoxalyl)indole **749**, gave the diastereomeric alcohols **750**. Removal of the indole protecting group, followed by coupling with vindoline (1% HCl/MeOH) gave the diastereomers, **751** (16′*S*), and its (16′*R*)-epimer, which were separated by HPLC. Deprotection to the free amine **752**, followed by an intramolecular Mannich reaction (formaldehyde/acetic acid), gave deethyldihydronavelbine (**745**) and its (16′*R*)-epimer. The natural C(16′)-epimer showed *in vitro* antitumor activity comparable to vinblastine, while the 16′*R* epimer was inactive.

Two new analogues of catharanthine, **753** and **754**, differing from catharanthine in the fusion of the indole ring to the non-aromatic portion of the iboga skeleton have been synthesized in racemic form and their reactivity toward coupling with vindoline examined. The [2,1] fused analogues (e.g., **753**) were found to give low

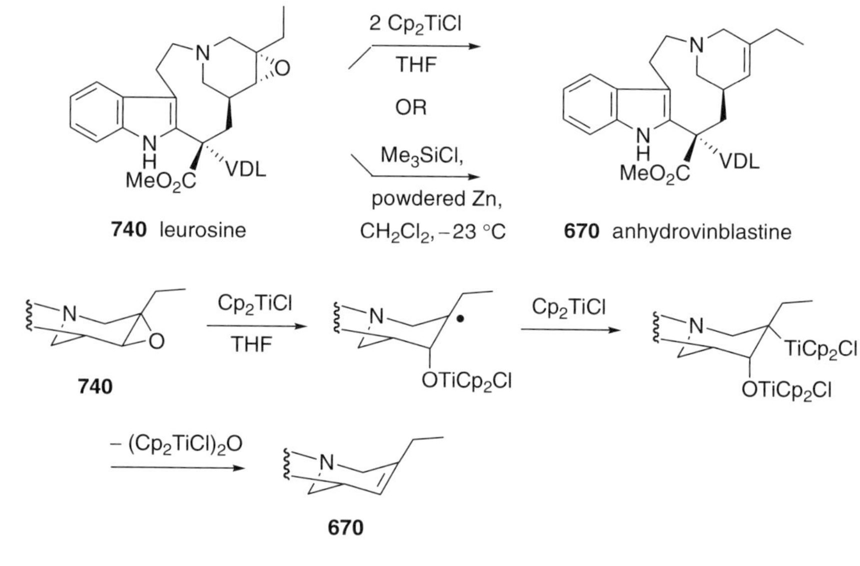


Scheme 47

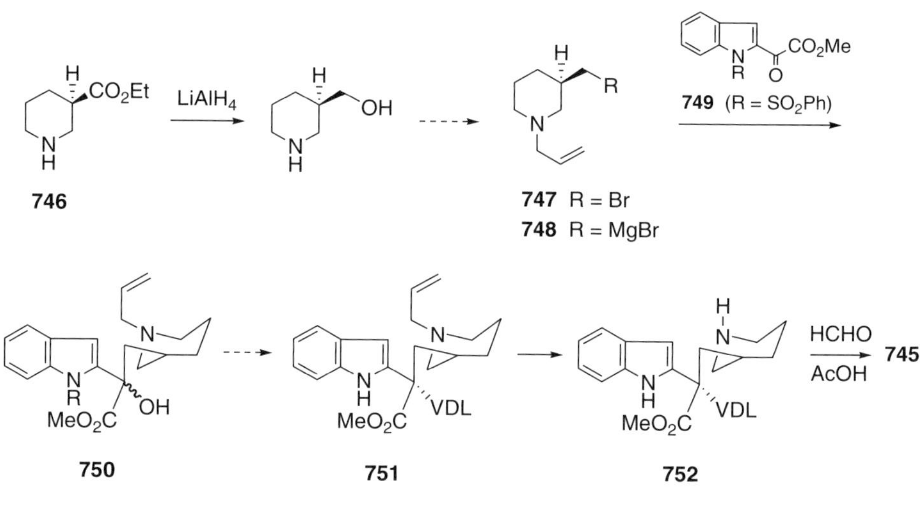


Scheme 48

yields of coupling products, which were the 3′-10 bisindoles (e.g., **755**), under Potier–Polonovski conditions, while the [3,2] isomers (e.g., **754**) underwent fragmentation of the C(16′)-C(21′) bond as in catharanthine, but no coupling to vindoline occurred (*470*).

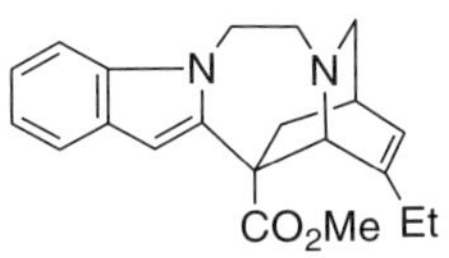


**753**

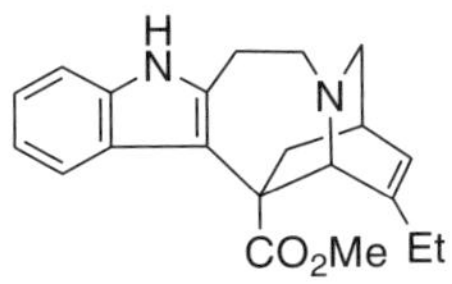


**754**

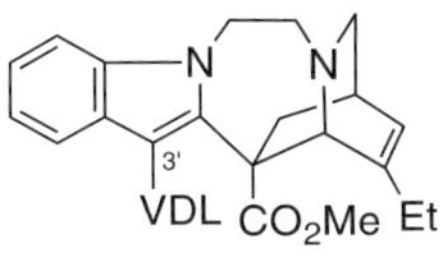


**755**

It was previously known that oxidation of vinblastine sulfate with chromyl acetate leads to vincristine (*471*). Szantay and coworkers have found that a similar oxidation carried out on the free base **665** led instead to the cyclic derivative, **756**, in which the original velbanamine part has undergone a 3′, 7′-transannular cyclization into a ψ-aspidosperma-type skeleton. Similar oxidation of vincristine (**666**) led to the corresponding cyclized product **757**. Acid treatment of the cyclized derivatives resulted in an aspidospermane→eburnane skeletal rearrangement leading to novel bisindoles with the ψ-eburnea–aspidosperma-type skeleton, e.g., **756** → **758** (*472*).

Nitration of vincristine yielded the non-toxic 7′-nitro derivative **759** (31%), in addition to the 11′-nitro and 9′-nitro derivatives (65%). Reduction of the 7′-nitro derivative **759** regenerated the original compound. It was proposed that if the non-toxic **759** could be enzymatically reduced to vincristine at the site of the tumor, it would constitute a new approach to avoiding the serious side effects of vincristine (*473*).

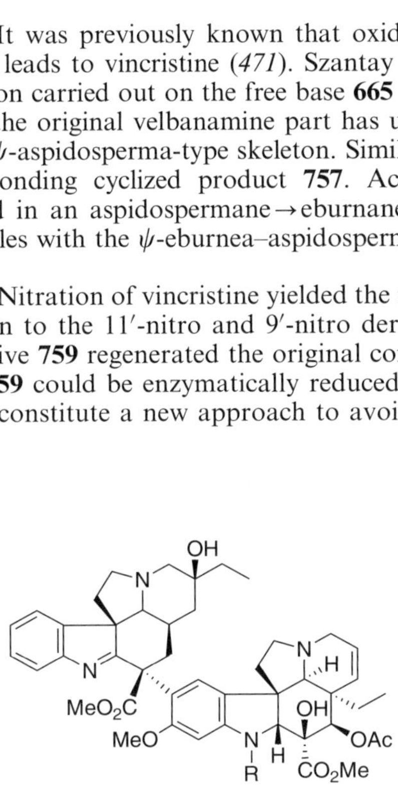


**756** R = Me
**757** R = CHO

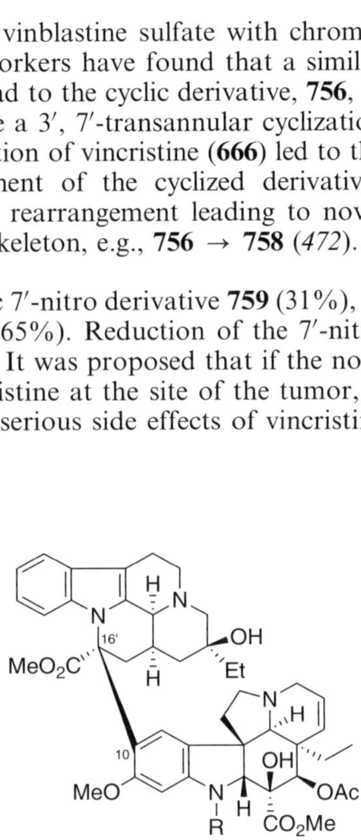


**758** R = Me

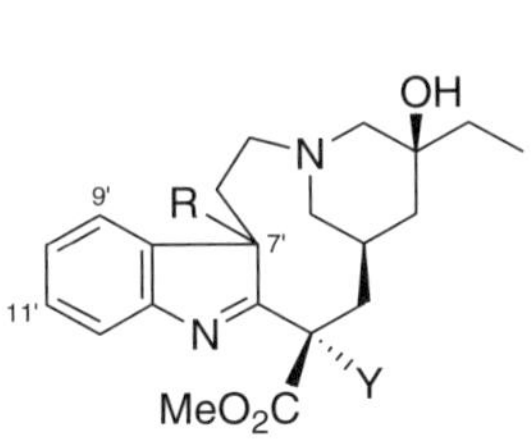


**759** R = $NO_2$

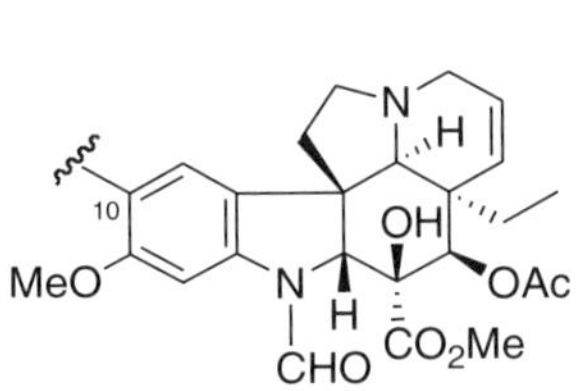


Y

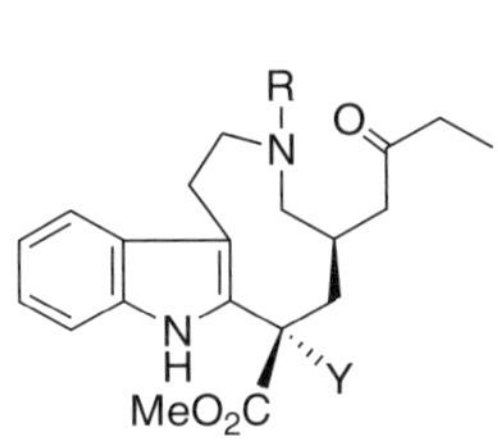


**760** R = CHO

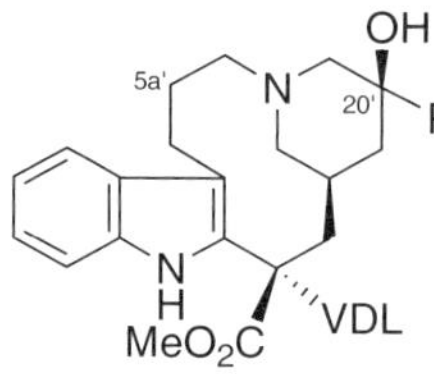

**761** R = Et
**762** R = Me

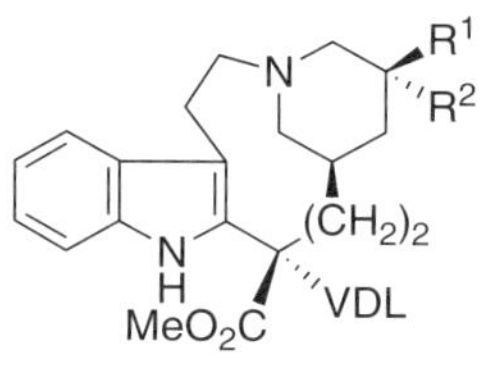

**763** $R^1$ = Et, $R^2$ = OH
**764** $R^1$ = OH, $R^2$ = Et

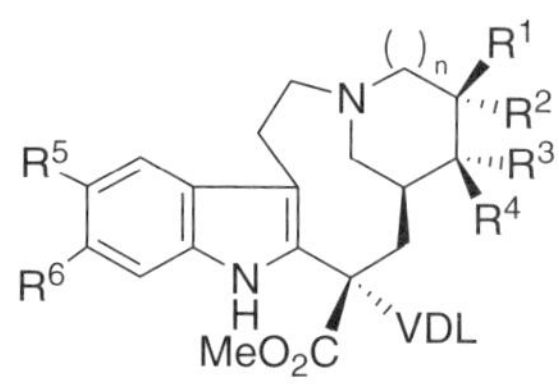

**765**

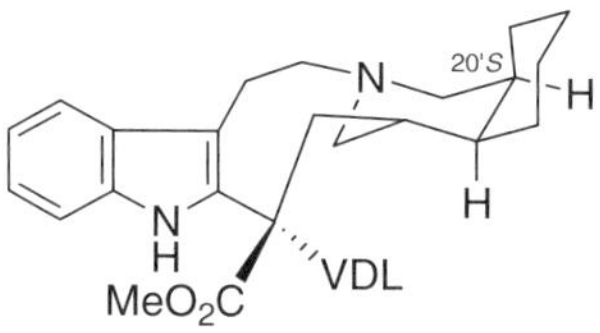

**766**

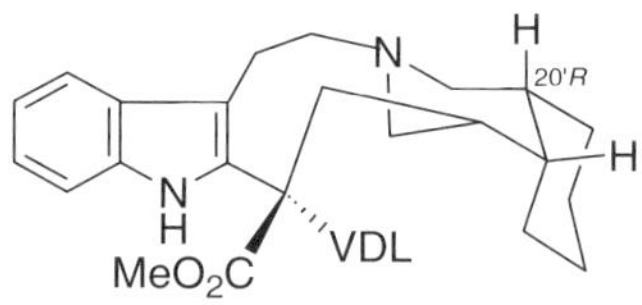

**767**

Vincristine was oxidized to the ring-cleavage product, *N*-formylcatharinine (**760**) in incubations with either horseradish peroxidase or the human serum copper oxidase ceruloplasmin. The horseradish peroxidase-catalyzed oxidation required hydrogen peroxide whereas the ceruloplasmin-catalyzed oxidation required chlorpromazine as a 'shuttle oxidant'. *N*-Formylcatharinine (**760**) was found to be 118 times less active than vincristine in an *in vivo* test against a human T-cell leukemic cell line (*474*).

The Kuehne group has recently reported syntheses of 5a′-*homo*-vinblastine (**761**), its C(20′) methyl congener **762** (*475*), 16a′-*homo*-leurosidine (**763**), and 16a′-*homo*-vinblastine (**764**) (*476*). The 5a′-*homo*-vinblastine congeners showed vinblastine-like inhibition of tubulin polymerization, as well as cytotoxicity to L1210 leukemia cells, but with lower potency for the latter effect compared to the corresponding compounds in the vinblastine series. No inhibition of tubulin polymerization or cytotoxicity against L1210 were observed for 16a′-*homo*-vinblastine. In common with leurosidine, 16a′-*homo*-leurosidine also did not display any activity at concentrations below $1 \times 10^{-5}$ M. In an expanded study, the Kuehne group reported the preparation of 62 vinblastine congeners, mostly with modifications in the piperidine ring of the cleavamine moiety (e.g., **765**), for evaluation of various biological activities, including cytotoxicity against murine L1210 leukemia and RCC-2 rat colon cancer cells, inhibition of polymerization of microtubular protein at $<10^{-6}$ M, induction of spiralization of microtubular protein, and microtubular disassembly at $<10^{-4}$ M (*477*). An $ID_{50}$ range of $>10^7$ M concentrations was found for L1210 inhibition by these compounds. In addition to confirming the earlier findings that the C(20′)-substituents play a critical role in modulating the *in vitro* biological activity (*433,434,436,455,478*), another notable finding was that in the vinblastine series, *cis*-fusion of a six-membered ring onto the D′-piperidine ring resulted in an appreciable increase in potency against L1210 leukemia. Thus the C(20′*S*) six-membered ring compound, **766**, was found to be 500 times more potent than vinblastine, while the C(20′*R*) compound, **767**, was 1000 times more potent than vinblastine (*477*).

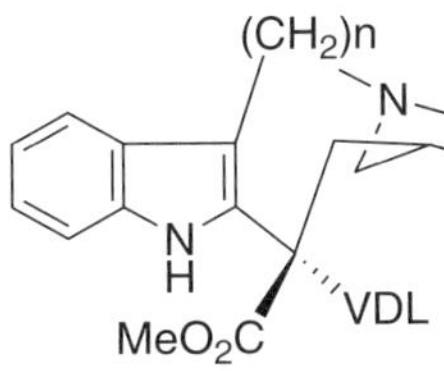

**768** n = 1
**769** n = 2

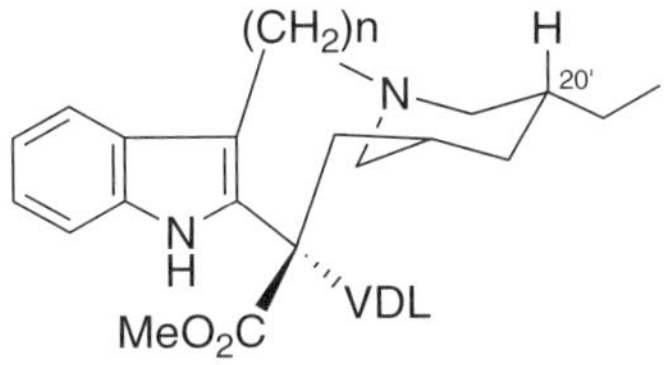

**770** n = 2
**771** n = 1

Another approach to novel vinblastine-type analogues was that developed by the French Pierre Fabre group, who explored the transformations of these alkaloids when exposed to superacid media (*479,480*). Vinorelbine (**737**), anhydrovinblastine (**670**), and vinblastine (**665**), upon treatment with $CCl_4$ in HF-$SbF_5$ at –40°C, yielded after workup, the corresponding *gem*-difluoro derivatives, **768** (for **737**), and **769** (for **670**, **665**). In the case of vinblastine (**665**), dehydration of the tertiary alcohol function probably occurred in the first step. These reactions represent the first examples of a one-step fluorination at an allylic position, C(20′), without affecting any of the other functionalities within the molecule. A mechanism was proposed as shown in Scheme 49 (*479*). It was subsequently shown that stereoselective reduction of vinblastine, anhydrovinblastine, and vinorelbine to the corresponding 20′(*R*)-20′-deoxy analogues could be achieved by reaction in HF-$SbF_5$ in the presence of a hydride donor such as methylcyclopentane at –20°C (*481*). Thus, reaction of

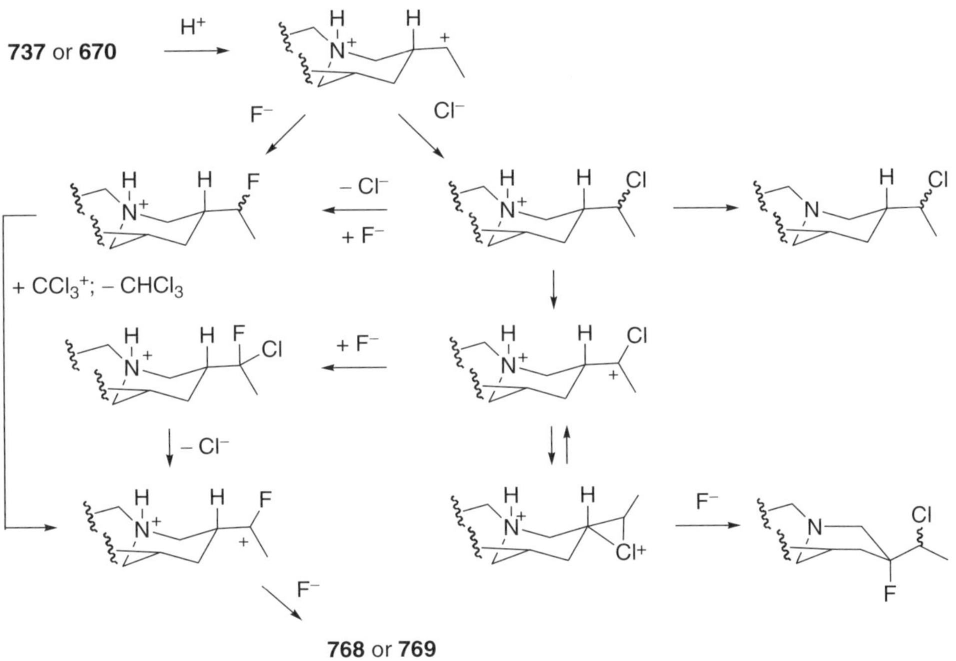

Scheme 49

vinblastine and anhydrovinblastine gave **770** while similar reaction of vinorelbine gave **771**. Deuterium labeling experiments showed that reduction occurred at C(19′).

Using superacid chemistry, a new family of derivatives based on modifications at C(20′) and C(19′) of vinorelbine, anhydrovinblastine, and vinblastine were prepared for *in vitro*, as well as *in vivo*, SAR studies (*456,458*). The data, however, showed a lack of correlation between the *in vitro* and *in vivo* results. The tubulin polymerization assay was marked by a narrow range of the $IC_{50}$ values for a majority of the compounds tested, while there were some clear contradictions for some compounds between the *in vitro* and the *in vivo* activity. Vinflunine (20′*R*-19′,19′-difluoro-15′,20′-dihydrovinorelbine)

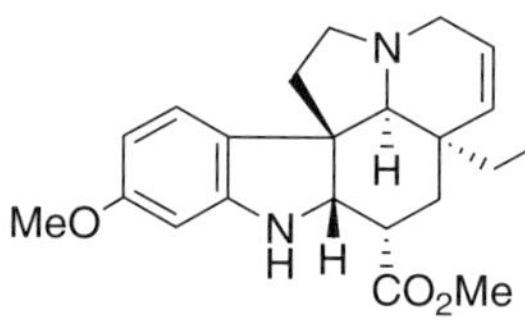

**772**

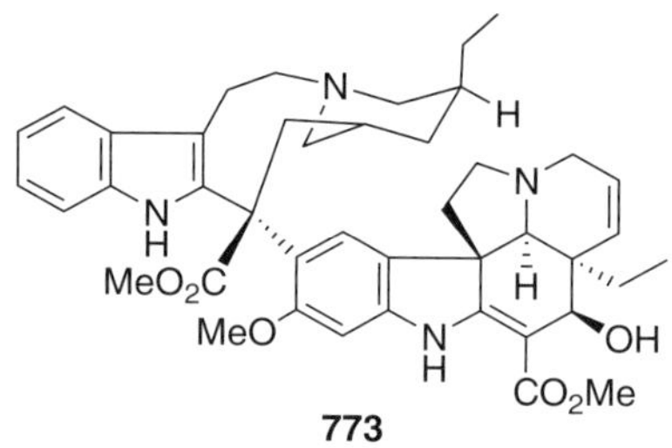

**773**

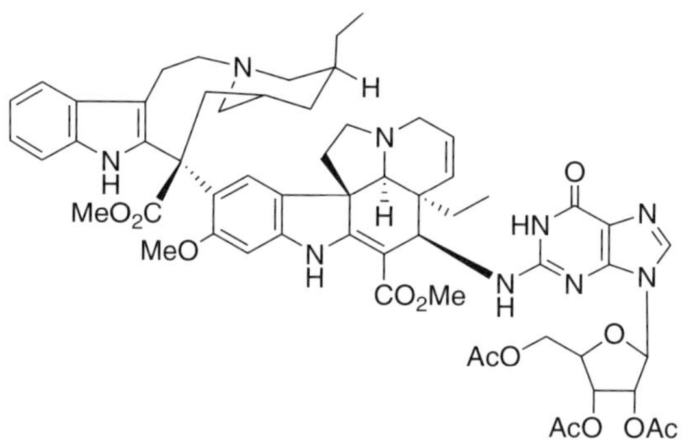

**774**

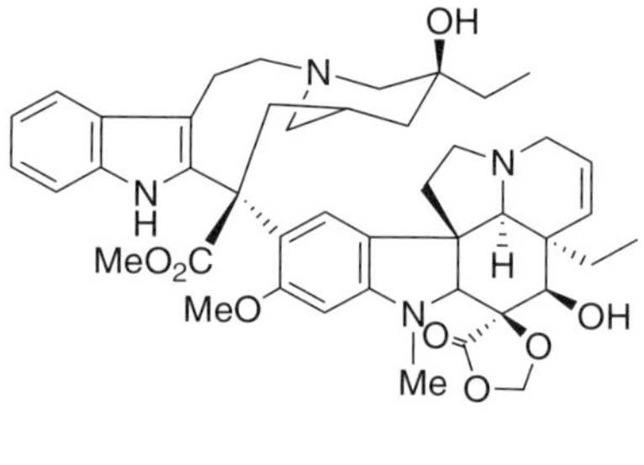

**775**

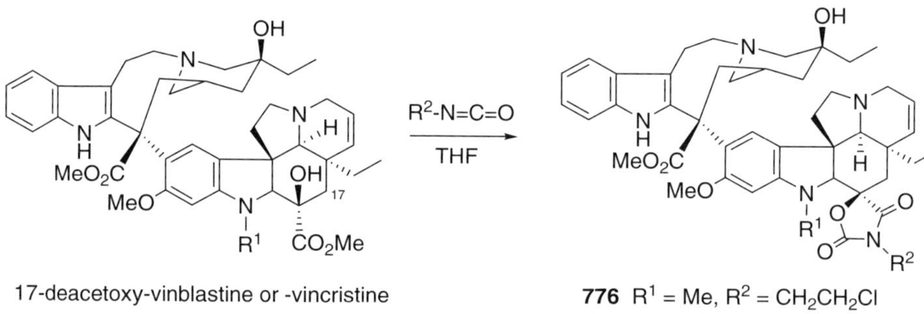


17-deacetoxy-vinblastine or -vincristine

**776** $R^1$ = Me, $R^2$ = $CH_2CH_2Cl$
**777** $R^1$ = CHO, $R^2$ = $CH_2CH_2Cl$
**778** $R^1$ = Me, $R^2$ = $CH_2CH{=}CH_2$

Scheme 50

(**768**), however, showed impressive *in vivo* antitumor activity in the murine P388 leukemia model which was superior to that shown by vinorelbine, and was therefore duly selected for preclinical investigations (*457,458*). The results obtained showed a distinct superiority for vinflunine over vinorelbine against s.c. implanted B16 melanoma (*482*), and against a panel of human tumour xenografts (*483*). Vinflunine is currently undergoing phase II clinical trials (*480*).

The 17-deacetyl derivatives of vinblastine and vincristine were conveniently prepared in high yields by base-catalyzed hydrolysis of the parent alkaloids (*484*). Danieli and coworkers synthesized C-17 modified analogues of vinblastine by coupling the tabersonine derivative **772** with catharanthine **667**, followed by subsequent hydrogenation and stereoselective hydroxylation of the bisindole to **773**. Reaction of **773** with tri-*O*-acetylguanosine provided the novel adduct, **774** (*485*). There has, however, been no report of the biological activity of these derivatives. A spirolactone **775**, however, obtained by the reaction of 17-deacetylvinblastine with paraformaldehyde, was found to be more effective compared to vinblastine in *in vivo* studies using $DBA_2$ mice with P388 leukemia (*486*).

Several semisynthetic derivatives of 17-deacetoxyvinblastine and 17-deacetoxyvincristine (KAR-2, KAR-3, KAR-4, **776**–**778**, respectively) with a spirofused

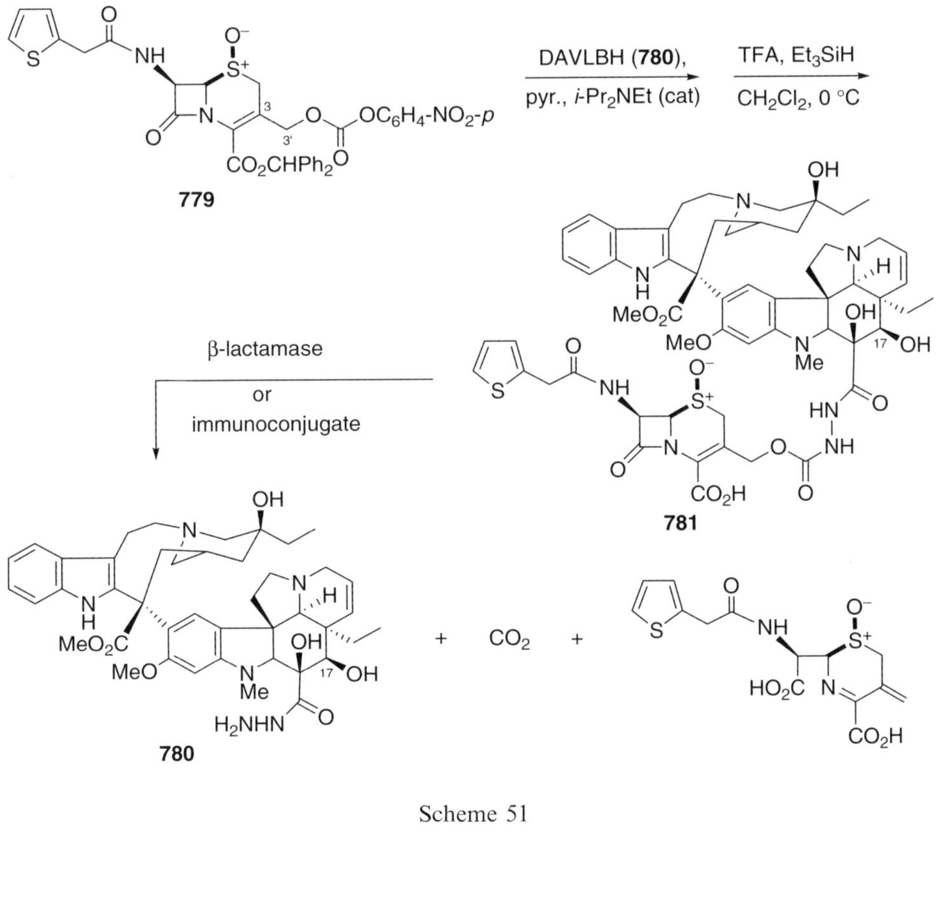


Scheme 51

oxazolidino ring at C(16) of the vindoline unit (reminiscent of vinzolidine), have been prepared by a Hungarian group (Scheme 50) (*487–489*).

These compounds in particular KAR-2 (**776**), showed interesting pharmacological behaviour compared to vinblastine. KAR-2 inhibited tubulin assembly *in vitro*, but this effect was found to be dependent on the ultrastructure of the microtubules, bundled microtubules appeared to be resistant to the drug. KAR-2 was effective against murine P388 leukemia *in vivo* with significantly lower toxicity compared to vinblastine. A significant difference compared to vinblastine, however, was the absence of anticalmodulin activity shown by KAR-2 and KAR-4, despite binding to calmodulin, which was attributed to the presence of the spirooxazolidino ring. KAR-3 on the other hand did show anticalmodulin activity, but with diminished potency compared to vincristine, suggesting that the substitution of a methyl group by a formyl can selectively modify the anticalmodulin potency of these spirooxazolidino-containing derivatives (*489*).

In a different approach, cephalosporin **781**, substituted at the C(3′) position with the potent cytotoxic agent, deacetylvinblastine hydrazide (DAVLBH **780**), was synthesized as a potential prodrug for the treatment of solid tumors (*490*). Such a design was based on the knowledge that an enzyme-mediated hydrolysis of the cephalosporin $\beta$-lactam bond can result in expulsion of the C(3′) substituent. It was envisaged that release of the free alkaloid will be mediated by an immunoconjugate, comprising a $\beta$-lactamase enzyme covalently linked to a monoclonal antibody which has been prelocalized at the tumor. The synthesis of **781** was readily achieved *via* condensation of deacetylvinblastine hydrazide (**780**) with the cephalothin-derived cephem, the 3′-*p*-nitrophenyl carbonate, **779** (Scheme 51).

The prodrug **781** was found to be an excellent substrate for the MoAb-enzyme conjugate, as shown by the facile release of free **780**, which was monitored by HPLC. Mouse xenograft studies using the antibody-enzyme immunoconjugate with prodrug **781** caused regression of established tumors. Such a combination was found to be significantly more active than either the prodrug **781**, or the free alkaloid **780**, administered alone.

## References

1. G. A. Cordell and J. E. Saxton, *in* "The Alkaloids", (R. H. F. Manske and R. G. A. Rodrigo, eds.), vol. 20, p. 1. Academic Press, New York, 1981.
2. G. A. Cordell, *in* "Indoles: The Monoterpenoid Indole Alkaloids", (J. E. Saxton, ed.), p. 539. Wiley-Interscience, New York, 1983.
3. J. Sapi and G. Massiot, *in* "Indoles: The Monoterpenoid Indole Alkaloids", (J. E. Saxton, ed.), p. 523. Wiley-Interscience, New York, 1994.
4. A. Brossi and M. Suffness (eds.). "The Alkaloids, Vol. 37: Antitumor Bisindole Alkaloids from *Catharanthus roseus* (L.)", vol. 37. Academic Press, San Diego, 1990.
5. H. Achenbach and C. Renner, *Heterocycles* **23**, 2075 (1985).
6. H. Achenbach, *Pure Appl. Chem.* **58**, 653 (1986).
7. H. Achenbach and D. Franke, *Arch. Pharm.* (Weinheim, Ger.). **320**, 91 (1987).
8. M. Somei, T. Funamoto, and T. Ohta, *Heterocycles* **26**, 1783 (1987).

9. F. Delle Monache, G. Delle Monache, M. A. De Moraes, E. Souza, M. D. S. Cavalcanti, and A. Chiappeta, *Gazz. Chim. Ital.* **119**, 435 (1989).
10. M. H. H. Nkunya, J. J. Makangara, and S. A. Jonker, *Nat. Prod. Res.* **18**, 253 (2004).
11. I. Muhammad and C. M. Hasan, *J. Bangladesh Acad. Sci.* **11**, 1 (1987); *Chem. Abstr.* **107**, 172452 (1987).
12. J. J. Makangara, L. Henry, S. A. Jonker, and M. H. H. Nkunya, *Phytochemistry* **65**, 227 (2004).
13. A. T. McPhail, T. S. Wu, T. Ohta, and H. Furukawa, *Tetrahedron Lett.* **24**, 5377 (1983).
14. H. Furukawa, C. Ito, T. S. Wu, and A. T. McPhail, *Chem. Pharm. Bull.* **41**, 1249 (1993).
15. H. Furukuwa, T. S. Wu, and T. Ohta, *Chem. Pharm. Bull.* **31**, 4202 (1983).
16. C. Ito, M. Nakagawa, T. S. Wu, and H. Furukawa, *Chem. Pharm. Bull.* **39**, 2525 (1991).
17. M. T. H. Nutan, C. M. Hasan, and M. A. Rashid, *Fitoterapia* **70**, 130 (1999).
18. H. Furukawa, T. S. Wu, and C. S. Kuoh, *Chem. Pharm. Bull.* **33**, 2611 (1985).
19. C. Ito, T. S. Wu, and H. Furukawa, *Chem. Pharm. Bull.* **36**, 2377 (1988).
20. C. Ito, T. S. Wu, and H. Furukawa, *Chem. Pharm. Bull.* **38**, 1143 (1990).
21. C. Ito and H. Furukawa, *Chem. Pharm. Bull.* **38**, 1548 (1990).
22. C. Ito, Y. Thoyama, M. Omura, I. Kajiura, and H. Furukawa, *Chem. Pharm. Bull.* **41**, 2096 (1993).
23. C. Ito, N. Okahana, T. S. Wu, M. L. Wang, J. S. Lai, C. S. Kuoh, and H. Furukawa, *Chem. Pharm. Bull.* **40**, 230 (1992).
24. C. Ito, T. S. Wu, and H. Furukawa, *Chem. Pharm. Bull.* **35**, 450 (1987).
25. C. Ito and H. Furukawa, *Chem. Pharm. Bull.* **39**, 1355 (1991).
26. T. S. Wu, M. L. Wang, J. S. Lai, C. Ito, and H. Furukawa, *Phytochemistry* **30**, 1052 (1991).
27. E. K. Desoky and D. W. Bishay, *Bull. Facul. Pharm.* (Cairo University) **30**, 231 (1992); *Chem. Abstr.* **119**, 177557 (1992).
28. Y. S. Wang, H. P. He, Y. M. Shen, X. Hong, and X. J. Hao, *J. Nat. Prod.* **66**, 416 (2003).
29. Y. Tachibana, H. Kikuzaki, N. H. Lajis, and N. Nakatani, *J. Agric. Food Chem.* **51**, 6461 (2003).
30. M. M. Rahman and A. I. Gray, *Phytochemistry* **66**, 1601 (2005).
31. V. Kumar, D. B. M. Wickramaratne, and U. Jacobsson, *Tetrahedron Lett.* **31**, 5217 (1990).
32. N. M. Cuong, T. Q. Hung, T. Van Sung, and W. C. Taylor, *Chem. Pharm. Bull.* **52**, 1175 (2004).
33. T. S. Wu, S. C. Huang, and P. L. Wu, *Tetrahedron Lett.* **37**, 7819 (1996).
34. G. Bringmann, S. Tasler, H. Endress, J. Kraus, K. Messer, M. Wohlfarth, and W. Lobin, *J. Am. Chem. Soc.* **123**, 2703 (2001).
35. G. Bringmann, S. Tasler, H. Endress, K. Peters, and E. M. Peters, *Synthesis* 1501 (1998).
36. G. Bringmann and S. Tasler, *Tetrahedron* **57**, 2337 (2001).
37. G. Bringmann, A. Ledermann, M. Stahl, and K. P. Gulden, *Tetrahedron* **51**, 9353 (1995).
38. G. Bringmann, A. Ledermann, and G. Francois, *Heterocycles* **40**, 293 (1995).
39. W. S. Murphy and M. Bertrand, *J. Chem. Soc., Perkin Trans. 1* 4115 (1998).
40. M. Itoigawa, Y. Kashiwada, C. Ito, H. Furukawa, Y. Tachibana, K. F. Bastow, and K. H. Lee, *J. Nat. Prod.* **63**, 893 (2000).
41. Y. C. Kong, K. F. Cheng, R. C. Cambie, and P. G. Waterman, *J. Chem. Soc., Chem. Commun.* 47 (1985).
42. Y. C. Kong, K. H. Ng, K. H. Wat, A. Wong, I. F. Saxena, K. F. Cheng, P. P. H. But, and H. T. Chang, *Planta Med.* 304 (1985).

43. Y. C. Kong, P. P. H. But, H. T. Chang, and W. S. Kan, *Fitoterapia* **57**, 387 (1986).
44. Y. C. Kong, P. P. H. But, K. H. Ng, Q. Li, K. F. Cheng, and P. G. Waterman, *Biochem. Syst. Ecol.* **16**, 485 (1988).
45. Y. C. Kong, K. H. Ng, P. P. H. But, Q. Li, S. X. Yu, H. T. Zhang, K. F. Cheng, D. D. Soejarto, W. S. Kan, and P. G. Waterman, *J. Ethnopharmacol.* **15**, 195 (1986).
46. Y. C. Kong, C. N. Lam, K. F. Cheng, and T. C. W. Mak, *Jiegou Huaxue* **4**, 30 (1985); *Chem. Abstr.* **104**, 31728 (1986).
47. K. F. Cheng, Y. C. Kong, and T. Y. Chan, *J. Chem. Soc., Chem. Commun.* 48 (1985).
48. J. H. Sheu, Y. K. Chen, and Y. L. V. Hong, *Tetrahedron Lett.* **32**, 1045 (1991).
49. J. H. Sheu, Y. K. Chen, and Y. L. V. Hong, *J. Org. Chem.* **58**, 5784 (1993).
50. Y. K. Chen, H. F. Chung, and J. H. Sheu, *Nat. Prod. Lett.* **5**, 225 (1994).
51. J. H. Sheu, Y. K. Chen, H. F. Chung, S. F. Lin, and P. J. Sung, *J. Chem. Soc., Perkin Trans. 1* 1959 (1998).
52. J. H. Sheu, C. A. Chen, and B. H. Chen, *J. Chem. Soc., Chem. Commun.* 203 (1999).
53. E. Wenkert, E. C. Angell, V. F. Ferreira, E. L. Michelotti, S. R. Piettre, J. H. Sheu, and C. S. Swindell, *J. Org. Chem.* **51**, 2343 (1986).
54. E. Wenkert, P. D. R. Moeller, S. R. Piettre, and A. T. McPhail, *J. Org. Chem.* **53**, 3170 (1988).
55. J. Bergman and L. Venemalm, *Tetrahedron Lett.* **29**, 2993 (1988).
56. J. Bergman and L. Venemalm, *Tetrahedron* **48**, 759 (1992).
57. J. P. Kutney, F. J. Lopez, S. P. Huang, and H. Kurobe, *Heterocycles* **28**, 565 (1989).
58. J. P. Kutney, F. J. Lopez, S. P. Huang, H. Kurobe, R. Flogaus, K. Piotrowska, and S. J. Rettig, *Can. J. Chem.* **69**, 949 (1991).
59. K. J. Henry Jr. and P. A. Grieco, *J. Chem. Soc., Chem. Commun.* 510 (1993).
60. M. Ishikura, K. Imaizumi, and N. Katagiri, *Heterocycles* **53**, 553 (2000).
61. M. Ishikura, K. Imaizumi, and N. Katagiri, *Heterocycles* **53**, 2201 (2000).
62. M. Ishikura, *Heterocycles* **41**, 1385 (1995).
63. K. F. Cheng, T. T. Chan, T. F. Lai, and Y. C. Kong, *J. Chem. Soc., Perkin Trans. 1* 3317 (1988).
64. G. A. Cao and K. F. Cheng, *Synth. Commun.* **26**, 1525 (1996).
65. K. F. Cheng, T. Y. Chan, T. T. Wong, and T. F. Lai, *J. Chem. Soc., Perkin Trans. 1* 1555 (1990).
66. K. F. Cheng, K. P. Chan, and T. F. Lai, *J. Chem. Soc., Perkin Trans. 1* 2461 (1991).
67. K. F. Cheng, K. P. Chan, Y. C. Kong, and D. D. Ho, *J. Chem. Soc., Perkin Trans. 1* 2955 (1991).
68. D. D. Ho, C. P. Lau, K. H. Ng, Y. C. Kong, K. F. Cheng, and K. P. Chan, *Eur. J. Pharmacol.* **205**, 209 (1991).
69. K. F. Cheng, G. A. Cao, Y. W. Yu, and Y. C. Kong, *Synth. Commun.* **24**, 65 (1994).
70. K. F. Cheng, M. K. Cheung, and Y. C. Kong, *Aust. J. Chem.* **50**, 349 (1997).
71. K. F. Cheng and M. K. Cheung, *J. Chem. Soc., Perkin Trans. 1* 1213 (1996).
72. W. L. Chan, D. D. Ho, C. P. Lau, K. H. Wat, Y. C. Kong, K. F. Cheng, T. T. Wong, and T. Y. Chan, *Eur. J. Med. Chem.* **26**, 387 (1991).
73. K. F. Cheng, T. T. Wong, K. P. Chan, and Y. C. Kong, *Eur. J. Med. Chem.* **27**, 121 (1992).
74. D. C. C. Wong, W. P. Fong, S. S. T. Lee, Y. C. Kong, K. F. Cheng, and G. Stone, *Eur. J. Pharmacol.* **362**, 87 (1998).
75. P. C. Ng, D. D. Ho, K. H. Ng, Y. C. Kong, K. F. Cheng, and G. Stone, *Eur. J. Pharmacol.* **264**, 1 (1994).
76. T. W. T. Leung, G. Cheng, C. H. Chui, S. K. W. Ho, F. Y. Lau, J. K. J. Tjong, T. C. C. Poon, J. C. O. Tang, W. C. P. Tse, K. F. Cheng, and Y. C. Kong, *Chemotherapy* **46**, 62 (2000).
77. V. U. Khuzhaev, S. F. Aripova, and R. S. Shakirov, *Khim. Prir. Soedin.* 685 (1994); *Chem. Abstr.* **124**, 4912 (1995).

78. V. U. Khuzhaev and S. F. Aripova, *Khim. Prir. Soedin.* 328 (1995); *Chem. Abstr.* **124**, 226550 (1996).
79. I. Zhalolov, V. U. Khuzhaev, B. Tashkhodzhaev, M. G. Levkovich, S. F. Aripova, and N. D. Abdullaev, *Chem. Nat. Compd.* **34**, 706 (1998).
80. I. Z. Zhalolov, B. Tashkhodzhaev, V. U. Khuzhaev, and S. F. Aripova, *Chem. Nat. Compd.* **38**, 83 (2002).
81. I. Z. Zhalolov, V. U. Khuzhaev, M. G. Levkovich, S. F. Aripova, and A. S. Shashkov, *Chem. Nat. Compd.* **38**, 276 (2002).
82. V. U. Khuzhaev, I. Z. Zhalolov, M. G. Levkovich, S. F. Aripova, and A. S. Shashkov, *Chem. Nat. Compd.* **38**, 280 (2002).
83. V. U. Khuzhaev, I. Z. Zhalolov, M. G. Levkovich, and S. F. Aripova, *Russ. Chem. Bull.* **52**, 745 (2003).
84. V. U. Khuzhaev, I. Zhalolov, K. K. Turgunov, B. Tashkhodzhaev, M. G. Levkovich, S. F. Aripova, and A. S. Shashkov, *Chem. Nat. Compd.* **40**, 261 (2004).
85. V. U. Khuzhaev, I. Zhalolov, K. K. Turgunov, B. Tashkhodzhaev, M. G. Levkovich, S. F. Aripova, and A. S. Shashkov, *Chem. Nat. Compd.* **40**, 269 (2004).
86. V. U. Khuzhaev, I. Z. Zhalolov, M. G. Levkovich, S. F. Aripova, and A. S. Shashkov, *Russ. Chem. Bull.* **53**, 1765 (2004).
87. T. Ohmoto and K. Koike, *Chem. Pharm. Bull.* **30**, 1204 (1982).
88. T. Ohmoto and K. Koike, *Chem. Pharm. Bull.* **31**, 3198 (1983).
89. K. Koike, T. Ohmoto, and K. Ogata, *Chem. Pharm. Bull.* **34**, 3228 (1986).
90. K. Koike and T. Ohmoto, *Chem. Pharm. Bull.* **35**, 3305 (1987).
91. K. Koike and T. Ohmoto, *Chem. Pharm. Bull.* **34**, 2090 (1986).
92. K. Koike, T. Ohmoto, and T. Higuchi, *Phytochemistry* **26**, 3375 (1987).
93. T. Ohmoto and K. Koike, *Chem. Pharm. Bull.* **33**, 3847 (1985).
94. T. Ohmoto and K. Koike, *Chem. Pharm. Bull.* **33**, 4901 (1985).
95. K. Koike and T. Ohmoto, *Phytochemistry* **27**, 3029 (1988).
96. M. Yoshikawa, E. Harada, S. Aoki, J. Yamahara, N. Murakami, H. Shibuya, and I. Kitagawa, *Chem. Pharm. Bull.* **41**, 2101 (1993).
97. J. Liu, R. S. Davison, and O. W. Howarth, *Magn. Reson. Chem.* **31**, 1091 (1993).
98. J. S. Yang, D. Q. Yu, and X. T. Liang, *Yaoxue Xuebao* **23**, 267 (1988); *Chem. Abstr.* **109**, 107726 (1988).
99. H. C. Krebs, J. V. Rakotoarimanga, P. Rasoanaivo, F. Frappier, and M. T. Martin, *J. Nat. Prod.* **60**, 1183 (1997).
100. H. Suzuki, Y. Ebihara, Y. Yokoyama, and Y. Murakami, *Heterocycles* **46**, 57 (1997).
101. Y. Adjibade, B. Weniger, J. C. Quirion, B. Kuballa, P. Cabalion, and R. Anton, *Phytochemistry* **31**, 317 (1992).
102. M. Leboeuf, A. Cave, P. Mangeney, and A. Bouquet, *Planta Med.* **41**, 374 (1981).
103. T. Tokuyama and J. W. Daly, *Tetrahedron* **39**, 41 (1983).
104. P. Solis, A. G. Ravelo, J. A. Palenzuela, M. P. Gupta, A. Gonzalez, and J. D. Phillipson, *Phytochemistry* **44**, 963 (1997).
105. L. Verotta, T. Pilati, M. Tato, E. Elisabetsky, T. A. Amador, and D. S. Nunes, *J. Nat. Prod.* **61**, 392 (1998).
106. A. Balayer, T. Sevenet, H. Schaller, A. H. A. Hadi, A. Chiaroni, C. Riche, and M. Pais, *Nat. Prod. Lett.* **2**, 61 (1993).
107. Dachriyanus, M. V. Sargent, and F. S. Wahyuni, *Aust. J. Chem.* **53**, 159 (2000).
108. H. Takayama, Y. Matsuda, K. Masubuchi, A. Ishida, M. Kitajima, and N. Aimi, *Tetrahedron* **60**, 893 (2004).
109. F. Libot, C. Miet, N. Kunesch, J. E. Poisson, J. Pusset, and T. Sevenet, *J. Nat. Prod.* **50**, 468 (1987).
110. F. Gueritte-Voegelein, T. Sevenet, J. Pusset, M. T. Adeline, B. Gillet, J. C. Beloeil, D. Guenard, and P. Potier, *J. Nat. Prod.* **55**, 923 (1992).
111. F. Libot, N. Kunesch, J. Poisson, M. Kaiser, and H. Duddeck, *Heterocycles* **27**, 2381 (1988).

112. V. Jannic, F. Gueritte, O. Laprevote, L. Serani, M. T. Martin, T. Sevenet, and P. Potier, *J. Nat. Prod.* **62**, 838 (1999).
113. R. K. Duke, R. D. Allan, G. A. R. Johnston, K. N. Mewett, A. D. Mitrovic, C. C. Duke, and T. W. Hambley, *J. Nat. Prod.* **58**, 1200 (1995).
114. N. H. Lajis, Z. Mahmud, and R. F. Toia, *Planta Med.* **59**, 383 (1993).
115. H. Takayama, I. Mori, M. Kitajima, N. Aimi, and N. H. Lajis, *Org. Lett.* **6**, 2945 (2004).
116. Y. Adjibade, H. Saad, B. Kuballa, J. P. Beck, T. Sevenet, P. Cabalion, and R. Anton, *J. Ethnopharmacol.* **29**, 127 (1990).
117. Y. Adjibade, H. Saad, T. Sevenet, B. Kuballa, J. C. Quirion, and R. Anton, *Planta Med.* **56**, 212 (1990).
118. H. E. A. Saad, S. H. El-Sharkawy, and W. T. Shier, *Planta Med.* **61**, 313 (1995).
119. A. Roth, B. Kuballa, P. Cabalion, and R. Anton, *Planta Med.* 289 (1985).
120. C. L. Fang, S. Horne, N. Taylor, and R. Rodrigo, *J. Am. Chem. Soc.* **116**, 9480 (1994).
121. H. Ishikawa, H. Takayama, and N. Aimi, *Tetrahedron Lett.* **43**, 5637 (2002).
122. L. Verotta, F. Orsini, M. Sbacchi, M. A. Scheildler, T. A. Amador, and E. Elisabetsky, *Bioorg. Med. Chem.* **10**, 2133 (2002).
123. J. T. Link and L. E. Overman, *J. Am. Chem. Soc.* **118**, 8166 (1996).
124. L. E. Overman, D. V. Paone, and B. A. Stearns, *J. Am. Chem. Soc.* **121**, 7702 (1999).
125. L. E. Overman, J. F. Larrow, B. A. Stearns, and J. M. Vance, *Angew. Chem. Int. Ed.* **39**, 213 (2000).
126. A. D. Lebsack, J. T. Link, L. E. Overman, and B. A. Stearns, *J. Am. Chem. Soc.* **124**, 9008 (2002).
127. L. E. Overman and E. A. Peterson, *Angew. Chem. Int. Ed.* **42**, 2525 (2003).
128. L. E. Overman and E. A. Peterson, *Tetrahedron* **59**, 6905 (2003).
129. J. J. Kodanko and L. E. Overman, *Angew. Chem. Int. Ed.* **42**, 2528 (2003).
130. L. Verotta, F. Peterlongo, E. Elisabetsky, T. A. Amador, and D. S. Nunes, *J. Chromatogr. A* **841**, 165 (1999).
131. R. Rasolonjanahary, T. Sevenet, F. Gueritte Voegelein, and C. Kordon, *Eur. J. Pharmacol.* **285**, 19 (1995).
132. E. Elisabetsky, T. A. Amador, R. R. Albuquerque, D. S. Nunes, and A. C. T. Carvalho, *J. Ethnopharmacol.* **48**, 77 (1995).
133. T. A. Amador, L. Verotta, D. S. Nunes, and E. Elisabetsky, *Planta Med.* **66**, 770 (2000).
134. A. Roth, B. Kuballa, C. Bounthanh, P. Cabalion, T. Sevenet, J. P. Beck, and R. Anton, *Planta Med.* **52**, 450 (1986).
135. Y. Adjibade, B. Kuballa, P. Cabalion, M. L. Jung, J. P. Beck, and R. Anton, *Planta Med.* **55**, 567 (1989).
136. A. Beretz, A. Roth-Georger, G. Corre, B. Kuballa, R. Anton, and J. P. Cazenave, *Planta Med.* 300 (1985).
137. A. M. Balde, L. A. Pieters, A. Gergely, V. Wray, M. Claeys, and A. J. Vlietinck, *Phytochemistry* **30**, 997 (1991).
138. J. C. Quirion, C. Kan, I. R. C. Bick, and H. P. Husson, *J. Org. Chem.* **52**, 4527 (1987).
139. J. C. Quirion, H. P. Husson, C. Kan, and I. R. C. Bick, *Nat. Prod. Lett.* **2**, 41 (1993).
140. J. C. Quirion, H. P. Husson, C. Kan, O. Laprevote, A. Chiaroni, C. Riche, S. Burkard, H. J. Borschberg, and I. R. C. Bick, *J. Org. Chem.* **57**, 5848 (1992).
141. F. Tillequin, M. Koch, and A. Rabaron, *J. Nat. Prod.* **48**, 120 (1985).
142. M. Caprasse, D. Tavernier, and L. Angenot, *J. Pharm. Belg.* **38**, 211 (1983).
143. M. Frederich, M. Tits, M. P. Hayette, V. Brandt, J. Penelle, P. DeMol, G. Llabres, and L. Angenot, *J. Nat. Prod.* **62**, 619 (1999).
144. G. Massiot, P. Thepenier, M. J. Jacquier, J. Lounkokobi, C. Mirand, M. Zeches, and L. Le Men-Olivier, *Tetrahedron* **39**, 3645 (1983).

145. G. M. T. Robert, A. Ahond, C. Poupat, P. Potier, C. Jolles, A. Jousselin, and H. Jacquemin, *J. Nat. Prod.* **46**, 694 (1983).
146. C. A. Coune, L. J. G. Angenot, and J. Denoel, *Phytochemistry* **19**, 2009 (1980).
147. C. Coune, M. Tits, and L. Angenot, *J. Pharm. Belg.* **37**, 189 (1982).
148. M. Nicoletti, J. U. Oguakwa, and I. Messana, *Fitoterapia* **51**, 131 (1980).
149. C. Mirand-Richard, L. Le Men-Olivier, J. Levy, and J. Le Men, *Heterocycles* **12**, 1409 (1979).
150. K. Yamada, K. Aoki, and D. Uemura, *J. Org. Chem.* **40**, 2572 (1975).
151. R. Verpoorte, G. Massiot, M. J. Jacquier, P. Thepenier, and L. Le Men-Olivier, *Tetrahedron Lett.* **27**, 239 (1986).
152. M. F. S. Marques, L. Kato, H. F. Leitao Filho, and F. D. A. M. Reis, *Phytochemistry* **41**, 963 (1996).
153. M. Azoug, A. Loukaci, B. Richard, J. M. Nuzillard, C. Moreti, M. Zeches-Hanrot, and L. Le Men-Olivier, *Phytochemistry* **39**, 1223 (1995).
154. T. S. Kam and K. M. Sim, *Phytochemistry* **47**, 145 (1998).
155. T. S. Kam, K. M. Sim, T. Koyano, and K. Komiyama, *Phytochemistry* **50**, 75 (1999).
156. M. Zeches, F. Sigaut, L. Le Men-Olivier, J. Levy, and J. Le Men, *Bull. Soc. Chim. Fr.* 75 (1981).
157. P. J. Guilhem, *Acta Crystallogr.* **B30**, 742 (1974).
158. M. Zeches, B. Richard, P. Thepenier, L. Le Men-Olivier, and J. Le Men, *Phytochemistry* **19**, 2451 (1980).
159. L. Dupont, J. Lamotte-Brasseur, O. Dideberg, H. Campsteyn, M. Vermeire, and L. Angenot, *Acta Crystallogr.* **B33**, 1801 (1977).
160. D. Tavernier, W. Zhang, L. Angenot, M. Chierici-Tits, and J. Leclercq, *Phytochemistry* **26**, 557 (1987).
161. M. Frederich, A. Cristino, Y. H. Choi, R. Verpoorte, M. Tits, L. Angenot, E. Prost, J. M. Nuzillard, and M. Zeches-Hanrot, *Planta Med.* **70**, 72 (2004).
162. C. Mirand, L. Le Men-Olivier, J. Le Men, and C. Delaude, *Phytochemistry* **22**, 577 (1983).
163. G. Massiot, P. Thepenier, M. J. Jacquier, C. Delaude, L. Le Men-Olivier, and R. Verpoorte, *Tetrahedron Lett.* **26**, 2441 (1985).
164. G. Massiot, P. Thepenier, M. J. Jacquier, L. Le Men-Olivier, R. Verpoorte, and C. Delaude, *Phytochemistry* **26**, 2839 (1987).
165. L. Angenot, *Plant Med. Phytother.* **12**, 123 (1978).
166. A. Lerchner and E. M. Carreira, *J. Am. Chem. Soc.* **124**, 14826 (2002).
167. A. T. Keene, J. D. Phillipson, D. C. Warhurst, M. Koch, and E. Sequin, *Planta Med.* **53**, 201 (1987).
168. J. D. Phillipson and C. W. Wright, *Planta Med.* **57**, S53 (1991).
169. C. W. Wright, D. H. Bray, M. J. O'Neill, D. C. Warhurst, J. D. Phillipson, J. Quetin-Leclercq, and L. Angenot, *Planta Med.* **57**, 337 (1991).
170. C. W. Wright, D. Allen, Y. Cai, Z. Chen, J. D. Phillipson, G. C. Kirby, D. C. Warhurst, M. Tits, and L. Angenot, *Phytother. Res.* **8**, 149 (1994).
171. M. Frederich, M. P. Hayette, M. Tits, P. De Mol, and L. Angenot, *Antimicrob. Agents Chemother.* **43**, 2328 (1999).
172. M. Frederich, M. J. Jacquier, P. Thepenier, P. De Mol, M. Tits, G. Philippe, C. Delaude, L. Angenot, and M. Zeches-Hanrot, *J. Nat. Prod.* **65**, 1381 (2002).
173. M. Frederich, M. Tits, E. Goffin, G. Philippe, P. Grellier, P. De Mol, M. P. Hayette, and L. Angenot, *Planta Med.* **70**, 520 (2004).
174. C. Caron, M. J. Hoizey, L. Le Men-Olivier, G. Massiot, M. Zeches, C. Choisy, E. Le Magrex, and R. Verpoorte, *Planta Med.* **54**, 409 (1988).
175. J. Leclercq, M. C. De Pauw-Gillet, R. Bassleer, and L. Angenot, *J. Ethnopharmacol.* **15**, 305 (1986).
176. R. Bassleer, M. C. Depauw-Gillet, B. Massart, J. M. Marnette, P. Wiliquet, M. Caprasse, and L. Angenot, *Planta Med.* **45**, 123 (1982).

177. M. Tits, C. Desaive, J. M. Marnette, R. Bassleer, and L. Angenot, *J. Ethnopharmacol.* **12**, 287 (1984).
178. J. Quetin-Leclercq, M. C. De Pauw-Gillet, L. Angenot, and R. Bassleer, *Chem. Biol. Interact.* **80**, 203 (1991).
179. K. A. P. M. Bonjean, M. C. A. J. De Pauw-Gillet, J. Quetin-Leclercq, L. Angenot, and R. J. B. Bassleer, *Anticancer Res.* **16**, 1129 (1996).
180. J. Quetin-Leclercq, B. Bouzahzah, A. Pons, R. Greimers, L. Angenot, R. Bassleer, and H. Barbason, *Planta Med.* **59**, 59 (1993).
181. M. Frederich, M. Bentires-Alj, M. Tits, L. Angenot, R. Greimers, J. Gielen, V. Bours, and M. P. Merville, *J. Pharmacol. Exp. Ther.* **304**, 1103 (2003).
182. L. Dassonneville, N. Wattez, C. Mahieu, P. Colson, C. Houssier, M. Frederich, M. Tits, L. Angenot, and C. Bailly, *Anticancer Res.* **19**, 5245 (1999).
183. T. S. Kam, K. H. Lee, and S. H. Goh, *Phytochemistry* **30**, 3441 (1991).
184. T. S. Kam, K. H. Lee, and S. H. Goh, *Phytochemistry* **31**, 2031 (1992).
185. S. H. Goh and S. A. A. Junan, *Phytochemistry* **24**, 880 (1985).
186. S. H. Goh, E. Soepadmo, P. Chang, S. Aishah, Y. K. Koh, A. Nasrulhaq, C. E. Taylor, and A. K. Wong, *Malaysian J. Sci.* **8,** 109 (1986); *Chem. Abstr.* **107**, 214793 (1987).
187. A. Arnone, G. Nasini, and L. Merlini, *J. Chem. Soc. Perkin Trans. 1* 571 (1987).
188. G. Massiot, B. Massoussa, M. J. Jacquier, P. Thepenier, L. Le Men-Olivier, C. Delaude, and R. Verpoorte, *Phytochemistry* **27**, 3293 (1988).
189. G. Massiot, M. Zeches, C. Mirand, L. Le Men-Olivier, C. Delaude, K. H. C. Baser, R. Bavovada, N. G. Bisset, P. J. Hylands, J. Strombom, and R. Verpoorte, *J. Org. Chem.* **48**, 1869 (1983).
190. G. Massiot, P. Thepenier, M. J. Jacquier, L. Le Men-Olivier, and C. Delaude, *Phytochemistry* **31**, 2873 (1992).
191. R. Verpoorte, J. Aadewiel, J. Strombom, and A. Baerheim Svendsen, *J. Ethnopharmacol.* **10**, 243 (1984).
192. M. Frederich, J. Quetin-Leclercq, R. G. Biala, V. Brandt, J. Penelle, M. Tits, and L. Angenot, *Phytochemistry* **48**, 1263 (1998).
193. M. Caprasse, C. Coune, and L. Angenot, *J. Pharm. Belg.* **38**, 135 (1983).
194. L. Angenot, M. Dubois, C. Ginion, W. Van Dorsser, and A. Dresse, *Arch. Int. Pharmacodyn. Ther.* **215**, 246 (1975); *Chem. Abstr.* **84**, 130147 (1976).
195. M. Caprasse, L. Angenot, D. Tavernier, and M. J. O. Anteunis, *Planta Med.* **50**, 131 (1984).
196. R. G. Biala, M. Tits, J. Penelle, M. Frederich, V. Brandt, C. Prosperi, G. Llabres, and L. Angenot, *J. Nat. Prod.* **61**, 139 (1998).
197. J. Quetin-Leclercq, G. Llabres, R. Warin, M. L. Belem-Pinheiro, H. Mavar-Manga, and L. Angenot, *Phytochemistry* **40**, 1557 (1995).
198. J. Penelle, M. Tits, P. Christen, V. Brandt, M. Frederich, and L. Angenot, *J. Nat. Prod.* **62**, 898 (1999).
199. J. Penelle, M. Tits, P. Christen, J. Molgo, V. Brandt, M. Frederich, and L. Angenot, *Phytochemistry* **53**, 1057 (2000).
200. J. Penelle, P. Christen, J. Molgo, M. Tits, V. Brandt, M. Frederich, and L. Angenot, *Phytochemistry* **58**, 619 (2001).
201. P. Wins, I. Margineanu, J. Penelle, L. Angenot, T. Grisar, and L. Bettendorff, *Naunyn-Schmiedeberg's Arch. Pharmacol.* **367**, 253 (2003).
202. C. L. C. De Medeiros, G. Thomas, and R. Mukherjee, *Phytother. Res.* **5**, 24 (1991).
203. G. Baudouin, F. Tillequin, M. Bert, and M. Koch, *J. Chem. Soc., Chem. Commun.* 3 (1986).
204. S. Subhadhirasakul, H. Takayama, Y. Miyabe, M. Kitajima, D. Ponglux, S. I. Sakai, and N. Aimi, *Heterocycles* **41**, 2049 (1995).
205. S. Sakai, N. Aimi, K. Yamaguchi, E. Yamanaka, and J. Haginiwa, *Tetrahedron Lett.* 719 (1975).

206. S. Sakai, N. Aimi, K. Yamaguchi, E. Yamanaka, and J. Haginiwa, *J. Chem. Soc., Perkin Trans. 1* 1257 (1982).
207. J. V. Silverton and T. Akiyama, *J. Chem. Soc., Perkin Trans. 1* 1263 (1982).
208. M. Lin, B. Yang, and D. Yu, *Yaoxue Xuebao* **21**, 114 (1986); *Chem. Abstr.* **104**, 221983 (1986).
209. A. M. Bui, B. C. Das, E. Guittet, and V. Stoven, *Heterocycles* **38**, 1025 (1994).
210. T. S. Kam and K. M. Sim, *Helv. Chim. Acta* **85**, 1027 (2002).
211. T. S. Kam, K. M. Sim, and H. S. Pang, *J. Nat. Prod.* **66**, 11 (2003).
212. V. Munoz, C. Moretti, M. Sauvain, C. Caron, A. Porzel, G. Massiot, B. Richard, and L. Le Men-Olivier, *Planta Med.* **60**, 455 (1994).
213. J. M. Nuzillard, T. M. Pinchon, C. Caron, G. Massiot, and L. Le Men-Olivier, *C. R. Acad. Sci., Ser. II* **309**, 195 (1989).
214. T. M. Pinchon, J. M. Nuzillard, B. Richard, G. Massiot, L. Le Men-Olivier, and T. Sevenet, *Phytochemistry* **29**, 3341 (1990).
215. G. Massiot, J. M. Nuzillard, B. Richard, and L. Le Men-Olivier, *Tetrahedron Lett.* **31**, 2883 (1990).
216. H. Takayama, S. Subhadhirasakul, J. Mizuki, M. Kitajima, N. Aimi, D. Ponglux, and S. Sakai, *Chem. Pharm. Bull.* **42**, 1957 (1994).
217. K. Ghedira, M. Zeches-Hanrot, B. Richard, G. Massiot, L. Le Men-Olivier, T. Sevenet, and S. H. Goh, *Phytochemistry* **27**, 3955 (1988).
218. M. Lin, B. Yang, D. Yu, X. Lin, and Y. Zhang, *Yaoxue Xuebao* **22**, 833 (1987); *Chem. Abstr.* **108**, 164701 (1988).
219. X. Lin, Q. Zheng, Y. Zhang, M. Lin, D. Yu, and X. Liu, *Jiegou Huaxue* **6**, 89 (1987); *Chem. Abstr.* **108**, 167761 (1988).
220. N. Keawpradub and P. J. Houghton, *Phytochemistry* **46**, 757 (1997).
221. N. Keawpradub, P. J. Houghton, E. Eno-Amooquaye, and P. J. Burke, *Planta Med.* **63**, 97 (1997).
222. T. S. Kam, Y. M. Choo, and K. Komiyama, *Tetrahedron* **60**, 3957 (2004).
223. D. E. Burke, J. M. Cook, and P. W. LeQuesne, *J. Am. Chem. Soc.* **95**, 546 (1973).
224. L. K. Hamaker and J. M. Cook, *in* "Alkaloids: Chemical and Biological Perspectives", (S. W. Pelletier, ed.), vol. 9, pp. 23–25. Elsevier, New York, 1995.
225. Y. Bi, J. M. Cook, and P. W. LeQuesne, *Tetrahedron Lett.* **35**, 3877 (1994).
226. Y. Bi and J. M. Cook, *Tetrahedron Lett.* **34**, 4501 (1993).
227. Y. Bi, L. H. Zhang, L. K. Hamaker, and J. M. Cook, *J. Am. Chem. Soc.* **116**, 9027 (1994).
228. T. Gan and J. M. Cook, *Tetrahedron Lett.* **37**, 5033 (1996).
229. T. Gan and J. M. Cook, *Tetrahedron Lett.* **37**, 5037 (1996).
230. T. Gan and J. M. Cook, *J. Org. Chem.* **63**, 1478 (1998).
231. F. Mayerl and M. Hesse, *Helv. Chim. Acta* **61**, 337 (1978).
232. H. Takayama, C. Phisalaphong, M. Kitajima, N. Aimi, and S. Sakai, *Tetrahedron* **47**, 1383 (1991).
233. P. Yu, T. Wang, J. Li, and J. M. Cook, *J. Org. Chem.* **65**, 3173 (2000).
234. D. E. Burke, C. A. DeMarkey, P. W. LeQuesne, and J. M. Cook, *J. Chem. Soc., Chem. Commun.* 1346 (1972).
235. D. E. Burke, G. A. Cook, J. M. Cook, K. G. Haller, H. A. Lazar, and P. W. LeQuesne, *Phytochemistry* **12**, 1467 (1973).
236. R. L. Garnick and P. W. LeQuesne, *J. Am. Chem. Soc.* **100**, 4213 (1978).
237. S. Zhao, X. Liao, and J. M. Cook, *Org. Lett.* **4**, 687 (2002).
238. S. Zhao, X. Liao, T. Wang, J. Flippen-Anderson, and J. M. Cook, *J. Org. Chem.* **68**, 6279 (2003).
239. X. Liu, J. R. Deschamp, and J. M. Cook, *Org. Lett.* **4**, 3339 (2002).
240. C. W. Wright, D. Allen, Y. Cai, J. D. Phillipson, I. M. Said, G. C. Kirby, and D. C. Warhurst, *Phytother. Res.* **6**, 121 (1992).

241. N. Keawpradub, G. C. Kirby, J. C. P. Steele, and P. J. Houghton, *Planta Med.* **65**, 690 (1999).
242. C. W. Wright, D. Allen, J. D. Phillipson, G. C. Kirby, D. C. Warhurst, G. Massiot, and L. Le Men-Olivier, *J. Ethnopharmacol.* **40**, 41 (1993).
243. N. Keawpradub, E. Eno-Amooquaye, P. J. Burke, and P. J. Houghton, *Planta Med.* **65**, 311 (1999).
244. A. M. Morfaux, P. Mouton, G. Massiot, and L. Le Men-Olivier, *Phytochemistry* **29**, 3345 (1990).
245. A. M. Morfaux, P. Mouton, G. Massiot, and L. Le-Men Olivier, *Phytochemistry* **31**, 1079 (1992).
246. A. Chatterjee, A. K. Ghosh, and E. W. Hagaman, *J. Org. Chem.* **47**, 1732 (1982).
247. G. Massiot, J. Vercauteren, B. Richard, M. J. Jacquier, and L. Le Men-Olivier, *C. R. Acad. Sci., Ser. II* **294**, 579 (1982).
248. S. Subhadhirasakul, H. Takayama, N. Aimi, D. Ponglux, and S. Sakai, *Chem. Pharm. Bull.* **42**, 1427 (1994).
249. M. J. Jacquier, J. Vercauteren, G. Massiot, L. Le Men-Olivier, J. Pusset, and T. Sevenet, *Phytochemistry* **21**, 2973 (1982).
250. C. Caron, Y. Yachaoui, G. Massiot, L. Le Men-Olivier, J. Pusset, and T. Sevenet, *Phytochemistry* **23**, 2355 (1984).
251. A. Jossang, P. Fodor, and B. Bodo, *J. Org. Chem.* **63**, 7162 (1998).
252. G. Massiot, J. Vercauteren, M. J. Jacquier, J. Levy, and L. Le Men-Olivier, *C. R. Acad. Sci., Ser. II* **292**, 191 (1981).
253. D. Guillaume, A. M. Morfaux, B. Richard, G. Massiot, L. Le Men-Olivier, J. Pusset, and T. Sevenet, *Phytochemistry* **23**, 2407 (1984).
254. N. Kunesch, A. Cave, E. W. Hagaman, and E. Wenkert, *Tetrahedron Lett.* **21**, 1727 (1980).
255. A. Cave, N. Kunesch, J. Bruneton, R. Goutarel, and G. P. Wannigama, *J. Nat. Prod.* **50**, 1178 (1987).
256. Atta-ur-Rahman, A. Pervin, A. Muzaffar, K. T. D. DeSilva, and W. S. J. Silva, *Heterocycles* **27**, 2051 (1988).
257. Atta-ur-Rahman, A. Pervin, I. Ali, A. Muzaffar, K. T. D. DeSilva, and W. S. J. Silva, *Planta Med.* **54**, 37 (1988).
258. A. A. Gorman, N. J. Dastoor, M. Hesse, W. Von Philipsborn, U. Renner, and H. Schmid, *Helv. Chim. Acta* **52**, 33 (1969).
259. R. Mukherjee, B. A. Da Silva, B. C. Das, P. A. Keifer, and J. N. Shoolery, *Heterocycles* **32**, 985 (1991).
260. G. Massiot, B. Massoussa, P. Thepenier, M. J. Jacquier, L. Le Men-Olivier, and C. Delaude, *Heterocycles* **20**, 2339 (1983).
261. P. Thepenier, M. J. Jacquier, G. Massiot, L. Le Men-Olivier, and C. Delaude, *Phytochemistry* **23**, 2659 (1984).
262. R. Verpoorte, M. J. Verzijil, and A. Baerheim Svendsen, *Planta Med.* **44**, 21 (1982).
263. R. Verpoorte and A. Baerheim Svendsen, *Lloydia* **39**, 357 (1976).
264. R. Mukherjee, B. C. Das, P. A. Keifer, and J. N. Shoolery, *Heterocycles* **38**, 1965 (1994).
265. M. J. G. Tits and L. Angenot, *Planta Med.* **34**, 57 (1978).
266. M. Tits, D. Tavernier, and L. Angenot, *Phytochemistry* **18**, 515 (1979).
267. M. Tits, L. Angenot, and D. Tavernier, *J. Nat. Prod.* **46**, 638 (1983).
268. J. Lamotte, L. Dupont, O. Dideberg, K. Kambu, and L. Angenot, *Tetrahedron Lett.* 4227 (1979).
269. K. Kambu, C. Coune, and L. Angenot, *Planta Med.* **37**, 161 (1979).
270. M. Frederich, M. C. De Pauw, G. Llabres, M. Tits, M. P. Hayette, V. Brandt, J. Penelle, P. De Mol, and L. Angenot, *Planta Med.* **66**, 262 (2000).
271. M. Frederich, M. C. De Pauw, C. Prosperi, M. Tits, V. Brandt, J. Penelle, M. P. Hayette, P. DeMol, and L. Angenot, *J. Nat. Prod.* **64**, 12 (2001).

272. G. Philippe, P. DeMol, M. Zeches-Hanrot, J. M. Nuzillard, M. H. Tits, L. Angenot, and M. Frederich, *Phytochemistry* **62**, 623 (2003).
273. G. Philippe, E. Prost, J. M. Nuzillard, M. Zeches-Hanrot, M. Tits, L. Angenot, and M. Frederich, *Tetrahedron Lett.* **43**, 3387 (2002).
274. J. M. Nuzillard, P. Thepenier, M. J. Jacquier, G. Massiot, L. Le Men-Olivier, and C. Delaude, *Phytochemistry* **43**, 897 (1996).
275. A. Lansiaux, C. Bailly, C. Houssier, P. Colson, M. C. De Pauw-Gillet, M. Frederich, M. Tits, and L. Angenot, *Planta Med.* **68**, 591 (2002).
276. M. D. F. F. Melo, C. A. D. M. Santos, A. D. A. Chiappeta, J. F. D. Mello, and R. Mukherjee, *J. Ethnopharmacol.* **19**, 319 (1987).
277. G. Thomas, M. D. F. EMelo Diniz, and R. Mukherjee, *Phytother. Res.* **6**, 84 (1992).
278. B. A. Da Silva, A. P. De Araujo Filho, R. Mukherjee, and A. D. A. Chiappeta, *Phytother. Res.* **7**, 419 (1993).
279. R. Verpoorte, E. Kos-Kuyck, A. Tjin A Tsoi, C. L. M. Ruigrok, G. De Jong, and A. Baerheim Svendsen, *Planta Med.* **48**, 283 (1983).
280. R. Verpoorte, C. L. M. Ruigrok, and A. Baerheim Svendsen, *Planta Med.* **46**, 149 (1982).
281. Atta-ur-Rahman, K. Zaman, Habib-ur-Rahman, and S. Malik, *J. Nat. Prod.* **49**, 1138 (1986).
282. Atta-ur-Rahman and K. Zaman, *Phytochemistry* **27**, 1926 (1988).
283. Atta-ur-Rahman and K. Zaman, *Phytochemistry* **28**, 1788 (1989).
284. S. Mukhopadhyay, G. A. Handy, S. Funayama, and G. A. Cordell, *J. Nat. Prod.* **44**, 696 (1981).
285. T. A. Van Beek, R. Verpoorte, and A. Baerheim Svendsen, *Tetrahedron Lett.* **25**, 2057 (1984).
286. T. A. Van Beek, R. Verpoorte, A. Baerheim Svendsen, and R. Fokkens, *J. Nat. Prod.* **48**, 400 (1985).
287. T. A. Van Beek, R. Verpoorte, and P. Q. Kinh, *Planta Med.* 277 (1985).
288. T. A. Van Beek, P. P. Lankhorst, R. Verpoorte, G. Massiot, R. Fokkens, C. Erkelens, P. Perera, and C. Tibell, *Z. Naturforsch.* **40B**, 693 (1985).
289. S. Michel, F. Tillequin, and M. Koch, *J. Chem. Soc., Chem. Commun.* 229 (1987).
290. X. Z. Feng, C. Kan, H. P. Husson, P. Potier, S. K. Kan, and M. Lounasmaa, *J. Nat. Prod.* **44**, 670 (1981).
291. X. Z. Feng, G. Liu, C. Kan, P. Potier, and S. K. Kan, *J. Nat. Prod.* **52**, 928 (1989).
292. A. T. Henriques, A. A. Melo, P. R. H. Moreno, L. L. Ene, J. A. P. Henriques, and E. E. S. Schapoval, *J. Ethnopharmacol.* **50**, 19 (1996).
293. T. S. Kam and K. M. Sim, *Phytochemistry* **63**, 625 (2003).
294. E. Wenkert, D. W. Cochran, H. E. Gottlieb, E. W. Hagaman, R. B. Filho, F. J. D. A. Matos, and M. I. L. M. Madruga, *Helv. Chim. Acta* **59**, 2437 (1976).
295. P. Clivio, B. Richard, J. R. Deverre, T. Sevenet, M. Zeches, and L. Le Men-Olivier, *Phytochemistry* **30**, 3785 (1991).
296. L. Huang, Y. Zhou, C. Li, and C. Wang, *Zhongcaoyao* **28**, 451 (1997); *Chem. Abstr.* **128**, 228465 (1998).
297. P. Perera, F. Sandberg, T. A. Van Beek, and R. Verpoorte, *Phytochemistry* **24**, 2097 (1985).
298. T. A. Van Beek, F. L. C. Kuijlaars, P. H. A. M. Thomassen, R. Verpoorte, and A. Baerheim Svendsen, *Phytochemistry* **23**, 1771 (1984).
299. H. Takayama, S. Suda, I. S. Chen, M. Kitajima, N. Aimi, and S. Sakai, *Chem. Pharm. Bull.* **42**, 280 (1994).
300. T. S. Kam and K. M. Sim, *Heterocycles* **57**, 2137 (2002).
301. E. Bombardelli, A. Bonati, B. Gabetta, E. M. Martinelli, G. Mustich, and B. Danieli, *J. Chem. Soc., Perkin Trans. 1* 1432 (1976).
302. M. Urrea, A. Ahond, A. M. Bui, and P. Potier, *Bull. Soc. Chim. Fr.* 147 (1981).

303. R. Van der Heijden, R. L. Brouwer, R. Verpoorte, T. A. Van Beek, P. A. A. Harkes, and A. Baerheim Svendsen, *Planta Med.* 144 (1986).
304. R. Van der Heijden, R. L. Brouwer, R. Verpoorte, R. Wijnsma, T. A. Van Beek, P. A. A. Harkes, and A. B. Svendsen, *Phytochemistry* **25**, 843 (1986).
305. T. S. Kam, K. M. Sim, T. Koyano, M. Toyoshima, M. Hayashi, and K. Komiyama, *Bioorg. Med. Chem. Lett.* **8**, 1693 (1998).
306. T. S. Kam and K. M. Sim, *Helv. Chim. Acta* **86**, 122 (2003).
307. T. S. Kam and K. M. Sim, *Heterocycles* **55**, 2405 (2001).
308. M. Bert, G. Baudouin, F. Tillequin, and M. Koch, *Heterocycles* **23**, 2505 (1985).
309. T. S. Kam and H. S. Pang, *Heterocycles* **63**, 845 (2004).
310. M. You, X. Ma, R. Mukherjee, N. R. Farnsworth, G. A. Cordell, A. D. Kinghorn, and J. M. Pezzuto, *J. Nat. Prod.* **57**, 1517 (1994).
311. E. Federici, G. Palazzino, M. Nicoletti, and C. Galeffi, *Planta Med.* **66**, 93 (2000).
312. D. Ramanitrahasimbola, P. Rasoanaivo, S. Ratsimamanga-Urverg, E. Federici, G. Palazzino, C. Galeffi, and M. Nicoletti, *Phytother. Res.* **15**, 30 (2001).
313. M. Damak, C. Poupat, and A. Ahond, *Tetrahedron Lett.* 3531 (1976).
314. E. Valencia, A. Madinaveitia, J. Bermejo, and A. G. Gonzalez, *J. Nat. Prod.* **58**, 134 (1995).
315. I. Chardon-Loriaux, M. M. Debray, and H. P. Husson, *Phytochemistry* **17**, 1605 (1978).
316. H. Ripperger, C. Kamperdick, G. Adam, T. P. Lien, and T. Van Sung, *J. Prakt. Chem.* **341**, 506 (1999).
317. T. P. Lien, T. Van Sung, H. Ripperger, C. Kamperdick, and G. Adam, *J. Prakt. Chem.* **342**, 725 (2000).
318. B. Proksa, D. Uhrin, E. Grossmann, and Z. Voticky, *Tetrahedron Lett.* **27**, 5413 (1986).
319. B. Proksa, D. Uhrin, E. Grossmann, Z. Voticky, and J. Fuska, *Planta Med.* **54**, 214 (1988).
320. B. Proksa, D. Uhrin, E. Grossmann, and Z. Voticky, *Planta Med.* **53**, 120 (1987).
321. C. Kan-Fan, G. Massiot, B. C. Das, and P. Potier, *J. Org. Chem.* **46**, 1481 (1981).
322. M. Damak, A. Ahond, and P. Potier, *Bull. Soc. Chim. Fr.* 213 (1981).
323. J. Abaul, E. Philogene, P. Bourgeois, A. Ahond, C. Poupat, and P. Potier, *J. Nat. Prod.* **52**, 1279 (1989).
324. J. Abaul, P. Bourgeois, E. Philogene, M. Damak, A. Ahond, C. Poupat, and P. Potier, *C. R. Acad. Sci., Ser. II* **298**, 627 (1984).
325. R. Torrenegra, J. A. Pedrozo P., H. Achenbach, and P. Bauereib, *Phytochemistry* **27**, 1843 (1988).
326. T. R. Govindachari, G. Sandhya, S. Chandrasekharan, and K. Rajagopalan, *J. Chem. Soc., Chem. Commun.* 1137 (1987).
327. T. P. Lien, C. Kamperdick, T. Van Sung, G. Adam, and H. Ripperger, *Phytochemistry* **49**, 1797 (1998).
328. G. Massiot, B. Richard, L. Le Men-Olivier, J. De Graeve, and C. Delaude, *Phytochemistry* **27**, 1085 (1988).
329. Y. L. He, W. M. Chen, and X. Z. Feng, *J. Nat. Prod.* **57**, 411 (1994).
330. A. M. Bui, B. C. Das, E. Guittet, J. Y. Lallemand, and P. Potier, *J. Nat. Prod.* **49**, 321 (1986).
331. A. M. Bui, B. C. Das, E. Guittet, and P. Potier, *J. Nat. Prod.* **54**, 514 (1991).
332. T. S. Kam, K. Y. Loh, L. H. Lim, W. L. Loong, C. H. Chuah, and W. Chen, *Tetrahedron Lett.* **33**, 969 (1992).
333. T. S. Kam, K. Y. Loh, and W. Chen, *J. Nat. Prod.* **56**, 1865 (1993).
334. T. S. Kam, H. S. Pang, and T. M. Lim, *Org. Biomol. Chem.* **1**, 1292 (2003).
335. A. Cavé, J. Bruneton, A. Ahond, A. M. Bui, H. P. Husson, C. Kan, G. Lukacs, and P. Potier, *Tetrahedron Lett.* 5081 (1973).
336. H. Achenbach, R. Waibel, and M. Zwanzger, *Phytochemistry* **37**, 1737 (1994).

337. K. Umezawa, T. Ohse, T. Yamamoto, T. Koyano, and Y. Takahashi, *Anticancer Res.* **14**, 2413 (1994).
338. P. Clivio, D. Guillaume, J. Vercauteren, B. Richard, J. M. Nuzillard, M. Zeches-Hanrot, and L. Le Men-Olivier, *Phytochemistry* **40**, 953 (1995).
339. T. S. Kam and S. Anuradha, *Phytochemistry* **40**, 313 (1995).
340. W. Chen, S. Anuradha, and T. S. Kam, *Zeitschrift f Krist* **213**, 183 (1998).
341. M. Zeches-Hanrot, J. M. Nuzillard, B. Richard, H. Schaller, H. A. Hadi, T. Sevenet, and L. Le Men-Olivier, *Phytochemistry* **40**, 587 (1995).
342. T. P. Lien, H. Ripperger, A. Porzel, K. Merzweiler, T. Van Sung, and G. Adam, *Phytochemistry* **49**, 1457 (1998).
343. N. Amino, T. Ohse, T. Koyano, and K. Umezawa, *Anticancer Res.* **16**, 55 (1996).
344. K. Umezawa, T. Taniguchi, M. Toi, T. Ohse, N. Tsutsumi, T. Yamamoto, T. Koyano, and M. Ishizuka, *Drugs Expt. Clin. Res.* **22**, 35 (1996).
345. T. Irie, K. Kubushiro, K. Suzuki, K. Tsukazaki, K. Umezawa, and S. Nozawa, *Anticancer Res.* **19**, 3061 (1999).
346. S. Atsumi, A. Nagasawa, T. Koyano, T. Kowithayakornd, and K. Umezawa, *Cell. Mol. Life Sci.* **60**, 2516 (2003).
347. J. Gohda, J. I. Inoue, and K. Umezawa, *Int. J. Oncol.* **23**, 1373 (2003).
348. K. Umezawa, A. Hiroki, M. Kawakami, H. Naka, I. Takei, T. Ogata, I. Kojima, T. Koyano, T. Kowithayakorn, H. S. Pang, and T. S. Kam, *Biomed. Pharmacother.* **57**, 341 (2003).
349. H. Takatsuna and K. Umezawa, *Biomed. Pharmacother.* **58**, 610 (2004).
350. T. Ogata, L. Li, S. Yamada, Y. Yamamoto, Y. Tanaka, I. Takei, K. Umezawa, and I. Kojima, *Diabetes* **53**, 2596 (2004).
351. M. Daudon, H. Mehri, M. Plat, E. W. Hagaman, and E. Wenkert, *J. Org. Chem.* **41**, 3275 (1976).
352. H. Mehri and M. Plat, *J. Nat. Prod.* **55**, 241 (1992).
353. J. Stöckigt, K. H. Pawelka, T. Tanahashi, B. Danieli, and W. E. Hull, *Helv. Chim. Acta* **66**, 2525 (1983).
354. W. Fahn, V. Kaiser, H. Schubel, J. Stöckigt, and B. Danieli, *Phytochemistry* **29**, 127 (1990).
355. H. Schubel, W. Fahn, and J. Stöckigt, *Helv. Chim. Acta* **72**, 147 (1989).
356. G. Palmisano, B. Danieli, G. Lesma, M. Santagostino, G. Fiori, and L. Toma, *Helv. Chim. Acta* **75**, 813 (1992).
357. B. Danieli, G. Lesma, D. Passarella, and A. Silvani, *Nat. Prod. Lett.* **7**, 141 (1995).
358. T. S. Kam, T. M. Lim, G. Subramaniam, Y. M. Tee, and K. Yoganathan, *Phytochemistry* **50**, 171 (1999).
359. T. S. Kam and K. Yoganathan, *Nat. Prod. Lett.* **10**, 69 (1997).
360. T. S. Kam, K. Yoganathan, H. Y. Li, and N. Harada, *Tetrahedron* **53**, 12661 (1997).
361. T. S. Kam and Y. M. Choo, *Tetrahedron Lett.* **44**, 1317 (2003).
362. T. S. Kam and Y. M. Choo, *Helv. Chim. Acta* **87**, 991 (2004).
363. T. S. Kam, T. M. Lim, and G. H. Tan, *J. Chem. Soc., Perkin Trans. 1* 1594 (2001).
364. R. Erra-Balsells, *Phytochemistry* **27**, 3945 (1988).
365. F. S. Sariaslani, M. W. Duffel, and J. P. Rosazza, *J. Med. Chem.* **28**, 629 (1985).
366. A. Henriques, C. Kan-Fan, A. Ahond, C. Riche, and H. P. Husson, *Tetrahedron Lett.* 3707 (1978).
367. A. Henriques, S. K. Kan, and M. Lounasmaa, *Acta Chem. Scand.* **B33**, 775 (1979).
368. A. Henriques, C. Kan, H. P. Husson, S. K. Kan, and M. Lounasmaa, *Acta Chem. Scand.* **B34**, 509 (1980).
369. A. Henriques, C. Kan, A. Chiaroni, C. Riche, H. P. Husson, S. K. Kan, and M. Lounasmaa, *J. Org. Chem.* **47**, 803 (1982).
370. M. Zeches, G. Lukacs, G. Massiot, L. Le Men-Olivier, and M. Debray, *J. Nat. Prod.* **45**, 707 (1982).
371. A. Henriques and H. P. Husson, *Tetrahedron Lett.* **22**, 567 (1981).

372. A. T. McPhail, E. W. Hagaman, N. Kunesch, E. Wenkert, and J. Poisson, *Tetrahedron* **39**, 3629 (1983).
373. B. Danieli, G. Lesma, G. Palmisano, S. Tollari, and B. Gabetta, *J. Org. Chem.* **48**, 381 (1983).
374. O. Lefebvre-Soubeyran, *Acta Crystallogr.* **29B**, 2855 (1973).
375. C. Lavaud, G. Massiot, J. Vercauteren, and L. Le Men-Olivier, *Phytochemistry* **21**, 445 (1982).
376. P. Magnus and P. Brown, *J. Chem. Soc., Chem. Commun.* 184 (1985).
377. X. Z. Feng, C. Kan, H. P. Husson, P. Potier, S. K. Kan, and M. Lounasmaa, *J. Nat. Prod.* **47**, 117 (1984).
378. C. Kan-Fan, T. Sevenet, H. P. Husson, and K. C. Chan, *J. Nat. Prod.* **48**, 124 (1985).
379. T. Sevenet, L. Allorge, B. David, K. Awang, A. H. A. Hadi, C. Kan-Fan, J. C. Quirion, F. Remy, H. Schaller, and L. E. Teo, *J. Ethnopharmacol.* **41**, 147 (1994).
380. T. S. Kam, L. Arasu, and K. Yoganathan, *Phytochemistry* **43**, 1385 (1996).
381. T. S. Kam, G. Subramaniam, and W. Chen, *Phytochemistry* **51**, 159 (1999).
382. T. S. Kam, G. Subramaniam, and W. Chen, *Nat. Prod. Lett.* **12**, 293 (1998).
383. T. S. Kam, P. S. Tan, and W. Chen, *Phytochemistry* **33**, 921 (1993).
384. Y. L. Zhou, J. H. Ye, Z. M. Li, and Z. H. Huang, *Planta Med.* **54**, 315 (1988).
385. K. Yan, S. Hong, and X. Feng, *Yaoxue Xuebao* **33**, 597 (1998); *Chem. Abstr.* **130**, 252520 (1998).
386. F. Batchily, S. Baassou, H. Mehri, M. Plat, T. Sevenet, and J. Pusset, *Ann. Pharm. Fr.* **43**, 359 (1985).
387. A. Laguna, L. Novotny, L. Dolejs, and M. Budesinsky, *Planta Med.* **50**, 285 (1984).
388. P. Yates, F. N. MacLachlan, I. D. Rae, M. Rosenberger, A. G. Szabo, C. R. Willis, M. P. Cava, M. Behforouz, M. V. Lakshmikantham, and W. Zeiger, *J. Am. Chem. Soc.* **95**, 7842 (1973).
389. M. V. Lakshmikantham, M. J. Mitchell, and M. P. Cava, *Heterocycles* **9**, 1009 (1978).
390. A. A. Adesomoju, V. H. Rawal, M. V. Lakshmikantham, and M. P. Cava, *J. Org. Chem.* **48**, 3015 (1983).
391. A. A. Adesomoju, M. V. Lakshmikantham, and M. P. Cava, *Heterocycles* **20**, 1511 (1983).
392. A. A. Adesomoju, M. V. Lakshmikantham, and M. P. Cava, *Heterocycles* **32**, 1461 (1991).
393. M. Mroue and M. Alam, *Phytochemistry* **30**, 1741 (1991).
394. H. Mehri, S. Baassou, and M. Plat, *J. Nat. Prod.* **54**, 372 (1991).
395. I. Moldvai, C. Szantay, Jr., G. Tarkanyi, and C. Szantay, *Tetrahedron* **51**, 9103 (1995).
396. I. Moldvai, A. Vedres, G. Toth, C. Szantay, Jr., and C. Szantay, *Tetrahedron Lett.* **27**, 2775 (1986).
397. I. Moldvai, G. Toth, C. Szantay, Jr., H. Kuhne, A. Vedres, and C. Szantay, *J. Chem. Res.* **10**, 314 (1988).
398. J. Vercauteren, G. Massiot, L. Le Men-Olivier, J. Levy, and C. Delaude, *Bull. Soc. Chim. Fr.* 291 (1982).
399. A. Chatterjee, A. K. Ghosh, P. C. Das, T. Bhowmik, and J. Banerji, *J. Indian Chem. Soc.* **75**, 695 (1998).
400. T. S. Kam, K. Yoganathan, and H. Y. Li, *Tetrahedron Lett.* **37**, 8811 (1996).
401. T. S. Kam, K. H. Lim, K. Yoganathan, M. Hayashi, and K. Komiyama, *Tetrahedron* **60**, 10739 (2004).
402. K. X. Yan and X. Z. Feng, *Chinese Chem. Lett.* **8**, 313 (1997); *Chem. Abstr.* **127**, 106600 (1997).
403. J. W. Moncrief and W. N. Lipscomb, *J. Am. Chem. Soc.* **87**, 4963 (1965).
404. J. W. Moncrief and W. N. Lipscomb, *Acta Crystallogr.* **21**, 322 (1966).
405. R. Bau and K. K. Jin, *J. Chem. Soc., Perkin Trans. 1* 2079 (2000).

406. V. M. Lynch, A. Stamford, P. Magnus, and B. E. Davis, *Acta Crystallogr.* **C47**, 1563 (1991).
407. J. G. Dong, W. Bornmann, K. Nakanishi, and N. Berova, *Phytochemistry* **40**, 1821 (1995).
408. J. P. Kutney, D. E. Gregonis, R. Imhof, I. Itoh, E. Jahngen, A. I. Scott, and W. K. Chan, *J. Am. Chem. Soc.* **97**, 5031 (1975).
409. C. A. Parish, J. G. Dong, W. G. Bornmann, J. Chang, K. Nakanishi, and N. Berova, *Tetrahedron* **54**, 15739 (1998).
410. T. D. Spitzer, R. C. Crouch, and G. E. Martin, *J. Heterocyclic Chem.* **29**, 265 (1992).
411. T. D. Spitzer, D. W. Reynolds, R. C. Crouch, J. P. Shockcor, R. L. Johnson, R. W. Morrison, Jr., and G. E. Martin, *J. Heterocyclic Chem.* **32**, 1255 (1995).
412. C. W. Andrews, J. Wisowaty, A. O. Davis, R. C. Crouch, and G. E. Martin, *J. Heterocyclic Chem.* **32**, 1011 (1995).
413. G. Blaskó and G. A. Cordell, *in* "The Alkaloids", (A. Brossi and M. Suffness, eds.), vol. 37, p. 1. Academic Press, San Diego, 1990.
414. R. Verpoorte, R. Van der Heijden, and P. R. H. Moreno, *in* "The Alkaloids", (A. Brossi, ed.), vol. 49, p. 221. Academic Press, San Diego, 1997.
415. M. Sottomayor, I. L. Cardoso, L. G. Pereira, and A. R. Barcelo, *Phytochem. Rev.* **3**, 159 (2004).
416. R. Van der Heijden, D. I. Jacobs, W. Snoeijer, D. Hallard, and R. Verpoorte, *Curr. Med. Chem.* **11**, 607 (2004).
417. V. De Luca, N. Brisson, J. Balsevich, and W. G. W. Kurz, *in* "Primary and Secondary Metabolism of Plant Cell Cultures II", (W. G. W. Kurz, ed.), p. 154. Springer, Berlin, 1989.
418. V. De Luca, B. St. Pierre, F. Vazquez Flota, and P. Laflamme, *in* "Towards Natural Medicine Research in the 21st Century", (H. Agetha, N. Aimi, Y. Ebizuka, T. Fujita, and G. Honda, eds.), p. 363. Elsevier Science B. V., Amsterdam, 1998.
419. J. P. Kutney, *Heterocycles* **25**, 617 (1987).
420. J. P. Kutney, *Nat. Prod. Rep.* **7**, 85 (1990).
421. M. Sottomayor, M. Lopez-Serrano, F. DiCosmo, and A. Ros Barcelo, *FEBS Lett.* **428**, 299 (1998).
422. M. Sottomayor and A. Ros Barcelo, *Protoplasma* **222**, 97 (2003).
423. M. E. Kuehne and I. Marko, *in* "The Alkaloids", (A. Brossi and M. Suffness, eds.), Vol. 37, p. 77. Academic Press, San Diego, 1990.
424. Atta-ur-Rahman, Z. Iqbal, and H. Nasir, *in* "Studies in Natural Product Chemistry" (Atta-ur-Rahman, ed.), vol. 14, p. 805. Elsevier Science B. V., Amsterdam, 1994.
425. P. Potier, N. Langlois, Y. Langlois, and F. Gueritte, *J. Chem. Soc., Chem. Commun.* 670 (1975).
426. N. Langlois, F. Gueritte, Y. Langlois, and P. Potier, *J. Am. Chem. Soc.* **98**, 7017 (1976).
427. P. Potier, *J. Nat. Prod.* **43**, 72 (1980).
428. J. P. Kutney, A. H. Ratcliffe, A. M. Treasurywala, and S. Wunderly, *Heterocycles* **3**, 639 (1975).
429. J. P. Kutney, T. Hibino, E. Jahngen, T. Okutani, A. H. Ratcliffe, A. M. Treasurywala, and S. Wunderly, *Helv. Chim. Acta* **59**, 2858 (1976).
430. J. P. Kutney, *Lloydia* **40**, 107 (1977).
431. J. Vucovic, A. E. Goodbody, J. P. Kutney, and M. Misawa, *Tetrahedron* **44**, 325 (1988).
432. R. J. Sundberg, K. G. Gadamasetti, and P. J. Hunt, *Tetrahedron* **48**, 277 (1992).
433. M. E. Kuehne and T. C. Zebovitz, *J. Org. Chem.* **52**, 4331 (1987).
434. M. E. Kuehne, T. C. Zebovitz, W. G. Bornmann, and I. Marko, *J. Org. Chem.* **52**, 4340 (1987).
435. M. E. Kuehne and W. G. Bornmann, *J. Org. Chem.* **54**, 3407 (1989).
436. M. E. Kuehne, P. A. Matson, and W. G. Bornmann, *J. Org. Chem.* **56**, 513 (1991).

437. M. E. Kuehne and U. K. Bandarage, *J. Org. Chem.* **61**, 1175 (1996).
438. G. Schill, C. U. Priester, U. F. Windhovel, and H. Fritz, *Helv. Chim. Acta* **69**, 438 (1986).
439. G. Schill, H. Lower, C. U. Priester, U. F. Windhovel, and H. Fritz, *Tetrahedron* **43**, 3729 (1987).
440. G. Schill, C. U. Priester, U. F. Windhovel, and H. Fritz, *Tetrahedron* **43**, 3747 (1987).
441. G. Schill, C. U. Priester, U. F. Windhovel, and H. Fritz, *Tetrahedron* **43**, 3765 (1987).
442. G. Schill, C. U. Priester, U. F. Windhovel, and H. Fritz, *Tetrahedron* **46**, 1211 (1990).
443. G. Schill, C. U. Priester, U. F. Windhovel, and H. Fritz, *Tetrahedron* **46**, 1221 (1990).
444. J. Harley-Mason and Atta-ur-Rahman, *J. Chem. Soc., Chem. Commun.* 1048 (1967).
445. J. Harley-Mason and Atta-ur-Rahman, *Tetrahedron* **36**, 1057 (1980).
446. P. Magnus, J. S. Mendoza, A. Stamford, M. Ladlow, and P. Willis, *J. Am. Chem. Soc.* **114**, 10232 (1992).
447. P. Magnus, A. Stamford, and M. Ladlow, *J. Am. Chem. Soc.* **112**, 8210 (1990).
448. P. Magnus, M. Ladlow, J. Elliott, and C. S. Kim, *J. Chem. Soc., Chem. Commun.* 518 (1989).
449. P. Magnus, M. Ladlow, and J. Elliott, *J. Am. Chem. Soc.* **109**, 7929 (1987).
450. P. Magnus and J. Mendoza, *Tetrahedron Lett.* **33**, 899 (1992).
451. S. Yokoshima, T. Ueda, S. Kobayashi, A. Sato, T. Kuboyama, H. Tokuyama, and T. Fukuyama, *J. Am. Chem. Soc.* **124**, 2137 (2002).
452. H. Tokuyama, T. Yamashita, M. T. Reding, Y. Kaburagi, and T. Fukuyama, *J. Am. Chem. Soc.* **121**, 3791 (1999).
453. S. Kobayashi, T. Ueda, and T. Fukuyama, *Synlett* 883 (2000).
454. D. A. Evans, M. T. Bilodeau, T. C. Somers, J. Clardy, D. Cherry, and Y. Kato, *J. Org. Chem.* **56**, 5750 (1991).
455. L. S. Borman and M. E. Kuehne, *in* "The Alkaloids" (A. Brossi and M. Suffness eds.), vol. 37, p. 133. Academic Press, San Diego, 1990; H. L. Pearce, *ibid.*, p. 145.
456. J. Fahy, *Curr. Pharm. Design* **7**, 1181 (2001).
457. B. T. Hill, *Curr. Pharm. Design* **7**, 1199 (2001).
458. A. Duflos, A. Kruczynski, and J. M. Barret, *Curr. Med. Chem.: Anticancer Agents* **2**, 55 (2002).
459. E. Gunic, I. Tabakovic, and M. J. Gasic, *J. Chem. Soc., Chem Commun.* 1496 (1993).
460. I. Tabakovic, E. Gunic, and I. Juranic, *J. Org. Chem.* **62**, 947 (1997).
461. S. Duangteraprecha, K. Hirata, E. Morihara, M. Nakae, H. Katayama, M. Honda, and K. Miyamoto, *J. Ferment. Bioengin.* **83**, 227 (1997).
462. K. Hirata, S. Duangteraprecha, E. Morihara, M. Honda, T. Akagi, M. Nakae, H. Katayama, and K. Miyamoto, *Biotech. Lett.* **19**, 53 (1997).
463. C. Hardouin, E. Doris, B. Rousseau, and C. Mioskowski, *Org. Lett.* **4**, 1151 (2002).
464. C. Hardouin, E. Doris, B. Rousseau, and C. Mioskowski, *J. Org. Chem.* **67**, 6571 (2002).
465. Atta-ur-Rahman and S. Perveen, *J. Nat. Prod.* **51**, 1271 (1988).
466. B. Schmidt, J. Kutney, and L. Mayer *PCT Int. Appl., WO98390 04* (1998).
467. G. Schwartz, D. Bong, C. G. Leichman, and L. L. Leichman, *Proc. Am. Assoc. Cancer Res.* **41**, 3895 (2000).
468. R. J. Sundberg, J. L. Bettiol, K. G. Gadamasetti, M. Marshalla, and L. Kelsh, *Bioorg. Med. Chem. Lett.* **4**, 1999 (1994).
469. P. Magnus and L. S. Thurston, *J. Org. Chem.* **56**, 1166 (1991).
470. R. J. Sundberg, J. Hong, S. Q. Smith, M. Sabat, and I. Tabakovic, *Tetrahedron* **54**, 6259 (1998).
471. G. Richter Ltd, US Pat. 3899493; Chem. Abstr., **83**, 17960 (1975).
472. K. Honty, A. Demeter, C. Szantay, Jr., M. Hollosi, P. Kolonits, and C. Szantay, *Heterocycles* **50**, 169 (1999).
473. L. Szabo, C. Szantay, E. G. Baitz, and M. Mak, *Tetrahedron Lett.* **36**, 5265 (1995).
474. S. H. Ahn, M. W. Duffel, and J. P. N. Rosazza, *J. Nat. Prod.* **60**, 1125 (1997).

475. M. E. Kuehne, S. D. Cowen, F. Xu, and L. S. Borman, *J. Org. Chem.* **66**, 5303 (2001).
476. M. E. Kuehne, Y. Qin, A. E. Huot, and S. L. Bane, *J. Org. Chem.* **66**, 5317 (2001).
477. M. E. Kuehne, W. G. Bornmann, I. Marko, Y. Qin, K. L. LeBoulluec, D. A. Frasier, F. Xu, T. Mulamba, C. L. Ensinger, L. S. Borman, A. E. Huot, C. Exon, F. T. Bizzarro, J. B. Cheung, and S. L. Bane, *Org. Biomol. Chem.* **1**, 2120 (2003).
478. L. S. Borman and M. E. Kuehne, *Biochem. Pharmacol.* **38**, 715 (1989).
479. J. Fahy, A. Duflos, J. P. Ribet, J. C. Jacquesy, C. Berrier, M. P. Jouannetaud, and F. Zunino, *J. Am. Chem. Soc.* **119**, 8576 (1997).
480. J. C. Jacquesy, C. Berrier, M. P. Jouannetaud, F. Zunino, J. Fahy, A. Duflos, and J. P. Ribet, *J. Fluorine Chem.* **114**, 139 (2002).
481. C. Lafitte, M. P. Jounannetaud, J. C. Jacquesy, J. Fahy, and A. Duflos, *Tetrahedron Lett.* **39**, 8281 (1998).
482. A. Kruczynski, F. Colpaert, J. P. Tarayre, P. Mouillard, J. Fahy, and B. T. Hill, *Cancer Chemother. Pharmacol.* **41**, 437 (1998).
483. B. T. Hill, H. H. Fiebig, W. R. Waud, M. F. Poupon, F. Colpaert, and A. Kruczynski, *Eur. J. Cancer* **35**, 512 (1999).
484. K. N. Thimmaiah, W. D. Lloyd, and V. S. Sethi, *Indian J. Chem.* **29B**, 678 (1990).
485. B. Danieli, G. Lesma, M. Martinelli, D. Passarella, A. Silvani, B. Pyuskyulev, and M. N. Tam, *J. Org. Chem.* **63**, 8586 (1998).
486. A. De Bruyn, M. Verzele, J. P. Dejonghe, K. S. P. B. Roa, M. P. Collard, A. Trouet, and J. Hannart, *Planta Med.* **55**, 364 (1989).
487. F. Orosz, B. G. Vertessy, C. Salerno, C. Crifo, E. Capuozzo, and J. Ovadi, *Br. J. Pharmacol.* **121**, 955 (1997).
488. F. Orosz, J. Kovacs, P. Low, B. G. Vertessy, Z. Urbanyi, T. Acs, T. Keve, and J. Ovadi, *Br. J. Pharmacol.* **121**, 947 (1997).
489. F. Orosz, B. Comin, B. Rais, J. Puigjaner, J. Kovacs, G. Tarkanyi, T. Acs, T. Keve, M. Cascante, and J. Ovadi, *Br. J. Cancer* **79**, 1356 (1999).
490. L. N. Jungheim, T. A. Shepherd, and D. L. Meyer, *J. Org. Chem.* **57**, 2334 (1992).

# CUMULATIVE INDEX OF TITLES

# INDEX